					0
					2 Helium **He** 4.00

IIIA	IVA	VA	VIA	VIIA
5 Boron **B** 10.81	6 Carbon **C** 12.01	7 Nitrogen **N** 14.01	8 Oxygen **O** 16.00	9 Fluorine **F** 19.00
13 Aluminum **Al** 26.98	14 Silicon **Si** 28.09	15 Phosphorus **P** 30.97	16 Sulfur **S** 32.06	17 Chlorine **Cl** 35.45

IB	IIB

IIIA	IVA	VA	VIA	VIIA
10 Neon **Ne** 20.18
18 Argon **Ar** 39.95

IB	IIB	IIIA	IVA	VA	VIA	VIIA	0
29 Copper **Cu** 63.55	30 Zinc **Zn** 65.37	31 Gallium **Ga** 69.72	32 Germanium **Ge** 72.59	33 Arsenic **As** 74.92	34 Selenium **Se** 78.96	35 Bromine **Br** 79.90	36 Krypton **Kr** 83.80
47 Silver **Ag** 107.87	48 Cadmium **Cd** 112.40	49 Indium **In** 114.82	50 Tin **Sn** 118.69	51 Antimony **Sb** 121.75	52 Tellurium **Te** 127.60	53 Iodine **I** 126.90	54 Xenon **Xe** 131.30
79 Gold **Au** 196.97	80 Mercury **Hg** 200.59	81 Thallium **Tl** 204.37	82 Lead **Pb** 207.2	83 Bismuth **Bi** 208.98	84 Polonium **Po** (210)	85 Astatine **At** (210)	86 Radon **Rn** (222)

64 Gadolinium **Gd** 157.25	65 Terbium **Tb** 158.93	66 Dysprosium **Dy** 162.50	67 Holmium **Ho** 164.93	68 Erbium **Er** 167.26	69 Thulium **Tm** 168.93	70 Ytterbium **Yb** 173.04	71 Lutetium **Lu** 174.97

96 Curium **Cm** (247)	97 Berkelium **Bk** (247)	98 Californium **Cf** (251)	99 Einsteinium **Es** (254)	100 Fermium **Fm** (253)	101 Mendelevium **Md** (256)	102 Nobelium **No** (254)	103 Lawrencium **Lr** (257)

Periodic Table of the Elements

Chemistry for the Health Sciences

Chemistry for the

George I. Sackheim *Associate Professor of Chemistry, and Executive Secretary of the Department, University of Illinois at Chicago Circle; Lecturer, School of Health Sciences, Michael Reese Hospital and Medical Center, Chicago. Formerly, Coordinator of Biological and Physical Sciences, School of Nursing, Michael Reese Hospital and Medical Center, Chicago; Science Instructor, School of Nursing, Presbyterian—St. Luke's Hospital, Chicago*

Health Sciences

Third Edition

Ronald M. Schultz *Formerly, Assistant Professor of Chemistry, Crane Campus, The Chicago City College; Science Instructor, Schools of Nursing, St. Ann's Hospital, St. Anthony de Padua Hospital, and St. Mary of Nazareth Hospital, Chicago*

Macmillan Publishing Co., Inc.
New York

Collier Macmillan Publishers
London

MACMILLAN PUBLISHING CO., INC.
866 Third Avenue, New York, New York 10022

COLLIER MACMILLAN CANADA, LTD.

Library of Congress Cataloging in Publication Data

Sackheim, George I
  Chemistry for the health sciences.

  Includes bibliographies and index.
  1. Chemistry. I. Schultz, Ronald M., joint author.
II. Title.
QD33.S13   1977   540   76-15269
ISBN 0-02-405040-7

Printing: 4 5 6 7 8   Year: 9 0 1 2 3

Preface to the Third Edition

This textbook of chemistry is designed primarily for first-year students in various health-related programs—nursing, dietetics, laboratory technology, inhalation therapy, dental hygiene, dental assisting, medical assisting, dental technology, and other health-service technologies. Emphasis is placed on *practical* aspects of inorganic chemistry, organic chemistry, and biochemistry. Theoretic topics are dealt with only as an aid to understanding bodily processes in the human.

Unit One, "Inorganic Chemistry," stresses relationships with the life processes dealt with in Unit Three, "Biochemistry." Included are discussions of the following topics:

> acids, bases, salts, and electrolytes, as related to acid-base balance and electrolyte balance in the body
>
> oxidation-reduction, as related to biologic oxidation-reduction reactions taking place in the mitochondria of cells
>
> solutions, for understanding the solvent action involved in digestion
>
> colloids, for understanding the particular nature and properties of proteins, amino acids, and nucleic acids
>
> covalent compounds, as related to the types of bonds that must be broken and rearranged in the formation of high-energy phosphate bonds
>
> emulsions, for understanding the need for emulsification of fats before digestion
>
> nuclear chemistry and radioactivity, for understanding the biologic effects of radiation on cells and organs

Unit Two, "Organic Chemistry," introduces the various classes of organic compounds—hydrocarbons, alcohols, ethers, cyclic compounds, and heterocyclic compounds. In addition, the text discussion relates such compounds to carbohydrates, fats, proteins, vitamins, and hormones.

Unit Three, "Biochemistry," deals with the chemical and molecular basis of life itself. It endeavors to explain the various chemical processes taking place in the body during its normal as well as its abnormal metabolism. The role of ATP, the principal direct source of energy for the body, is stressed throughout the various chapters on metabolism. The formation and decomposition of this compound, and the energies involved, serve to indicate the very complex processes that take place "normally." The role of coenzymes, such as CoA, is also emphasized, to show the role these substances play in metabolic cycles. Discussions of excesses and deficiencies of vitamins and hormones are designed to demonstrate the involved inter-relationships in the body's metabolic processes. In the chapter on heredity the authors have endeavored to combine chemistry with the molecular basis of life, as evidenced by the many recent advances in our understanding of DNA structure and of the replication of DNA and RNA.

An outline appears at the beginning of each chapter, indicating the topics to be discussed and, in general, how they are interrelated. The summary at the end of each chapter is also designed to help the reader identify the particularly important aspects of the subject matter. The student who wishes to know more about any specific topic will find additional information in the references appended to each chapter. Also at the end of each chapter is a set of questions, which may be used for oral review or assigned as homework.

New to the third edition are:

> introduction of the Kelvin temperature scale
> problems involving gas laws
> discussion of newer radioisotopes in various types of diagnostic and therapeutic uses
> introduction of material on xeroradiography, ultrasonography, and x-ray scans
> sections on air and water pollution, and noxious gases
> introduction of the IUPAC system of naming, along with the common system
> sections on mind drugs, chemotherapy, and chromatography
> additional material on peptide linkages and amides
> expanded section on isomerism, including *cis–trans* and optical activity
> introduction of primary, secondary, tertiary, and quaternary structures of proteins
> additional material on phospholipids, including their part in cell membranes
> discussion of enzyme activators and inhibitors
> section on electron transport system in the citric acid cycle
> newer material on cytoplasmic lipogenesis, immunoglobulins, and the role of vitamin A in the visual cycle
> section on Starling's hypothesis of capillary diffusion and water balance
> additional material on electrolytes, including such ions as lithium, iron, and copper
> discussion of role of vitamins as coenzymes, and hypervitaminosis diseases
> additional material on the role of cyclic-AMP in hormonal action; release and release-inhibiting factors of the hypothalamus
> discussion of some causes of genetic disease
> list of organic functional groups

A comprehensive set of laboratory experiments (Sackheim-Schultz, *Laboratory Chemistry for the Health Sciences*, Macmillan Publishing Co., Inc., New York) has been designed to implement and supplement the various topics in the textbook. These experiments are keyed to specific chapters and, when performed, will aid the student in understanding the fundamental concepts involved. Also available to the instructor is a helpful *Answer and Test Manual for Chemistry for the Health Sciences* (Macmillan Publishing Co., Inc., New York).

G. I. S.
R. M. S.

Acknowledgments

For their helpful comments, criticisms, and suggestions the authors wish to thank Don Hoster, Associate Professor, Community College of Baltimore; Dennis D. Lehman, Associate Professor, Loop College, and Cook County Hospital School of Nursing, Chicago; John G. McGrew, Assistant Professor, Alderson-Broaddus College (Philippi, West Virginia); Don Roach, Chairman, Department of Chemistry, Miami-Dade Community College (Miami, Florida); and M. L. Rodenburg, Associate Professor, Kettering College of Medical Arts (Kettering, Ohio). In addition, the authors are indebted to the following persons for their critical review of the third-edition manuscript: David Gorenstein, Associate Professor of Chemistry, University of Illinois at Chicago Circle; Elaine Herzog, Assistant Professor of Chemistry, University of Illinois at Chicago Circle; Louise B. Katz, Chairman, Department of Life Sciences, Sinclair Community College (Dayton, Ohio); David McCormick, Assistant Professor of Biology, Sinclair Community College (Dayton, Ohio); and Edward G. Rietz, Professor of Chemistry, Wright City College (Chicago, Illinois).

Contents

unit one

Inorganic Chemistry

Units of Measurement

Temperature Scales

When we speak of the outside temperature as being 80° or 40° or even 2° below zero, we are normally speaking in terms of degrees Fahrenheit (F), even though we may not express these units as such. Now that the metric system of measurement is being used more and more widely, we must specify the unit, since more than one temperature scale may be utilized. In chemistry, the Celsius or centigrade (C) and the Kelvin (K) temperature scales are commonly used. Let us see first how the Celsius and the Fahrenheit temperature scales compare with one another.

If a Fahrenheit and a Celsius thermometer are placed in a mixture of ice and water, the reading on the Fahrenheit scale will be 32° and the reading on the Celsius scale 0° (see Figure 1–1). That is, 32°F corresponds to 0°C. These temperatures indicate the freezing point of water. They also indicate the melting point of ice.

Next, if the same two thermometers are placed in boiling water at one atmosphere pressure, the reading on the Fahrenheit scale will be 212° and the reading on the Celsius scale will be 100°. These temperatures indicate the boiling point of water.

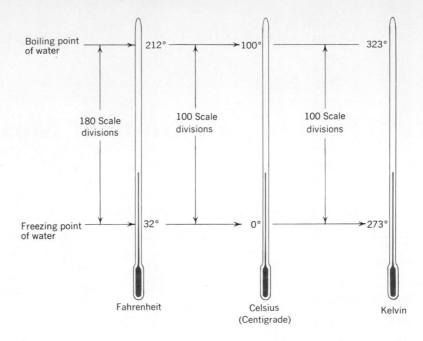

Figure 1–1. Comparison of Fahrenheit, Celsius, and Kelvin temperature scales.

It should be noted that there is a 180-degree difference between the boiling and the freezing points of water on the Fahrenheit scale (212° to 32°) and a 100-degree difference between the boiling and the freezing points of water on the Celsius scale (100° to 0°). The difference between the freezing points of water on the Fahrenheit and Celsius scales is 32° (32° − 0°).

This information may be combined into the following formulas

$$°F = 9/5\,°C + 32° \quad \text{and} \quad °C = 5/9\,(°F − 32°)$$

The first formula is used to change Celsius temperatures to the corresponding Fahrenheit readings; the second formula is used to do the reverse, to change Fahrenheit readings to Celsius. Note that the first formula may be changed to the second by subtracting 32° from both sides and then dividing by 9/5.

Example 1.   Change 80°C to °F.
Using the formula °F = 9/5 °C + 32°

$$°F = 9/5(80°) + 32°$$

$$= 144° + 32°$$

$$= 176°$$

so 80°C = 176°F

Example 2.   Change 50°F to °C.
Using the formula °C = 5/9 (°F − 32)

$$°C = 5/9(50° − 32°)$$

$$= 5/9(18°)$$

$$= 10°$$

so 50°F = 10°C

A Kelvin thermometer will indicate the freezing point of water as 273°K and the boiling point of water as 373°K. Note that there are 100 degrees between the boiling and freezing points of water on the Kelvin scale (373°K − 273°K), just as there were on the Celsius scale. The difference between the freezing points of water on the Kelvin and Celsius scales is 273° (273°K − 0°K). This information may be combined into the following formula:

$$°K = °C + 273°$$

Example 3.   Change 37°C to °K.
Using the formula °K = °C + 273°

$$°K = 37° + 273°$$

$$°K = 310°$$

so 37°C = 310° K

The Metric System

While we are all familiar with the English system of measurement, which uses the foot as the unit of length, the pound as the unit of weight, and the gallon as the unit of volume, we are gradually becoming used to the metric system of measurement in our everyday life. In chemistry the metric system is used exclusively. The main advantage of the metric system is that it is a decimal system. That is, the units are all multiples of ten of larger or smaller units.

Units

The unit of length in the metric system is the meter. A *meter* is a little longer than a yard. The meter was originally defined as one ten-millionth of the distance of a meridian from the equator to the North Pole on a line passing through Paris. The standard meter was later defined as the length between two marks on a platinum-irridium bar kept at the temperature of melting ice and stored at the International Bureau of Weights and Measures near Paris. The standard meter is now defined in terms of the orange-red wavelengths of light emitted by the element krypton-86 (see Chapter 3). The word meter is abbreviated as m.

A unit of weight (mass) in the metric system is the *gram*, abbreviated as g. A gram is about 1/450th of a pound.

The unit of volume in the metric system is the *liter* (usually spelled out or abbreviated as L or l). A liter is approximately equal to a quart.

Derived Units

In the metric system, prefixes are used to designate various multiples or submultiples. The most commonly used prefixes are as follows:

milli- (abbreviated m), which means one thousandth
centi- (abbreviated c), which means one hundredth
deci- (abbreviated d), which means one tenth
kilo- (abbreviated k), which means one thousand

Thus, a millimeter (mm) is one thousandth of a meter, or 0.001 m; a centigram (cg) is one hundredth of a gram, or 0.01 g. Likewise, a kiloliter (kl) is 1000 liters (see Table 1–1). Other prefixes used occasionally are

TABLE 1–1

Metric Conversions

Weight	
1 g = 1000 mg	one gram = one thousand milligrams
1 g = 100 cg	one gram = one hundred centigrams
1 g = 10 dg	one gram = ten decigrams
1 kg = 1000 g	one kilogram = one thousand grams

Length	
1 m = 1000 mm	one meter = one thousand millimeters
1 m = 100 cm	one meter = one hundred centimeters
1 m = 10 dm	one meter = ten decimeters
1 km = 1000 m	one kilometer = one thousand meters

Volume	
1 L = 1000 ml	one liter = one thousand milliliters
1 L = 100 cl	one liter = one hundred centiliters
1 L = 10 dl	one liter = ten deciliters
1 kl = 1000 l	one kiloliter = one thousand liters

Another unit of volume is the cubic centimeter, abbreviated cc or cm^3. One cc is equal to 1 ml so that these two units may be used interchangeably.

mega- (abbreviated M), which means one million, and micro- (abbreviated μ), which means one-millionth. Note that the letter m indicates both meter and milli-; however, when used by itself, it indicates meters.

Conversion Units

Frequently the student has difficulty in converting from one system of measurement to another (see Figure 1–2). The procedure is the same as that used when converting within the English system. Consider, for example, the following English units:

$$12 \text{ in.} = 1 \text{ ft}$$

$$3 \text{ ft} = 1 \text{ yd}$$

$$1760 \text{ yd} = 1 \text{ mile}$$

Figure 1–2. Comparison of English and metric units (1/2 scale).

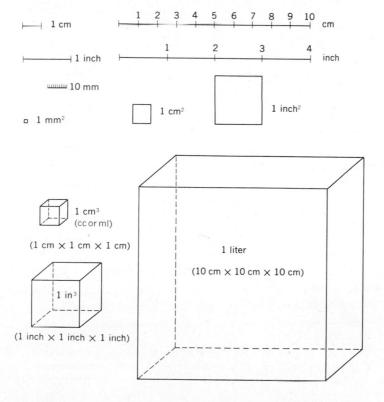

When asked to convert 6 ft to inches, many students merely multiply by 12. However, mathematically, the correct procedure is to multiply by the conversion unit 12 in./ft so that the result is correct both numerically and in units.

$$6 \text{ ft} = 6 \text{ ft} \times 12 \frac{\text{in.}}{\text{ft}} = 72 \text{ in.}$$

Likewise, when converting 3.58 m to millimeters (mm), and recalling that there are 1000 mm in a meter, the procedure is to multiply by 1000 mm/m. Note that the unwanted units cancel leaving the answer in millimeters, as was desired.

$$3.58 \text{ m} = 3.58 \text{ m} \times 1000 \frac{\text{mm}}{\text{m}} = 3580 \text{ mm}$$

(Note that if the 3.58 m were divided by the conversion unit of 1000 mm/m, the units would not cancel, indicating an error in computation.)

When converting 50 cm to inches, the conversion factor 2.5 cm = 1 in. should be used (see Table 1–2). Thus

$$50 \text{ cm} = 50 \text{ cm} \times 1 \text{ in.}/2.5 \text{ cm} = 20 \text{ in.}$$

TABLE 1–2

Approximate English–Metric Conversions

1 in. = 2.5 cm
1 ft = 30 cm
1 lb = 450 g
2.2 lb = 1 kg
1.06 qt = 1 liter

Density

Density is defined as weight (or mass) per volume (see Figure 1–3). It may be expressed as the following formula

$$D = \frac{W}{V}$$

where D is the density, W is the weight or mass, and V the volume. Density must have units of weight and volume such as g/ml, lb/ft^3, mg/liter, or others.

Example 1. What is the density of the mercury in a thermometer if 31.2 g of it occupy 2.29 ml?

$$D = \frac{W}{V}$$

$$= 31.2 \text{ g}/2.29 \text{ ml}$$

$$= 13.6 \text{ g/ml}$$

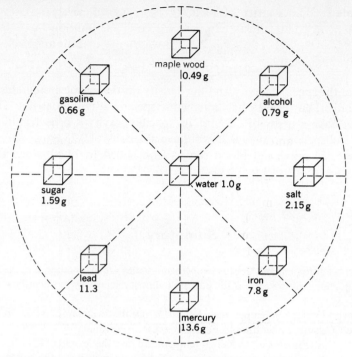

Figure 1–3. A comparison of the weights of 1 ml of various substances.

Example 2.   Alcohol has a density of 0.80 g/ml. How much will 100 ml of it weigh?

$$D = \frac{W}{V}$$

$$W = D \times V$$

$$= 0.80 \text{ g/ml} \times 100 \text{ ml}$$

$$= 80 \text{ g}$$

Specific Gravity

Specific gravity is defined as the weight of an object compared to the weight of an equal volume of water, or also as the density of an object compared to the density of water (which is 1 g/ml). Specific gravity may be expressed mathematically as:

$$\text{sp gr} = \frac{\text{density of object}}{\text{density of water}}$$

Example 1. The density of mercury is 13.6 g/ml. What is its specific gravity?

$$\text{sp gr} = \frac{\text{density of object}}{\text{density of water}} = \frac{13.6 \text{ g/ml}}{1 \text{ g/ml}} = 13.6$$

Note that the specific gravity and the density (in the metric system) have the same number, but density has units while specific gravity does not. Thus, if an object has a density of 1.5 g/ml, its specific gravity will be 1.5; likewise, an object of specific gravity 0.75 will have a density of 0.75 g/ml.

The specific gravity of blood ranges from 1.054 to 1.060, while that of urine ranges from 1.003 to 1.030.

Summary

Although Fahrenheit and Celsius temperature scales are in common use, the Celsius (centigrade) temperature scale is the one used almost exclusively in scientific measurements.

To convert Fahrenheit temperatures to Celsius, use the formula, $°C = 5/9\,(°F − 32°)$. To convert Celsius temperature to Fahrenheit, use the formula, $°F = 9/5°C + 32°$. To convert Celsius temperature to Kelvin temperature, use the formula $°K = °C + 273°$.

The units of length, weight, and volume in the metric system are the meter (m), the gram (g), and the liter (L), respectively. Prefixes in common use in the metric system are the following: milli- (1/1000), centi- (1/100), deci- (1/10), kilo- (1000).

Density is defined as weight per unit volume, or $D = W/V$.

Specific gravity is defined as the density of a substance compared to the density of water.

Questions and Problems

1. Body temperature is 98.6°F. What is it in degrees Celsius?
2. Which patient has a higher fever, one with a temperature of 101°F or one with a temperature of 38.2°C?
3. At the top of a certain mountain, water boils at 85°C. What is the boiling point in degrees Fahrenheit? In degrees Kelvin?
4. If the water in a pressure cooker boils at 250°F, what will be the boiling point in degrees Celsius? In degrees Kelvin?
5. What is the standard of mass in the metric system? In the English system?
6. What is the standard of length in the metric system? In the English system?
7. What do the following prefixes indicate: milli-, deci-, kilo-, centi-?
8. What do the following abbreviations indicate: g, cm, mm, kl, m?
9. Convert the following

    (a) 87.6 mm =   m   (c) 394 cg =   g
    (b) 2.5 liters =   ml   (d) 16.4 cm =   m

10. Convert the following

    (a) 20 in. =   cm   (c) 10 cm =   in.

    (b) 22 lb =   kg   (d) 50 kg =   lb

11. A patient is 5 ft 10 in. tall and weighs 160 lb. What is his height in centimeters and his weight in kilograms?

12. The following information was recorded for a patient: height 152.5 cm, weight 68 kg, temperature 37.2°C. What would be the readings in inches, pounds, and degrees Fahrenheit?

13. What is the density of a medication if the contents of a filled 2-ml syringe weigh 2.50 g?

14. Normal urine has a density between 1.003 g/ml and 1.030 g/ml. Assuming a value of 1.020 g/ml, what will be the weight of a 250-ml sample of urine?

15. An object has a specific gravity of 3.06. What is its density in the metric system?

16. Mercury has a density of 13.6 g/ml. What volume will 200 g of it occupy?

17. An order for a medication reads: "Give 1.5 mg per kilogram of body weight." How much medication should be given to a patient weighing 165 lb?

18. What do the following mean in terms of known units: kilowatts, milliseconds, megacuries, microamperes?

19. An infant weighs 7000 g. How many kilograms does the infant weigh? How many pounds?

20. A rectangular object is 25 cm tall, 15 cm wide, and 10 cm deep. What is its volume in cc, L, kl?

21. Using Figure 1–3, what is the specific gravity of lead? Salt?

References

Baum, S. J., and Scaife, C. W.: *Chemistry: A Life Science Approach*. Macmillan Publishing Co., Inc., New York, 1975, Chap. 1.

Fernandez, J. E., and Whitaker, R. D.: *An Introduction to Chemical Principles*. Macmillan Publishing Co., Inc., New York, 1975, Chap. 2.

Hein, M., and Best, L. R.: *College Chemistry*. Dickenson Publishing Co., Inc., Encino, Calif., 1976, Chap. 2.

Longo, F. R.: *General Chemistry*. McGraw-Hill Book Co., New York, 1974, Chap. 2.

Nebergall, W. H.; Schmidt, F. C.; and Holtzclaw, H. F., Jr.: *College Chemistry*, 5th ed. D. C. Heath & Co., Lexington, Mass., 1976, Chap. 2.

Peters, E. I.: *Introduction to Chemical Principles*. W. B. Saunders Co., Philadelphia, 1974, Chap. 2.

Sackheim, G. I.: *Chemical Calculations*, 10th ed., series B. Stipes Publishing Co., Champaign, Ill., 1976, Chap. 2.

Chapter 2

Properties of Matter

What Is Matter?

Matter is anything that occupies space and has weight. Everything we see or feel is matter, as well as many things we cannot see or feel. Such things as trees, food, machinery, and soil are examples of matter we can see and feel. Air is an example of matter we cannot see, yet we know that it is all around us. However, not all matter is of the same type. Matter can be classified into three types or states.

States of Matter

Matter can exist in the *solid* state, in the *liquid* state, or in the *gaseous* state. A piece of iron is an example of matter in the solid state. A bar of iron has a definite shape, a shape that cannot be easily changed. The volume of the piece of iron (the amount of space that it occupies) is a definite volume; it cannot easily be changed. Also, the piece of iron does not flow.

When a pint of water is poured from a container of one shape into a container of another shape, the water (a liquid) assumes the shape of the new container as far as it fills it. However, the volume of the water remains the same—1 pint—regardless of the shape of the container into which it is poured.

When air is pumped into an empty bottle, the air occupies all of the space and also takes the shape of the container. Forcing more air into the bottle will increase the pressure but the air will still occupy all of the space and take the shape of the container. If the air is transferred to another bottle of different shape and size, again the air will occupy all of the space and take the shape of the container.

In solids, the particles are closely adhering and tightly packed; in liquids, the particles are mobile and relatively close to one another; in gases, the particles are independent and relatively far apart.

Some solids have a high density (gold, 19.3 g/ml), while others have a low density (cork, 0.2 g/ml). Most liquids have a relatively low density (water, 1 g/ml; gasoline, 0.66 g/ml), but mercury has a high density (13.6 g/ml). All gases have a very low density, expressed in the units g/L (air, 1.3 g/L; hydrogen, 0.09 g/L). In summary:

1. Solids have a definite shape, a definite volume, do not flow, have particles that are closely adhering and tightly packed, and may exhibit high or low density.
2. Liquids have no definite shape, do have a definite volume, flow, have particles that are mobile and relatively close to one another, and may have high or low density.
3. Gases have no definite shape, no definite volume, flow, have particles that are independent and relatively far apart, and have a low density.

Changes of State

When water reaches the freezing point it changes to ice (from the liquid to the solid state). When water boils, it changes from a liquid to a gas, steam. These are examples of matter changing from one state to another. Likewise, a piece of metal can be heated enough to melt it (change it from the solid to the liquid state). Further heating can change the liquid metal to vaporized metal, a gas. The reverse steps are also possible. Oxygen is a gas at room temperature. Upon sufficient cooling, it becomes a liquid. Further cooling then changes liquid oxygen to solid oxygen.

In general, then, matter can usually be changed from one state to another merely by changing its temperature. Such changes are called *physical changes*. A physical change is one in which no new substance is produced, although there may be a change of state or a change of color or both. Note that not all substances can be changed from solid to liquid to gas merely by changing the temperature. What about a piece of wood?

The other type of change that matter can undergo is called a *chemical change*. A chemical change is one in which a new substance or new substances are produced that have entirely different properties from the original substance. Burning a piece of wood is an example of a chemical change. New substances—ash and smoke (gas)—are produced. Note that these substances have properties different from those of the original piece of wood.

Properties of Matter

One portion of matter can be distinguished from another by means of its properties. These distinguishing properties of matter may be classified into two main types—*physical properties* and *chemical properties*.

Physical Properties

Physical properties include color, odor, taste, solubility in water, density, hardness, melting point, and boiling point. These physical properties can serve to identify a substance, although not all of these properties may be necessary for the identification. For example, when we say that the color of a substance is white, we automatically eliminate all substances that are not white. Next, if we say that the white substance is odorless, we can eliminate all white objects that have an odor, leaving a smaller number of substances that are both white and odorless. If we continue to eliminate in this manner by using additional physical properties, such as density, hardness, melting point, and boiling point, eventually only one substance will fit all of these properties—the substance we are trying to identify.

Chemical Properties

Properties such as reacting (or not reacting) in air, or reacting (or not reacting) with an acid, or burning (or not burning) in a flame are chemical properties. An object can be identified by means of its chemical properties, but it is usually much simpler to do so by means of the physical properties.

Comparison of Physical and Chemical Properties

A physical property tells what a substance *is*—it is white, or it is green; it is odorless, or it has a sharp odor; it is hard, or it is soft. A chemical property tells what a substance *does*—it burns, or it does not burn; it reacts with an acid, or it does not react with an acid, and so on.

Energy is defined as the ability to do work. The muscles in our bodies get their energy from the chemical reactions that take place in the muscle cells. The heat energy necessary to keep our bodies at a temperature of 98.6°F or 37°C comes from the oxidation of the foods we eat. The electrical energy we use in our homes comes from the burning of a fuel or from atomic energy. Energy exists in several forms—heat, light, electrical, mechanical, sound, chemical, and atomic. Energy may be classified into two categories: *kinetic energy* and *potential energy*.

Kinetic Energy

Kinetic energy is energy associated with motion; that is, energy which is doing something now, such as heat energy obtained from burning wood, light energy from an incandescent lamp, mechanical energy from a motor, and atomic energy from a nuclear reactor.

Potential Energy

Potential energy is stored energy, energy not associated with motion. Examples of potential energy are a dry cell (which can supply electrical energy when it is connected to something), food (which supplies energy to our bodies when it is metabolized), and water at the top of a waterfall (which can supply mechanical energy as it falls to the bottom).

Transformation of Energy

Energy may be transformed from one form into another. Thus, burning a piece of coal changes its potential energy into heat (kinetic) energy. The heat energy thus produced might be used to boil water, which produces large amounts of steam. The steam might be used to drive a generator to produce electrical energy. In turn, this electrical energy might be used to drive a motor (mechanical energy), to produce light in a fluorescent lamp (light energy), to operate a radio (sound energy), or to operate a toaster (heat energy).

The sun produces energy by nuclear reactions and radiates this energy to the earth. Plants on the earth pick up the light energy from the sun during the process of photosynthesis and produce compounds that contain chemical energy. When man eats these compounds in the plants, his body converts the chemical energy into heat energy and mechanical energy.

Conservation of Energy and Matter

The *law of conservation of energy* states that energy is neither created nor destroyed during a chemical reaction. Energy can be changed from one form to another, but the total amount of energy remains the same regardless of what form the energy is changed to.

The *law of conservation of matter* states that during a chemical reaction matter is neither created nor destroyed. This means that the weights of substances before they react should be the same as the weights after they react. For example, if a candle is placed in a sealed container, a certain weight will be obtained for both the candle and the container. If the candle is lit and allowed to burn (inside the sealed container) and if it is weighed as it continues to burn, the weight will be found to remain the same even though part of the candle is disappearing and even though several gaseous products are being produced. The same is true for a camera flashbulb before and after being fired. Other experiments performed under very carefully controlled conditions have produced similar results. That is, the sum of the weights after a reaction are the same as the sum of the weights before a reaction.

In the early twentieth century Albert Einstein stated that matter and energy were interchangeable. That is, under certain conditions, matter could be changed into energy or energy into matter. These changes do not occur under the conditions of an ordinary chemical reaction so that the laws of conservation of energy and of the conservation of matter are still used. However, these laws may be combined into one overall law, which states that matter and energy cannot be created or destroyed, but they can be converted from one to the other.

Measurement of Energy

Heat is the most common form of energy; all other forms of energy can be converted into heat energy. The unit of heat energy is the *calorie*, which is defined as the amount of heat required to raise the temperature of one gram of water one degree Celsius. The calorie is abbreviated as cal. It is a rather small unit of heat.

A larger unit of heat, the kilocalorie or the large calorie, is equal to 1000 small calories. The kilocalorie is abbreviated as Cal, with a capital "C," or as kcal. The large calorie is used when measuring the heat energy of the body and for nutritional values of foods.

There are three principal kinds of foods that produce energy in the body: carbohydrates, fats, and proteins. The oxidation of 1 g of carbohydrate produces 4 Cal (kcal), the oxidation of 1 g of fat produces 9 Cal (kcal), and that of 1 g of protein 4 Cal (kcal).

The number of calories produced by a chemical reaction may be calculated in terms of the amount of water that can be heated from one temperature to another, by using the following formula

No. calories = No. grams of water × change in temperature in °C

Example 1. How many calories are produced during the oxidizing (burning) of a piece of food if the heat is sufficient to warm 2000 g of water from 20°C to 38°C?

No. calories = No. grams of water × change in temperature in °C

$$= 2000 \times (38 - 20)$$

$$= 36{,}000 \, \text{cal} = 36 \, \text{Cal(kcal)}$$

Example 2. How much carbohydrate must be oxidized in the body to produce 36 Cal? How much fat? Protein?

Since the oxidation of 1 g of carbohydrate produces 4 Cal, it will require 36/4 g, or 9 g, of carbohydrate to produce 36 Cal.

Since the oxidation of 1 g of fat produces 9 Cal, it will require 36/9 g or 4 g of fat.

Since the oxidation of 1 g of protein produces 4 Cal, as did carbohydrates, it will require 9 g of protein.

Composition of Matter

All matter can be divided into three classes, depending upon the properties of the material being considered. These three classes are *elements*, *compounds*, and *mixtures*.

Elements

Elements are the building blocks of all matter. An element can be defined as a substance that cannot be broken down into any simpler substance by ordinary chemical means. There are more than 100 elements known to man. The names of some of these elements, such as oxygen, hydrogen, iron, carbon, copper, mercury, and uranium, are probably familiar. The names of some of the rarer elements, such as cesium, einsteinium, molybdenum, and xenon, may not be familiar. Many elements are in common industrial use. Consider iron in steel, germanium in transistors, and tungsten in incandescent light bulbs.

Because an element cannot be broken down into anything simpler by ordinary chemical means, it must contain only one type of substance.

Classification. Elements can be classified into three main types: metals, nonmetals, and noble gases. Each type—the metal, the nonmetal, and the noble gas—has its own specific properties.

Metals conduct heat and electricity. They have a luster (a shiny surface) similar to that of silver and aluminum. Metals reflect light. Some metals are ductile (they can be drawn out into a thin wire). Some metals are malleable (they can be pounded out into thin sheets). Some metals have a high tensile strength. Metals such as iron, copper, and silver are solid at room temperature, but mercury is a liquid.

Nonmetals do not conduct heat and electricity very well. They have little luster and seldom reflect light. Nonmetals frequently are brittle. They cannot be pounded into thin sheets (they are not malleable) and cannot be drawn into thin wires (they are not ductile). Nonmetals may be solids at room temperature (carbon and sulfur), liquid at room temperature (bromine), or gaseous at room temperature (oxygen and nitrogen). There are many more metallic elements than nonmetallic; see the periodic table inside the front cover.

Rare gases are gases at room temperature. They are relatively unreactive and were formerly called inert or noble gases. Among the rare gases are the elements helium, neon, and krypton.

Symbols for the Elements. Each element can be identified by a symbol that represents that element. The symbol C stands for the element carbon, S for sulfur, and O for oxygen. In these instances, the symbol is the first letter of the name of the element. But conflicts would arise if the first letter were to be used for every element, so the first two letters are used for some elements, such as Ca for calcium and Al for aluminum. Or the symbol may use the first letter and one other letter to suggest a sound that is apparent in the name, such as Zn for zinc and Cl for chlorine. Note that when two letters are used to form a symbol, the first letter is capitalized and the second letter is small.

The symbols for some elements are based upon their Latin names. Ag, the symbol for silver, comes from the Latin word *argentum*. Fe, the symbol for iron, comes from the Latin word *ferrum*.

Elements Present in the Human Body. Table 2–1 lists the elements necessary for life, the symbols for these elements, their functions, and the percentage present in the body.

Compounds

When an electrical current is passed through a container of water, the water is decomposed into two gases, each having its own set of properties. One gas is hydrogen and the other is oxygen. If sugar is placed in a test tube and heated strongly, drops of water will collect at the top of the test tube while pieces of carbon will remain at the bottom. Both of these changes are chemical changes; that is, some new products have been formed. Both sugar and water are classified as compounds, as are such other substances as salt, boric acid, carbon dioxide, and ether. Both water and sugar can be broken down into other substances by chemical means. Compounds, then, are substances that can be broken down into simpler substances by chemical means. Compare this definition with that of elements, which cannot be broken down into simpler substances by ordinary means.

Take a crystal of sugar from a sugar bowl. Examine it carefully. List its physical and chemical properties. Take another crystal of sugar from the sugar bowl and again list its physical and chemical properties. The properties are identical, as are the pieces of sugar. That is, the sugar in a bowl of sugar is homogeneous; it is the same throughout.

TABLE 2–1

Elements of Life

Element	Symbol	Percentage in Human Body	Function
Oxygen	O	65.00	Required for water and organic compounds
Carbon	C	18.00	Required for organic compounds
Hydrogen	H	10.00	Required for water and organic compounds
Nitrogen	N	3.00	Required for many organic compounds, particularly protein
Calcium	Ca	2.00	Required for bones, teeth, and certain enzymes; necessary for clotting of blood
Phosphorus	P	1.00	Required for bones, teeth, and energy transfer
Potassium	K	0.35	Principal positive cellular ion
Sulfur	S	0.25	Required for some protein and organic compounds
Sodium	Na	0.15	Principal positive extracellular ion
Chlorine	Cl	0.15	Principal negative ion in body
Iodine	I		Required for thyroid hormones
Fluorine	F		Required for bones and teeth; inhibitor of some enzymes
Iron	Fe		Required for hemoglobin and many enzymes
Magnesium	Mg		Required for many enzymes
Copper	Cu		Required for many enzymes and for the synthesis of hemoglobin
Zinc	Zn	Trace amounts	Required for many enzymes; related to action of insulin
Cadmium	Cd		Found in renal cortex; function unknown
Manganese	Mn		Required for some enzymes acting in the mitochondria
Cobalt	Co		Required for vitamin B_{12}
Molybdenum	Mo		Required for some enzymes
Chromium	Cr		Related to action of insulin
Selenium	Se		Essential to liver function
Vanadium	V		Essential in some plants; function unknown
Tin	Sn		Essential in rats; function unknown
Silicon	Si		Essential in chicks
Nickel	Ni		Possibly essential in chicks
Boron	B		Essential in some plants; function unknown

If a sample of sugar is analyzed, it will be found to contain carbon, hydrogen, and oxygen. The percentage of carbon in a portion of sugar can be calculated by weighing the piece of sugar as it is and again after heating it until nothing is left but carbon. Repeating this procedure with another portion of sugar will yield the same percentage of carbon. Likewise, analysis of several portions of sugar would give identical results for the percentage of oxygen and also for the percentage of hydrogen. The analysis of many samples of water gives a constant percentage of oxygen and hydrogen. The result of examining many different compounds for the percentages of their components can be stated as the *law of definite proportions*—compounds have a definite proportion or percentage by weight of the substances from which they were made.

It has already been mentioned that compounds have properties that are entirely different from those of the substances from which they were made. That is, water, a compound, has entirely different properties than does the oxygen and hydrogen from which it is made.

Compounds, then, have the following characteristics:

1. They can be separated into their component substances by chemical means.
2. They are homogeneous in composition.
3. They have a definite proportion by weight of the substances from which they were made.
4. They have different properties from those of the substances from which they were made.

Mixtures

If a few crystals of salt are dissolved in a cup full of water, a mixture of salt and water—or salt water—is produced. If a few more crystals of salt are added to that same salt water, a little stronger salt solution is produced. If a teaspoonful of salt is added to the cup and stirred until it is dissolved, an even stronger salt solution will be produced. The salt and water formed a mixture because they were placed together and stirred. The salt water made by dissolving different amounts of salt in water will have a varying composition, depending upon how much salt was added to the water. Thus, one property of a mixture is that it can have a variable composition or variable proportions. More water, more salt, or both salt and water can be added to change the strength of the mixture, but in each case salt water is produced.

The property of variable proportions is characteristic of any mixture. If sugar and sand are stirred together, a mixture is produced regardless of the amounts of each used. Likewise, sugar and iron filings can be mixed, and again, the proportion of iron filings to sugar does not alter the fact that it is a mixture.

The ingredients of any mixture can be separated from each other by such physical processes as evaporation and filtering. Salt can be separated from a salt-water mixture merely by evaporating the water. The sugar-sand mixture can be separated by placing both of them in water, stirring to dissolve the sugar, filtering the solution, and recovering the sand. The sugar can be recovered from the filtered solution by evaporating the water. A sugar-iron filings mixture can be separated by passing a magnet over the mixture. The magnet will attract the iron filings, leaving the sugar behind. Mixtures can also be separated by a process known as chromotography (see page 316).

Evaporation, chromotography, and separation by means of a magnet are examples of a physical change—one in which no new substance is produced. In each of these mixtures, the individual substances retained their own properties. There is no evidence of a chemical reaction because no new substance is produced. Thus, in the salt-water mixture, the salt retains its own properties, as does the water. In the sugar-sand mixture each retains its own properties. The sugar and the sand can be recognized separately under a microscope.

In summary, then, mixtures have the following characteristics:

1. They have no definite proportion or composition.
2. They can be separated into their component substances by physical means.
3. They retain the properties of the individual substances from which they were made.

Mixtures may be either homogeneous or heterogeneous in composition. *Homogeneous* means the same composition throughout. *Heterogeneous* means different composition throughout. Let us consider two of the mixtures already discussed, the salt-water and the sugar-sand mixtures.

Samples of salt-water are found to be identical from the top, from the middle, and from the bottom of a container of salt water. That is, the salt-water mixture is homogeneous. It is of the same composition throughout.

Now consider a mixture of sugar and sand. Samples from different parts of the mixture, when examined carefully under a microscope or with a magnifying glass, would not appear to have the same composition. That is, a mixture of sugar and sand is heterogeneous.

A mixture can be either homogeneous or heterogeneous. What determines which it shall be? The answer is quite simple. A mixture of two or more solids is heterogeneous. One solid can be distinguished from another, even if they are ground together in making the mixture. On the other hand, a solution is a homogeneous mixture. When a substance is dissolved in a liquid such as water, the mixture becomes the same throughout; it is homogeneous.

Although both the salt-water and sugar-sand mixtures contained two different substances, mixtures can be made from any number of substances. Note the word "substances." It can mean either elements or compounds. That is, a mixture can be composed of two (or more) elements (powdered iron and powdered sulfur), two (or more) compounds (sugar and salt), or both elements and compounds (iodine and water).

Summary

Matter is anything that occupies space and has weight. Matter can exist in the solid, the liquid, or the gaseous state. These states of matter are interchangeable and depend primarily upon temperature. Physical properties of matter describe what a substance is; chemical properties describe what a substance does.

Energy is the ability to do work. The two types of energy are kinetic, the energy associated with motion, and potential, or stored, energy. The law of conservation of energy states that energy may be transformed from one type to another but cannot be created or destroyed. The law of conservation of matter states that matter cannot be created or destroyed. These two laws may be combined into one stating that matter and energy cannot be created or destroyed but can be converted from one to the other.

The calorie is the unit of heat energy and is the amount of heat required to raise the temperature of 1 g of water 1°C. One large calorie, 1 Cal (kcal), equals 1000 cal.

Matter may be divided into three categories: elements, compounds, and mixtures.

Elements are substances that cannot be broken down into simpler substances by ordinary means. Elements are homogeneous in composition. Elements are either metals or nonmetals. Each element may be represented by a symbol.

Compounds can be separated into their component substances by chemical means, are homogeneous in composition, have a definite proportion by weight of their component substances, and have different properties from those of the substances from which they were made.

Mixtures have no definite proportions or composition, can be separated into their component substances by physical means, and retain the properties of the individual substances from which they were made. Mixtures can be made from two or more substances. Mixtures of solids are heterogeneous; mixtures made by dissolving a substance in a liquid are homogeneous in composition.

Questions and Problems

1. Compare the properties of solids, liquids, and gases as to definite shape and volume and also as to density.
2. Which of the following properties are physical and which are chemical?
   (a) Odor
   (b) Reactivity
   (c) Taste
   (d) Boiling point
   (e) Flammability
   (f) Density
   (g) Color
   (h) Melting point
3. Which of the following substances are elements, which are compounds, and which are mixtures?
   (a) Milk
   (b) Zinc
   (c) Mercury
   (d) Paint
   (e) Carbon
   (f) Water
   (g) Air
   (h) Table salt
4. Distinguish between potential energy and kinetic energy.
5. State the laws of conservation of energy, conservation of matter, conservation of matter and energy.
6. How many calories are required to change the temperature of 500 g of water from 15°C to 27°C? How many grams of protein must be oxidized to produce this energy?
7. Which of the following processes involves a physical change and which a chemical change?
   (a) Breaking glass
   (b) Burning wood
   (c) Boiling water
   (d) Digesting food

(e) Distilling mercury   (g) Winding a clock
(f) Electrolysis of water   (h) Souring of milk

8. Which *must* be homogeneous: elements, compounds, or mixtures? Which *may* be?
9. What elements do the following symbols represent?

   (a) O   (e) Na   (h) F
   (b) N   (f) Cl   (i) S
   (c) Fe   (g) H   (j) Ag
   (d) K

10. What are the symbols for the following elements?

    (a) Magnesium   (e) Phosphorus   (h) Bromine
    (b) Calcium   (f) Zinc   (i) Mercury
    (c) Iodine   (g) Copper   (j) Boron
    (d) Carbon

11. Why are gases compressible?
12. Can all substances be changed from solid to liquid to gas? Explain.

References

Ault, F. K., and Lawrence, R. M.: *Chemistry: A Conceptual Introduction.* Scott, Foresman & Co., Glenview, Ill., 1976, Chap. 2.

Baum, S. J., and Scaife, C. W.: *Chemistry: A Life Science Approach.* Macmillan Publishing Co., Inc., New York, 1975, Chap. 1.

Fernandez, J. E., and Whitaker, R. D.: *An Introduction to Chemical Principles.* Macmillan Publishing Co., Inc., New York, 1975, Chap. 2.

Frieden, E.: The chemical elements of life. *Scientific American,* **227**: 52–60 (July), 1972.

Hein, M., and Best, L. R.: *College Chemistry.* Dickenson Publishing Co., Inc., Encino, Calif., 1976, Chap. 3.

Masterson, W. L., and Slowinski, E. J.: *Chemical Principles*, 3rd ed. W. B. Saunders Co., Philadelphia, 1973, Chap. 1.

Nebergall, W. H.; Schmidt, F. C.; and Holtzclaw, H. F., Jr.: *College Chemistry*, 5th ed. D. C. Heath & Co., Lexington, Mass., 1976, Chap. 1.

Pauling, L., and Pauling, P.: *Chemistry.* W. H. Freeman & Co., San Francisco, 1975, Chap. 1.

Peters, E. I.: *Introduction to Chemical Principles.* W. B. Saunders Co., Philadelphia, 1974, Chap. 3.

Sienko, M. J., and Plane, R. A.: *Chemistry*, 5th ed. McGraw-Hill Book Co., New York, 1976, Chap. 1.

Chapter 3

Structure of Matter

The Atom

The symbol of an element not only represents that element, it also represents one atom of that element. But what is an atom? We have all heard of atoms in connection with atomic bombs, "splitting the atom," and atomic power.

Consider a bar of iron. Iron is an element. It has certain properties. Cutting the bar in half produces two pieces of iron. Both pieces have the same properties as the original bar. Continued cutting produces smaller and smaller pieces, all with identical properties.

But, in time, we could theoretically arrive at the smallest piece of iron attainable. This smallest piece of iron is an atom—an atom of iron. If this atom of iron were cut into two, particles with different properties would be produced. It would no longer be iron. Thus, an atom can be defined as the smallest portion of an element that retains all of the properties of the element.

A piece of iron is made up of many atoms of iron, a piece of copper of many atoms of copper, and a piece of silver of many atoms of silver. The atoms of one element differ from those of another and so give characteristic properties to each element. Atoms are called building blocks of the universe. A chemist uses different kinds of atoms to build chemical compounds just as the different letters of the alphabet are used to form words. Since there are more than 100 elements, there are more than 100 different kinds of atoms.

Size

Although an atom is extremely small, its size can be accurately measured. An atom has a diameter of approximately one hundred-millionth of a centimeter (1/100,000,000 cm). Because an atom is so small, one hundred trillion of them (100,000,000,000,000) could be placed on the head of a pin. And the head of a pin itself is a tiny object; it measures only about 1/25 inch in diameter.

Weight

An atom is extremely small; therefore, it is not surprising that an atom weighs very little. In fact, it would take 18 million billion billion (18,000,000,000,000,000,000,000,000) hydrogen atoms to weigh 1 oz.

Atomic Weight

One atom cannot be weighed even on the most sensitive weighing device. However, the weights of individual atoms can be determined accurately by weighing large numbers of them. As would be expected, the weights of individual atoms are infinitesimal. But the chemist is not as interested in the exact weights of atoms as he is in their relative weights. The relative weight of an atom is called its atomic weight. In his work the chemist uses atomic weights rather than exact weights. Atomic weights are easier to use than the exact weights and they are just as accurate because they can be determined very precisely.

What does the term *relative weight* mean? The chemist has arbitrarily given the carbon-12 atom a weight of 12.0000 units.[1] He then compares the weights of atoms of all other elements with this weight. Thus if an atom of an element is exactly twice as heavy as the carbon-12 atom, that element is

[1] In 1961, by international agreement, the atomic weights of the elements were based upon the carbon-12 atom having a relative weight of 12.0000. Previously, the atomic weights had been based upon oxygen at 16.0000.

assigned a relative weight, or atomic weight, of 24.0000. For our purposes, we will usually use the atomic weights as whole numbers, ignoring the decimal values. Thus, the atomic weight of carbon-12 is 12, that of oxygen is 16, and that of sodium is 23. But for precise work, the exact atomic weights must be used.

The atomic weights of all the elements are listed in the periodic table inside the front cover and also in the chart inside the back cover.

Inside the Atom

The early chemists believed that the atom was indivisible, that it could not be broken down into any simpler substances. Modern theory states that the atom is composed of a small, heavy nucleus with particles revolving around it at relatively great distances. Thus an atom is composed mostly of empty space, that space being between the nucleus and the revolving particles.

If the nucleus of an atom could be expanded so that it was about 400 ft in diameter, the closest revolving particle would be 4000 miles away. Between that nucleus and the closest revolving particle would be empty space, 4000 miles of it. Actually, over 99.9 per cent of the volume of an atom is empty space.

Fundamental Particles

Atoms are considered to be made primarily of three fundamental particles, the proton, the electron, and the neutron.[2] The proton (p) has a charge of positive one ($+1$) and weight of 1 atomic mass unit (amu) (relative weight, compared to 12 for carbon-12). Protons are located inside the nucleus of the atom. The electron (e) has a negative charge (-1) and a weight of 1/1837 amu. Electrons are located outside the nucleus of the atom. The neutron (n) has no charge; it is neutral, as the name implies. It has a weight of 1 amu. Neutrons are located inside the nucleus (see Table 3–1). In addition to these three

TABLE 3–1

Particle	Symbol	Charge	Weight, amu	Location in the Atom
Proton	p	$+1$	1	Inside nucleus
Electron	e	-1	1/1837	Outside nucleus
Neutron	n	0	1	Inside nucleus

[2] Theoretic physicists now believe that electrons are fundamental particles but protons and neutrons are made of even smaller particles called quarks. However, in this discussion we shall assume that protons, neutrons, and electrons are all fundamental particles.

fundamental particles there are many, many more particles—among them
the positron, the meson, and the neutrino—but a discussion of these is beyond
the scope of this book.

Atomic Number

Each element has a given atomic number which represents that element
and no other. The atomic number indicates the number of protons in an atom
of that element. Because the protons are located in the nucleus, the atomic
number indicates the number of protons that are in the nucleus of that atom.
However, since all atoms are electrically neutral, there must be as many
electrons (negative charges) as protons (positive charges). Therefore, the
atomic number also tells the number of electrons in the atom, these electrons
being located outside the nucleus.

Number of Neutrons in the Nucleus

Because the weight of the electron is quite small (1/1837th of that of the
proton or the neutron) and because only the electrons are located outside the
nucleus of the atom, practically all the weight of the atom is located in its
nucleus. The atomic number of an element indicates the number of protons
in its nucleus. Knowing that each proton weighs 1 amu, you can calculate the
total weight of these protons. The rest of the weight of the atom must be
due to the neutrons in the nucleus. Knowing that each neutron weighs 1 amu,
you can calculate the number of neutrons present in the nucleus.

For example, if the atomic number of an element is 5, there must be five
protons in the nucleus of the atom and also five electrons outside that nucleus.
If the mass number (relative weight of one atom) of that element is 11, the
number of neutrons may be calculated as follows: the five electrons outside
the nucleus weigh practically nothing so their weight may be ignored; as the
five protons weigh 5 units and the whole atom weighs 11 units, the neutrons in
it must weigh 6 units. Knowing that each neutron weighs 1 amu, the number of
neutrons must be 6. *The number of neutrons can be found by subtracting the
atomic number of an element from its mass number.*

Structure of the Atom

What does the atom look like? The simplest atom, the hydrogen atom, has
the atomic number 1 and a mass number of 1. The atomic number indicates
one proton inside the nucleus of this atom; it also indicates one electron outside that nucleus. The number of neutrons can be calculated by subtracting
the atomic number (1) from the mass number (1); thus, there are no neutrons

in the nucleus of a hydrogen atom. The hydrogen atom can be represented as follows

$1\,e$

Hydrogen, atomic number 1, mass number 1

The circle represents the nucleus of the atom, the letter p the protons, e the electrons, and n the neutrons.

The element helium—atomic number 2 and mass number 4— has two protons in the nucleus and two electrons outside that nucleus. The number of neutrons is two (atomic number subtracted from mass number). The helium atom can be represented as follows

$2\,e$

Helium, atomic number 2, mass number 4

The sodium atom—atomic number 11 and mass number 23—has 11 protons in its nucleus, 11 electrons outside its nucleus, and 12 neutrons ($23 - 11$) in its nucleus.

$11\,e$

Sodium, atomic number 11, mass number 23

The uranium atom—atomic number 92 and mass number 238—has in its nucleus 92 protons and 146 neutrons. Outside the nucleus are 92 electrons.

$$\left(\begin{array}{c} 92\,p \\ 146\,n \end{array}\right) \qquad 92\,e$$

Uranium, atomic number 92, mass number 238

Isotopes

The periodic chart at the front of the book indicates that the element chlorine has an atomic number of 17 and an atomic weight of 35.5 (to one decimal place). According to the discussion in the previous paragraphs, the chlorine atom should have 17 protons in its nucleus, 17 electrons outside its nucleus, and 18.5 neutrons ($35.5 - 17$) in its nucleus. However, a neutron is a fundamental particle so that there can never be a fraction of a neutron. How can this problem be resolved?

The answer to this problem is that there are two types of chlorine atoms.

One chlorine atom has a mass number of 35 and the other has a mass number of 37. (Chlorine with a mass number of 36 does not exist in nature.) Both of these types of atoms of chlorine have an atomic number of 17. These two varieties of chlorine are termed *isotopes*. Isotopes are defined as atoms of an element having the same atomic number but different mass numbers. The first isotope of chlorine—atomic number 17 and mass number 35—has 17 protons in its nucleus, 17 electrons outside its nucleus, and 18 neutrons (35 − 17) in its nucleus. The second isotope of chlorine—atomic number 17 and mass number 37—has 17 protons in its nucleus, 17 electrons outside its nucleus, and 20 neutrons (37 − 17) in its nucleus.

Chlorine, mass number 35   Chlorine, mass number 37

The atomic weight is the average weight of all of the isotopes. If the two isotopes of chlorine, mass numbers 35 and 37, were present in equal amounts, the atomic (average) weight would be 36. However, since the atomic weight is listed as 35.5, the isotope of mass number 35 must be the predominant one because the atomic weight is closer to 35 than to 37.

The element carbon—atomic number 6—has three isotopes. Their mass numbers are 12, 13, and 14. They all have atomic number 6, which means that they all have six protons in their nucleus and six electrons outside their nucleus. The isotope of mass number 12 has six neutrons in its nucleus; the istotope of mass number 13 has seven neutrons in its nucleus; and the isotope of mass number 14 has eight neutrons in its nucleus. Carbon-12 is the most abundant since the atomic weight of carbon is 12.011, indicating small amounts of the other isotopes.

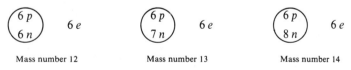

Mass number 12   Mass number 13   Mass number 14

Three isotopes of carbon, atomic number 6

Isotopes have been defined as atoms of an element having the same atomic number but different mass numbers; therefore isotopes of an element must have the same number of protons and electrons but different numbers of neutrons.

Most of the known elements have isotopes. Some have only two, whereas others have many more. In addition to the naturally occurring isotopes, there are many more artificially prepared isotopes. These isotopes will be discussed in the chapter on radioactivity (Chapter 4).

Arrangement of the Electrons in the Atom

The electrons are not located just anywhere outside the nucleus. There are a definite order and arrangement for them as they revolve about the nucleus. The electrons travel around the nucleus in *energy levels* or *shells*. A shell represents a volume occupied by an electron cloud. Formerly, the electrons were compared to planets revolving around the sun and were called planetary electrons. The electrons do travel around the nucleus, but they do so at tremendous speeds. The electron in the hydrogen atom revolves around its nucleus at the rate of almost 6 million billion revolutions per second. This traveling electron moves so rapidly that it appears to be everywhere at the same time; that is, there appears to be an electron cloud around the nucleus of the atom.

Where there are many electrons in an atom, they revolve around the nucleus in different shells or electron clouds. For our purpose we shall use the terms electron cloud and electron shell interchangeably.

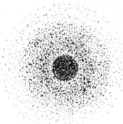 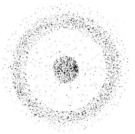

Electron clouds

The maximum number of electrons in each shell may be calculated from the formula $2n^2$, where n is the number of the shell counting out from the nucleus. Thus, the first shell holds a maximum of two electrons (if $n = 1$, $2n^2 = 2$); the second shell holds a maximum of eight electrons (if $n = 2$, $2n^2 = 8$), and the third shell has a maximum of 18 electrons. *The first shell must be completely filled before electrons can begin filling the second shell; the second shell must be completely filled before electrons can begin filling the third shell.*

The element hydrogen—atomic number 1 and mass number 1—has one proton and no neutrons in its nucleus and one electron outside of its nucleus. This one electron can go into the first shell so that the hydrogen atom may be represented as follows

Hydrogen, atomic number 1, mass number 1

The curved line indicates the first shell or energy level.

The element helium—atomic number 2 and mass number 4—has two protons and two neutrons in its nucleus. The helium atom has two electrons out-

side its nucleus. These two electrons can go into the first shell, which can hold
a maximum of two electrons, as shown below.

$$\left(\begin{array}{c} 2\,p \\ 2\,n \end{array}\right) \quad \Big) 2\,e$$

Helium, atomic number 2, mass number 4

'The lithium atom—atomic number 3 and mass number 7—has three protons and four neutrons in its nucleus. Outside of its nucleus it has three electrons. Two of these electrons can go into the first shell or energy level. The third electron must go into the second shell because the first shell can hold only two electrons.

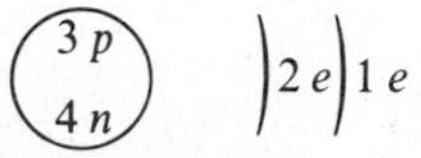

$$\left(\begin{array}{c} 3\,p \\ 4\,n \end{array}\right) \quad \Big) 2\,e \Big) 1\,e$$

Lithium, atomic number 3, mass number 7

The element sodium—atomic number 11 and mass number 23—has 11 protons and 12 neutrons in its nucleus and 11 electrons outside its nucleus. The first shell can hold two electrons and the second shell eight electrons, so that the one remaining electron must go into the third shell. Therefore, the structure of the sodium atom is as follows:

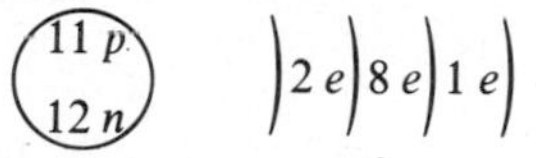

$$\left(\begin{array}{c} 11\,p \\ 12\,n \end{array}\right) \quad \Big) 2\,e \Big) 8\,e \Big) 1\,e \Big)$$

Sodium, atomic number 11, mass number 23

Table 3–2 lists the progression of atomic numbers and the electron arrangements in the first 18 elements. The arrangement of the electrons for the heavier elements does not follow the simple rules mentioned above and will not be discussed in this text. For a discussion of the electron arrangements of the elements with larger atomic numbers, consult one of the references listed at the end of this chapter.

The Periodic Table

Toward the end of the nineteenth century, chemists tried to group the elements into certain families because of their similarities in properties and reactions. This was the beginning of the periodic table.

The modern periodic table places the elements according to the number and arrangement of the electrons in the atom. Look at the periodic table inside the front cover of this book. Note that it is divided into horizontal and vertical columns. Each box in the table represents one element. There are over 100 elements known to man; thus, there are over 100 different kinds of

TABLE 3–2

Electron Arrangements for the First 18 Elements

| Element | Symbol | Atomic Number | Electron Arrangement | | |
			First Shell	Second Shell	Third Shell
Hydrogen	H	1	1		
Helium	He	2	2		
Lithium	Li	3	2	1	
Beryllium	Be	4	2	2	
Boron	B	5	2	3	
Carbon	C	6	2	4	
Nitrogen	N	7	2	5	
Oxygen	O	8	2	6	
Fluorine	F	9	2	7	
Neon	Ne	10	2	8	
Sodium	Na	11	2	8	1
Magnesium	Mg	12	2	8	2
Aluminum	Al	13	2	8	3
Silicon	Si	14	2	8	4
Phosphorus	P	15	2	8	5
Sulfur	S	16	2	8	6
Chlorine	Cl	17	2	8	7
Argon	Ar	18	2	8	8

atoms, each having its own place in the periodic table. The vertical columns are called *groups* and are numbered with Roman numerals. The horizontal rows are called *periods* and are numbered with Arabic numbers.

Each box on the chart contains a symbol with a number above and below that symbol. The name of the element is given above the symbol. Consider the box with the symbol O in it:

| 8 |
| Oxygen |
| O |
| 16.00 |

The symbol O represents the element oxygen. The atomic number of the element (8) is given above the name. Below the symbol is the atomic weight, 16.00.

Under group IA of the periodic chart are such elements as hydrogen (H), lithium (Li), and sodium (Na). The electron arrangement of these atoms is given in Table 3–3.

TABLE 3–3

	First Shell	Second Shell	Third Shell
Hydrogen	1		
Lithium	2	1	
Sodium	2	8	1

Note that they all have one electron in their outermost shell. All are listed under group IA. (A-group elements are commonly called main-group elements.)

Consider the elements beryllium (Be) and magnesium (Mg). Each of these elements has two electrons in its outermost shell; they are in group IIA. Likewise, in group IIIA the elements aluminum (Al) and boron (B) have three electrons in their outermost shell. In group VIIA the elements fluorine (F) and chlorine (Cl) each have seven electrons in their outermost shell.

In general, the A-group number corresponds to the number of electrons in the outermost shell. The principal exception to this rule is group 0 at the far right of the table. This group contains the rare gases. The rare gases all have eight electrons in their outer shell except for helium which contains only two (because the first shell is filled when it contains two electrons).

The B-group elements are called "transition" elements. They are all metals and usually have two electrons in their outer shell.

Reading horizontally, in period 1 there are only two elements, hydrogen and helium. Both of these elements have an electron or electrons in the first shell only. In period 2 there are eight elements, all of which have one or more electrons in their outer shell, the second shell. In period 3 there are 18 elements, all having electrons in the third shell.

Another generalization, then, is that the period corresponds to the number of shells of electrons in the atom. Thus, the element oxygen—group VIA and period 2—has six electrons in its outer shell (from group VIA) and two shells of electrons (from period 2). The element vanadium (symbol V) is in group VB, period 4. Vanadium has two electrons in its outer shell (B-group elements usually have two electrons in their outer shell) and 4 shells of electrons (from period 4).

Summary

Atoms have an extremely small size and weight. The weights of all atoms are compared to that of the carbon-12 atom, which has been assigned a weight of 12.0000 atomic mass units (amu). These relative weights are called atomic weights.

The atom is composed of a nucleus containing protons and neutrons and of electrons revolving around that nucleus. Both protons and neutrons weigh 1 amu; electrons weigh practically nothing. Almost the entire weight of the atom is therefore in its nucleus.

The atomic number of an element indicates the number of protons inside the nucleus of an atom of that element and also the number of electrons outside that nucleus.

The number of neutrons may be found by subtracting the atomic number of an element from its mass number.

The electrons revolve around the nucleus in what are called electron clouds or shells. The first shell holds a maximum of two electrons, the second a maximum of eight electrons, and the third shell a maximum of 18 electrons.

The first electron shell must be filled before electrons can begin filling the second shell. The second electron shell must be filled before electrons can begin filling the third shell.

Isotopes are atoms having the same atomic number but different mass numbers.

Most elements have isotopes; some have two, others have several.

The periodic table lists all of the elements in the order of their atomic numbers. The vertical columns are called groups and the horizontal rows are called periods. The A-group number indicates the number of electrons in the outer shell; the B-group elements are called transition elements. They all are metals and usually have two electrons in their outer shell; the period indicates the number of electron shells.

Questions and Problems

1. Diagram the following atoms showing protons, neutrons, and electrons in each shell.

Symbol	Atomic Number	Mass Number
N	7	14
Be	4	9
F	9	19
S	16	32
Ne	10	20
Si	14	28

2. Diagram the structures of the following isotopes.

Symbol	Atomic Number	Mass Number
H	1	1,2,3
O	8	16,17,18
Mg	12	24,25,26
N	7	13,14
Ne	10	20,22

3. Using the periodic table inside the front cover of this book, predict the number of electrons in the outer shell and also the number of electron shells in the following elements

(a) Beryllium   (Be)
(b) Phosphorus (P)
(c) Bromine   (Br)
(d) Calcium   (Ca)
(e) Rubidium (Rb)
(f) Arsenic   (As)
(g) Radium   (Ra)
(h) Zirconium (Zr)
(i) Manganese (Mn)
(j) Actinium   (Ac)
(k) Radon   (Rn)

4. How do isotopes differ from one another with respect to chemical properties? Physical properties?

5. What is an atom?

6. State the mass, charge, and location in the atom of the three fundamental particles.
7. Where is the mass of the atom concentrated?
8. What are the vertical columns in the periodic table called?
9. What are the horizontal rows in the periodic table called?
10. All the elements in the periodic table are not listed in the order of increasing atomic weights. Give two examples of exceptions.
11. Are all A-group elements metals? All B-group elements?
12. Group O contains what type of elements?
13. In which part of the periodic chart are the metals located? The nonmetals? The rare gases?
14. Elements 93–106 have been prepared synthetically. Are they all transition elements?

References

Ault, F. K., and Lawrence, R. M.: *Chemistry: A Conceptual Introduction.* Scott, Foresman & Co., Glenview, Ill., 1976, Chap. 3.

Baum, S. J., and Scaife, C. W.: *Chemistry: A Life Science Approach.* Macmillan Publishing Co., Inc., New York, 1975, Chap. 2.

Fernandez, J. E., and Whitaker, R. D.: *An Introduction to Chemical Principles.* Macmillan Publishing Co., Inc., New York, 1975, Chap. 8.

Hein, M., and Best, L. R.: *College Chemistry.* Dickenson Publishing Co., Inc., Encino, Calif., 1976, Chaps. 4, 5, 6.

Longo, F. R.: *General Chemistry.* McGraw-Hill Book Co., New York, 1974, Chaps. 3, 4.

Masterson, W. L., and Slowinski, E. J.: *Chemical Principles,* 3rd ed. W. B. Saunders Co., Philadelphia, 1973, Chap. 2.

Nebergall, W. H.; Schmidt, F. C.; and Holtzclaw, H. F., Jr.: *College Chemistry,* 5th ed. D. C. Heath & Co., Lexington, Mass., 1976, Chap. 3.

Pauling, L., and Pauling, P.: *Chemistry.* W. H. Freeman & Co., San Francisco, 1975, Chaps. 2, 3.

Peters, E. I.: *Introduction to Chemical Principles.* W. B. Saunders Co., Philadelphia, 1974, Chap. 4.

Sackheim, G. I.: *Chemical Calculations,* 10th ed., series B. Stipes Publishing Co., Champaign, Ill., 1976, Chap. 9.

Sienko, M. J., and Plane, R. A.: *Chemistry,* 5th ed. McGraw-Hill Book Co., New York, 1976, Chaps. 2, 3.

Chapter 4

Radioactivity

Discovery of Radioactivity

In 1896 a French physicist, Henri Becquerel (1852–1908), found that uranium crystals had the property of "fogging" a photographic plate that had been placed near those crystals. This fogging took place even though the photographic plate was wrapped in black paper. By placing crystals of uranium on a photographic plate covered with black paper and then developing the plate, he obtained a self-photograph of the crystals. Becquerel concluded that the uranium gave off some kind of radiation or rays that affected the photographic plate. Figure 4–1 illustrates an autoradiograph (self-picture) of a radioactive bone section.

Substances like uranium that spontaneously give off radiation are called *radioactive*. *Radioactivity* is the property that causes an element to emit radiation. This radiation comes from the nucleus of the atom.

Types of Radiation Produced by a Radioactive Substance

The following experiment was performed to study the radiations produced by a radioactive element (Figure 4–2). A piece of radium was placed at the bottom of a thick lead well. The purpose of the lead was to absorb all the radiation except that going directly upward. The escaping radiation was allowed to fall on a photographic plate. When these radiations were passed through a strong electrostatic field, three different areas showed up on the photographic plate. This indicated that there were actually three different kinds of radiation. These were called alpha, beta, and gamma.

Alpha Particles

Alpha particles (α particles) are attracted toward the negative field, which indicates that they are positively charged. Alpha particles, or alpha rays as they are sometimes called, consist of positively charged helium nuclei; that is, they consist of the nuclei of helium atoms (each of which contains two protons and two neutrons). Alpha particles have a very low penetrating power and so are relatively harmless when they strike the body because they do not penetrate the outer layer of the skin.

Beta Particles

Beta particles (β particles) are attracted toward the positive electrostatic field, which indicates that they consist of negatively charged particles. Beta particles, or beta rays, consist of high-speed electrons that travel at a speed in excess of 100,000 miles per second. Beta particles have a slight penetrating power. When they strike the body, they penetrate only a few millimeters of skin and do not reach any vital organs.

Note that beta particles are deflected by the electrostatic field to a much greater extent than are the alpha particles. This indicates that beta particles have a much smaller mass than alpha particles.

Gamma Rays

Gamma rays (γ rays) are not affected by an electrostatic field because they have no charge. They are not particles at all but are a form of electromagnetic

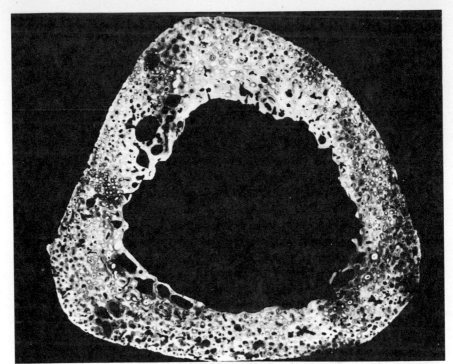

A

B

38

Figure 4–1. A section of bone from the body of a former radium watch-dial painter, who, in order to maintain a fine brush tip, was in the habit of touching the tip with his tongue. *A.* Darkened areas of damaged bone. *B.* An autoradiograph, in which the bone "took its own picture" by being held against film, showing areas exposed by the radium alpha particles; note that the areas of high alpha activity correspond to the areas of maximum damage in *A.* (Reproduced from Frigerio, N.: *Your Body and Radiation.* U.S. Atomic Energy Commission, Washington, D.C., 1967. Illustrations from Argonne National Laboratory.)

radiation similar to x-rays. Gamma rays are very penetrating; they will pass through the body, causing cellular damage as they travel through (see page 57).

Nuclear Reactions

When the nucleus of an atom emits a ray, its atomic number and mass number may change. Such changes in the nucleus are called nuclear reactions. In writing equations for nuclear reactions the following symbols are used:

1. The atomic number and the mass number are written to the left of the symbol of the element.

$$\text{(mass number)} \quad {}^{238}_{92}\text{U}$$
$$\text{(atomic number)}$$

Figure 4–2. Radium emits radiation that an electrostatic field separates into alpha (α) and beta (β) particles and gamma (γ) rays.

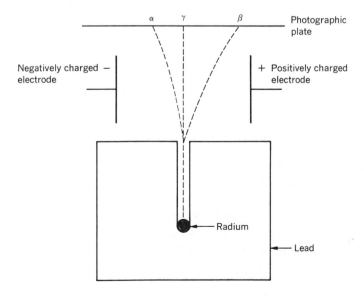

element because chemical properties are based upon electrons only, and isotopes of an element have identical electron structures. Thus, an organism cannot distinguish between normal carbon (^{12}C) and radioactive carbon (^{14}C).

If a plant is allowed to breathe carbon dioxide containing radioactive ^{14}C, the resulting carbohydrates will contain radioactive carbon. The movement of the radioactive atoms as they proceed through the plant may be followed by means of a Geiger counter, a device which registers the presence of radioactive substances. Radioisotopes that are introduced into living organisms are called "tagged" atoms because their path can be followed readily as they move through the organism.

A radioisotope is usually indicated by its symbol and mass number only. The atomic number is not necessary because the symbol itself serves to identify the element. Some radioisotopes commonly used in medicine and in biochemistry are ^{131}I, ^{60}Co, ^{99m}Tc, ^{14}C, and ^{59}Fe (see page 47).

Radioisotopes are used medicinally in the diagnosis and treatment of various disorders of the human body.

A medical tool called a scanner helps to locate malignancies (see Figure 4–3). A patient is given a selected radioisotope that will accumulate in the body area being studied. The scanner moves back and forth across that

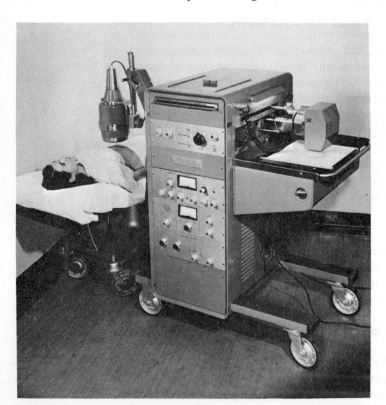

Figure 4–3. A scanner, an instrument used for depicting the distribution of radioactive material in the human body. (Courtesy of Picker Nuclear, Division of Picker X-Ray Corp., White Plains, N.Y.)

site and detects the radiation in each area over which it travels. This information is fed to a receiver and recorder, which then produces a picture called a scan (see Figure 4–4). Proper interpretation of a scan can tell not only if there is a malignancy but where it is located and also its exact dimensions.

Units of Radiation

Radiation is measured in terms of several different units, depending upon whether the measurement relates to a physical effect or a biologic effect.

The physical unit of radiation is a measure of the number of nuclear disintegrations occurring per second in a radioactive source. The standard unit is the *curie*, which is defined as the number of nuclear disintegrations occurring per second in 1 g of radium. One curie equals 37 billion disintegrations per second. Smaller units are the millicurie (37 million disintegrations per second) and the microcurie (37 thousand disintegrations per second). These smaller units are frequently used in describing the amount of radioactive fallout. This unit, the curie, is not useful in biologic work because it simply indicates the number of disintegrations per second regardless of the type of radiation and regardless of the effect of that radiation upon tissue.

The *roentgen* (abbreviated r) is a unit of radiation generally applied to x-rays and gamma rays only. X-rays and gamma rays produce ionization in air and also in tissue. The roentgen is defined as the intensity of x-rays or gamma rays which produces 1 electrostatic unit (esu) of positive or negative ions in 1 cc of air. This is not the same for tissue as it is for air, so that the roentgen does not accurately indicate the amount of radiation on tissue.

The *rad* (radiation absorbed dosage) refers to the amount of radiation energy absorbed by tissue that has been radiated. One rad corresponds to the absorption of 100 *ergs* of energy per gram of tissue. An erg is a very small unit of energy. It requires more than 40 million ergs to equal 1 cal. However, even though the erg is an extremely small unit of energy, the effect of 1 rad (100 ergs per gram) is important because of the ionization that the radiation produces in the cells.

The *rem* (rad equivalent to man) represents the amount of radiation absorbed by a human being. This unit of measure takes into consideration the difference in energy for various radioactive sources. A rem is the amount of ionizing radiation that, when absorbed by man, has an effect equal to the absorption of 1 r.

Detection and Measurement of Radiation

The problem of detecting and measuring radiation is very important in medical work, particularly in the protection of personnel. One device frequently used to detect radiation is the Geiger counter (see Figure 4–5). This

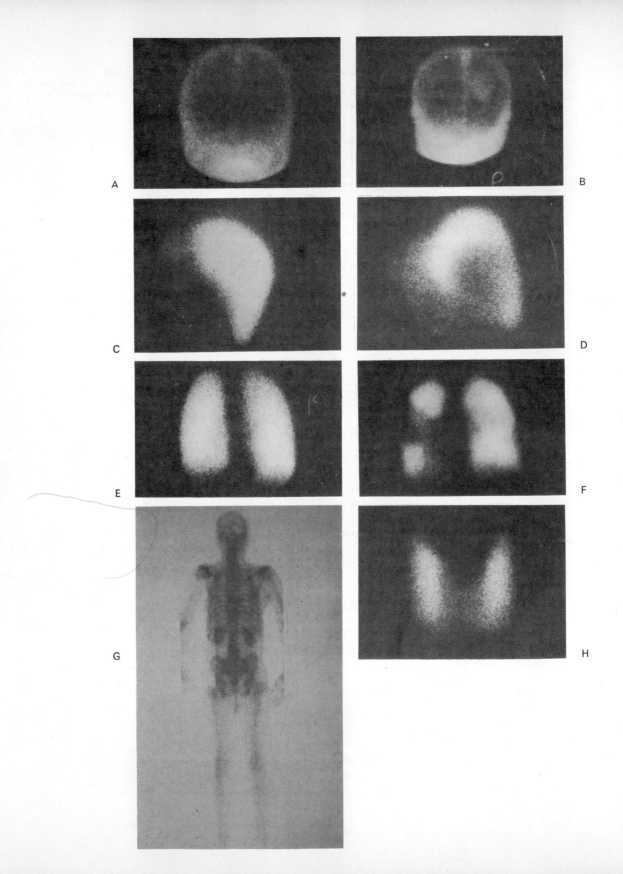

Figure 4–4. Typical scans. *A.* Normal brain, anterior view. *B.* Brain with cardiovascular aneurysm, anterior view. *C.* Normal liver, posterior view. *D.* Abnormal liver, posterior view. *E.* Normal lungs, posterior view. *F.* Lungs with pulmonary embolism, posterior view. *G.* Normal bone, posterior view. *H.* Normal thyroid, anterior view. (Courtesy of X-ray Department, Michael Reese Hospital and Medical Center, Chicago, Ill.)

45

Radioactivity

device consists of a glass tube containing a gas at low pressure through which runs a wire connected to a high voltage power supply. When the device is brought close to a radioactive substance, the radiation causes a momentary current to flow through the tube. A speaker is usually placed in the circuit to produce a click indicating a momentary flow of current. Sometimes a counting device is connected to the tube to indicate the amount of radiation.

X-ray technicians and others who work around radiation usually are required to wear film badges (see Figure 4–6). These badges indicate the accumulated amount of radiation to which they have been exposed. They

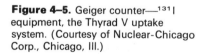
Figure 4–5. Geiger counter—^{131}I equipment, the Thyrad V uptake system. (Courtesy of Nuclear-Chicago Corp., Chicago, Ill.)

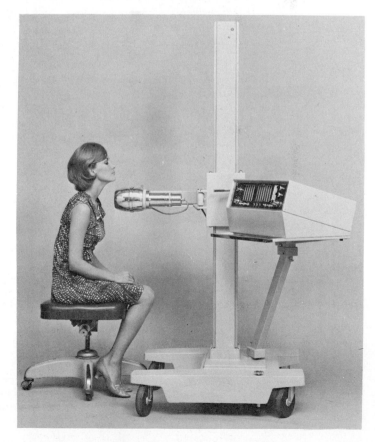

Figure 4–6. Film badge worn by x-ray technician. (Courtesy of R. S. Landauer, Jr. & Co., Glenwood, Ill.)

contain a piece of photographic film whose darkening is directly proportional to the amount of radiation received. This film must be checked frequently to see how much radiation has been absorbed.

Another device for detecting radiation is the Wilson cloud chamber. Essentially this device consists of enclosed air saturated with water vapor. When the rubber bulb at the base of the equipment is compressed and then suddenly released, the air inside becomes supersaturated with water vapor. This air, supersaturated with water vapor, condenses about any particle or ion in that air. The particles emitted from a radioactive source produce a visible fog track as the water vapor condenses on the ions formed along that path.

Half-Life

When a radioactive element gives off a particle, it changes or decays into another element. The rate of decay of all radioactive elements is not the same. Some elements decay rapidly, whereas others decay at an extremely slow rate. The half-life of a radioactive element is defined as the amount of time required

TABLE 4–1

Half-Life Periods of Several Common Radioisotopes

Radioisotope	Half-Life	Radiation
^{99m}Tc	6 hours	γ
^{59}Fe	45 days	$\beta + \gamma$
^{198}Au	2.7 days	$\beta + \gamma$
^{131}I	8.0 days	$\beta + \gamma$
^{32}P	14.3 days	β

for half of the atoms in a given sample to decay. Some radioactive elements have a half-life measured in terms of billions of years, whereas others are measured in fractions of a second.

The radioisotope ^{131}I has a half-life of approximately 8 days. Consider a 100-mg sample of ^{131}I. After 8 days (one half-life period of time) only half as much, or 50 mg, would be left. After another half-life period (after a total of 16 days), only half of that amount, or 25 mg, of ^{131}I would be left. Every half-life period, half of the remaining amount decays.

For medical work, a radioisotope must have a half-life such that it will remain in the body long enough to supply the radiation needed and yet will decrease in radioactivity within a reasonable period of time so that the body does not receive excess radiation.

The half-life periods of several commonly used radioisotopes are listed in Table 4–1.

Radioisotopes of long half-life are very dangerous to the body. For example, radium has a half-life of 1590 years; therefore, if it is taken into the body, it continues to give off its radiation with practically no change in amount during the lifetime of that person.

Radioisotopes in Medicine

When a radioisotope is to be used for diagnostic purposes, it must meet several criteria. Among them are the following: The radioactive element must be contained in a compound that will tend to concentrate in the area under study or in certain abnormal tissues. Since the presence of radiation is usually determined by an external counter or by a scan, a radioisotope emitting alpha particles is not generally used because such particles have too low a penetrating power to be detected outside the body. Radioisotopes with gamma radiation are preferred, although those emitting beta particles are also used.

The radioisotope selected should have a short half-life and should be in the form of a compound that will be eliminated from the body shortly after its diagnostic use is completed. Thus, the body will receive a minimum amount of radiation after the test is completed. In addition, the amount of radio-isotope used should be as small as practicable.

When a radioisotope is to be used for therapy, external measurement is not as necessary, so that alpha emitters may be used as well as beta and gamma. In radiation therapy, selected cells or tissues are to be destroyed without damage to nearby healthy tissues. Thus, the given radioisotope should have the property of concentrating in the desired area and preferably should emit alpha or beta particles because these have limited penetrating power and will not damage adjacent tissues.

Among the many radioisotopes in common medical use are those discussed below.

Iodine-131

This isotope is used in the diagnosis and treatment of thyroid conditions. The thyroid gland requires iodine to function normally. If a patient is suspected of having a thyroid disorder, he is given a drink of water containing a small amount of ^{131}I in the form of sodium iodide. If the thyroid is functioning normally, it should take up about 12 per cent of the radioactive iodine within a few hours. This iodine uptake may be measured with a Geiger counter (Figure 4–5) or with a scan. If less than the normal amount of ^{131}I is taken up, then the patient may have a hypothyroid condition. If the amount of ^{131}I taken up is greater than normal, then a condition known as hyperthyroidism may exist.

For a typical thyroid scan, a dose of 50 microcuries is administered orally. This amount of radiation gives 50 rads to the thyroid gland and 0.02 rad to the total body. Another method for a thyroid scan involves the use of technetium-99m. This radioisotope has the following advantages: (1) 5 millicuries of technetium-99m administered intravenously give a dose of only 1 rad to the thyroid gland and 0.02 rad to the whole body; and (2) the scan can be performed 20 minutes after injection, rather than after a 24-hour wait as when ^{131}I is used.

In addition, ^{131}I is used for scans of the adrenal glands, the gallbladder, and the liver. ^{131}I is employed therapeutically in the treatment of hyperthyroidism and in cancer of the thyroid gland.

^{125}I injected intravenously in the form of radioisotope-labeled fibrinogen is being used to detect deep thromboses.

Technetium-99m

Technetium-99m, ^{99m}Tc (*m* for metastable), is one of the most widely used radioisotopes for various types of scans. Technetium pertechnetate is used for brain scans and thyroid scans (see page 44); technetium sulfur colloid, for liver scans and bone-marrow scans; technetium macroaggregates of albumin, for lung-perfusion scans; technetium pyrophosphate, for bone scans; technetium diethylenetriamine pentaacetic acid (DTPA), for renal scans; and technetium-labeled red blood cells, for pericardial studies.

^{99m}Tc is also used to measure blood volume. A sample of this radioisotope is injected into the bloodstream. A short while later a sample of blood is withdrawn and its radioactivity measured. By knowing the radioactivity of the ^{99m}Tc and the radioactivity present in the sample withdrawn, an accurate determination of the total blood volume may be obtained.

Cobalt-60

^{60}Co gives off powerful gamma rays as well as beta particles. For treatment purposes, the beta particles are shielded out and only the gamma rays are used. ^{60}Co has a half-life of 5.3 years and is used as a substitute for radium because it is much cheaper and easier to handle. This radioisotope is employed in the treatment of many different types of cancer. Hospitals use large machines containing a cobalt "bomb" to supply gamma radiation directly to the cancer site (see Figure 4–7).

Figure 4–7. Cobalt machine in use. (Courtesy of University of Illinois Hospitals, Chicago, Ill.)

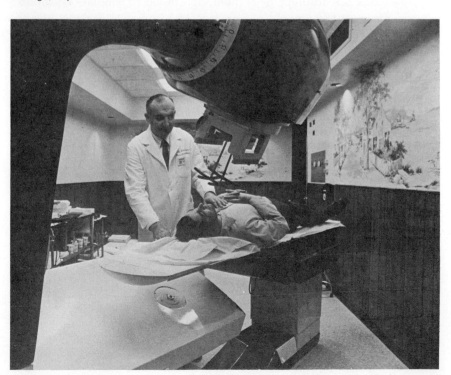

Among the many other radioisotopes in medical use are ^{133}Xe and ^{81m}Kr for ventilation lung scans, ^{75}Se for pancreas scans, and ^{67}Ga for whole-body scans for tumors. ^{3}H is used to determine the total amount of water present in the body; ^{14}C has been employed to study the path of carbohydrates, fats, and protein in the body and their conversion into other substances; ^{59}Fe is used to measure absorption of iron from the digestive tract.

^{32}P has been used for the treatment of bone metastases, and ^{198}Au has been employed for the treatment of pleural and peritoneal metastases.

Unlike technetium-99m, which concentrates in damaged heart cells, thallium-201 (^{201}Tl) concentrates in normal heart muscle but not in abnormal tissue. Thus, damaged heart muscle areas show up on a scan as "cold spots."

Cesium-129 (^{129}Cs) is used for scans on patients suspected of having acute myocardial infarction.

Radioactive plutonium batteries are being recommended for use in cardiac pacemakers since they will last the life of the patient.

X-Ray Therapy

X-rays are a penetrating type of radiation, similar to gamma rays but of a lower frequency. The amount of radiation and how deep it penetrates the tissue are adjustable in an x-ray machine, whereas these factors are generally fixed in a radioactive source such as radium or cobalt-60.

X-rays may be used for treatment of superficial skin conditions by adjusting the voltage of the machine so that it produces a "soft" or nonpenetrating radiation (see Figure 4–8).

Although more penetrating x-rays have been used for the treatment of deep-seated malignancies, the cobalt machine (see Figure 4–7) is now widely utilized for the treatment of different types of cancer.

Another form of x-ray therapy involves the use of oxygen at pressures higher than atmospheric. This treatment is based upon the fact that cancer cells are almost three times as sensitive to destruction by x-rays when the tissues are under 3 atmospheres (atm) pressure as they are at normal atmospheric pressure. This is not true for normal cells. The patient is placed in a chamber containing air at 3 atm pressure and then given x-ray treatments. This type of treatment is called hybaroxic cancer radiation.

X-rays will not pass through bone and teeth as easily as through tissue, so dental x-rays will show the presence of cavities, advanced bone destruction, and abscesses as well as the positions of normal and impacted teeth (see Figure 4–9).

A new type of x-ray instrument is the ACTA-scanner (Automatic Computerized Transverse Axial-scanner; see Figure 4–10). The ACTA scanner rotates in a circle around the body and makes sharp, detailed records of

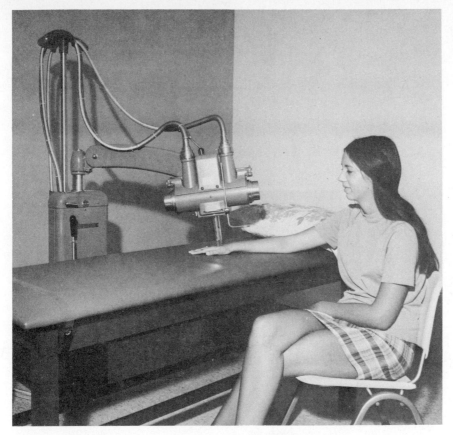

Figure 4–8. Superficial x-ray therapy unit. (Courtesy of Picker Corp., Cleveland, Ohio.)

narrow strips of a cross section of the body, each record yielding thousands of bits of information. All of the information is fed into a computer, which then produces a picture of the body section scanned. The results are available in minutes on television screens in black and white, or in color (see Figure 4–11). Even though the scanner makes many passes, the patient is exposed to about the same amount of radiation as with traditional x-ray equipment.

The use of the ACTA-scanner will make possible, for the first time, a detailed examination of any part of a patient's body, for the detection of tumors of the brain, breast, kidney, lung, or pancreas. It can also detect abnormal cavities in the spinal cord and enlargements of such organs as the liver, spleen, and heart. In addition, this equipment will allow examinations to be performed on an outpatient basis, since no preparation is necessary. The examination may even be performed with the patient wearing his clothing.

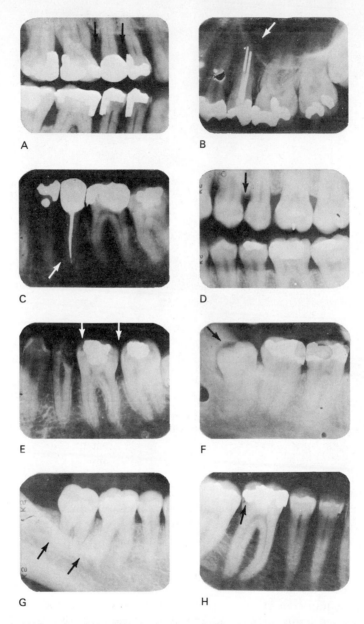

Figure 4–9. Dental x-rays. *A.* Advanced bone destruction. *B.* Root canal. *C.* Root canal. *D.* Bone destruction. *E.* Caries. *F.* Impacted molar. *G.* Bone destruction. *H.* Recurrent decay. (Courtesy of Dr. Norman Cohen, Chicago, Ill.)

Figure 4–10. The ACTA-scanner. (Courtesy of Pfizer Medical Systems, Inc., Columbia, Md.)

Xeroradiography

In xeroradiography, conventional x-ray equipment is used to make the exposure (see Figure 4–12, *A*). However, the image is produced by an entirely different process. In x-ray work, the image is produced on a piece of film by a photochemical process. In xeroradiography the image is produced on opaque paper by a photoelectric process. In contrast to x-rays, the xeroradiographic process is dry and requires no dark room.

Xeroradiography is applied principally to mammography for early detection of breast cancer. It produces better resolution and makes the interpretation of soft tissue studies easier and more accurate (see Figure 4–12, *B*). Xeroradiography is also useful in the search for foreign bodies, such as plastic, wood, and glass, that may not show up on a regular x-ray.

Ultrasonography

Ultrasonography employs very high-frequency sound waves in place of x-rays. Normal sound waves have a frequency range from 20 to 20,000 hertz. Ultrasound frequencies are those above 20,000 hertz. Those used in diagnostic ultrasound range from 1,000,000 to 15,000,000 hertz (1 to 15 megahertz).

Ultrasound waves are sent into the body, and the reflected waves are picked up, analyzed, and displayed on an oscilloscope. Photos of the image may also be taken (see Figure 4–13).

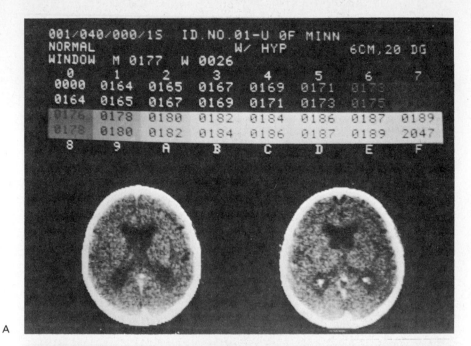

A

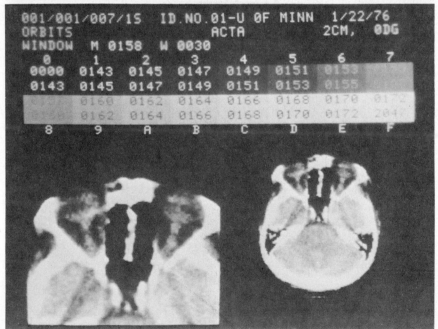

B

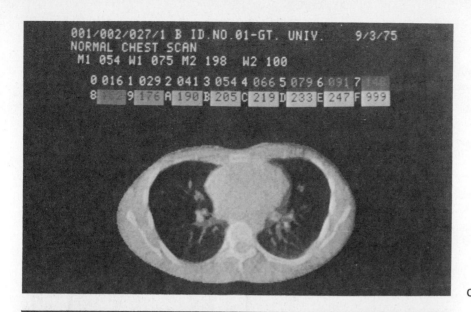

```
001/002/027/1 B ID.NO.01-GT. UNIV.      9/3/75
NORMAL CHEST SCAN
 M1 054 W1 075 M2 198  W2 100

  0 016 1 029 2 041 3 054 4 066 5 079 6 091 7
 8      9 178 A 190 B 205 C 219 D 233 E 247 F 999
```

C

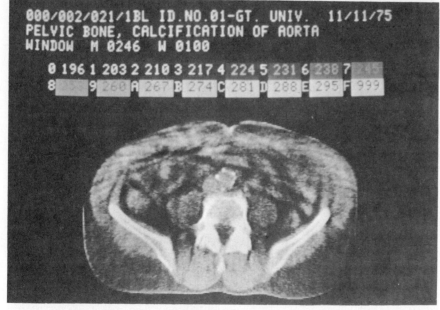

```
000/002/021/1BL ID.NO.01-GT. UNIV.   11/11/75
PELVIC BONE, CALCIFICATION OF AORTA
WINDOW  M 0246  W 0100

  0 196 1 203 2 210 3 217 4 224 5 231 6 238 7
 8      9 260 A 267 B 274 C 281 D 288 E 295 F 999
```

D

Figure 4–11. *A*. Scan of a head showing two adjacent slices, 1 mm apart. Structures visible include mildly enlarged ventricles (both slices), calcified pineal gland (both slices), and calcified choroid plexus (right slice only). *B*. Head scan through the orbits. Visible are the eyeballs, optic nerves, rectus muscles, and sinuses. A computer-generated enlargement is on the left. *C*. Chest scan showing the lungs, heart, ribs, scapulae, sternum, vertebra, and pulmonary vasculature. *D*. Pelvic scan showing the pelvic bones, lower vertebra, and a calcified aorta immediately in front of the vertebra. (Courtesy of Pfizer Medical Systems, Inc., Silver Spring, Md.)

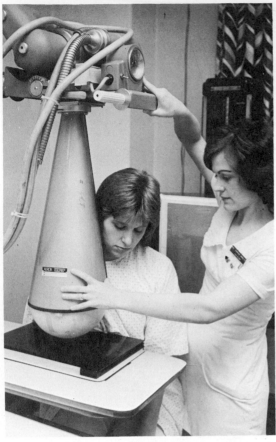 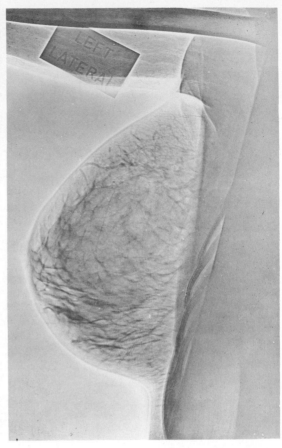

A   B

Figure 4–12. *A*. Xeroradiography in use. *B*. A mammograph. (*A*, courtesy of Miami Valley Hospital, Dayton, Ohio; *B*, courtesy of X-ray Department, Michael Reese Hospital and Medical Center, Chicago, Ill.)

N-Radiography

N-radiography is similar to x-radiography except that it uses a beam of neutrons rather than x-rays. However, unlike x-rays, which interact with the electron cloud around the atomic nucleus, neutrons are absorbed and scattered by atomic nuclei. While x-rays go through tissue like a needle, fast neutrons because of their great mass smash their way through, producing a wide track of particles, which in turn create ionization in their paths. The effect of neutron bombardment on tumors is greater than with x-rays, while the effect on normal tissue is not significantly different. In addition, patients receive less radiation with neutrons than with x-rays.

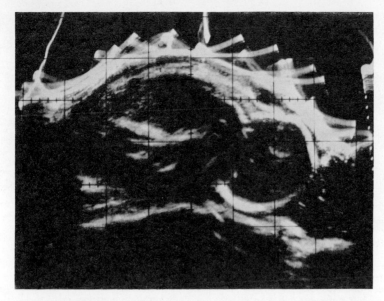

Figure 4–13. Ultrasonograph of a fetus in womb, showing head at right and trunk to left. (Courtesy of X-ray Department, Michael Reese Hospital and Medical Center, Chicago, Ill.)

N-radiation has another advantage over x-rays. Recall that the more oxygenated a tissue, the greater the effect of radiation on it (see page 50). As a tumor grows, it tends to become short of oxygen because it is outgrowing its supply of that vital element. This hypoxia, in turn, diminishes the effectiveness of x-ray treatment. However, lack of oxygen does not protect the tumor from the effects of neutron radiation.

N-radiation is most effective against those cancers that resist x-rays and gamma rays from cobalt machines. Such cancers frequently occur in the mouth, stomach, colon, and uterus. Like x-rays, neutrons are less effective in treating deep-seated malignancies.

Another method of detecting and treating cancerous tissues involves the use of a proton beam. However, this is still in the preliminary testing stages.

Biologic Effects of Radiation

Externally, alpha and beta rays are relatively harmless to man since they have slight penetrating power (see Figure 4–14). The gamma rays, with their great penetrating power, have a very definite effect upon the body. If a radioactive substance is taken inside the body, it is the gamma rays that are most harmful. Alpha and beta rays (particles) are also quite harmful internally— alpha rays because of their great size and beta rays because of their greater penetrating power.

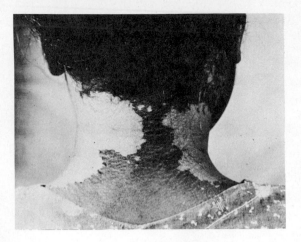
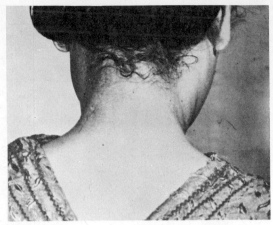

A   B

Figure 4–14. Changes in pigmentation of skin due to beta burns. *A*. A burn area on the neck of a Rongelap native one month after accidental exposure to radiation in 1954. *B*. The same burn area one year after the accident, showing complete recovery. (Reproduced from Frigerio, N.: *Your Body and Radiation*. U.S. Atomic Energy Commission, Washington, D.C., 1967. Illustrations from Brookhaven National Laboratory.)

Radiation of tissues causes ionization of the atoms inside the cells and also produces substances that impair cellular metabolism. Ionization in the cell disrupts the chemical processes within the cell. It may alter the genes, causing the cell to grow abnormally or to die.

Exposure to external radiation may be controlled by increasing the distance between the body and the source of the radiation. The amount of radiation received varies inversely as the square of the distance; therefore doubling the distance from a radioactive source permits the body to receive only one fourth as much radiation. Shielding material, such as lead, when placed between the body and the radioactive source will also protect the body against radiation.

Radiation causes ionization in cells. This ionization, in turn, can alter DNA. As a result, some cells may die or fail to multiply but other cells will continue to live and reproduce. Because of the alteration in DNA, there may be genetic changes (mutations) that can show up in future generations. Large amounts of radiation may also produce cataracts, sterility, and leukemia. Exposure to radiation lessens life expectancy. Rapidly dividing tissue is highly susceptible to radiation. For this reason unnecessary radiation during pregnancy is to be avoided.

Exposure to large amounts of radiation can cause "radiation sickness." The symptoms are gastrointestinal disturbances (nausea, vomiting, diarrhea, general body weakness), a drop in red blood cell and white blood cell counts,

58

loss of hair, extensive skin damage, and ulcerative sores that are difficult to heal. Extremely large doses of radiation are fatal very quickly.

The National Council on Radiation Protection and Measurement and the International Commission on Radiological Protection have set the following radiation standards:

1. A dose not exceeding 0.5 rem (500 millirems, mrem) per year of whole-body exposure for individual members of the general population.
2. An average dose to the general population not exceeding 0.17 rem (170 mrem) per year, whole-body radiation.
3. A dose not exceeding 5 rem per year, whole-body exposure for radiation workers.

It is also recommended that actual exposures to radiation be kept as low as possible.

Sources of Radiation

The body receives radiation externally from three principal sources: natural background radiation, medical radiation, and fallout.

The body normally receives radiation from space and from radioactive material present in the soil, in the air, in water, and in the building materials of our houses. The average natural background radiation in the U.S. is about 150 mrem per year.

The amount of medical radiation varies with the type and frequency of medical treatment. The average in the U.S. is 70 mrem per year. However, x-ray photographs of specific parts of the body may give that body part a very high amount of radiation. Fluoroscopy produces an even greater amount of radiation.

Summary

Radioactivity is the property of emitting radiation from the nucleus of an atom. The three types of radiation are alpha, beta, and gamma. Alpha particles are positively charged helium nuclei. Beta particles are high-speed electrons and are negatively charged. Gamma rays are a form of electromagnetic radiation and have no charge.

In nuclear reactions both the sum of the atomic numbers and mass numbers on both sides of an equation are the same. In addition to naturally occurring radioactive substances, artificially radioactive substances may be prepared by bombardment with such particles as protons, neutrons, and alpha particles.

A scanner helps to locate malignancies by moving back and forth across the site being studied and detecting the radiation in each area over which it travels. The radiation comes from radioisotopes, which are selected to accumulate at the desired body part.

Radiation may be detected by means of a Geiger counter or with a film badge.

The half-life of a radioisotope is the amount of time required for half of its atoms to decay. For medical work, a radioisotope must have a half-life long enough to give the body part the radiation it needs and short enough so that it will not give the body too much radiation within a reasonable period of time.

Radioisotopes are used in the diagnosis and treatment of various disorders in the body—^{131}I for thyroid conditions, ^{32}P for brain tumors, ^{99m}Tc for scans, and ^{60}Co for radiation therapy.

X-rays are a type of radiation similar to gamma rays. The penetration powers of x-rays may be controlled, whereas those of gamma rays may not. X-ray treatment, combined with high-pressure oxygen, has had some success in the treatment of cancer.

The units of radiation are the curie, the roentgen, the rad, and the rem. The roentgen applies primarily to x-rays. The rem is the unit most commonly used in terms of the body.

Radiation produces ionization within the cells, causing some type of damage. Small amounts of radiation produce genetic changes, larger amounts cause a shortened life-span. Very large amounts of radiation may cause death within a short period of time.

X-rays are harmful to the body because of the effects produced by the radiation so care should be taken to avoid unnecessary exposure. Fluoroscopy produces even more radiation than x-ray photographs.

Questions and Problems

1. What are alpha particles? Beta particles? Gamma rays?
2. Define the term *half-life*.
3. Balance the following equations

    (a) $^{9}_{4}Be + ^{4}_{2}He \longrightarrow ^{1}_{0}n + ?$

    (b) $^{27}_{13}Al + ^{1}_{0}n \longrightarrow ^{4}_{2}He + ?$

    (c) $^{30}_{15}P \longrightarrow ^{0}_{-1}e + ?$

4. What is artificial radioactivity?
5. What are radioisotopes? Give one use for ^{14}C, ^{131}I, and ^{59}Fe.
6. If 100 mg of ^{32}P is present on a certain day, approximately how much will be present 2 months later?
7. If 2 mg of ^{99m}Tc are present at 8 A.M. Monday, approximately how much was present at 2 A.M. on the preceding Sunday?
8. How may radiation be detected?
9. What are x-rays used for?
10. What are the units of radiation?
11. List some of the physiologic effects of radiation.
12. What may be done to minimize the effects of radiation on the body?
13. From where does the body receive external radiation?
14. Explain how a Wilson cloud chamber works.
15. Describe the use of a scanner. Name one radioisotope used in scans and indicate where it may be used.
16. How may blood volume be determined?
17. Compare x-rays with n-rays in terms of their interaction with the atom.
18. What is xeroradiography? For what purposes is it used?
19. What is ultrasonography?

20. Compare criteria for selection of radioisotopes for diagnostic and therapeutic uses.
21. How does the ACTA-scanner work?
22. What are the values of radiation dosage for the general population? For radiation workers?
23. Explain changes in atomic number and mass number when an atom emits an alpha particle. A beta particle.

References

Baum, S. J., and Scaife, C. W.: *Chemistry: A Life Science Approach*. Macmillan Publishing Co., Inc., New York, 1975, Chap. 11.

Fernandez, J. E., and Whitaker, R. D.: *An Introduction to Chemical Principles*. Macmillan Publishing Co., Inc., New York, 1975, Chap. 28.

Gordon, R.; Herman, G. T.; and Johnson, S. A.: Image reconstruction from projections. *Scientific American*, **233**:56–68 (Oct.), 1975.

Hein, M., and Best, L. R.: *College Chemistry*. Dickenson Publishing Co., Inc., Encino, Calif., 1976, Chap. 18.

Pauling, L., and Pauling, P.: *Chemistry*. W. H. Freeman & Co., San Francisco, 1975, Chap. 20.

Sackheim, G. I.: *Chemical Calculations*, 10th ed., series B. Stipes Publishing Co., Champaign, Ill., 1976, Chap. 24.

Chapter 5

Chemical Bonding

Molecules

Atoms may remain by themselves or they may combine with other atoms to form molecules. A molecule is a combination of two or more atoms. These atoms may be of the same elements, as in the oxygen molecules (O_2), or may be of different elements, as in the hydrogen chloride molecule (HCl). A more complicated molecule is that of glucose, $C_6H_{12}O_6$. What holds the atoms together in a molecule? Atoms are held together by bonds. The type of bond depends primarily upon the number of electrons in the outer shell. Bonds are classified into two main types—electrovalent and covalent.

Stability of the Atom

Atoms are considered stable (nonreactive) when their outer shell (the valence shell) is filled to eight. The rare gases—neon, argon, krypton, xenon, and radon—all have eight electrons in their valence shell. They are stable. One exception to this rule of eight (the octet rule) is helium, which is stable even though it has only two electrons in its valence shell. Helium has only one shell, the first shell, and that shell can hold only two electrons.

Atoms that do not have eight electrons in their outer shell may lose, gain, or share their valence electrons with other atoms in order to reach a more stable structure of eight. This process of rearrangement of the valence electrons is responsible for the chemical reactions between atoms.

Electron-Dot Structures

The electron-dot structure of an atom (also called a Lewis structure) is an abbreviated representation for the structure of that atom. In this system, the nucleus and all of the electron shells except the outer shell are represented by the symbol for that element. Each electron in the outer shell is indicated by a dot. For example, the element sodium (symbol Na, atomic number 11) has its nucleus surrounded by 11 electrons—2 in the first shell (energy level), 8 in the second shell (energy level), and 1 in the third (outer) shell. The electron-dot structure for the sodium atom is Na·, with the dot representing the one electron in the outer shell and the symbol Na representing the remainder of the atom. Carbon, atomic number 6, has the electron configuration 2e) 4e). The electron-dot representation for carbon is

Argon, atomic number 18, has the electron configuration of 2e) 8e) 8e). The electron-dot structure for argon is

Formation of Ions

Consider the sodium atom with the electron structure 2e) 8e) 1e). If the sodium atom loses the one electron in its valence shell, then it will reach a stable structure of eight electrons in its outer shell. When a sodium atom loses an electron, it becomes a positively charged particle called a sodium ion. This reaction may be written as:

$$\text{Na·} \ -1e \ \longrightarrow \ \text{Na}^+$$

or, with the electron-dot being understood,

$$Na - 1e \longrightarrow Na^+$$

where the positive sign indicates a charge of $+1$ on the sodium ion. (Note that the number 1 is understood and not written.) The charge on the sodium ion is a positive one because the sodium ion still has eleven protons in its nucleus but now has only ten electrons outside that nucleus.

Likewise, the aluminum atom, which has the electron structure of 2e) 8e) 3e), will lose all three electrons from its valence shell to form an aluminum ion with a charge of $+3$ (written above the symbol as $3+$).

$$Al - 3e \longrightarrow Al^{3+} \quad \text{or simply} \quad Al - 3e \longrightarrow Al^{3+}$$

In general, metals have one, two, or three electrons in their valence shell. Also, metals lose all the electrons in their valence shell to form positively charged ions. The positive charge on a metallic ion is equal to the number of electrons lost by that metal.

Elements that have six or seven electrons in their outer shell tend to gain electrons to reach a stable configuration of eight. Such elements are called nonmetals. (Elements having four or five electrons in their valence shell are also nonmetals. These shall be discussed separately under covalent bonds.)

Consider the element chlorine, 2e) 8e) 7e). The chlorine atom will tend to gain one electron to fill its outer shell to eight and thus will form an ion with a charge of -1, or, omitting the dots,

$$Cl + 1e \longrightarrow Cl^-$$

Since the ion has one more electron than the atom, it will have a charge of -1.

Likewise, the sulfur atom, 2e) 6e), can gain two electrons to form an ion with a charge of -2.

$$S + 2e \longrightarrow S^{2-}$$

Therefore, an atom that has either lost or gained electrons in its valence shell is called an *ion*. Ions formed from a metal will have a positive charge equal to the number of electrons lost. Ions formed from nonmetals will have a negative charge equal to the number of electrons gained.

Electrovalence

When a sodium atom, Na, combines with a chlorine atom, Cl, to form a sodium chloride molecule (NaCl), the sodium atom loses one electron to form a positively charged sodium ion (Na^+). At the same time the chlorine atom gains that one electron to form a negatively charged chloride ion (Cl^-).

The reaction is

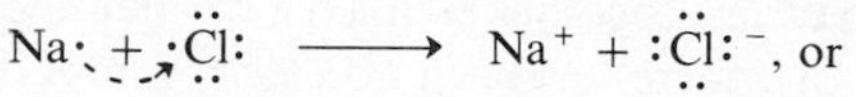

$$Na\cdot + \cdot\ddot{\underset{\cdot\cdot}{Cl}}: \longrightarrow Na^+ + :\ddot{\underset{\cdot\cdot}{Cl}}:^-, \text{ or}$$

$$Na + Cl \longrightarrow Na^+ + Cl^-$$

The positively charged sodium ion and the negatively charged chloride ion will be attracted to each other and these two ions will be held together by the electrostatic attraction of their charges (opposite charges attract each other). This type of bonding is called *electrovalent*. Electrovalence is a transfer of an electron or electrons from one atom to another with the formation of ions which attract each other.

Another example of a transfer of electrons from a metal to a nonmetal is in the reaction between magnesium (Mg) and chlorine (Cl_2 here written as two Cl's).

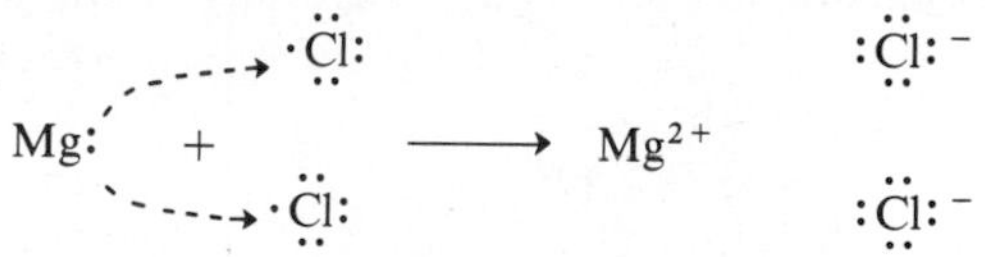

$$Mg\text{:} + \quad \begin{array}{c} \cdot\ddot{\underset{\cdot\cdot}{Cl}}: \\ \\ \cdot\ddot{\underset{\cdot\cdot}{Cl}}: \end{array} \longrightarrow Mg^{2+} \qquad \begin{array}{c} :\ddot{\underset{\cdot\cdot}{Cl}}:^- \\ \\ :\ddot{\underset{\cdot\cdot}{Cl}}:^- \end{array}$$

$$\text{or} \quad Mg + 2Cl \longrightarrow Mg^{2+} + 2Cl^-$$

where the positively charged magnesium ion and the negatively charged chlorine (chloride) ions are held together by electrovalent bonds. (Note that again each ion has a completed outer shell of eight.)

Ionic Compounds

Compounds that contain ions are called ionic compounds. Therefore, ionic compounds are electrovalent. As shall be discussed later, ionic compounds fall into three categories—acids, bases, and salts.

Covalence

In an ionic compound, electrovalent bonding results from the loss or gain of electrons. However, there is another method by which atoms can be bonded together. This is by the sharing of electrons (*covalence*).

In the chlorine molecule, Cl_2, each of the two chlorine atoms has seven electrons in its valence shell. In this case both atoms will share electrons so that each will have a completed valence shell of eight electrons. The following diagram shows two chlorine atoms with their electrons so situated that each has a completed valence shell of eight electrons around it.

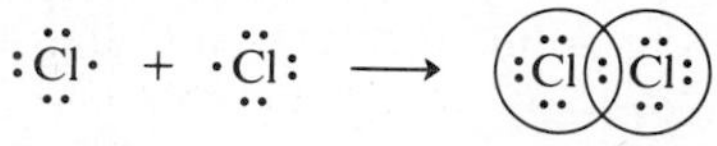

$$:\ddot{\underset{\cdot\cdot}{Cl}}\cdot + \cdot\ddot{\underset{\cdot\cdot}{Cl}}: \longrightarrow \left(:\ddot{\underset{\cdot\cdot}{Cl}}\middle(:\middle)\ddot{\underset{\cdot\cdot}{Cl}}:\right)$$

Each of the chlorine atoms is sharing one electron with the other. The bond that holds these two atoms together is called a covalent bond. Note that in the chlorine molecule, Cl_2, there has been no electron loss or gain and so there are no chloride ions present. This is one of the primary differences between electrovalence and covalence. In electrovalence ions are formed, whereas no ions are formed in covalence.

The covalent bond between the two chlorine atoms may be indicated by a short line, or Cl—Cl, with the electrons being understood and not written.

Covalent bonds may also be formed between atoms of different elements. In compounds containing covalent bonds each atom must have eight electrons around it, since eight electrons in the outer shell represent a stable structure. An exception to this rule (see page 63) is hydrogen, which in compounds has only two electrons around it (recall that the first shell can hold only two electrons).

The compound carbon tetrachloride, CCl_4, may be diagrammed as:

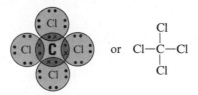

The compound ammonia, NH_3, may be represented as:

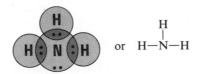

In carbon tetrachloride there are four covalent bonds, one between each of the chlorines and the carbon. In ammonia there are three covalent bonds. Note that in the above structures each element has eight electrons around it, except for hydrogen, which has only two.

Since metals tend to lose electrons, they generally do not form covalent compounds. Thus, we can say that covalent compounds are formed between nonmetals having four, five, six, or seven electrons in their valence shell.

Radicals

A group of atoms that stay together and act as a unit in a chemical reaction is called a *radical*. A radical acts as if it were one ion. Since radicals are ions, they must have a charge.

Table 5–1 indicates the name, formula, and charge of several common radicals.

TABLE 5–1

Common Radicals

Name	Formula and Charge
Sulfate	SO_4^{2-}
Nitrate	NO_3^-
Phosphate	PO_4^{3-}
Carbonate	CO_3^{2-}
Ammonium	NH_4^+
Hydroxide	OH^-
Bicarbonate	HCO_3^-

Oxidation Numbers

For ionic compounds, the oxidation number of an element is equal to the number of electrons lost or gained and so is the same as the charge on the ion. That is, in sodium chloride, Na^+Cl^-, the oxidation number of sodium is $+1$ and that of chlorine -1. In the compound $MgBr_2$, where the magnesium ion has a charge of $2+$ and each bromide ion a charge of $1-$, the oxidation number of magnesium is $+2$ and that of each bromide -1.

For covalent compounds, where electrons are shared and not transferred, oxidation numbers are arbitrarily assigned to elements using the following rules:

1. All elements in their free state have an oxidation number of zero.
2. The oxidation number of oxygen is -2.
3. The oxidation number of hydrogen is $+1$.
4. The sum of oxidation numbers in all compounds must equal zero.

These arbitrarily assigned oxidation numbers are based upon unequal sharing of electrons between atoms. For example, in the compound NH_3, shown on page 66, the electrons shared between the nitrogen and each of the hydrogens are not uniformly distributed between the nitrogen and the hydrogens. Each shared pair is actually closer to the nitrogen (because nitrogen has a greater attraction for electrons than does hydrogen). Thus, there is an unequal sharing of electrons in this compound, NH_3.

Table 5–2 indicates the oxidation numbers of various elements and radicals.

Calculating Oxidation Numbers from Formulas

As has been previously mentioned, the sum of the oxidation numbers in any compound must equal zero. Let us find the oxidation number of zinc in the compound ZnO. Note that the oxidation number of oxygen, as listed in Table 5–2, is -2. Writing the formula of the compound with the known

TABLE 5–2

Oxidation Numbers of Elements and Radicals*

Positive Oxidation Numbers		Negative Oxidation Numbers	
Name and Symbol	Oxidation Number	Name and Symbol	Oxidation Number
Hydrogen H^+	$+1$	Chloride Cl^-	-1
Sodium Na^+	$+1$	Bromide Br^-	-1
Potassium K^+	$+1$	Iodide I^-	-1
Silver Ag^+	$+1$	Sulfate SO_4^{2-}	-2
Calcium Ca^{2+}	$+2$	Nitrate NO_3^-	-1
Magnesium Mg^{2+}	$+2$	Phosphate PO_4^{3-}	-3
Aluminum Al^{3+}	$+3$	Carbonate CO_3^{2-}	-2
Ammonium NH_4^+	$+1$	Bicarbonate HCO_3^-	-1
Iron Fe^{2+} and Fe^{3+}	$+2$ and $+3$	Sulfide S^{2-}	-2
Copper Cu^+ and Cu^{2+}	$+1$ and $+2$	Oxide O^{2-}	-2
Tin Sn^{2+} and Sn^{4+}	$+2$ and $+4$	Hydroxide OH^-	-1

* Note that some elements, such as copper, tin, and iron, have more than one oxidation number.

oxidation number above

$$? \quad -2 = 0$$
$$Zn\,O$$

we see that the oxidation number of the Zn must be $+2$ in order for the sum of the oxidation numbers to be zero.

Next let us find the oxidation number of Mn in $KMnO_4$. From the table we see that the oxidation number of K is $+1$ and that of O is -2. Therefore, four oxygens will have a total oxidation number of $4(-2)$, or

$$+1 + (?) + 4(-2) = 0$$
$$K\,Mn \quad O_4$$

In order for the sum of the oxidation numbers to be zero, the oxidation number of Mn must be $+7$.

Now consider the compound As_2S_3. The oxidation number of S from the table is -2. Therefore, three S's will have a total oxidation number of $3(-2)$

$$? \quad 3(-2) = 0$$
$$As_2 \quad S_3$$

In order for the sum of the oxidation numbers to be zero, the total oxidation number of the As's must be $+6$. However, this value of $+6$ applies to two As atoms, so that the oxidation number of each As is $+3$.

Writing Formulas from Oxidation Numbers

To write the formula of a compound formed between calcium and chlorine, look up the oxidation numbers of these elements in Table 5–2 and write

$$\overset{+2\ -1}{Ca\ Cl}$$

Note that the sum of the oxidation numbers is not zero so that this is not the correct formula for the compound. An easy method for obtaining the correct formula is to use a system of crisscrossing the oxidation numbers

$$\overset{+2}{Ca}\diagup\overset{-1}{Cl}$$

Thus, the formula of the compound between calcium and chlorine is $CaCl_2$. (Note that the subscript 1 is always understood and never written.) In this compound the sum of the oxidation numbers ($+2$ for the calcium and -2 for the two chlorines) does equal zero.

To write the formula for the compound formed between magnesium and the phosphate radical, we first write the oxidation numbers above the symbols and then crisscross them, or

$$\overset{+2}{Mg}\diagup\overset{-3}{PO_4}$$

so that the formula is $Mg_3(PO_4)_2$, where the sum of the oxidation numbers now equals zero ($+6$ from three Mg's and -6 from two PO_4's). The parentheses around the $PO_4{}^{3-}$ radical indicate that the entire radical occurs more than once in the formula. If the radical occurs only once in a compound, parentheses are not necessary. Thus, in the compound between the sodium ion (oxidation number $+1$) and the nitrate radical (oxidation number -1), the formula is simply written as $NaNO_3$.

When both positive and negative oxidation numbers are even numbers, the formula may be correct as written. The compound formed between calcium (oxidation number $+2$) and the sulfate radical (oxidation number -2) is $CaSO_4$ since the sum of the oxidation numbers is already zero.

Occasionally, when both positive and negative oxidation numbers are even numbers, the formula of the compound can be simplified by dividing by 2. Thus, the compound formed between tin (oxidation number $+4$) and the sulfate radical (oxidation number -2) may be written as

$$\overset{+4}{Sn}\diagup\overset{-2}{SO_4}$$

or $Sn_2(SO_4)_4$, which may be simplified to $Sn(SO_4)_2$.

Summary

Molecules are combinations of two or more atoms. Atoms are held together in molecules by electrovalent or covalent bonds.

The valence shell of an atom is the outermost shell of electrons. The number of electrons in the valence shell determines the chemical properties of the atom. Atoms are most stable when they have eight electrons in their valence shell.

The electron-dot structure of an element uses the symbol of that element to represent the nucleus and all of the electrons except those in the valence shell. Each electron in the valence shell is represented by a dot placed near the symbol.

Metals have one, two, or three electrons in their outer shell and tend to lose all those electrons to form positively charged ions. Nonmetals with six or seven electrons in their valence shell tend to gain electrons to fill their valence shell to eight, thereby forming ions with a negative charge.

When a metal loses an electron to form a positively charged ion and a nonmetal gains that electron to form a negatively charged ion, these ions are held together by the attraction of their charges. This type of bonding is called electrovalent. This type of electron transfer is called electrovalence.

Nonmetals may also share electrons to complete their valence shell. Such a bond is called a covalent bond. In a covalent bond there are no ions formed. In forming covalent bonds, each element has eight electrons around it, except for hydrogen, which has only two.

A radical is a group of atoms that acts as a unit in a chemical reaction. Radicals are ions and have a charge.

For ionic compounds, the oxidation number is the same as the charge on the ion. For covalent compounds the oxidation number is an arbitrary number based upon unequal sharing of electrons. The oxidation number of oxygen is -2 and that of hydrogen $+1$. In all compounds the sum of the oxidation numbers must equal zero.

In writing the formula of a compound from a table of oxidation numbers, the values of the oxidation numbers are crisscrossed. Care must be taken to note when the oxidation numbers are divisible by two so that the formula can be simplified.

Questions and Problems

1. What is a molecule?
2. What is the valence shell of an atom? What effect does it have on an atom?
3. What is meant by the term "electron-dot structure"?
4. Give the electron-dot structures for the following elements (use periodic table)

   (a) Phosphorus   (d) Neon
   (b) Oxygen   (e) Nitrogen
   (c) Hydrogen   (f) Aluminum

5. What electron configuration does an element usually need to reach maximum stability?
6. What is an ion? What type of elements form positively charged ions? Negatively charged?
7. What type of bond consists of ions held together by the attraction of their charges?
8. Define "covalent bond." Compare electrovalent and covalent bonds.
9. Draw the electron-dot structures for the following covalent compounds.

   (a) H_2 (d) PH_3
   (b) HCl   (e) CH_4
   (c) H_2S (f) N_2

10. Write the formulas for the compounds formed from the following ion pairs

    (a) H^+ and SO_4^{2-} (c) Cu^+ and I^-
    (b) Fe^{2+} and Cl^- (d) Ca^{2+} and PO_4^{3-}

(e) Ba^{2+} and HCO_3^-

(f) Mg^{2+} and NO_3^-

(g) Ca^{2+} and S^{2-}

(h) Al^{3+} and CO_3^{2-}

11. Write the formulas for the compounds formed between the following: silver ions and sulfate ions; potassium ions and bicarbonate ions; ammonium ions and sulfide ions; hydrogen ions and nitrate ions.

12. Calculate the oxidation number for *each* atom of the underlined element (consult Table 5–2 for oxidation numbers).

(a) $K\underline{Cl}O_4$

(b) $Na\underline{N}O_2$

(c) $\underline{Fe}_2O_3$

(d) $\underline{C}_{12}H_{22}O_{11}$

(e) $H_4\underline{P}_2O_7$

(f) $\underline{C}_6H_6$

(g) $\underline{Ra}(HCO_3)_2$

(h) $(\underline{N}H_4)_2S$

References

Ault, F. K., and Lawrence, R. M.: *Chemistry: A Conceptual Introduction.* Scott, Foresman & Co., Glenview, Ill., 1976, Chap. 4.

Baum, S. J., and Scaife, C. W.: *Chemistry: A Life Science Approach.* Macmillan Publishing Co., Inc., New York, 1975, Chap. 3.

Fernandez, J. E., and Whitaker, R. D.: *An Introduction to Chemical Principles.* Macmillan Publishing Co., Inc., New York, 1975, Chap. 2.

Hein, M., and Best, L. R.: *College Chemistry.* Dickenson Publishing Co., Inc., Encino, Calif., 1976, Chap. 7.

Longo, F. R.: *General Chemistry.* McGraw-Hill Book Co., New York, 1974, Chap. 5.

Masterson, W. L., and Slowincki, E. J.: *Chemical Principles*, 3rd ed. W. B. Saunders Co., Philadelphia, 1973, Chap. 7.

Nebergall, W. H.; Schmidt, F. C.; and Holtzclaw, H. F., Jr.: *College Chemistry*, 5th ed. D. C. Heath & Co., Lexington, Mass., 1976, Chap. 4.

Pauling, L., and Pauling, P.: *Chemistry.* W. H. Freeman & Co., San Francisco, 1975, Chap. 6.

Peters, E. I.: *Introduction to Chemical Principles.* W. B. Saunders Co., Philadelphia, 1974, Chap. 8.

Sackheim, G. I.: *Chemical Calculations*, 10th ed. series B. Stipes Publishing Co., Champaign, Ill., 1976, Chap. 11.

Sienko, M. J., and Plane, R. A.: *Chemistry*, 5th ed. McGraw-Hill Book Co., New York, 1976, Chap. 4.

Chapter 6

Chemical Equations

Symbols and Formulas

A symbol not only identifies an element but also represents one atom of that element. Thus, the symbol Cu designates the element copper and also indicates one atom of copper (the number 1 being understood and not written). Two atoms of copper are designated as 2Cu.

A formula consists of a group of symbols that represent the elements present in a substance. It also indicates one molecule of that substance. Thus, the formula NaCl indicates that the compound (sodium chloride) consists of one atom of sodium (Na) and one atom of chlorine (Cl).

If there is more than one atom of an element present in a compound, then subscripts (lower numbers) are used to indicate how many atoms of each element are present. In the compound HNO_3 (nitric acid) there is one atom of hydrogen (H), one atom of nitrogen (N), and three atoms of oxygen (O), all of which make up one molecule of nitric acid. In the compound $K_2Cr_2O_7$ there are two atoms of potassium (K), two atoms of chromium (Cr), and seven atoms of oxygen making up one molecule of potassium dichromate.

To designate more than one molecule of that substance, a number (a coefficient) is placed in front of the formula for that substance. For example, $2HNO_3$ indicates two molecules of HNO_3; $6K_2Cr_2O_7$ indicates six molecules of $K_2Cr_2O_7$.

The formula O_2 indicates one molecule of oxygen. This molecule consists of two atoms of oxygen. The formula H_2 indicates one molecule of hydrogen. This molecule consists of two atoms of hydrogen. Both O_2 and H_2 are called *diatomic molecules* since they are made up of two atoms of those elements.

Not all molecules are diatomic. Some are *monatomic*; that is, they consist of only one atom. Examples of such monatomic molecules are neon, Ne, and argon, Ar. Molecules of other elements such as sulfur, S_8, are *polyatomic*; they contain several atoms in their molecules.

Be very careful in distinguishing between 2H and H_2. 2H represents two atoms of hydrogen that are not combined; they are separate, independent atoms. H_2 represents one molecule of hydrogen, which consists of two atoms of hydrogen that are chemically combined with a covalent bond between them. (Two molecules of hydrogen would be shown as $2H_2$.)

Molecular Weight

The *molecular weight* of any compound is the sum of the atomic weights of all of the atoms present in that compound. The molecular weight of sodium chloride (NaCl), is 58.5, which represents the sum of the atomic weight of sodium, 23, plus that of chlorine, 35.5. (See Table of Atomic Weights inside back cover.)

Calculating the Molecular Weight (Formula Weight) of a Compound

To find the molecular weight of a compound, add the atomic weights of all of the atoms that are present in that compound. In the compound H_2O, the molecular weight can be calculated by adding the weight of two atoms of hydrogen and one atom of oxygen.

$$
\begin{array}{llr}
\text{2 hydrogen atoms (atomic weight 1)} = 2 \times 1 &=& 2 \\
\text{1 oxygen atom (atomic weight 16)} = 1 \times 16 &=& 16 \\
\hline
\text{molecular weight} &=& 18
\end{array}
$$

The molecular weight of glucose, $C_6H_{12}O_6$, can be calculated as follows

$$
\begin{array}{llll}
\text{6 carbon atoms (atomic weight 12)} = & 6 \times 12 = & 72 \\
\text{12 hydrogen atoms (atomic weight 1)} = & 12 \times 1 = & 12 \\
\text{6 oxygen atoms (atomic weight 16)} = & 6 \times 16 = & 96 \\
\hline
& \text{molecular weight} = & 180
\end{array}
$$

The molecular weight of calcium phosphate, $Ca_3(PO_4)_2$, can be calculated as follows:

$$
\begin{array}{llll}
\text{3 calcium atoms (atomic weight 40)} = & 3 \times 40 = & 120 \\
\text{2 phosphorus atoms (atomic weight 31)} = & 2 \times 31 = & 62 \\
\text{8 oxygen atoms (atomic weight 16)} = & 8 \times 16 = & 128 \\
\hline
& \text{molecular weight} = & 310
\end{array}
$$

Percentage Composition

The *percentage composition* of a compound can be calculated from the relative atomic weights of the elements present in that compound. Consider the compound $Ca_3(PO_4)_2$, whose molecular weight was found to be 310 (see previous paragraph). Of this weight, 120 was calcium, 62 phosphorus and 128 oxygen. Then

$$\%Ca = \frac{\text{weight of calcium in compound}}{\text{weight of compound}} \times 100 = \frac{120}{310} \times 100 = 38.7\%$$

$$\%P = \frac{\text{weight of phosphorus in compound}}{\text{weight of compound}} \times 100 = \frac{62}{310} \times 100 = 20.0\%$$

$$\%O = \frac{\text{weight of oxygen in compound}}{\text{weight of compound}} \times 100 = \frac{128}{310} \times 100 = 41.3\%$$

$$\text{total} = \overline{100\%}$$

Chemical Equations

When an electrical current is passed through water (a process known as *electrolysis*) hydrogen gas and oxygen gas are produced. The chemist uses symbols and formulas to describe this chemical reaction.

$$H_2O \xrightarrow{\text{electrical current}} H_2\uparrow + O_2\uparrow$$

The arrow is used instead of an equal sign and is read as "yields" or "produces." The plus sign on the right side of the equation is read as "and." A plus sign on the left side of the equation is read as "reacts with." The materials that react are called the *reactants*. The reactants are written on

the left side of the equation. The substances that are produced are called
the *products*. They are written on the right side of the equation. An arrow
pointing downward after a substance indicates that that substance is a
precipitate; that is, it is an insoluble solid, one that does not dissolve.
An arrow pointing upward after a substance indicates that that substance is
a gas.

Note that the previous equation does not contain the same number of
hydrogen and oxygen atoms on each side of the arrow. To be balanced, a
chemical equation must contain the same number of atoms of each element
on both sides of that equation. Thus the following equation is *not*
balanced.

$$H_2O \xrightarrow{\text{electrical current}} H_2\uparrow + O_2\uparrow \quad \text{(unbalanced)}$$

In balancing a chemical equation, do not change the subscripts (the
lower numbers after the symbols) because doing so would change either
the reactants or the products, thus changing the meaning of the reaction.
Instead, place coefficients (numbers) in front of the symbols and formulas to
indicate how many of each is needed.

In the equation $H_2O \rightarrow H_2 + O_2$, there are two hydrogen atoms on each
side of the equation. There is one oxygen atom on the left side and two oxygen
atoms on the right side of the equation. To get two atoms of oxygen on the
left side of the equation (in order to balance the two oxygen atoms on the
right side), place a 2 in front of the H_2O. The 2 cannot be placed as a subscript
after the O in H_2O because then another substance would be represented,
not water. The 2 cannot be placed between the H and the O because this also
would change the meaning of the formula. Therefore, place the 2 in front of
the H_2O.

$$2H_2O \xrightarrow{\text{electrical current}} H_2\uparrow + O_2\uparrow \quad \text{(unbalanced)}$$

However, now there are four hydrogen atoms on the left side of the equation
(2 H_2's). In order to get four hydrogen atoms on the right side of the equation,
place a 2 in front of the H_2. There are already two oxygen atoms on each side
of the equation. Thus the balanced equation is as follows:

$$2H_2O \xrightarrow{\text{electrical current}} 2H_2\uparrow + O_2\uparrow \quad \text{(balanced)}$$

This balanced equation now shows that two molecules of water, on electrolysis, yield two molecules of hydrogen gas and one molecule of oxygen gas.

When aluminum metal is reacted with sulfuric acid, the products are
hydrogen gas, H_2, and aluminum sulfate, $Al_2(SO_4)_3$. The unbalanced equation for this reaction is as follows

$$Al + H_2SO_4 \longrightarrow Al_2(SO_4)_3 + H_2\uparrow \quad \text{(unbalanced)}$$

This equation is not balanced because there are more aluminum atoms on the right side of the equation than on the left side. The same is true for the sulfur and oxygen atoms. To balance an equation of this type, pick out the most complicated-looking formula and assume that one molecule of it is present. The most complicated-looking formula in the above equation is $Al_2(SO_4)_3$. Assuming that one molecule of it is produced, then there are 2 atoms of aluminum on the right side of the equation. To balance this, place a 2 in front of the Al on the left side of the equation.

$$2Al + H_2SO_4 \longrightarrow Al_2(SO_4)_3 + H_2\uparrow \quad \text{(unbalanced)}$$

Next note that there are three SO_4 groups in the molecule of $Al_2(SO_4)_3$. There then must be three SO_4 groups on the left side of the equation. To get these three groups place a 3 in front of the H_2SO_4.

$$2Al + 3H_2SO_4 \longrightarrow Al_2(SO_4)_3 + H_2\uparrow \quad \text{(unbalanced)}$$

To complete the equation, note that there are now six hydrogen atoms on the left side of the equation (in the $3H_2$'s). Therefore, there must be six hydrogen atoms on the right side, so place a 3 in front of the H_2.

$$2Al + 3H_2SO_4 \longrightarrow Al_2(SO_4)_3 + 3H_2\uparrow \quad \text{(balanced)}$$

Now the equation is balanced. There are two aluminums, six hydrogens, three sulfurs, and twelve oxygens (or 3 SO_4's) on each side of the equation.

When sulfur is burned in excess oxygen, sulfur trioxide is produced. When this reaction is written in equation form, it becomes

$$S + O_2 \longrightarrow SO_3 \quad \text{(unbalanced)}$$

Following the balancing procedure, pick out the most complicated compound and take one molecule of it. Thus, in this equation take one molecule of SO_3. This molecule contains one atom of sulfur. There is already one atom of sulfur on the left side of the equation. There are three atoms of oxygen on the right side of the equation and only two on the left side. However, there is no *whole* number that can be placed in front of the O_2 to make three oxygen atoms on that side of the equation. If a 2 is placed there, there will be four atoms of oxygen. In this case, then, instead of selecting one molecule of the most complicated compound, select two molecules of it.

$$S + O_2 \longrightarrow 2SO_3 \quad \text{(unbalanced)}$$

Then, in order to balance two sulfur atoms on the right side of the equation, start with two sulfur atoms on the left side.

$$2S + O_2 \longrightarrow 2SO_3 \quad \text{(unbalanced)}$$

Next, the right side of the equation contains six oxygen atoms and so must
the left side. Place a 3 in front of the O_2 in order to have six oxygen atoms on
that side of the equation. The equation then becomes balanced.

$$2S + 3O_2 \longrightarrow 2SO_3 \quad \text{(balanced)}$$

Types of Chemical Reactions

In a beginning chemistry course, chemical reactions are divided into four
main types. (In more advanced courses, there are several different methods of
subdividing chemical reactions into various types.) These four main types are
combination reactions, decomposition reactions, single replacement reactions,
and *double displacement reactions.*

Combination Reactions

Combination reactions (also called composition reactions) take place when
two or more substances combine to form a more complex substance. An
example of such a reaction is that between carbon and oxygen to form carbon
dioxide. The carbon dioxide formed is more complex than either the carbon
or the oxygen that combined to produce it. The equation for the reaction
can be written as

$$C + O_2 \longrightarrow CO_2$$

Another example of a combination reaction is that of sulfur and oxygen to
form sulfur trioxide. The balanced equation is as follows

$$2S + 3O_2 \longrightarrow 2SO_3$$

Combination reactions can occur between compounds as well as between
elements. For example

$$H_2O + CO_2 \longrightarrow H_2CO_3$$

Combination reactions have the general equation

$$A + B \longrightarrow AB$$

where A and B are the reactants that combine to form the product AB.

Decomposition Reactions

A decomposition reaction occurs when a compound breaks down into
two or more simpler substances. Decomposition is the reverse of combination.

One decomposition reaction described is water being electrolyzed to yield hydrogen and oxygen. Other examples of decomposition reaction are mercuric oxide (HgO) yielding mercury and oxygen when heated

$$2HgO \xrightarrow{\text{heat}} 2Hg + O_2\uparrow$$

and sulfurous acid yielding water and sulfur dioxide

$$H_2SO_3 \longrightarrow H_2O + SO_2\uparrow$$

The general equation for a decomposition reaction is as follows

$$AB \longrightarrow A + B$$

where AB is the reactant (the starting material) which breaks down into A and B. The products A and B are simpler substances than AB. A and B may be either elements or compounds.

Single Replacement Reactions

Single replacement reactions (also called single displacement reactions) occur when one element in a compound is replaced by another element. In the following reaction

$$Zn + CuSO_4 \longrightarrow ZnSO_4 + Cu$$

the zinc replaces the copper in the $CuSO_4$, producing the element copper. One element (zinc) has taken the place of another element (copper) in a compound. In the following single replacement reaction

$$Mg + 2HCl \longrightarrow MgCl_2 + H_2\uparrow$$

the magnesium has taken the place of the hydrogen in the HCl to form $MgCl_2$, freeing the hydrogen as gas.

This same type of reaction occurs with nonmetals. In the following reaction one nonmetal (chlorine) replaces another nonmetal (bromine) in its compound.

$$Cl_2 + 2NaBr \longrightarrow 2NaCl + Br_2$$

The general equation for a single replacement reaction where one metal (A) replaces another metal (B) in its compound (BC) is as follows

$$A + BC \longrightarrow AC + B$$

A second equation for a single replacement reaction where a nonmetal (D) replaces another nonmetal (F) in its compound (EF) is as follows

$$D + EF \longrightarrow ED + F$$

Double Displacement Reactions

Double displacement reactions (also called double replacement, double decomposition, or metathesis reactions) occur when substances in two different compounds displace each other. In the following reaction

$$Na_2SO_4 + BaCl_2 \longrightarrow 2NaCl + BaSO_4\downarrow$$

the Na from the Na_2SO_4 has taken the place of the Ba in the $BaCl_2$. At the same time the Ba from the $BaCl_2$ has taken the place of the Na in the Na_2SO_4. The arrow beside $BaSO_4$ indicates that it is a precipitate. Diagrammatically this can be shown as follows

$$Na_2SO_4 + BaCl_2 \longrightarrow 2NaCl + BaSO_4\downarrow$$

In another example of a double displacement reaction

$$FeS + 2HCl \longrightarrow FeCl_2 + H_2S\uparrow$$

the Fe has taken the place of the H in the HCl and the H in turn has taken the place of the Fe in its compound forming the gas, hydrogen sulfide.

The general equation for a double displacement reaction is as follows

$$AB + CD \longrightarrow AD + CB$$

where A has taken the place of C and likewise C has taken the place of A in their corresponding compounds.

Reaction Rates

Some chemical reactions proceed at a slow rate. Iron, for example, rusts very slowly. Wood takes years to decay. On the other hand, some chemical reactions proceed more rapidly. Coal burns steadily and quickly. Concrete begins to set within a few hours. Some chemical reactions not only occur rapidly, they take place almost instantaneously. Consider the violent explosion of dynamite. Within a fraction of a second, the complete reaction has taken place.

What determines the speed of a chemical reaction? The speed of a chemical reaction depends upon several factors. These are (1) the nature of the reacting substances; (2) the temperature; (3) the concentration of the reacting substances; (4) the presence of a catalyst; and (5) the surface area and the intimacy of contact of the reacting substances.

Nature of Reacting Substances

When a solution of sodium sulfate (Na_2SO_4) is mixed with a solution of barium chloride ($BaCl_2$), a white precipitate of barium sulfate ($BaSO_4$) is formed immediately.

$$Na_2SO_4 + BaCl_2 \longrightarrow 2NaCl + BaSO_4\downarrow$$

This equation may be rewritten to show the ions of which these salts consist.

$$2Na^+ + SO_4^{2-} + Ba^{2+} + 2Cl^- \longrightarrow 2Na^+ + 2Cl^- + BaSO_4\downarrow$$

Next, as in any algebraic equation, cancel the sodium ions and the chloride ions from both sides of the equation leaving the net equation

$$Ba^{2+} + SO_4^{2-} \longrightarrow BaSO_4\downarrow$$

This is an example of an ionic reaction—the reaction between ions. Many of the reactions taking place in the body are of this type.

Consider, however, the reaction between hydrogen (H_2) and oxygen (O_2) to form water (H_2O). This reaction proceeds very slowly, even at a temperature of 200°C unless a spark is introduced into the mixture.

$$2H_2 + O_2 \xrightarrow{\text{spark}} 2H_2O$$

In this reaction, it is necessary for the bonds between the hydrogen atoms in the hydrogen molecules to be broken. Also, the bonds between the oxygen atoms must be broken before the reaction can occur. This is an example of a reaction in which covalent bonds must be broken and new ones formed. Such reactions proceed much more slowly than ionic reactions.

Thus, the speed of a reaction is affected by the nature of the reacting substances, whether their bonding is ionic or covalent.

Temperature

As the temperature rises, the speed of a chemical reaction increases. Actually, for every 10°C rise in temperature, the speed of a given reaction approximately doubles. Thus, a reaction that proceeds at a certain speed at 0°C will proceed twice as fast at 10°C and four times as fast at 20°C.

Since a change in temperature of 10°C doubles the speed of a chemical reaction, even a slight change in temperature may affect the speed of a reaction with definitely noticeable results.

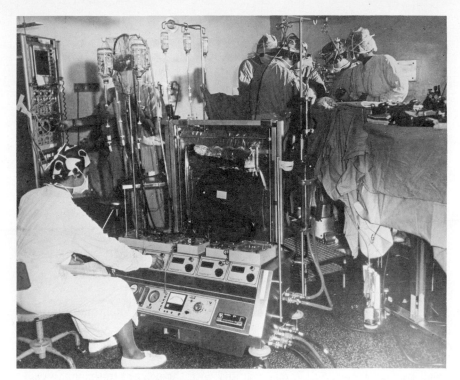

Figure 6–1. Open-heart surgery. (Courtesy of Massachusetts General Hospital, Boston, Mass.)

When a patient has a fever of only a few degrees, he has an increased pulse rate and also an increased respiratory rate. Reactions taking place throughout the body proceed at an accelerated rate.

When the temperature of the human body drops, the various metabolic processes slow down considerably. This fact is of great importance, for example, during open-heart surgery when the temperature of the body is lowered considerably (see Figure 6–1).

Concentration

The concentration of a reactant is the amount present in a given unit of volume. The more of a given material present in a certain volume, the greater its concentration. Greater concentration produces faster reactions because there are more molecules that can react.

A patient with a respiratory disease can breathe more easily when using a nasal catheter with oxygen because the concentration of the oxygen in the lungs is increased. This increased concentration increases the speed of oxygen uptake, making breathing easier for the patient (see Figure 8–4).

When a protein substance is placed in water and heated, a hydrolysis reaction (see page 309) proceeds at an extremely slow speed. If a strong acid is added to the mixture, the reaction proceeds at a much faster rate. The acid is not used up (that is, it is not changed chemically). Its presence merely increases the speed of the reaction. Any substance that changes the speed of a reaction without itself being changed chemically is called a catalyst. Many of the chemical reactions used in industry would not be practical without a catalyst. They would take too long to be of commercial use.

The body uses catalysts to enable its chemical reactions to proceed at a rapid pace. Those catalysts present in the body are called *enzymes*. During digestion, for example, the food undergoes many chemical changes, each under the influence of a specific enzyme. There are also catalysts (and enzymes) that slow down rather than speed up chemical reactions.

Surface Area

The speed of a chemical reaction also depends upon the amount of surface area present in the reacting substances. Although a pile of flour is quite harmless, the same flour in the form of dust can cause a dangerous explosion. This effect is due to the tremendous amount of surface area of the dust. This large surface area can then react rapidly with the oxygen in the air to cause an explosion.

Many medications are given in the form of finely divided suspended solids. In this manner, more surface area means more rapid absorption in the body.

Increasing the concentration of the reacting substances may also be considered as increasing the amount of surface area.

Summary

A symbol for an element not only identifies that element but also represents one atom of that element. A formula consists of a group of symbols that represent the elements present in a substance.

The molecular weight of a compound is equal to the sum of the atomic weights of the atoms present in that compound.

A chemical equation uses symbols and formulas to represent a chemical reaction. The substances on the left side of the equation are called reactants and those on the right side products. An arrow pointing upward indicates a gas; an arrow pointing downward indicates a precipitate.

To balance a chemical equation, pick out the most complicated-looking compound and assume that one molecule of it is present. Then proceed back and forth adding coefficients in front of the reactants and products until the number of atoms of each type on each side of the equation is the same.

The four types of chemical reactions and their general equations are:

Combination	$A + B \longrightarrow AB$
Decomposition	$AB \longrightarrow A + B$
Single replacement	$A + BC \longrightarrow AC + B$

or

$$D + EF \longrightarrow ED + F$$

Double displacement $\quad AB + CD \longrightarrow AD + CB$

The speed of a chemical reaction depends upon the nature of the reacting substances, upon the temperature, upon the concentration of the reacting substances, upon the presence of a catalyst, and upon the surface area of the reacting substances.

Questions and Problems

1. What do the following symbols or formulas indicate: $O_2, 2O, CO_2, CO$?
2. Calculate the molecular weight of each of the following compounds (use atomic weights as whole numbers)
   (a) $NaNO_3$
   (b) KBr
   (c) $Ca(HCO_3)_2$
   (d) $C_{12}H_{22}O_{11}$
   (e) $C_{57}H_{110}O_6$
3. Balance the following equations
   (a) $Mg + O_2 \longrightarrow MgO$
   (b) $Zn + HCl \longrightarrow ZnCl_2 + H_2$
   (c) $C + O_2 \longrightarrow CO_2$
   (d) $NaCl + AgNO_3 \longrightarrow AgCl + NaNO_3$
   (e) $ZnSO_4 + NaOH \longrightarrow Zn(OH)_2 + Na_2SO_4$
   (f) $Fe + O_2 \longrightarrow Fe_3O_4$
   (g) $Mg + AgNO_3 \longrightarrow Mg(NO_3)_2 + Ag$
   (h) $Al(OH)_3 + H_2SO_4 \longrightarrow Al_2(SO_4)_3 + H_2O$
4. Label each of the reactions in question 3 as to type.
5. What factors determine the rate of a chemical reaction? Indicate at least one practical application of each factor.
6. Calculate the percentage composition of each compound in question 2 (answers to one decimal place).

References

Baum, S. J., and Scaife, C. W.: *Chemistry: A Life Science Approach.* Macmillan Publishing Co., Inc., New York, 1975, Chaps. 4, 9.

Hein, M., and Best, L. R.: *College Chemistry.* Dickenson Publishing Co., Inc., Encino, Calif., 1976, Chap. 10.

Peters, E. I.: *Introduction to Chemical Principles.* W. B. Saunders Co., Philadelphia, 1974, Chap. 5.

Sackheim, G. I.: *Chemical Calculations*, 10th ed., series B. Stipes Publishing Co., Champaign, Ill., 1976, Chaps. 4, 5, 7.

Sienko, M. J., and Plane, R. A.: *Chemistry*, 5th ed. McGraw-Hill Book Co., New York, 1976, Chap. 5.

Chapter 7

The Gaseous State

General Properties

In Chapter 2 we saw that gases have no definite shape, no definite volume, and a low density. Gases have a much greater volume than an equal mass of solid or liquid. For example, 1 gram of liquid water occupies 1 ml and 1 gram of solid water (ice) occupies nearly 1 ml, but 1 gram of water vapor (at 0°C and 1 atmosphere pressure) occupies nearly 1250 ml. These and other properties of gases can be explained in terms of the kinetic molecular theory.

The Kinetic Molecular Theory

The principal assumptions of this theory are:

1. Gases consist of tiny particles called molecules.
2. The distance between molecules of a gas is very great compared to the size of the molecules themselves (that is, the volume occupied by a gas is mostly empty space).
3. Gas molecules are in rapid motion and move in straight lines, frequently colliding with each other and with the walls of the container.

4. Gas molecules do not attract each other.

5. When molecules of a gas collide with each other or with the walls of the container, they bounce back with no loss of energy. Such collisions are said to be perfectly elastic.
6. The average kinetic energy of the molecules is the same for all gases at the same temperature. The average kinetic energy increases as the temperature increases and decreases as the temperature decreases.

Let us see how the various properties of gases can be explained in terms of this theory.

1. If the distance between molecules is very great compared to the size of the molecules themselves, the molecules will occupy only a small fraction of that volume and so the density will be very low.

2. If the molecules are moving rapidly in all directions, they can fill any size container. They can keep on moving until they hit a wall of a container or until they hit each other and bounce back; that is, the gas will have no definite volume and no definite shape.

3. If the molecules of a gas strike the walls of a container, they should exert a pressure on that wall. However, since the molecules are moving in all directions, they should exert a pressure equally in all directions. And, gases do so.

4. If a bottle of ether is opened in a room, the odor is soon apparent in all parts of that room; that is, the molecules of ether gas diffuse into the air (a mixture of gases). According to the kinetic molecular theory, there is a great deal of empty space between the molecules of a gas so that the ether molecules can diffuse into the spaces between the air molecules.

5. We can show that the collisions must be perfectly elastic by means of a reverse type of reasoning. Suppose that the collisions between molecules are not perfectly elastic; that is, some energy was lost upon collisions with each other and with the walls of the container. Eventually, the gas molecules would have so little energy left that they would settle to the bottom of the container. However, gases never settle. Therefore, the collisions between the molecules themselves and the walls of the container must be perfectly elastic.

6. Since gas molecules are so far apart from one another, they should be able to be forced closer together by increasing the pressure; that is, gases should be compressible, as indeed they are.

The Gas Laws

Boyle's Law

If the volume of a gas is reduced, the molecules will have less space in which to move. Therefore, they will strike the walls of the container more often and cause a greater pressure.

The relationship between the volume of a given quantity of a gas and its pressure is expressed by Boyle's law, which states that the volume occupied by a gas is inversely proportional to the pressure, if the temperature remains constant.

A direct application of Boyle's law may be seen in a respirator, a machine that has been used in the treatment of patients with polio or other respiratory difficulties. When pressure inside of the respirator is decreased, the air in the lungs expands, forcing the diaphragm down. When the pressure in the respirator is increased the volume of air in the lungs is decreased, allowing the diaphragm to move upward again. This alternate increase and decrease in pressure enables the patient to breathe even though his own muscles cannot control the movement of the diaphragm.

Boyle's law may be stated mathematically as

$$P_1 V_1 = P_2 V_2$$

where P_1 and P_2 are the initial and final pressure, respectively, and V_1 and V_2 the initial and final volumes. Pressures are usually expressed in the unit torr, where 760 torr equals 1 atmosphere pressure.

Example 1. 500 ml gas at 25°C and 750 torr will occupy what volume at 25°C and 650 torr?

Note first that temperature is constant so that we can use Boyle's law. Note also that P_1 is 750 torr, V_1 is 500 ml, and P_2 is 650 torr. Then, using $P_1 V_1 = P_2 V_2$, we have

$$750 \text{ torr} \times 500 \text{ ml} = 650 \text{ torr} \times V_2$$

Dividing by 650 torr, we have

$$\frac{750 \text{ torr} \times 500 \text{ ml}}{650 \text{ torr}} = V_2 \quad \text{and} \quad V_2 = 577 \text{ ml}$$

Example 2. A gas exerts a pressure of 858 torr when confined in a 5-liter container. What will be the pressure if the gas is confined in a 10-liter container at constant temperature?

Again using $P_1 V_1 = P_2 V_2$

$$858 \text{ torr} \times 5\text{L} = P_2 \times 10\text{L}$$

$$P_2 = 429 \text{ torr}$$

Charles' Law

When gases are heated, they expand; when gases are cooled, they contract. The relationship between volume and temperature is expressed by Charles' law, which states that the volume of a fixed quantity of a gas is directly proportional to its Kelvin temperature, if the pressure remains constant.

Recall that Kelvin (absolute) temperature is Celsius temperature plus 273°
(see page 5).

Charles' law may be explained in terms of the kinetic molecular theory.
As the temperature of a gas is increased, the molecules move faster and strike
the walls of the container more often. However, if the pressure is to be kept
constant, then the volume must increase; that is, the higher the temperature,
the greater the volume of a gas, at constant pressure, and vice versa.

A direct application of Charles' law may be seen in such equipment as an
incubator. When air comes in contact with the heating element, it expands
and becomes lighter. This lighter air rises, causing a circulation of warm air
throughout the incubator.

Charles' law may be expressed mathematically as:

$$\frac{V_1}{T_1} = \frac{V_2}{T_2}$$

where V_1 and V_2 are the initial and final volumes and T_1 and T_2 the initial
and final temperatures, respectively, all temperature in degrees Kelvin.

Example 3. A gas occupies 368 ml at 27°C and 600 torr. What will be the
volume of that gas at 127°C and 600 torr?

We note that pressure is constant so that we can use Charles' law. Also
note that V_1 is 368 ml, T_1 is 27° + 273° or 300°K, and T_2 is 127° + 273° or
400°K. Then, using

$$\frac{V_1}{T_1} = \frac{V_2}{T_2}$$

$$\frac{368 \text{ ml}}{300°\text{K}} = \frac{V_2}{400°\text{K}}$$

$$V_2 = 491 \text{ ml}$$

Example 4. 200 ml of gas at 35°C is cooled at constant pressure to −20°C.
What will be its new volume?

Here V_1 is 200 ml, T_1 is 35° + 273° or 308°K, and T_2 is −20° + 273° or
253°K. Then, using

$$\frac{V_1}{T_1} = \frac{V_2}{T_2}$$

$$\frac{200 \text{ ml}}{308°\text{K}} = \frac{V_2}{253°\text{K}}$$

$$V_2 = 164 \text{ ml}$$

Pressure–Temperature Relationship

Another relationship may be expressed between the pressure exerted by
a gas and its temperature. These two factors are directly proportional. That is,

Figure 7–1. Autoclave being loaded. (Courtesy of Castle Co., Rochester, N.Y.)

as the temperature on a gas increases, the pressure increases, and vice versa, if the volume remains constant.

A common application of this relationship is the autoclave, a device used for sterilization in the hospital (see Figure 7–1). As the pressure of the incoming steam is increased, the temperature of the gas inside the autoclave is also increased. The normal temperature of steam is 100°C but in an autoclave it may rise as high as 120°C because of the increased pressure. This higher temperature is sufficient to destroy any microorganisms that may exist in the material being autoclaved.

Combined Gas Laws

Boyle's law refers to the volume of a fixed quantity of a gas at constant temperature; Charles' law, to the volume of such a gas at constant pressure. However, frequently none of these factors is constant. In such a case we use the combined gas laws, which may be stated mathematically as:

$$\frac{P_1 V_1}{T_1} = \frac{P_2 V_2}{T_2}$$

Example 5. 250 ml of gas at 27°C and 800 torr will occupy what volume at STP? (STP means standard temperature and pressure, 0°C and 760 torr or 760 mm, where the units "torr" and "mm" are commonly used interchangeably.) Using

$$\frac{P_1 V_1}{T_1} = \frac{P_2 V_2}{T_2},$$

$$\frac{800 \text{ torr} \times 250 \text{ ml}}{300° \text{K}} = \frac{760 \text{ torr} \times V_2}{273° \text{K}}$$

whence $V_2 = 239$ ml.

Dalton's Law

Dalton's law refers to a mixture of gases rather than to a pure gas. Dalton's law states that, in a mixture of gases, each gas exerts a partial pressure proportional to its concentration. For example, if air contains 21 per cent oxygen, then 21 per cent of the total air pressure is exerted by the oxygen. Normal air pressure of 1 atm will support a mercury column 760 mm high. The partial pressure of the oxygen in the air would be 21 per cent of 760 mm, or 0.21×760 mm, or 160 mm.

Dalton's law may be explained in terms of the kinetic molecular theory. Since there is no attraction between gas molecules, each kind of molecule strikes the walls of the container the same number of times per second as if it were the only kind of molecule present. That is, the pressure exerted by each gas (its partial pressure) is not affected by the presence of other gases. Each gas exerts a partial pressure proportional to the number of molecules of that gas (proportional to its concentration).

Gases always flow from an area of higher partial pressure to one of lower partial pressure. An example of the flow of gases caused by a difference in partial pressures is found in our own body (see Figure 7–2). The partial pressure of oxygen in the inspired air is 158 mm. The partial pressure of oxygen in the alveoli is 100 mm. Therefore, oxygen passes from the lungs into the alveoli (from a higher partial pressure to a lower one). From the alveoli the oxygen flows into the venous blood (from a partial pressure of 100 mm to one of 40 mm). This flow of oxygen into the venous blood in the lungs changes the venous blood into arterial blood, in which the partial pressure of oxygen is 90 mm. When the arterial blood reaches the tissues where the partial pressure of oxygen is 30 mm, oxygen flows to those tissues (again from a higher partial pressure to a lower one). When the arterial blood loses oxygen to the tissues, its oxygen partial pressure drops to 40 mm and it becomes venous blood, which returns to the lungs to begin the cycle anew.

Conversely, in the tissues the partial pressure of the carbon dioxide is 50 mm and in the arterial blood it is 40 mm so that carbon dioxide flows out of the tissues into the blood. When the arterial blood picks up carbon dioxide (and at the same time loses oxygen) it becomes venous blood with a carbon dioxide partial pressure of 46 mm. This venous blood, in turn, loses carbon

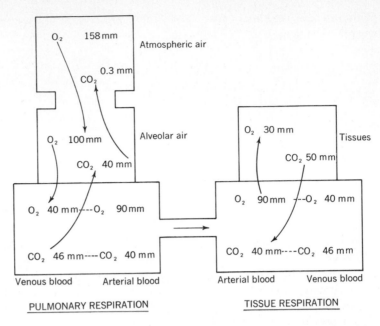

Figure 7–2. Diagram showing oxygen and carbon dioxide (CO_2) flow and partial pressures in body.

dioxide to the alveoli where the carbon dioxide partial pressure is 40 mm. From the alveoli the carbon dioxide passes into the lungs (from a partial pressure of 40 mm to one of 0.3 mm) and then is exhaled.

Graham's Law

If a container of ammonia and a container of ether are opened simultaneously, a person standing at the far end of the room will notice the odor of ammonia before that of the ether. Both gases will diffuse through the air but one will diffuse at a faster rate than the other. Ammonia has a molecular weight of 17 and ether a molecular weight of 74. We should expect the lighter molecule to diffuse faster than the heavier one, as indeed it does. This relationship was stated by Graham as, "The rates of diffusion of gases are inversely proportional to the square roots of the molecular weights" (or inversely proportional to the square roots of their densities).

Air Pollution

Air pollutants may be gases such as sulfur dioxide, nitrogen oxides, ozone, hydrocarbons, and carbon monoxide, or they may be particulate matter such as smoke particles, asbestos, and lead aerosols.

Concentrations of pollutants in the air are often expressed as parts per million (ppm), where 1 ppm corresponds to one part pollutant to one million parts air.

Concentrations of pollutants at levels far below 1 ppm can have an adverse effect upon human life. For example, 0.2 ppm sulfur dioxide in the atmosphere leads to an increased death rate and 0.02 ppm peroxybenzoyl nitrate (a constituent of smog) causes severe eye irritation.

The word "smog" is derived from the words "smoke" and "fog" and is characterized by air that contains lung and eye irritants along with reduced visibility. Smog frequently occurs during a thermal inversion. Normally, warm air near the ground surface rises and carries away pollutants. However, during a thermal inversion, the air near the surface is cooler than the air above it and so remains at the surface keeping the air pollutants down at that level.

Gases

The principal source of three gaseous pollutants—carbon monoxide, hydrocarbons, and nitrogen oxides—is the automobile, which is also the source of such particulate matter in the air as asbestos (from brake linings) and lead aerosols (from leaded gasoline). Tobacco smoke is a source of carbon monoxide as well as particulate matter (ash): fossil fuel–powered electrical generating stations are also a major source of air pollutants.

Each pollutant poses a different threat to human life.

1. *Carbon monoxide* is a deadly poison that interferes with the transportation of oxygen by the blood. Low concentrations of this gas, such as are found in automobiles, garages, downtown streets during "rush hours," and space-heated rooms, cause impairments of judgment and vision. Evidence has shown that intermittent exposure to carbon monoxide at low levels of concentration may cause strokes and hypertension in susceptible individuals. High concentrations of carbon monoxide may cause headache, drowsiness, coma, respiratory failure, and death.

2. There is no evidence that *hydrocarbons*, at present levels, exert any direct undesirable effects upon humans.

3. *Nitrogen oxides* are just as dangerous a pollutant as carbon monoxide, even though environmental groups do not stress them equally. The first effects of nitrogen oxides upon humans is an irritation of the eyes and respiratory passages. Concentrations of 1.6 to 5 ppm of nitrogen dioxide for a one-hour exposure cause increased airway resistance and diminish diffusing capacity of the lungs. Concentrations of 25 to 100 ppm cause acute but reversible bronchitis and pneumonitis. Concentrations above 100 ppm are usually fatal, with death resulting from pulmonary edema.

4. *Oxides of sulfur* cause acute airway spasm and poor airway clearance in all people and exert a deadly effect upon patients already disabled by lung disease. Concentrations of 8 to 10 ppm cause immediate throat irritation, and concentrations of 20 ppm cause immediate coughing.

Particulate matter absorbs sulfur dioxide, and the resulting tiny particles may get into the small air passages in the lungs, causing spasms and destruction of cells. Particles of the smallest size penetrate deepest into the lungs and remain there the longest.

1. *Cigarette and tobacco smoke* is one of the most dangerous types of air pollution. Smokers inhale large amounts of carbon monoxide, tars, and particulate matter. People around smokers are also exposed to these same pollutants. Evidence has shown that expectant mothers who smoke have smaller babies who have a higher infant mortality rate than those of non-smokers. Smokers also have a greater chance of developing lung cancer, emphysema, and cardiovascular disease.

2. *Ozone* is normally present in the atmosphere in extremely small amounts. However, it is also produced by the action of sunlight upon the oxides of nitrogen given off by automobile engines. Ozone, in turn, reacts with hydrocarbon emissions to form peroxyacyl nitrates, which are the eye irritants of smog. Since ozone is produced by the action of sunlight, its levels in the air are usually lower at night than during the day.

Los Angeles–type smog is primarily a photochemical phenomenon since many of the changes that take place in the air pollutants occur because of the action of sunlight upon the substances mentioned above.

In some sections of North America ozone is such a problem that when levels reach 0.1 ppm, a yellow alert is sounded. During this yellow alert people having respiratory or heart problems are warned to stay indoors and to reduce their activity. If the ozone level should reach 0.4 ppm, a red alert is sounded. Under these conditions certain factories are forced to shut down and motorists are asked to restrict their use of automobiles. In some areas a red alert is sounded if the ozone level remains at 0.1 ppm for 24 hours at any reporting station.

Summary

The properties of gases may be explained in terms of the kinetic molecular theory. which states that (1) gases are composed of tiny particles called molecules, (2) the distances between molecules of a gas are very great compared to the size of the molecules themselves, (3) gas molecules move rapidly in straight lines, (4) gas molecules do not attract each other, (5) collisions between molecules and between the molecules and the walls of the container are perfectly elastic, and (6) the average kinetic energy of the molecules is the same for all gases at the same temperature.

Boyle's law states that the volume of a fixed quantity of a gas is inversely proportional to the pressure if the temperature remains constant. Boyle's law may be expressed mathematically as $P_1 V_1 = P_2 V_2$.

Charles' law states that the volume of a fixed quantity of a gas is directly proportional to its Kelvin temperature if the pressure remains constant. Charles' law may be stated mathematically as $V_1/T_1 = V_2/T_2$.

The combined gas laws may be stated mathematically as $P_1 V_1/T_1 = P_2 V_2/T_2$.

Dalton's law states that in a mixture of gases each gas exerts a partial pressure proportional to its concentration.

Gases flow from an area of high partial pressure to one of lower partial pressure.

Graham's law states that the rates of diffusion of gases are inversely proportional to the square roots of their molecular weights.

Air pollutants may be gases such as carbon monoxide, ozone, nitrogen oxides, sulfur oxides, and hydrocarbons, and they also may be particulate matter such as smoke particles, asbestos, and lead aerosols. Each pollutant poses a different threat to human life.

Questions and Problems

1. Explain in terms of the kinetic molecular theory (a) why gases have a low density, (b) why gases are compressible, (c) why gases do not settle, (d) why gases diffuse into each other, (e) why gases have no definite shape or volume, (f) Charles' law, and (g) Boyle's law.
2. 800 ml of neon gas at 25°C and 720 torr will occupy what volume at 25°C and 680 torr?
3. 1.50 liters of oxygen at 13°C and 675 torr occupy 1.75 liters at 13°C and what pressure?
4. 200 ml of oxygen at 27°C and 600 torr occupy what volume at 227°C and 600 torr?
5. 3.00 ft^3 of hydrogen at 40°C and 1.15 atmosphere pressure will occupy what volume at −40°C and 1.15 atmosphere pressure?
6. 250 ml of carbon dioxide gas at 15°C and 900 torr will occupy what volume at STP (0°C and 760 torr)?
7. Explain how a respirator enables a person whose lungs are paralyzed to breathe.
8. Explain how an autoclave works in terms of the appropriate gas law.
9. Why does carbon dioxide pass from the blood to the lungs instead of vice versa?
10. Explain why oxygen flows to the tissues and carbon dioxide from the tissues instead of vice versa?
11. Which gas will diffuse faster: methane, CH_4, or ammonia, NH_3?
12. Arrange the following gases in order of increasing rates of diffusion: H_2, O_2, NH_3, CO_2, SO_2, C_2H_6, Ar, and He.
13. What are the principal sources of air pollution?
14. What are the hazards of cigarette smoking?
15. What threats to life are caused by (a) oxides of nitrogen, (b) ozone, (c) carbon monoxide, (d) oxides of sulfur, (e) particulate matter?
16. When is an air pollution yellow alert called? A red alert? What precautions should be taken?

References

Baum, S. J., and Scaife, C. W.: *Chemistry: A Life Science Approach.* Macmillan Publishing Co., Inc., New York, 1975, Chap. 5.

Eastman, R. D.: *Essentials of Modern Chemistry.* Holt, Rinehart & Winston, Inc., New York, 1975, Chap. 5.

Fernandez, J. E., and Whitaker, R. D.: *An Introduction to Chemical Principles.* Macmillan Publishing Co., Inc., New York, 1975, Chaps. 4, 5.

Hein, M., and Best, L. R.: *College Chemistry.* Dickenson Publishing Co., Inc., Encino, Calif., 1976, Chaps. 12, 21.

Hodges, L.: *Environmental Pollution.* Holt, Rinehart & Winston, Inc., New York, 1973, Chaps. 3, 4, 5, 6.

Longo, F. R.: *General Chemistry.* McGraw-Hill Book Co., New York, 1974, Chap. 6.

Nebergall, W. H.; Schmidt, F. C.; and Holtzclaw, H. F., Jr.: *College Chemistry*, 5th ed. D. C. Heath & Co., Lexington, Mass., 1976, Chaps. 10, 24.

Peters, E. I.: *Introduction to Chemical Principles.* W. B. Saunders Co., Philadelphia, 1974, Chap. 10.

Sackheim, G. I.: *Chemical Calculations*, 10th ed., series B. Stipes Publishing Co., Champaign, Ill., 1976, Chap. 6.

Sienko, M. J., and Plane, R. A.: *Chemistry*, 5th ed. McGraw-Hill Book Co., New York, 1976, Chap. 6.

Stocker, H. S., and Seager, S. L.: *Environmental Chemistry: Air and Water Pollution*, 2nd ed. Scott, Foresman & Co., Glenview, Ill., 1976, Part 1.

Chapter 8

Oxygen and Other Gases

CHAPTER OUTLINE

Oxygen
Occurrence / *Properties* / *Combustion* / *Preparation* / *Uses*

Ozone
Preparation / *Properties and Uses*

Nitrous Oxide

Noxious Gases

Oxygen

Occurrence

Oxygen is the most abundant element on the earth's surface. Air contains about 21 per cent free oxygen. The oceans and lakes on the earth's surface consist of about 80 per cent oxygen in the combined state. The earth's crust consists of about 50 per cent oxygen, combined mostly with silicon. Oxygen compounds comprise most of the weight of plants and animals.

Properties

Physical Properties. At room temperature oxygen is a colorless, odorless, tasteless gas. It is slightly heavier than air and is slightly soluble in water.

The method of preparation illustrated in Figure 8–1 shows that oxygen does not dissolve appreciably in water. If it were very soluble in the water, it could not be collected by this method. However, a small amount of oxygen does dissolve in water. This amount, even though small, is of very definite importance to marine life. It is this small amount of dissolved oxygen that enables fish and other marine animals to "breathe."

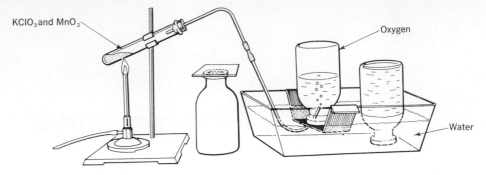

KClO$_3$ and MnO$_2$

Oxygen

Water

Figure 8–1. Laboratory preparation of oxygen.

The density of oxygen is 1.43 g per liter. The density of air is 1.29 g per liter. Therefore, oxygen is slightly heavier than air. When oxygen is collected in the laboratory, it is kept in covered bottles with the mouth upward.

When oxygen is cooled sufficiently, it forms a pale blue liquid which boils at −182.5°C. Further cooling produces a pale blue solid, freezing at −218.4°C.

Chemical Properties. Oxygen is a moderately active element at room temperature but is extremely active at higher temperatures. Oxygen combines with almost all elements to produce a class of compounds called oxides.

$$2Mg \; + \; O_2 \longrightarrow 2MgO$$

magnesium   oxygen   magnesium
oxide

$$C + O_2 \longrightarrow CO_2$$

carbon   oxygen   carbon
dioxide

The reaction between oxygen and some other substance is an example of *oxidation*. Common examples of oxidation are the rusting of iron, the burning of a candle, and the decaying of wood. Oxidation also occurs in living plant and animal tissues. These oxidation reactions are able to occur rapidly at relatively low temperatures because of the presence of specific catalysts called enzymes. An example of such a reaction occurring in the human body is the oxidation of glucose, a simple sugar, to carbon dioxide and water.

$$C_6H_{12}O_6 + 6O_2 \xrightarrow{\text{enzymes}} 6CO_2 + 6H_2O + \text{energy}$$

glucose   oxygen   carbon   water
dioxide

(This reaction is greatly oversimplified. As will be discussed later, when glucose is oxidized to carbon dioxide and water in the body, there are many intermediate steps involved, each with its own particular enzyme.)

Wood, coal, and gas burn in the presence of oxygen. This type of reaction is called combustion. Combustion may be defined as a rapid reaction with oxygen, in which heat and light are produced, usually accompanied by a flame. Oxygen supports combustion (that is, substances burn in oxygen), but oxygen itself does not burn.

Combustion is usually thought of in terms of a reaction with oxygen. However, oxygen is not absolutely necessary for a combustion reaction. When powdered iron and sulfur are heated together in a test tube, a rapid reaction occurs in which heat and light are given off. This is also a combustion reaction even though no oxygen is involved.

Spontaneous Combustion. When iron combines with the oxygen of the air, its rusts, continuously liberating a small amount of heat as the reaction continues. The total amount of liberated heat, however, will be the same as that which would have been liberated had the iron been burned in oxygen. That is, the total amount of heat produced is the same regardless of whether the reaction proceeds rapidly or slowly.

This is the principle underlying spontaneous combustion, which can be defined as a slow oxidation which develops by itself into combustion.

How is such a process possible? If oily rags are placed in a dry container without adequate ventilation, the oil will slowly combine with the oxygen in the air, liberating a small amount of heat during the process. Without ventilation the heat will not be dissipated, especially since rags are such poor conductors of heat. As the oxidation proceeds, more and more heat will be liberated until a sufficient amount accumulates to start the rags burning. This is spontaneous combustion.

If the oily rags had been placed in a closed container, the oxidation would not continue after the oxygen supply was used up. Likewise, if the rags were hung in a place where freely circulating air could carry away the heat, no spontaneous combustion could occur.

Fire Prevention and Control. Care should be taken in handling and storing flammable liquids such as ether and alcohol. Where smoking is allowed, nonflammable receptacles should be provided for the butts. One frequent cause of fires is throwing lighted cigarettes into a wastebasket. Oily rags and mops should be stored in well-ventilated fireproof lockers to avoid the danger of spontaneous combustion.

In case a fire should occur, how can it be extinguished? There are two methods of putting out a fire—removing the oxygen from the burning material or lowering the temperature of the burning substance below its kindling point.

One type of fire extinguisher contains carbon dioxide gas under pressure (see Figure 8–2). This type of extinguisher has several advantages. The carbon dioxide is extremely cold and lowers the temperature of the burning substance.

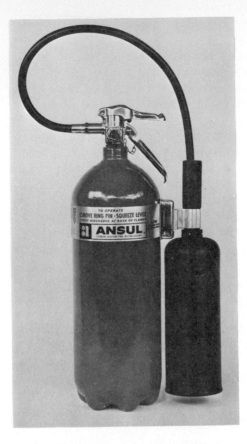

Figure 8–2. Carbon dioxide fire extinguisher. (Courtesy of The Ansul Co., Marinette, Wis.)

A large amount of carbon dioxide, which is heavier than oxygen, surrounds the burning area, keeping out the oxygen. For these reasons the carbon dioxide should be directed at the base of the flame. This type of extinguisher is recommended for electrical fires because the carbon dioxide does not conduct electricity. It is also used for oil and gasoline fires because it shuts off the oxygen supply and at the same time lowers the temperature. When a fire is extinguished, the carbon dioxide disappears into the air so that it does not have to be cleaned up along with the material that was burning. A cylinder of carbon dioxide holds a large volume of that gas under pressure and thus has a great capability in fire fighting.

Another type of fire extinguisher contains a solution of sodium bicarbonate and a vial of sulfuric acid (see Figure 8–3). When this type of extinguisher is inverted, the sulfuric acid reacts with the sodium bicarbonate to form sodium bisulfate, water, and carbon dioxide.

$$H_2SO_4 + NaHCO_3 \longrightarrow NaHSO_4 + H_2O + CO_2\uparrow$$

The carbon dioxide formed exerts enough pressure to force the water out of the extinguisher so that it may be directed at the fire. The water lowers the

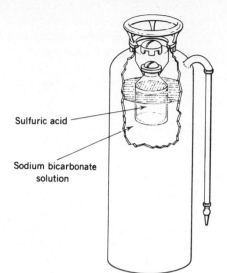

Sulfuric acid

Sodium bicarbonate
solution

Figure 8–3. Soda-acid type of fire
extinguisher.

temperature of the burning substance while the carbon dioxide surrounds it
with an inert atmosphere, keeping out the oxygen. Both of these effects help
to extinguish the flame.

However, this type of fire extinguisher has several disadvantages. It
cannot be used on a fire of electrical equipment because the liquid itself is a
conductor of electricity. Also, once this type of extinguisher has been used
on a fire, a residue of solid material remains which must then be removed.
Because of the weight involved, such an extinguisher cannot contain a large
amount of liquid, so that its use is limited. Finally, this type of fire extinguisher cannot be used for oil or gasoline fires because if it is sprayed on the
burning liquids, it sinks underneath them and actually helps spread the
fire.

Preparation

Laboratory Methods. One very common method for preparing oxygen in
the laboratory is by heating potassium chlorate, $KClO_3$ (see Figure 8–1).
The reaction is as follows

$$2KClO_3 \xrightarrow{\text{heat}} 2KCl + 3O_2\uparrow$$

It takes a considerable amount of heat to produce oxygen by this method
because the potassium chlorate must be heated to its melting point (370°C)
before it gives off oxygen.

However, when manganese dioxide (MnO_2) is added to the potassium
chlorate, oxygen is evolved from the heated mixture at a much lower temperature and at a more rapid rate. The manganese dioxide acts as a catalyst in

this reaction—it increases the speed of the reaction but does not take part in it. The presence of a catalyst in a reaction is indicated over the arrow in the equation for that reaction.

$$2KClO_3 \xrightarrow[\text{heat}]{\text{MnO}_2} 2KCl + 3O_2\uparrow$$

Oxygen may also be produced in the laboratory by the electrolysis of water.

$$2H_2O \xrightarrow{\text{electricity}} 2H_2\uparrow + O_2\uparrow$$

Commercial Method. The commercial source of oxygen is the air, an inexhaustible supply. When air is cooled to a low enough temperature under compression, it becomes a liquid. Liquid air, like ordinary air, consists mostly of nitrogen and oxygen. Liquid nitrogen boils at $-196°C$. Liquid oxygen boils at $-183°C$. When liquid air is allowed to stand, the nitrogen boils off first (because of its lower boiling point) leaving almost pure oxygen behind. Liquid oxygen is stored in steel cylinders under high pressure.

Uses

Medical Uses. Oxygen is necessary to life. When oxygen is taken into the lungs, it combines with the hemoglobin of the blood to form a compound called oxyhemoglobin (see page 425). The blood carries the oxyhemoglobin to the tissues where oxygen is released. This oxygen then reacts with the food and waste products in the cells, producing energy. At the same time, carbon dioxide is produced and carried back to the lungs where it is exhaled. Blood going to the tissues (arterial blood) contains oxyhemoglobin and has a characteristic bright red color. Blood coming from the tissues (venous blood) does not contain oxyhemoglobin, so it has a characteristic reddish-purple color. Patients with lung diseases such as pneumonia frequently do not have enough functioning lung tissue to pick up sufficient oxygen from the air. These patients are given a mixture of oxygen and air to breathe instead of air alone. Then, because of the higher partial pressure of oxygen, a small functioning area can pick up more oxygen than it could if air alone were breathed in. Therefore, the patient can breathe and live until the diseased area is cured and returns to normal. The oxygen may be administered by nasal catheter or mask (see Figures 8–4 and 8–5). Oxygen is given to patients with lung cancer to help them adjust to a decreased lung area.

Living tissues require oxygen. Without it they will soon die. However, if the temperature is lowered sufficiently, tissues can survive with very little oxygen. At normal body temperature, 98.6°F, the brain is extremely sensitive to a lack of oxygen. However, it too can live for a longer period of time without oxygen if its temperature is lowered. But how low is low when we are talking about body temperature?

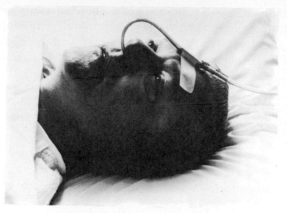

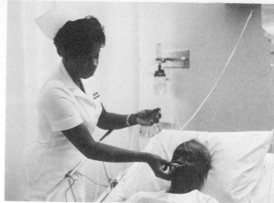

A   B

Figure 8–4. *A.* Nasal oxygen catheter. *B.* Nasal oxygen cannula. (*A,* courtesy of Puritan Disposables, Puritan–Bennett Corp., Kansas City, Mo.; *B,* courtesy of South Chicago Community Hospital, Chicago, Ill.)

In one form of surgery used to repair ruptured blood vessels in the brain, the body temperature is dropped from normal to 86°F by means of an ice bath. At this temperature the heart can be stopped for about 15 min without damage to the body. However, brain surgery requires more time than the 15 min allowed even at this low temperature. After the body temperature is lowered to 86°F and after the heart is stopped, a salt-water solution at 32°F

Figure 8–5. Aerosol mask. (Courtesy of Puritan Disposables, Puritan–Bennett Corp., Kansas City, Mo.)

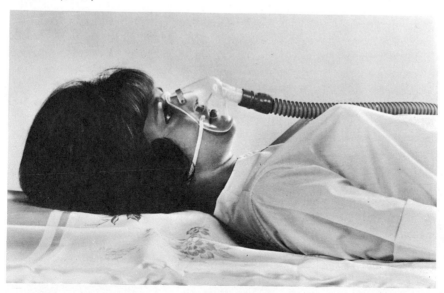

is pumped directly into the main artery that feeds the brain. This cold solution lowers the brain's temperature to approximately 60°F; at this temperature brain surgery may be completed without damage to either the brain or the body. After surgery, the heart is restarted and blood (carrying oxygen) again flows to the brain.

Since oxygen supports combustion, an object such as a candle which burns slowly in air, will burn very vigorously in oxygen. Therefore, a patient must not smoke or use matches in a room where oxygen is in use. The following precautions should also be taken

1. Electrical devices such as radios, televisions, electric shavers, and so on, are likewise banned because of the danger that a spark from the equipment could cause a fire.
2. Electric signal cords should be replaced by a hand bell because of the danger of a spark.
3. Patients should not be given back rubs with alcohol or oil because of the danger of a fire. Instead, lotion or powder should be used.
4. Oil or grease should never be applied to any part of the oxygen equipment. The nurse should take care that she does not have oil on her hands when she manipulates the regulator of the oxygen tank.

Hypoxia is a condition in which the body does not receive enough oxygen. In cases of hypoxia, oxygen must be administered to permit the body to function normally. Newborn babies who have difficulty in breathing are given oxygen containing a small amount of carbon dioxide. The carbon dioxide stimulates the respiratory center of the brain so that the rate of breathing increases, and the oxygen is picked up more rapidly and carried to the tissues. During dental surgery when nitrous oxide is used as the anesthetic, oxygen must be administered along with the nitrous oxide to prevent asphyxiation (lack of oxygen). Firemen who breathe large quantities of smoke may suffer from asphyxiation. They are treated by breathing from an oxygen mask. A person who is under water for several minutes and becomes unconscious because of lack of oxygen is given oxygen in an effort to build up the oxygen concentration in the blood rapidly. Pilots and astronauts breathe oxygen through a face mask because of the decrease or lack of oxygen in the atmosphere around them, as do miners, deep sea divers, and smoke fighters.

In the hospital, oxygen under pressure (hybaroxic treatment) is administered in the treatment of cancerous tissues (see page 50). Another use of oxygen in the hospital is in the determination of the basal metabolic rate (BMR). Basal metabolism tests measure the energy production of the body by measuring the amount of oxygen a patient breathes during a specified period of time (usually 6 min). The BMR has been used as an indicator of thyroid function, but its use for this purpose has been largely supplanted by tests using ^{131}I (radioactive iodine).

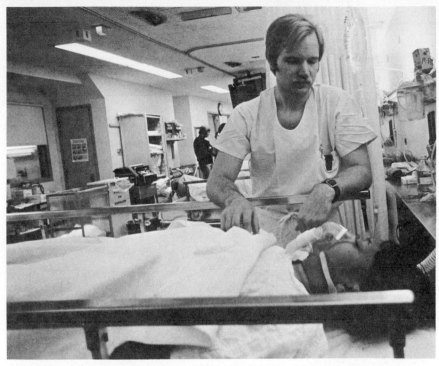

Figure 8–6. Administration of oxygen in the recovery room. (Courtesy of University of Illinois Hospitals, Chicago, Ill.)

Formerly, all premature infants were routinely given oxygen until respiratory sufficiency had been established (see Figure 8–7). Now it is known that, when the concentration of oxygen rises above 40 per cent in the inspired air, premature infants develop retrolental fibroplasia, a disease that affects the eyes. This disease produces complete or nearly complete blindness due to separation and fibrosis of the retina.

Commercial Uses. Deep-sea divers who worked at great depths formerly were supplied air under pressure to enable them to breathe. However, this process introduced a difficulty in that the nitrogen in the air became much more soluble in the blood under the higher pressure. When the diver was taken out of the water too rapidly, this dissolved nitrogen became less soluble because of the decrease in pressure. When this happened, bubbles of nitrogen formed in the blood, causing a condition called the "bends." Unless the diver could be placed immediately in a compression chamber under high pressure, which then was gradually reduced to normal pressure, the results were generally fatal.

A newer method enabling the diver to breathe a mixture of oxygen and helium is now used. Helium is less soluble in the blood than is nitrogen and so decreases the chances of the bends.

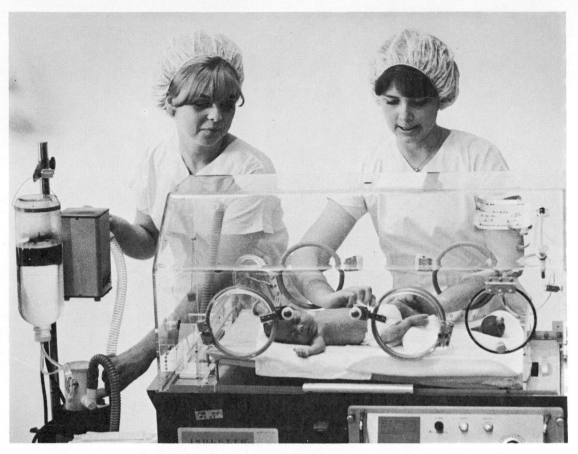

Figure 8–7. Infant in isolette. (Courtesy of Augustana Hospital, Chicago, Ill.)

Industrially, oxygen is used in the operation of furnaces for production of iron and steel. It is also used in welding torches. These torches frequently use acetylene as their fuel. When acetylene is burned with oxygen, the heat produced is high enough to melt steel. Liquid oxygen (LOX) is used in rockets and in launching satellites.

Ozone

Oxygen occurs most frequently as the diatomic molecule, O_2. However, it also exists as a triatomic molecule, O_3, called ozone. These two different forms of the element oxygen are called allotropic forms (see Figure 8–8). *Allotropic forms* of an element have different physical and chemical properties. Other elements that exist in allotropic forms are carbon, sulfur, and phosphorus.

Oxygen molecule   Ozone molecule

Figure 8–8. Scale models of oxygen and ozone molecules.

Preparation

When air or oxygen is passed between two electrically charged plates the volume decreases and a pale blue gas with a strong odor is formed. This gas is ozone. An equation for the formation of ozone is as follows

$$3O_2 \rightleftharpoons 2O_3$$

The double arrow indicates an *equilibrium reaction*, one that proceeds in both directions.

Ozone is formed in the air by the action of ultraviolet light from the sun upon the oxygen in the air. It is also formed by electrical discharges (lightning) in the air. Ozone may also be found around high-voltage machinery where sparks convert some of the oxygen of the air into ozone (see page 92).

Properties and Uses

Ozone is a colorless gas (or sometimes slightly blue) at room temperature. It has a very pungent odor, similar to that of garlic. It is heavier than oxygen and more soluble in water. Liquid ozone is blue in color. Ozone is a powerful oxidizing agent. The use of ozone is rather limited because of its poisonous nature and its instability. Ozone is irritating to the mucous membranes and it is quite toxic to the body except in extremely small amounts.

Ozone is not tolerated in industrial establishments in concentrations of more than 1 part per million (ppm), whereas carbon monoxide (CO) can be tolerated in strengths as high as 100 ppm. In other words, ozone is 100 times as poisonous as carbon monoxide. It is also 100 times as poisonous as hydrogen sulfide (H_2S) and approximately ten times as poisonous as hydrocyanic acid (HCN).

Ozone is not used medically for any purpose because it reaches a fatal concentration long before it can act in any useful way. However, the industrial use of ozone is increasing rapidly. It is a more powerful oxidizing agent than oxygen itself. Its usefulness as a bacteriocide, a decolorizer, and a deodorizer is great, when it is employed carefully and in sufficient strengths. Ozone is used in the rapid aging of wood, for rapid drying of varnishes and inks, in the treatment of water, and in the disinfection of swimming-pool water, but it should only be handled with great care by trained personnel.

Nitrous Oxide

Nitrous oxide, N_2O, is frequently used as a general anesthetic. It is a colorless gas with almost no odor or taste. It is heavier than air and has a relatively low solubility in blood. N_2O does not combine with hemoglobin and is carried dissolved in the bloodstream. It is excreted, unchanged, primarily through the lungs, but a small fraction may escape through the skin.

Although N_2O is not flammable, it does support combustion, as does oxygen, so that care must be taken with its use.

The highest concentration of N_2O that can be safely given for the maintenance of anesthesia is 70 per cent. Above this concentration hypoxia develops. However, at a concentration of 70 per cent or less, N_2O is not potent enough for most patients, so additional drugs such as halothane, methoxyflurane, or ether are used to complete the anesthesia.

Noxious Gases

Carbon monoxide, CO, is a colorless, odorless, tasteless gas. It is non-irritating to the body but has a great affinity for hemoglobin. CO is poisonous because the hemoglobin forms such a strong bond with the gas that the blood is unable to carry sufficient oxygen to the tissues.

Vapors of *carbon tetrachloride*, CCl_4, are very toxic to the body because of the damaging effects upon both the liver and the kidneys. CCl_4 has been used as an anesthetic, as a cleaning fluid, in fire extinguishers, and for the management of hookworm infestations; however, because of its great toxicity, its use has been banned by the United States Food and Drug Administration. Today carbon tetrachloride has no clinical use at all.

Chlorine is a pale yellow gas with a very pungent odor. In neutral or acid solutions chlorine acts not only as a bacteriocide but also as a virucide and amebicide. Chlorine is more effective as a germicide in acid and neutral solution than in basic solution. Chlorine kills microorganisms in water and so is used for water purification. Chlorine vapors, in large amounts, are toxic because they react with and destroy lung tissues.

Summary

Oxygen is the most abundant element on the earth's surface. It is colorless, odorless, tasteless, slightly heavier than air, and very slightly soluble in water.

Oxygen combines with most elements to form a class of compounds called oxides. The reaction between oxygen and some other substances is an example of oxidation. Oxidation also occurs in plant and animal tissues in the presence of specific catalysts called enzymes.

Combustion is a rapid reaction with oxygen, one in which heat and light are produced, usually accompanied by a flame. Spontaneous combustion is a slow oxidation which develops by itself into combustion.

Fire extinguishers accomplish their work either by removing the oxygen from the burning substance or by lowering its temperature below its kindling point, or both.

Oxygen may be prepared in the laboratory by heating such compounds as potassium chlorate ($KClO_3$) or by the electrolysis of water.

Oxygen is prepared commercially from liquid air.

Oxygen is used medically in the treatment of various respiratory diseases. Oxygen is inhaled and carried to the tissues by the hemoglobin in the blood. Care must be taken to prevent fires in the room of a patient receiving oxygen therapy. This means no smoking or use of electrical devices, no alcohol or oil backrubs.

Asphyxiation (hypoxia) may be overcome by the administration of oxygen. Oxygen is used in the treatment of cancer for some patients and in the determination of basal metabolic rate.

Industrially, oxygen is used in welding, in the production of steel, and in rockets. Ozone, O_3, is an allotropic form of oxygen, O_2. Ozone is a colorless gas with a pungent odor. It is very toxic to the body. Ozone is not used medically but has many industrial uses; it must be employed under carefully controlled conditions by trained personnel.

Nitrous oxide is used as a general anesthetic, usually in conjunction with another drug such as halothane, methoxyflurane, or ether.

Among the noxious gases are carbon monoxide, the vapor of carbon tetrachloride, and chlorine.

Questions and Problems

1. When oxygen is collected in the laboratory, why is the bottle kept mouth upward?
2. List several physical properties of oxygen.
3. When elements combine with oxygen, what type of compound is produced?
4. What is combustion? How does it differ from spontaneous combustion?
5. What precautions should the nurse take in preventing fires?
6. Describe the operation of two different types of fire extinguishers. List advantages and disadvantages of each.
7. How may oxygen be prepared in the laboratory? Commercially?
8. How is oxygen carried to the tissues?
9. Discuss the medical uses of oxygen and also the hazards involved in oxygen therapy.
10. What causes retrolental fibroplasia? How may this disease be prevented?
11. Why is carbon dioxide sometimes given along with oxygen to newborn babies?
12. What is hypoxia?
13. What is the "bends," and how may it be overcome?
14. What is ozone? List some properties of this substance. How may it be prepared?
15. List the general properties of nitrous oxide. What precautions should be taken with its use?
16. Why must nitrous oxide be used at concentrations less than 70 per cent? Why is it used in conjunction with other drugs?
17. Why is carbon monoxide poisonous?
18. Why is carbon tetrachloride poisonous?

19. Under what conditions is a solution of chlorine an effective germicide?
20. Why are chlorine vapors toxic to the body?

References

Cloud, P., and Gibor, A.: The oxygen cycle. *Scientific American*, **223**:110–23 (Sept.), 1970.

Delwiche, C. C.: The nitrogen cycle. *Scientific American*, **223**:136–46 (Sept.), 1970.

Hein, M., and Best, L. R.: *College Chemistry*. Dickenson Publishing Co., Inc., Encino, Calif., 1976, Chap. 20.

Klystra, J. A.: Experiments in water breathing. *Scientific American*, **219**:66–74 (Aug.), 1968.

Nebergall, W. H.; Schmidt, F. C.; and Holtzclaw, H. F., Jr.: *College Chemistry*, 5th ed. D. C. Heath & Co., Lexington, Mass., 1976, Chap. 8.

Nordman, J.: *What Is Chemistry?* Harper & Row, New York, 1974, Chap. 13.

Pauling, L., and Pauling, P.: *Chemistry*. W. H. Freeman & Co., San Francisco, 1975, pp. 165–167, 201–202, 644–645.

Sienko, M. J., and Plane, R. A.: *Chemistry*, 5th ed. McGraw-Hill Book Co., New York, 1976, Chap. 14.

Windle, W. F.: Brain damage by asphyxia at birth. *Scientific American*, **221**:76–84 (Oct.), 1969.

Chapter 9

Oxidation-Reduction

CHAPTER OUTLINE

Oxidation

Oxidation was illustrated in the last chapter in the combination of a substance with oxygen. A broader definition of oxidation states that it is an increase in oxidation number. Consider the following reaction.

$$2Na + Cl_2 \longrightarrow 2NaCl$$

An uncombined element has an oxidation number of zero. The oxidation number of sodium in sodium chloride (NaCl) is +1 and that of chlorine −1. Therefore, the above reaction may be written as follows

$$2\overset{0}{Na} + \overset{0}{Cl_2} \longrightarrow 2\overset{+1}{Na} + 2\overset{-1}{Cl}$$

where the upper numbers indicate the respective oxidation numbers of the substances. The sodium has changed in oxidation number from zero to +1, a gain. This is oxidation. The sodium atom was oxidized.

Oxidation may also be defined as "a loss of electrons." Consider the electron-dot structures given on the following page. The sodium atom has one electron in its valence shell. When the sodium atom loses this one electron, it forms a sodium ion with a $+1$ charge. This loss of an electron is defined as oxidation. Therefore, the sodium atom was oxidized.

$$Na\cdot + \cdot\ddot{\underset{..}{Cl}}: \longrightarrow Na^+ + :\ddot{\underset{..}{Cl}}:^-$$

The cells in the body "burn" glucose producing carbon dioxide, water, and energy.

$$\overset{0}{C_6H_{12}O_6} + \overset{0}{6O_2} \longrightarrow \overset{+4}{6CO_2} + \overset{-2}{6H_2O} + energy$$
$$\text{glucose}$$

The oxidation number of each carbon atom in glucose is zero. (Refer to page 67 for a discussion of oxidation numbers.) The oxidation number of the carbon atom in carbon dioxide, CO_2, is $+4$. Therefore, the carbon atom increased in oxidation number. A gain in oxidation number is oxidation; therefore, the carbon atom in glucose was oxidized, or it can be said that the glucose, which contains the carbon atom, was oxidized.

The following oxidation reactions take place in the body. They will be discussed in detail in the appropriate chapters on carbohydrates, fats, and proteins.

$$carbohydrate + O_2 \longrightarrow CO_2 + H_2O + energy$$
$$fat + O_2 \longrightarrow CO_2 + H_2O + energy$$
$$protein + O_2 \longrightarrow CO_2 + H_2O + urea + energy$$

Reduction

Oxidation was defined as an increase in oxidation number or as a loss of electrons. Reduction is the opposite of oxidation. Reduction is a decrease in oxidation number. *Oxidation can never take place without reduction* because something must be able to pick up the electrons lost by the oxidized atom, ion, or compound. Free electrons cannot exist by themselves for very long.

In the reaction of sodium with chlorine

$$\overset{0}{2Na} + \overset{0}{Cl_2} \longrightarrow \overset{+1}{2Na} + \overset{-1}{2Cl}$$

the sodium increased in oxidation number from 0 to $+1$. It was oxidized. At the same time, the chlorine decreased in oxidation number from 0 to -1. Therefore, the chlorine was reduced. Consider the structures at the top of this page. The chlorine atom has seven electrons in its valence shell. It gains one electron to form the chloride ion with a charge of -1. This gain of an electron is called reduction. Thus, by either definition, the chlorine was reduced.

In the reaction of glucose with oxygen

$$\overset{0}{C_6H_{12}O_6} + \overset{0}{6O_2} \longrightarrow \overset{+4}{6CO_2} + \overset{-2}{6H_2O} + energy$$

glucose was oxidized because the carbon atoms changed in oxidation number
from 0 to +4. However, at the same time, the oxygen changed from 0 to −2,
a decrease in oxidation number. Therefore, the oxygen was reduced.

Oxidizing Agents and Reducing Agents

In the reaction

$$H_2 + PbO \longrightarrow Pb + H_2O$$

$$\overset{0}{H_2} + \overset{+2\ -2}{Pb\ O} \longrightarrow \overset{0}{Pb} + \overset{2(+1)\ -2}{H_2\ O}$$

the lead decreases in oxidation number from +2 to 0. Therefore, the lead
was reduced. What reduced it? What supplied the electrons that it had to gain
in order to decrease in oxidation number? The answer is that the hydrogen
reduced it. The hydrogen supplied the electrons so that the lead could be
reduced; therefore, hydrogen is called *the reducing agent*. The substance
which causes the reduction of an element or compound is known as a
reducing agent.

At the same time, the hydrogen gained in oxidation number from 0 to +1.
Therefore, the hydrogen was oxidized. What oxidized it? What picked up
the electrons that the hydrogen must lose in being oxidized? The answer is
that the PbO picked up these electrons; thus, the PbO is called the *oxidizing
agent*. An oxidizing agent is defined as a substance that causes the oxidation
of some element or compound.

Rewriting the previous equation and indicating what was oxidized, what
was reduced, what the oxidizing agent was, and what the reducing agent was,
we have the following

$$\underset{\substack{\text{oxidized} \\ \text{(reducing} \\ \text{agent)}}}{H_2} + \underset{\substack{\text{reduced} \\ \text{(oxidizing} \\ \text{agent)}}}{PbO} \longrightarrow Pb + H_2O$$

Another way of stating the same thing is to say that *whatever is oxidized is
automatically the reducing agent* and *whatever is reduced is automatically the
oxidizing agent*.

Oxidation and reduction reactions produce the energy the body needs to
carry out its normal functions. Oxidation-reduction in the body involves
either oxygen or hydrogen or both. Whatever reacts with oxygen becomes
oxidized; whatever reacts with hydrogen becomes reduced. These facts can
be combined into the general statements. (1) Oxidation in the body takes
place when a substance combines with oxygen or loses hydrogen. (2) Reduction takes place in the body when a substance loses oxygen or combines with
hydrogen.

TABLE 9–1

Antiseptic Agents

Formula	Name	Use
3% H_2O_2	Hydrogen peroxide	Minor cuts and scratches
$KMnO_4$	Potassium permanganate	Treatment of infection in urethra and bladder
$KClO_3$	Potassium chlorate	Treatment of sore throat
I_2 in H_2O	Lugol's solution	Treatment of minor cuts
NaOCl	Sodium hypochlorite (Dakin's solution)	Treatment of wounds

Medical Importance of Oxidation-Reduction

Antiseptic Effects

Because they are oxidizing agents, many antiseptics have the property of killing bacteria. Among these is chlorine, which oxidizes organic matter and bacteria and so is used in the treatment of water to make it potable (see page 123). Table 9–1 lists some of the common antiseptics that are oxidizing agents.

Formaldehyde and sulfur dioxide are two reducing agents used in the disinfecting of rooms formerly occupied by patients with contagious diseases.

Another commonly used oxidizing agent and bleaching powder, $Ca(OCl)_2$, is used as a disinfectant for clothes and hospital beds.

Stain Removal

Oxidizing agents and reducing agents are used to remove most stains that cannot otherwise be removed. Table 9–2 lists some of the common stain removers and indicates where they may be safely used.

TABLE 9–2

Stain Removers

Substance	Name	Function	Use
H_2O_2	Hydrogen peroxide	Oxidizing agent	Blood stains on cotton or linen
$KMnO_4$	Potassium permanganate	Strong oxidizing agent	Almost all stains on white fabrics except rayon ($KMnO_4$ stain must be removed, usually with oxalic acid)
$(COOH)_2$	Oxalic acid	Reducing agent	Rust spots and $KMnO_4$ stains
NaOCl	Javelle water	Oxidizing agent	Effective on almost all stains on cotton and linen (not to be used on wool or silk)
$Na_2S_2O_3$	Sodium thiosulfate	Reducing agent	Iodine and silver stains

Summary

Oxidation is defined as a loss of electrons. Oxidation is also an increase in oxidation number, a combination with oxygen, or a loss of hydrogen.

Reduction is a gain of electrons. Reduction is also a loss in oxidation number, a gain of hydrogen, or a loss of oxygen. Oxidation can never take place without reduction and vice versa.

Whatever is oxidized is called a reducing agent.

Whatever is reduced is called an oxidizing agent.

Oxidizing and reducing agents may be useful as antiseptics and also as stain removers.

Questions and Problems

1. In the following equation, indicate the oxidizing agent and the reducing agent

    (a) $Cl_2 + 2KI \longrightarrow 2KCl + I_2$
    (b) $2Al + 3H_2SO_4 \longrightarrow Al_2(SO_4)_3 + 3H_2$
    (c) $Cu + 2AgNO_3 \longrightarrow Cu(NO_3)_2 + 2Ag$
    (d) $2H_2 + O_2 \longrightarrow 2H_2O$

2. In the following equations, what was oxidized and what was reduced?

    (a) $4Fe + 3O_2 \longrightarrow 2Fe_2O_3$
    (b) $2S + 3O_2 \longrightarrow 2SO_3$
    (c) $Zn + 2HCl \longrightarrow ZnCl_2 + H_2$

3. Why can oxidation never take place without reduction?
4. List several substances that can be used to remove stains.

References

Ault, F. K., and Lawrence, R. M.: *Chemistry: A Conceptual Introduction.* Scott, Foresman & Co., Glenview, Ill., 1976, Chap. 8.

Baum, S. J., and Scaife, C. W.: *Chemistry: A Life Science Approach.* Macmillan Publishing Co., Inc., New York, 1975, Chap. 8.

Fernandez, J. E., and Whitaker, R. D.: *An Introduction to Chemical Principles.* Macmillan Publishing Co., Inc., New York, 1975, Chap. 21.

Hein, M., and Best, L. R.: *College Chemistry.* Dickenson Publishing Co., Inc., Encino, Calif., 1976, Chap. 17.

Longo, F. R.: *General Chemistry.* McGraw-Hill Book Co., New York, 1974, Chap. 12.

Nebergall, W. H.; Schmidt, F. C.; and Holtzclaw, H. F., Jr.: *College Chemistry,* 5th ed. D. C. Heath & Co., Lexington, Mass., 1976, Chap. 16.

Pauling, L., and Pauling P.: *Chemistry.* W. H. Freeman & Co., San Francisco, 1975, Chap. 11.

Peters, E. I.: *Introduction to Chemical Principles.* W. B. Saunders Co., Philadelphia, 1974, Chap. 15.

Sackheim, G. I.: *Chemical Calculations,* 10th ed., series B. Stipes Publishing Co., Champaign, Ill., 1976, Chap. 16.

Chapter 10

Water

Importance

Water is one of the most important chemicals known to man. Without it neither animal nor plant life would exist. Man can live for a few weeks without food, but not without water. Almost three fourths of the earth's surface is covered by water. Rain, snow, sleet, hail, fog, and dew are manifestations of the water vapor present in the air.

Water is present on the solid part of the earth's surface as lakes, streams, waterfalls, and glaciers. The human body is approximately 50 per cent water.

Water is essential in the processes of digestion, circulation, elimination and the regulation of body temperature. Indeed, every activity of every cell in the body takes place in a watery environment.

Water is important as a solvent. Many substances dissolve in water—sugar, salt, and alcohol, for example.

Physical Properties

Pure water is a flat-tasting liquid. Tap water owes its taste to dissolved gases and minerals. Pure water and tap water are colorless and odorless. But large amounts of water such as lakes and oceans appear blue owing to the presence of finely divided solid material.

When water at room temperature is cooled, its volume contracts. However, water is an unusual liquid in that, after it is cooled to 4°C, further cooling produces an expansion in volume. When water freezes at 0°C, it expands even more, increasing in volume by almost 10 per cent. This expansion on freezing explains why ice floats and why water pipes may burst upon freezing.

The density of water changes with the temperature. At 4°C water has its smallest volume and its maximum density—1 g per milliliter. However, for all practical purposes, the density of water is given as 1 g per milliliter regardless of temperature, since the variation in density between 0° and 100°C is quite small.

Pure water boils at 100°C at 1 atm pressure. At lower pressures it boils at a lower temperature. In certain mountainous localities, water boils at 80°C because of the lower pressure. If more heat is applied to the water, it merely boils faster. The boiling point does not increase. When water boils at 80°C, there may not be sufficient heat to cook food. In this case, the food must be heated in a pressure cooker. This device increases pressure and so increases the boiling point of the water. The same principle is true in the autoclave where the increased pressure increases the boiling temperature.

Water is one of the best solvents known because it dissolves so many different compounds. It is used in making solutions for medications; it dissolves many of the materials found in the cells of the body; it brings about chemical changes in the substances dissolved in it. These changes, especially those taking place in the body cells, will not take place in the absence of water. Urine consists principally of water containing dissolved waste materials of the body.

For water to evaporate, a certain amount of heat is necessary. This amount of heat is approximately 540 cal per gram. When water is placed on the skin, the heat it needs to make it evaporate comes from that skin. Therefore, the skin loses the heat and so is cooled. Likewise, evaporation of perspiration is a cooling process.

For ice to melt, the amount of heat required is 80 cal per gram. When an ice pack is placed on the skin, the heat needed to melt the ice comes from the body, thus lowering the temperature of the body.

Physical Constants Based on Water

The freezing point of water (0°C or 32°F) and the boiling point of water at 1 atm pressure (100°C or 212°F) are the standard reference points for the measurement of temperature.

The weight of 1 ml of water at 4°C (its maximum density) is a standard of weight in the metric system—the gram. Specific gravity is based upon water. Specific gravity is defined as the weight of a substance compared with the weight of an equal volume of water.

The calorie is defined as the amount of heat required to change the temperature of 1 g of water 1°C.

Structure of the Water Molecule

Pure water does not conduct electricity. This indicates that water is a covalent compound in which the atoms share electrons. Thus each hydrogen atom of the water molecule shares its one electron with the oxygen atom.

We might expect the oxygen and the hydrogen atoms in the water molecule to be arranged in a straight line such as HOH. However, laboratory evidence indicates that atoms in the water molecule are arranged in a nonlinear manner with the angle between the hydrogen atoms being approximately 105°.

$$H \underset{X}{\overset{\cdot}{\cdot}} \overset{\cdot\cdot}{\underset{\cdot}{O}} \colon$$
$$\overset{X}{H}$$

The X's stand for the electrons of the hydrogen atom and the dots the electrons from the oxygen atom. However, the oxygen atom has a greater attraction for electrons than does the hydrogen atom. Therefore, the electrons will spend more of their time closer to the oxygen atom than to the hydrogen atom. This shifting of the electrons toward the oxygen atom will tend to give the oxygen atom a slight negative charge while the hydrogen atoms have a slight positive charge. The structure and distribution of relative charges in the water molecule may be represented as follows

$$^{+}H$$
$$\overset{\cdot\cdot}{\underset{x\cdot x}{O}}\colon^{-}$$
$$105° \quad H_{+}$$

Molecules in which there is an unequal or uneven *distribution* of charges are called *polar molecules*. Water is a polar molecule. The polar nature of the water molecule is responsible for its property of dissolving many materials. When sodium chloride (NaCl) is placed in water, it dissolves partly because of the water molecule attracting the ions and pulling them apart (see Figure

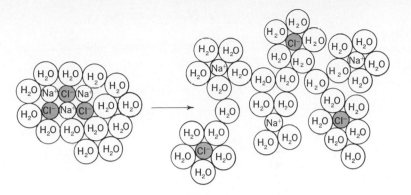

Figure 10–1. Model of the solution of sodium chloride (NaCl) in water. (Reproduced from Fernandez, J. E., and Whitaker, R. D.: *An Introduction to Chemical Principles.* Macmillan Publishing Co., Inc., New York, 1975.)

10–1). Sodium chloride is a polar compound because of the uneven distribution of charges, Na^+Cl^-, within the molecule. A general rule is that polar compounds dissolve in polar liquids and nonpolar compounds dissolve in nonpolar liquids. Nonpolar compounds generally do not dissolve in polar liquids. Carbon tetrachloride, CCl_4, and benzene, C_6H_6, are nonpolar liquids. Water is a polar liquid. As should be expected, carbon tetrachloride dissolves in benzene, but not in water. Likewise, benzene does not dissolve in water.

Chemical Properties

Electrolysis

When water undergoes electrolysis—that is, when an electric current is passed through it—hydrogen gas, H_2, and oxygen gas, O_2, are formed. The volume of hydrogen produced is twice that of oxygen.

$$2H_2O \xrightarrow{\substack{\text{electric} \\ \text{current}}} 2H_2\uparrow + O_2\uparrow$$

Stability

When water is heated to 100°C at 1 atm pressure, it boils and turns into a gas, steam. If the steam is heated to a high temperature, it does not decompose. The water molecule is extremely stable. Even at a temperature of 1600°C there is very little decomposition of the water molecule.

Reaction with Metal Oxides

Water reacts with soluble metal oxides to form a class of compounds called bases (see Chapter 12). For example

$$CaO + H_2O \longrightarrow Ca(OH)_2$$

calcium oxide   calcium hydroxide
                                a base

$$Na_2O + H_2O \longrightarrow 2NaOH$$

sodium oxide   sodium hydroxide
                                a base

Reaction with Nonmetal Oxides

Water reacts with soluble nonmetal oxides to form a class of compounds called acids (see Chapter 12). For example

$$CO_2 + H_2O \longrightarrow H_2CO_3$$

carbon   carbonic
dioxide   acid

$$SO_3 + H_2O \longrightarrow H_2SO_4$$

sulfur   sulfuric
trioxide   acid

Reaction with Active Metals

When an active metal such as sodium or potassium is placed in water, a vigorous reaction takes place, with the rapid evolution of hydrogen gas. At the same time, a base is formed.

$$2Na + 2H_2O \longrightarrow 2NaOH + H_2\uparrow$$

sodium   sodium hydroxide,
                                a base

Formation of Hydrates

When water solutions of some soluble compounds are evaporated, the substances separate as crystals that contain the given compound combined with water in a definite proportion by weight. Crystals that contain a definite proportion of water as part of their crystalline structure are called *hydrates*. The water contained in a hydrate is called *water of hydration* or *water of crystallization*.

Barium chloride crystallizes from solution as a hydrate containing two molecules of water. The formula for this hydrate is $BaCl_2 \cdot 2H_2O$. It is called barium chloride dihydrate. The dot in the middle of the compound indicates that two molecules of water have combined with one molecule of barium chloride. When the water of hydration is removed from a hydrate, the resulting compound is called *anhydrous* (meaning without water). When barium chloride dihydrate is heated the water of hydration is driven off, leaving anhydrous barium chloride.

$$BaCl_2 \cdot 2H_2O \xrightarrow{\text{heat}} BaCl_2 + 2H_2O\uparrow$$

barium chloride   anhydrous
dihydrate   barium chloride

Substances that lose their water of hydration on exposure to air are called *efflorescent*. An example of an efflorescent hydrate is washing soda ($Na_2CO_3 \cdot 10H_2O$), sodium carbonate decahydrate. Substances that pick up moisture from the air are called *hygroscopic*. An example of an hygroscopic compound is calcium chloride, $CaCl_2$. A hygroscopic compound may be used as a drying agent because it will pick up and remove moisture from the air.

A hydrate of particular importance in the medical field is plaster of paris. When plaster of paris is mixed with water, it forms a hard crystalline compound called gypsum.

$$(CaSO_4)_2 \cdot H_2O + 3H_2O \longrightarrow 2(CaSO_4 \cdot 2H_2O)$$

plaster of paris (soft)   gypsum (hard)

Plaster of paris was formerly used extensively in preparing surgical casts (see Figure 10–2). It is spread on crinoline to form a bandage. When the bandage is to be used, it is placed in water, wrung out, and quickly applied.

Figure 10–2. Plaster of paris cast. (Courtesy of University of Illinois Hospitals, Chicago, Ill.)

Plaster of paris expands on setting, so that the cast may appear comfortable when applied but become too tight when it is set. Therefore, it is very important for the nurse to check the circulation to the body part to which a plaster cast has been applied.

A newer method of preparing surgical casts involves the use of a webbed glass-fiber tape impregnated with an ultraviolet-sensitive plastic resin. The resin is hardened by a 3-min "cure" under an ultraviolet lamp. These casts have the advantages that they are lighter than comparable plaster casts, they can be immersed in water, they allow better air circulation, and they are resistant to breakage.

Neutralization

Neutralization is the reaction of an acid with a base to form a salt and water. For example

$$HNO_3 \; + \; KOH \longrightarrow \; KNO_3 \; + H_2O$$

nitric acid   potassium hydroxide   potassium nitrate   water
(acid)   (base)   (salt)

Note that regardless of the acid or base used, water is always one product of a neutralization reaction.

Hydrolysis

When some salts are placed in water, hydrolysis occurs. Hydrolysis is the reaction of a compound with water in which the water molecule is split. When ammonium chloride, a salt, is placed in water, hydrolysis occurs according to the following reaction.

$$NH_4Cl + H_2O \longrightarrow \; NH_4OH + \; HCl$$

ammonium   water   ammonium   hydrochloric
chloride   hydroxide   acid

Note that hydrolysis is the reverse of neutralization. Hydrolysis occurs during the process of digestion of foods. For example, sucrose is hydrolyzed into glucose and fructose through the action of an enzyme.

$$C_{12}H_{22}O_{11} + H_2O \; \xrightarrow{\text{enzyme}} \; C_6H_{12}O_6 + C_6H_{12}O_6$$

sucrose   glucose   fructose

Note that both glucose and fructose have the same formula, $C_6H_{12}O_6$. How two different compounds can have the same formula will be discussed in Chapter 21.

The hydrolysis of fats in the body yields fatty acids and glycerol. The hydrolysis of proteins yields amino acids. Specific enzymes are necessary

TABLE 10–1

Properties of Water*

Physical	Chemical
Colorless, except in deep layer, when it is greenish-blue	Stable compound—not easily decomposed at 1600°C about 0.3% dissociates
Odorless	Reacts violently with sodium and potassium
Tasteless	
At 1 atm, water freezes and ice melts at 0°C or 32°F	Reacts with *basic anhydrides* (metal oxides)—CaO, for example—to form bases
At 1 atm, water boils and steam condenses at 100°C or 212°F	Reacts with *acidic anhydrides* (non-metal oxides)—SO_2, for example—to form acids
Water expands when it freezes to ice	
Density at 4°C, 1.0 g/ml	When crystals are formed from aqueous solutions of certain substances, water and the solute combine to build crystals which are called hydrates
Specific heat, 1.0 cal/(g)(°C)	
Heat of vaporization, 540 cal/g	
Heat of fusion, 80 cal/g	

*Reproduced from Quick, F. J.: *Introductory College Chemistry.* Macmillan Publishing Co., Inc., New York, 1965.

for these hydrolytic reactions. They will be discussed in the appropriate chapters on fats and protein.

Purification of Water

Impurities Present in Water

Natural water contains many dissolved and suspended materials. Rain water contains dissolved gases—oxygen, nitrogen, and carbon dioxide—plus air pollutants (see page 91), suspended dust particles, and other particulate matter. Ground water contains minerals dissolved from the soil through which the water has passed. It also contains some suspended materials. Sea water contains over 3.5 per cent of dissolved matter, the principal compound being sodium chloride. Both sea and ground water also contain dissolved and undissolved pollutants.

Lake water or river water may appear clear when a glass full of it is held up to the light, or it may at first contain suspended clay or mud which tends to settle slowly, leaving what appears to be pure water. However, in both of these cases, the "clear" water may contain many bacteria and other microorganisms that can be quite harmful to the body. Their destruction or removal is necessary for the proper purification of water. Water may be purified by several processes. Among these are distillation, boiling, filtration, and aeration.

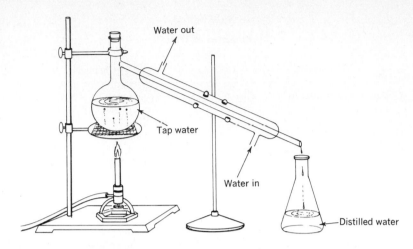

Figure 10–3. A laboratory distillation apparatus.

Distillation

Distillation is a process of converting water to steam and then changing the steam back to water again. In the laboratory, a "still" is used to prepare distilled water (see Figure 10–3). Impure water is placed in the flask at the left and then heated to boiling. As the water boils and changes into steam, it passes into a condenser, which consists of a long glass tube surrounded by another glass tube through which cold water runs. The cold water changes the steam in the condenser back into liquid water; this then runs out of the end of the tube into the receiving vessel at the right. The suspended and dissolved solids (including the bacteria) that were present in the impure water remain behind in the flask. They do not pass over into the condenser with the steam. The dissolved gases originally present in the impure water, however, do pass over with the steam. The usual practice is to discard the first few milliliters of distilled water coming from the condenser, since this volume will contain most of the dissolved gases.

Although distilled water is pure water, it is too expensive and the process too slow for large-scale use. The principal use of distilled water in the hospital is in the preparation of sterile solutions.

Boiling

Ground water or contaminated water may usually be made safe for drinking by boiling it for at least 15 min. The boiling does not remove the dissolved impurities but does kill any bacteria that might be present. Freshly boiled water has a flat taste because of the loss of dissolved gases. The taste may be brought back to normal by pouring the water back and forth from one clean vessel to another. This process, called aeration, allows air to dissolve in the water again.

Sedimentation and Filtration

For large-scale use, water is first allowed to stand in large reservoirs where most of the suspended dirt, clay, and mud settle out. This process is called sedimentation. However, sedimentation is a very slow process. Put some finely divided clay in a graduated cylinder full of water, shake, and see how slowly the clay settles out. Commercially, a mixture of aluminum sulfate and lime is added to the water. These two chemicals combine to form aluminum hydroxide, which precipitates as a gelatinous (sticky) substance. As the sticky aluminum hydroxide settles out, it carries down with it most of the suspended material. The main advantage of this material is that it settles much more rapidly than does the suspended material by itself and so increases the rate of sedimentation.

After sedimentation, the water is filtered through several beds of sand and gravel to remove the rest of the suspended material. The water then is essentially free of suspended material. However, this process does not remove the dissolved material nor much of the bacteria originally present. The water must then be treated with chlorine to kill the bacteria before it is fit to drink.

Aeration

Water may be purified by exposing it to air for a considerable period of time. The oxygen in the air dissolves in the water and destroys the bacteria by the process of oxidation. The oxygen also oxidizes the dissolved organic material in the water so that the bacteria have no source of food. However, this process is slow and expensive because of the long time involved in exposing water to the air. Commercially, aeration is accomplished by spraying filtered chlorinated water into the air. This additional process also removes objectionable odors from the water.

Hardness in Water

When a small amount of soap is added to soft water it forms copious suds. When a small amount of soap is added to hard water, it forms a precipitate or scum and no lather. What is the difference between soft water and hard water? Hard water contains dissolved compounds of calcium and magnesium. Soft water may contain other dissolved compounds but these compounds do not cause hardness. The calcium and magnesium compounds (which cause the hardness) react with soap to form a precipitate, thus removing the soap from the water. More and more soap must be added until all the hardness is removed. Only then will the soap cause a lather. The reactions involved are as follows

$$Ca^{2+} + Na \text{ (soap)} \longrightarrow Ca \text{ (soap)}\downarrow + Na^+$$

$$Mg^{2+} + Na \text{ (soap)} \longrightarrow Mg \text{ (soap)}\downarrow + Na^+$$

hardness   hardness   no hardness

The precipitated soap adheres to washed materials, making them rough so that they may irritate tender skin, or to washed hair, making it sticky and gummy (see Figure 10–4). Food cooked in hard water is likely to be tougher than that cooked in soft water because of the existence of additional minerals. When hard water is boiled, some of the salts form a deposit on the inside of the container in which it is heated. Look inside an old teakettle at home and see the "boiler scale." Hard water must never be used to sterilize surgical instruments because the precipitated hardness will dull the cutting edges. (If iron compounds are present in water, they also will cause hardness.)

Detergents have replaced soaps for washing clothes because they do not precipitate in the presence of hardness in water.

Home Methods for Water Softening

Temporary hardness is hardness that can be removed by boiling. It is caused by the bicarbonates of calcium and magnesium. Boiling converts these to insoluble carbonates, thus removing part of the hardness. Other soluble compounds of calcium and magnesium cause permanent hardness. Permanent hardness is not removed by boiling. Following is an equation for a method of removing temporary hardness

$$Ca(HCO_3)_2 \xrightarrow{heat} CaCO_3\downarrow + CO_2\uparrow + H_2O$$

Ammonia (ammonium hydroxide) is frequently used to soften water used in washing clothes and windows because it precipitates all temporary hardness and some of the permanent hardness. Borax, sodium tetraborate $(Na_2B_4O_7)$, is also frequently used in the home as a laundry water softener.

Figure 10–4. *A*. Taken through a powerful microscope, these photographs show how soap curd sticks to fabric fibers, shortening their life by as much as one third. The nylon fabric at the left was washed 15 times in hard water. Note how soap curd has accumulated on the fibers. The nylon fabric on the right was washed 15 times in soft water. Note how clean, soft, and free from curd the fibers are.

B. These photomicrographs show why the diaper on the left, washed 15 times in hard water, feels harsh and rough because it is matted with soap curd, whereas the diaper on the right, washed 15 times in softened water, remains fluffy and soft.

C. When the hair is shampooed in hard water, curd clings to the hair strands, dulling their natural luster and interfering with their ability to reflect light. The hair strands on the left are stringy and lack cleanliness, due to the clinging hard water curd. Those on the right, washed in soft water, are radiant and clean. (Courtesy of Culligan Water Institute, Northbrook, Ill.)

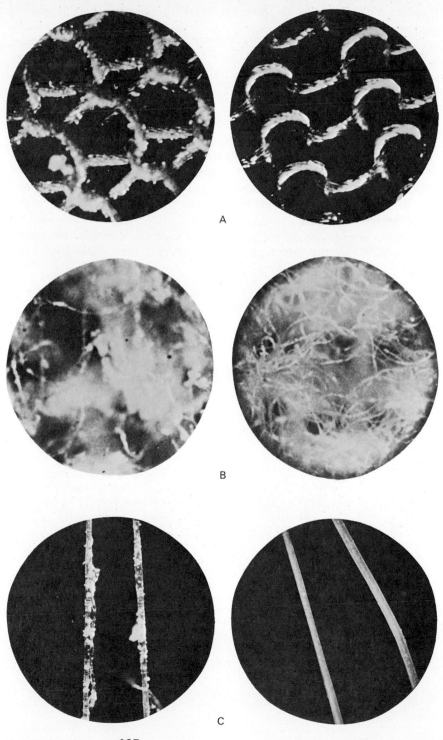

A

B

C

Its effect is similar to that of ammonia. Washing soda (Na_2CO_3), frequently used as a home-laundry water softener, removes both temporary and permanent hardness from the water. Trisodium phosphate, TSP (Na_3PO_4), which is another home-laundry water softener, has an action similar to that of washing soda. However, TSP is no longer recommended for laundry use because of the effects of phosphates on our lakes and streams.

Commercial Water Softeners

The preceding methods for water softening are practical for use in a home, but they are too expensive for large-scale commercial use. One commercial method used is the lime-soda process. In this process lime, $Ca(OH)_2$, and then soda, Na_2CO_3, are added to water to remove both temporary and permanent hardness.

Complex silicates called zeolites are also used commercially to remove both types of hardness from water. These zeolites in the form of large gran-

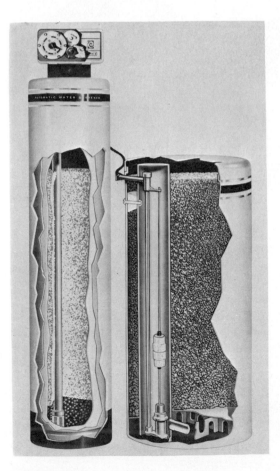

Figure 10–5. Zeolite tank. (Courtesy of Culligan Water Institute, Northbrook, Ill.)

ules are placed in a cylindrical container and hard water is allowed to flow through that container (see Figure 10–5). As the water passes over the zeolites, the sodium ions in the zeolite replace the calcium and magnesium ions in the water. This process, called ion exchange, takes the calcium and magnesium ions out of the water and replaces them with sodium ions. Sodium ions do not cause hardness. The reactions involved are as follows

$$Ca^{2+} + \underset{\substack{\text{sodium} \\ \text{zeolite}}}{Na_2Z} \longrightarrow \underset{\substack{\text{calcium} \\ \text{zeolite}}}{CaZ} + 2Na^+$$

$$Mg^{2+} + \underset{\substack{\text{sodium} \\ \text{zeolite}}}{Na_2Z} \longrightarrow \underset{\substack{\text{magnesium} \\ \text{zeolite}}}{MgZ} + 2Na^+$$

The zeolite in the cylinder will be gradually used up as more and more water is softened. However, the used zeolite may be regenerated and used over and over again. Many hospitals use this process for water softening.

Water that has been softened by the zeolite process still has dissolved minerals in it—the sodium compounds obtained by ion exchange. To prepare water comparable to distilled water, all soluble materials must be removed. This may be accomplished by using a set of two different resins, one to remove the soluble metal (positive) ions and the other to remove the negative ions present in the water. If a sample of hard water is assumed to contain calcium chloride, $CaCl_2$, the dissolved mineral may be removed as follows

$$CaCl_2 + \underset{\text{resin}}{2HY} \longrightarrow 2HCl + CaY$$

The HCl then reacts with the second resin, XNH_2

$$HCl + \underset{\text{resin}}{XNH_2} \longrightarrow XNH_3Cl$$

so that no ions are left in the water. Such water is called deionized water. Although this process is inexpensive and quick, it will not remove bacteria or dissolved nonelectrolytes.

Water Pollution

What is polluted water? Strictly speaking, it is any water that is not pure. However, tap water contains many dissolved and suspended substances. It is not pure, yet it is not called polluted water. Any substance that prevents or prohibits the normal use of water is termed a pollutant. The signs of polluted water are usually quite obvious—oil and dead fish floating on the surface of a body of water or deposited along the shores, a bad taste to drinking water, a foul odor along a waterfront, unchecked growth of aquatic weeds along the shore, or tainted fish that cannot be eaten.

Water pollutants may be classified into several categories. Among them are:

1. *Oxygen-demanding wastes.* Dissolved oxygen is required for both plant and animal life in a body of water. Anything that tends to decrease the supply of this vital element endangers the survival of the life forms. Oxygen-demanding wastes are acted upon by bacteria in the presence of oxygen, thus leading to a depletion of the dissolved oxygen. Oxygen-demanding pollutants include sewage and wastes from paper mills, food-processing plants, and other industrial processes that discharge organic materials into the water.

2. *Disease-causing agents.* Among the diseases that may be caused by pathogenic microorganisms present in polluted water are typhoid fever, cholera, infectious hepatitis, and poliomyelitis.

3. *Radioactive material.* Low-level radioactive wastes from nuclear power plants are sealed in concrete and buried underground. High-level radioactive wastes are initially stored as liquids in large underground tanks and later converted into solid form for burial in concrete. In either case, leakage can lead to pollution of nearby water supplies.

4. *Heat.* Although heat is not normally considered a pollutant of waterways, it does have a detrimental effect on the amount of dissolved oxygen. Thermal pollution results when water is used as a coolant for industrial plants and nuclear reactors and then returned to its source.

In addition to decreasing the amount of dissolved oxygen, thermal pollution also causes an increased rate of chemical reactions. The metabolic processes of fish and microorganisms are speeded up, increasing their need for oxygen, at a time when the supply of oxygen is diminishing. Higher water temperatures may also be fatal to certain forms of marine life.

5. *Plant nutrients.* Nutrients stimulate the growth of aquatic plants. This may lead to lower levels of dissolved oxygen. It may also lead to disagreeable odors when the large amount of plant material decays. Excessive plant growth is often unsightly and interferes with recreational use of water. Excess phosphorus in sewage comes from phosphate detergents and is one cause for this type of pollution.

6. *Synthetic organic chemicals.* In this category of pollutants are such substances as surfactants in detergents, pesticides, plastics, and food additives.

7. *Inorganic chemicals and minerals.* These pollutants come from industrial wastes as well as from run-off water from urban areas. One example of such a pollutant is mercury. It was once believed that metallic mercury was inert and settled to the bottom of a lake. It is now known that anaerobic bacteria in bottom muds are capable of converting this mercury into compounds that are poisonous. Another pollutant in this category is sulfuric acid, which is formed by the reaction of sulfur-containing ores with water and oxygen in the air. Salt is also a pollutant, which may occur when brine from oil wells is released into fresh water.

Water is one of the most important chemicals known to man. Pure water is a colorless, odorless, flat-tasting liquid which freezes at 0°C and boils at 100°C at 1 atm pressure. Water has its maximum density at 4°C. When water freezes, its volume increases by almost 10 per cent.

Water is used as a solvent for many substances. Many chemical reactions take place only when the reactants are dissolved in water. The evaporation of water or perspiration from the skin is a cooling process because the skin provides (loses) the heat required to change the liquid to the vapor state. Water is the standard of reference for such physical constants as the temperature scale, specific gravity, the calorie, and the standard of weight in the metric system (the gram).

The water molecule is a covalent one with a hydrogen-oxygen-hydrogen angle of about 105°. Water is a polar liquid. Polar liquids usually dissolve polar compounds and nonpolar liquids dissolve nonpolar compounds.

The electrolysis of water yields hydrogen and oxygen gases. Water does not otherwise appreciably decompose into these gases, even when heated to 1600°C.

Water reacts with certain metal oxides to form bases; water reacts with certain nonmetal oxides to form acids. Water reacts with sodium and potassium to form hydrogen gas and a base. Water reacts with certain salts to form hydrates. Hydrates that lose their water of hydration on standing are called efflorescent. Substances that pick up moisture from the air are called hygroscopic.

When an acid reacts with a base, a salt and water are produced. Such a reaction is called a neutralization reaction.

Hydrolysis of a salt is the reaction of that salt with water, whereby the water molecule is split.

Water may be purified by distillation, by boiling, by sedimentation and filtration, and by aeration.

Hardness in water is caused by ions of calcium and magnesium (and iron). Temporary hardness is due to bicarbonates of calcium and magnesium. Temporary hardness can be removed by boiling. Permanent hardness is not removed by boiling. Hardness in water can be removed by using ammonia, washing soda, trisodium phosphate, lime-soda, or zeolite. Deionized water contains no ions and is frequently used in place of distilled water.

Water pollution may be caused by oxygen-demanding wastes, disease-causing agents, radioactive materials, heat, plant nutrients, synthetic organic chemicals, and inorganic chemicals and minerals.

Questions and Problems

1. Why is pure water flat-tasting?
2. Why do lakes and oceans appear blue?
3. Why does ice float?
4. At what temperature does water have its maximum density?
5. Why use a pressure cooker at high altitudes?
6. How much heat is required to evaporate 10 g water from the skin?
7. Why is evaporation a cooling process?
8. Water is used as the standard of reference for which physical constants?
9. Diagram the structure of the water molecule. Why is the molecule polar?

10. Why does salt dissolve in water?
11. Will carbon tetrachloride, CCl_4, a nonpolar liquid, dissolve in water?
12. What type of compound is produced when water reacts with a metal oxide?
13. What type of compound is produced when water reacts with a nonmetal oxide?
14. What type of reaction occurs when an acid reacts with a base?
15. What type of reaction occurs when a salt reacts with water?
16. Why purify water?
17. Describe the preparation of distilled water.
18. What effect does boiling have on water as far as purification is concerned? As far as the removal of impurities is concerned?
19. Of what importance are filtration and aeration of water?
20. What is a hydrate?
21. What reaction occurs when a hydrate is heated?
22. What is efflorescence?
23. What is a hygroscopic compound? For what purpose may it be used?
24. What causes temporary hardness in water? How may it be removed?
25. What causes permanent hardness in water? How may it be removed?
26. Describe the zeolite process for water softening.
27. What is deionized water and how may it be prepared?
28. What is polluted water? What signs may indicate such a condition?
29. What diseases may be caused by polluted water?
30. A depletion of the oxygen content of water may be caused by what factors?
31. How are radioactive wastes disposed of? Why are precautions necessary?
32. What might be the effects of thermal pollution? What might cause this type of pollution?
33. Indicate several synthetic organic water pollutants. Where do they come from?
34. List three inorganic pollutants and indicate their source.

References

Freeman, H. L.: *Chemistry in the Environment.* W. H. Freeman & Co., San Francisco, 1973, pp. 27–30.

Hein, M., and Best, L. R.: *College Chemistry.* Dickenson Publishing Co., Inc., Encino, Calif., 1976, Chap. 13.

Hussian, F.: *Living Underwater.* Praeger Publishers, Inc., New York, 1971.

Nebergall, W. H.; Schmidt, F. C.; and Holtzclaw, H. F., Jr.: *College Chemistry,* 5th ed. D. C. Heath & Co., Lexington, Mass., 1976, Chap. 12.

Nordmann, J.: *What Is Chemistry?* Harper & Row, New York, 1974, Chap. 15.

Pauling, L., and Pauling, P.: *Chemistry.* W. H. Freeman & Co., San Francisco, 1975, Chap. 9.

Penman, H. C.: The water cycle. *Scientific American,* **223**:98–108 (Sept.), 1970.

Peters, E. I.: *Introduction to Chemical Principles.* W. B. Saunders Co., Philadelphia, 1974, Chap. 4.

Solomon, A. K.: The state of water within red blood cells. *Scientific American,* **224**:88–96 (Feb.), 1971.

Stoker, H. S., and Seager, S. L.: *Environmental Chemistry: Air and Water Pollution,* 2nd ed. Scott, Foresman & Co., Glenview, Ill., 1976, Part 2.

Liquid Mixtures

Liquid mixtures may be divided into four types: solutions, suspensions, colloids, and emulsions. Each type has its own specific properties and uses.

In a solution, a substance is dissolved in a liquid. In suspensions, colloids, and emulsions, one substance is not dissolved but rather is suspended in a liquid. In a suspension, the suspended material soon settles; in a colloid the suspended material does not settle; in an emulsion, the suspended material is a liquid which does not dissolve.

Solutions

General Properties

A solution consists of two parts: the material that has dissolved, the *solute*, and a material in which it has dissolved, the *solvent*.

When a crystal of salt is placed in water which is then stirred, the crystal dissolves and a clear solution is formed. When more salt is added to this salt-water solution, it too dissolves, making the solution stronger than the previous one. Even more salt can be dissolved in the water to make it a much stronger salt solution. Thus, one of the properties of solutions is that they have a variable composition. Varying amounts of salt and water can be mixed to form a salt-water solution.

When salt is dissolved in water to make a salt-water solution, the solution formed is perfectly clear. When sugar is dissolved in water, again a clear solution is formed. When copper sulfate is dissolved in water, it also forms a clear solution. However, the solutions formed with the salt and the sugar are colorless whereas that formed with copper sulfate is blue. Solutions are always clear. They may or may not have a color. Clear merely means that the solution is transparent to light.

When a salt solution is examined under a high-power microscope, it appears homogeneous. The same is true for a sugar solution. In general, all solutions are homogeneous. The solute cannot be distinguished from the solvent in a solution.

When a solution is allowed to stand undisturbed for a long period of time, no crystals of solute settle out, provided the solvent is not allowed to evaporate. This is another property of solutions—the solute does not settle out.

The salt in a salt-water solution may be recovered by allowing the water to evaporate. Likewise, the sugar from a sugar solution. In general, solutions may be separated by physical means.

If a solution (such as salt water) is poured into a funnel containing a piece of filter paper, the solution will pass through unchanged. Solutions also pass through membranes unchanged. That is, the particles in solution must be smaller than the openings in the filter paper and the even finer openings in the membranes.

The properties of solutions are summarized as follows. Solutions

1. Consist of a soluble material (the solute) dissolved in a liquid (the solvent).
2. Have a variable composition.
3. Are clear.
4. Are homogeneous.
5. Do not settle.
6. May be separated by physical means.
7. Pass through filter paper.
8. Pass through membranes.

Solvents Other Than Water

Solvents other than water are also used in the hospital. One common solvent is alcohol. An alcohol solution used medicinally is called a tincture. Tincture of iodine contains iodine dissolved in alcohol. Tincture of green soap contains potassium soap dissolved in alcohol.

Both carbon tetrachloride and ether are excellent solvents for fats and oils.

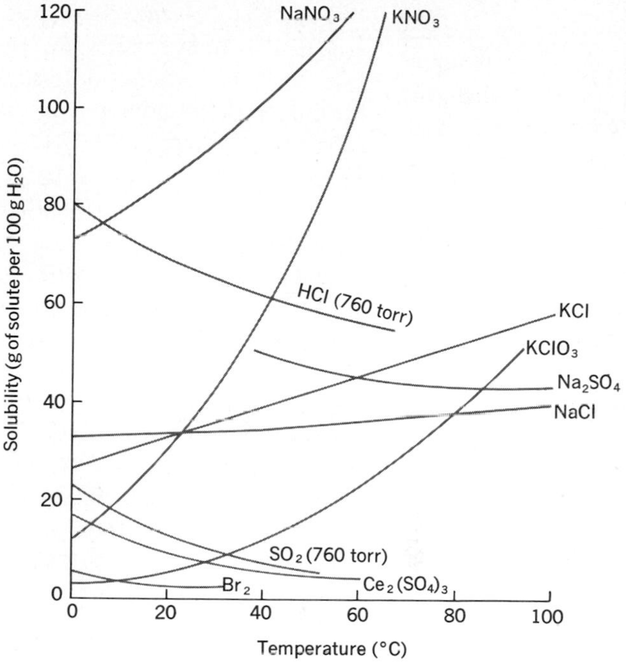

Figure 11–1. Variation of solubilities of various substances in water as a function of temperature. (Reproduced from Baum, S. J., and Scaife, C. W.: *Chemistry: A Life Science Approach.* Macmillan Publishing Co., Inc., New York, 1975.)

Factors Affecting Solubility of a Solid Solute

Temperature. Most solutes are more soluble in hot water than in cold water. Figure 11–1 shows that KNO_3 becomes much more soluble as the temperature increases, while $Ce_2(SO_4)_3$ becomes less soluble with an increase in temperature. NaCl shows little change in solubility. Note that gases such as HCl and SO_2 become less soluble with increasing temperature. The solubility of Br_2, a liquid, is practically unaffected by temperature.

Pressure. A change in pressure has no noticeable effect on the solubility of a solid or liquid solute in a given solvent but will affect the solubility of a gaseous solute. The greater the pressure, the greater the solubility of a gas in a liquid.

Surface Area. The greater the amount of surface area, the quicker a solute will dissolve in a solvent. Thus, to make a solute dissolve faster we frequently powder it, thereby increasing the surface area.

Stirring. The rate at which a solute dissolves can be increased by stirring the mixture. The process of stirring brings fresh solvent in contact with the solute and so permits more rapid solution.

Therefore, to dissolve most solutes rapidly, the solute should be powdered and the mixture should be heated while it is stirred.

Nature of Solvent. Sometimes in spite of all attempts, a solute does not appreciably dissolve in water. In this case, a different solvent should be used. For example, iodine is only slightly soluble in water but is quite soluble in alcohol so that a common solution of iodine contains alcohol as the solvent rather than water.

Importance of Solutions

During digestion, foods are changed to soluble substances so that they may pass into the bloodstream and be carried to all parts of the body. At the same time the waste products of the body are dissolved in the blood and carried to other parts of the body where they can be eliminated. Plants obtain minerals from the ground water in which those minerals have dissolved.

Many chemical reactions take place in solution. When solid silver nitrate is mixed with solid silver chloride, no reaction takes place because the movement of the ions in the solid state is highly restricted. However, when a solution of silver nitrate is mixed with a solution of sodium chloride, a precipitate of silver chloride is formed instantaneously. This reaction occurs because the ions in the solution are free to move and react with other ions.

Many medications are administered orally, subcutaneously, or intravenously as solutions.

Strength of Solutions

Dilute and Concentrated. When a few crystals of sugar are placed in a beaker of water, a *dilute* sugar solution is produced. As more and more sugar is added to the water, the solution becomes *concentrated*. But when does the solution change from dilute to concentrated? There is no sharp dividing line. Dilute merely means that the solution contains a small amount of solute. Concentrated merely means that the solution contains a large amount of solute.

However, both dilute and concentrated are relative terms. A dilute sugar solution may contain 5 g of sugar per 100 ml of solution, whereas 5 g (the same amount) of boric acid per 100 ml of solution will produce a concentrated boric acid solution. That is, the terms dilute and concentrated usually have no specific quantitative meaning and so are not generally used for medical applications.

Saturated. Suppose a small amount of sugar is placed in a beaker of water. When the mixture is stirred, all the sugar will dissolve. If more and more sugar is added with stirring, a point will soon be reached where some of the sugar settles to the bottom of the beaker. This excess sugar does not dissolve even upon more rapid agitation. This type of solution is called a *saturated solution.* Some of the crystals are continually dissolving and going into solution. At the same time, the same amount of solute crystallizes out of the solution. This type of interchange is called *equilibrium.*

Thus, a saturated solution may be defined as one in which there is an equilibrium between the solute and the solution. A saturated solution may also be defined as one that contains all the solute that it can hold under the given conditions.

Unsaturated. An unsaturated solution contains less of a solute than it could hold under normal conditions. Suppose that a saturated solution of glucose in a certain amount of water contains 25 g glucose. An unsaturated glucose solution would be one that contained an amount of glucose less than 25 g in the same amount of water. In an unsaturated solution, no equilibrium exists because there is no undissolved solute.

The terms saturated and unsaturated are relative terms. The same amount of two different solutes may produce entirely different types of solutions. For glucose, 5 g in 100 ml of water produces an unsaturated solution, whereas 5 g boric acid in 100 ml of water produces a saturated solution. Therefore, saturated and unsaturated solutions are relative terms and are not used for medical applications.

Percentage Solutions. The weight-volume method expresses the *exact* concentration of a solute by stating the weight of solute in a given volume of solvent, usually water. A 10 per cent glucose solution will contain 10 g glucose per 100 ml of solution. A 0.9 per cent saline solution will contain 0.9 g sodium chloride per 100 ml of solution. The percentage indicates the number of grams of solute per 100 ml of solution. How can we prepare 500 ml of a 2 per cent citric acid solution? A 2 per cent citric acid solution contains 2 g citric acid per 100 ml of solution. Therefore, in 500 ml solution there should be five times as much citric acid as in 100 ml solution—or 10 g (2 g × 5).

To prepare the solution, the procedure is as follows

1. Weigh out exactly 10 g citric acid.
2. Dissolve the 10 g citric acid in a small amount of water contained in a 500-ml graduated cylinder.
3. Add water to the 500-ml mark and stir.

Note the 10 g citric acid was dissolved in water and then diluted to the required volume—500 ml. It was *not* dissolved directly in 500 ml water.

Ratio Solutions. Another method of expressing concentration is a ratio solution. A 1:1000 merthiolate solution contains 1 g merthiolate in 1000 ml solution. A 1:10,000 $KMnO_4$ solution contains 1 g $KMnO_4$ in 10,000 ml solution. The first number in the ratio indicates the number of grams of solute and the second number gives the number of milliliters of solution.

As with percentage solutions, the solute is dissolved in a small amount of solvent (water) and then diluted to the desired volume.

Percentage and ratio solutions are frequently used by doctors, nurses, and pharmacists.

Molar Solutions. Molar solutions are used most frequently by chemists. A molar solution is defined as one that contains 1 mole, or 1 gram molecular weight, of solute per liter of solution. *Gram molecular weight* is the molecular weight expressed in grams. A 1 molar (1 M) solution of glucose, $C_6H_{12}O_6$, will contain 1 gram molecular weight glucose (180 g) in 1 liter solution. As before, the solute is dissolved in a small amount of water and then diluted to the desired volume.

To prepare 100 ml 2 M (2 molar) sulfuric acid solution, the following procedure may be used.

1. Calculate the molecular weight of sulfuric acid. The molecular weight of sulfuric acid, H_2SO_4, equals the sum of the atomic weights.

$$
\begin{array}{l}
2H = 2 \times 1 \ = \ \ 2 \\
1S = 1 \times 32 = 32 \\
4O = 4 \times 16 = \underline{64} \\
 98
\end{array}
$$

2. Expressed as gram molecular weight, this would be 98 g. Therefore 98 g sulfuric acid are required to prepare 1 liter of a 1 M solution.
3. To prepare 1 liter of a 2 M solution, 2 × 98 g or 196 g sulfuric acid are required. (To prepare twice as strong a solution in the same volume requires twice as much solute.)
4. To prepare 100 ml of a 2 M solution, 1/10 × 196 or 19.6 g sulfuric acid are required, since 100 ml is 1/10 of a liter.

Normal Solutions. A 1 normal (1 N) solution contains 1 gram equivalent weight of solute per liter of solution.

The gram equivalent weight of an acid can be calculated by dividing the gram molecular weight of that acid by the number of replaceable hydrogens it contains. Hydrochloric acid, HCl, contains one replaceable hydrogen atom. The gram equivalent weight of HCl is equal to the gram molecular weight divided by one: 36.5 g/1 or 36.5 g.

The gram equivalent weight of sulfuric acid, H_2SO_4, which has two replaceable hydrogen atoms, equals the gram molecular weight divided by two: 98 g/2 or 49 g.

The gram equivalent weight of a base can be calculated by dividing the gram molecular weight of that base by the number of hydroxide, OH^-, groups in that base. The gram equivalent weight of sodium hydroxide, NaOH, which contains one OH^- group, equals its gram molecular weight divided by one: 40 g/1 or 40 g.

The gram equivalent weight of calcium hydroxide, $Ca(OH)_2$, which contains two OH^- groups, equals its gram molecular weight divided by two: 74 g/2 or 37 g.

How much sulfuric acid is needed to prepare 3 liters of a 0.2 N solution? The steps in the calculation are as follows

1. The gram molecular weight of sulfuric acid, H_2SO_4, is 98 g.
2. The gram equivalent weight of sulfuric acid is 98 g/2 or 49 g.
3. To prepare 1 liter of 1 N solution, 49 g H_2SO_4 are needed.
4. To prepare 3 liters of 1 N solution 3 × 49 g or 147 g H_2SO_4 are needed. (For three times as much solution, three times as much solute is needed.)
5. To prepare 3 liters of 0.2 N solution, 0.2 × 147 g or 29.4 g H_2SO_4 are needed. (For 0.2 times as strong a solution, 0.2 times as much solute is needed.)

The word equivalent means equal to, so that one equivalent weight of one substance will react exactly with one equivalent weight of another. Also, a certain volume of a 1 N solution of an acid will react with exactly the same volume of a 1 N solution of a base.

A smaller unit of equivalent weight is the milliequivalent weight, which is one thousandth of an equivalent weight. One milliequivalent is one thousandth of an equivalent. Therefore, 1 milliequivalent of an acid will react exactly with 1 milliequivalent of a base. Milliequivalents are frequently used in measuring the concentration of various electrolytes present in body fluids (see Chapter 30).

Special Properties of Solutions

Effect of Solute on Boiling Point and Freezing Point. Whenever a nonvolatile solute is dissolved in a solvent, the boiling point of the solution thus prepared is always greater than that of the pure solvent. Water boils at 100°C. A solution of salt in water or a solution of sugar in water will boil at a temperature above 100°C.

Likewise, when a nonvolatile solute is dissolved in a solvent, the freezing point of the solution is always less than the freezing point of the pure solvent. The freezing point of water is 0°C. The freezing point of salt solution or sugar solution is always less than 0°C.

Use is made of these facts in the cooling system of automobiles. Permanent antifreeze is added to the water in the car radiator to lower the freezing point so that the liquid will not freeze when the temperature drops below 32°F. The same material is used in desert areas to raise the boiling point of the liquid in the automobile radiator—to prevent it from boiling over when the temperature rises.

Surface Tension. Consider a water molecule in the center of a beaker of water (see Figure 11–2). This water molecule will be attracted in all

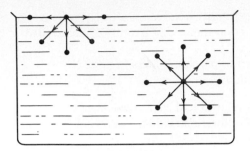

Figure 11–2. Surface tension of a liquid.

directions by the water molecules around it. Next, consider a molecule at the surface of the water. This molecule is attracted sideways and downward, but it is not attracted very much by the air molecules above it. Therefore there is a net downward attraction on the surface water molecules. This downward pull on the surface molecules causes them to form a surface film. Surface tension may be defined as the force that causes the surface of a liquid to contract. Surface tension also is the force necessary to break this surface film. All liquids exhibit surface tension; the surface tension of water is higher than that of most liquids.

Surface tension is responsible for the formation of drops of water on a greasy surface. The surface film holds the drop in a spherical shape rather than letting it spread over the surface as a sheet of water.

Some medications designed for use on the tissues in the throat contain a very special surface active agent—one that will reduce the surface tension of the water. This surface active agent, called a surfactant, lowers the surface tension of the liquid so that it spreads rapidly over the tissues rather than collecting in the form of droplets with less "active" surface area.

Diffusion. When a crystal of copper sulfate pentahydrate (a blue crystalline substance) is dropped into a cylinder of water, the blue color is soon observed in the water surrounding the crystal. After a while, the blue color may be seen extending upward from the crystal. The liquid at the bottom of the cylinder will be darker blue and the liquid above it will be lighter blue. After several hours the entire contents of the cylinder will be uniformly blue. This is an example of diffusion—the solute particles from the crystal are uniformly distributed into all parts of the solution.

Gases will diffuse into one another. When a bottle of ether is opened the odor may soon be detected at a distance. Diffusion into a gas takes place more rapidly than diffusion into a liquid.

Osmosis. Consider two sugar solutions (one weak and the other strong) separated by a semipermeable membrane, a selective membrane. The two solutions will tend to equalize in concentration. That is, the weak one will tend to become stronger while the strong one will tend to become weaker. How can they do this? There are two possibilities. First, the solvent can flow

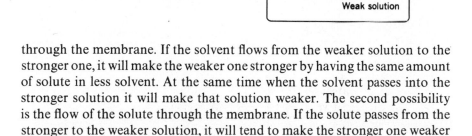

Figure 11-3. Osmosis; prune in water.

through the membrane. If the solvent flows from the weaker solution to the stronger one, it will make the weaker one stronger by having the same amount of solute in less solvent. At the same time when the solvent passes into the stronger solution it will make that solution weaker. The second possibility is the flow of the solute through the membrane. If the solute passes from the stronger to the weaker solution, it will tend to make the stronger one weaker and the weaker one stronger.

Which of these two possibilities is osmosis? The first one is. Osmosis is defined as the flow of solvent (water) through a semipermeable membrane (with the solvent passing from a dilute to a more concentrated solution). The flow of solute through the membrane is another example of diffusion; it is *not* osmosis. Both osmosis and diffusion may occur at the same time but not at the same rate. In general, osmosis occurs more rapidly than diffusion.

An example of osmosis that is quite common in the home can be observed by placing a dried prune in water (see Figure 11-3). The skin of the prune acts as a semipermeable membrane. Inside the prune are rather concentrated juices. The water surrounding the prune is certainly dilute in comparison to the juices inside. Thus there are two different concentrations of a solution separated by a semipermeable membrane, and osmosis can take place. In which direction? In osmosis the flow of solvent is from the weaker to the stronger solution. Therefore the water will flow into the prune, causing it to swell.

Another example of osmosis may be seen when a cucumber is placed in a strong salt solution (Figure 11-4). The skin of the cucumber acts as a semipermeable membrane. The liquid inside the cucumber is quite dilute in comparison to that of the salt solution. Therefore osmosis takes place because solutions of two different concentrations are separated by a semipermeable membrane. The flow, again from dilute to concentrated, takes

Figure 11-4. Osmosis; cucumber in salt water.

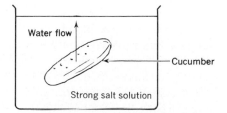

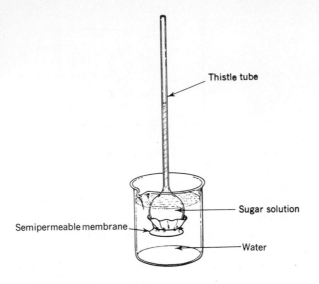

Thistle tube

Figure 11–5. Osmotic pressure as shown by a thistle tube containing sugar solution that is placed in water.

Sugar solution

Semipermeable membrane

Water

place from the cucumber into the solution. Thus the cucumber shrinks and becomes a pickle, again by the process of osmosis.

Osmotic Pressure. Consider a beaker of water into which is placed a thistle tube with a semipermeable membrane over its end. Assume that the thistle tube contains sugar solution (see Figure 11–5). In which direction will osmosis take place? Osmosis will take place with the water flowing into the thistle tube (from dilute to stronger solution). As the water flows into the thistle tube, the water level in the tube will rise. The rising water in the thistle tube will exert a certain amount of pressure, as does any column of liquid. The pressure exerted during the osmotic flow is called the osmotic pressure. This concept is of great importance in the regulation of fluid and electrolyte balance in the body (see pages 429, 432).

Isotonic Solutions

Two solutions that have the same solute concentration are said to be *isotonic*. The normal salt concentration of the blood is approximately equal to 0.9 per cent sodium chloride solution. The common name for 0.9 per cent sodium chloride solution is physiologic saline solution. The blood and physiologic saline solution are isotonic—they have the same salt concentration.

Suppose that a red blood cell were placed in a small amount of physiologic saline solution. What would happen? The red blood cell is surrounded by its cell wall, which acts as a semipermeable membrane. Will osmosis take place? The answer is no, because there is no difference in concentration on either side of the semipermeable membrane. Actually the water molecules move in both directions equally with no net change (see Figure 11–6).

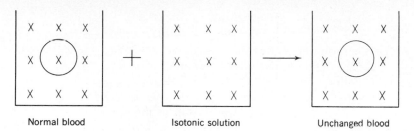

Normal blood   Isotonic solution   Unchanged blood

Figure 11–6. Red blood cell in isotonic solution.

Thus, physiologic saline solution can be given intravenously to a patient without any effect on the red blood cells (see Figure 11–7). Physiologic saline is administered under the following conditions:

1. When the patient has become dehydrated.
2. When the patient has lost considerable fluid, as in the case of hemorrhage.
3. To prevent postoperative shock.

A 5.5 per cent glucose solution is also approximately isotonic with body fluids.

Figure 11–7. Intravenous infusion with physiologic saline solution. (Reproduced from "The Hospital People," copyright 1967 by the Blue Cross Association.)

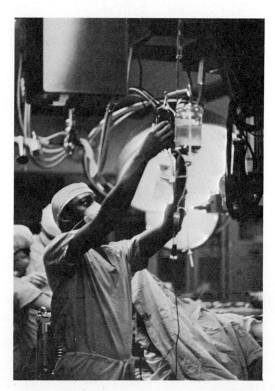

Hypotonic Solutions

A *hypotonic* solution is one that contains a lower solute concentration than that of another solution. Distilled water and tap water are hypotonic when compared with blood.

Suppose that a red blood cell is placed in water (a hypotonic solution). What will happen? The salt concentration in the red blood cell is higher than that of the water. Therefore, osmosis will take place with the water flowing into the red blood cell (from dilute to concentrated solution) (see Figure 11–8). The red blood cell thus enlarges until it bursts. This process of the bursting of a red blood cell because of a hypotonic solution is called *hemolysis*. During hemolysis the blood is said to be *laked*. Thus a hypotonic solution is *not* used for transfusions.

Hypertonic Solutions

A *hypertonic* solution is one that contains a higher solute concentration than that of another solution. A 5 per cent sodium chloride solution or a 10 per cent glucose solution is an example of a hypertonic solution when compared with blood.

Suppose a red blood cell is placed in a hypertonic solution. What will happen? The salt concentration in the red blood cell is less than that in the hypertonic solution. Therefore, osmosis will take place with the water flowing out of the red blood cell (from dilute to concentrated) (see Figure 11–9). The red blood cell thus shrinks. This shrinking of the red blood cell in a hypertonic solution is called *crenation* or *plasmolysis*.

Thus it should be noted that only isotonic solutions may be safely introduced into the bloodstream. Hypotonic solutions may cause hemolysis and hypertonic solutions may cause crenation.

Saline cathartics such as magnesium sulfate, milk of magnesia, and magnesium citrate are absorbed from the large intestine slowly and incom-

Figure 11–8. Red blood cell in hypotonic solution.

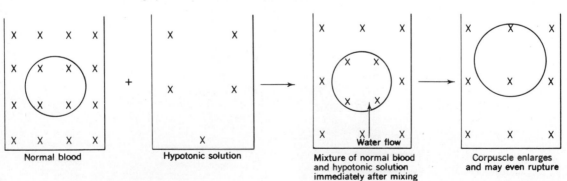

| Normal blood | Hypotonic solution | Mixture of normal blood and hypotonic solution immediately after mixing | Corpuscle enlarges and may even rupture |

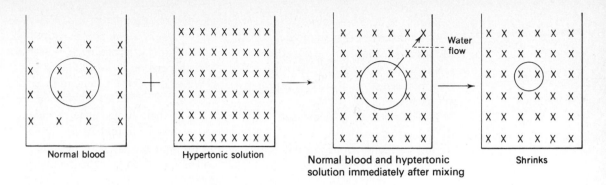

Normal blood   Hypertonic solution   Normal blood and hyptertonic solution immediately after mixing   Shrinks

Water flow

Figure 11–9. Red blood cell in hypertonic solution.

pletely. Thus, when these substances are ingested a hypertonic solution is produced in the large intestine, and water will flow into the intestinal tract until the solution is again isotonic with the body fluids. This additional water in the large intestine produces a watery stool that is easily evacuated. Therefore, by the continual use of cathartics a patient may become dehydrated. On the same basis, cathartics have been used to rid the body of excess fluid, although diuretics are now preferred for this purpose.

Suspensions

Suppose that some powdered clay is placed in water and then vigorously shaken. A suspension of clay in water will be produced. This suspension will not be clear; it will be opaque. Upon standing, the clay will slowly settle. The composition of the suspension is actually changing as the clay settles out so that it is a heterogeneous mixture.

The clay is not dissolved in the water; it is merely suspended in it. When the suspension is poured into a funnel containing a piece of filter paper, only the water passes through the filter paper; the clay does not. Evidently the suspended clay particles are too large to pass through the holes in a piece of filter paper. Undoubtedly they will also be too large to pass through a membrane that has even finer openings.

Properties of suspensions are summarized as follows. Suspensions

1. Consist of an insoluble substance suspended in a liquid.
2. Are heterogeneous.
3. Are not clear.
4. Settle.
5. Do not pass through filter paper.
6. Do not pass through membranes.

Some medications, such as milk of magnesia, are administered as a suspension. Many bottles of medication state on the label "shake before using." Most suspensions use water as the suspending medium but procaine penicillin G, for example, is usually administered as an oil suspension.

A mist is a suspension of a liquid in a gas. Water droplets suspended in air are one example of a mist.

Patients having a decreased water content in their lungs usually have thickened bronchial secretions. To increase the water content of the lungs requires a higher-than-normal water content in the inspired air. A nebulizer is a device that generates an aerosol mist consisting of large water particles that can penetrate into the trachea and large bronchi. An ultrasonic nebulizer produces a supersaturated mist by means of ultrasonic sound waves. Such a mist is very effective in inhalation therapy because the particle size is small enough to reach the smallest bronchioles (see Figure 11–10). Also, the concentration of the mist is greater than that produced by other means.

Figure 11–10. Electronic nebulizer. (Courtesy of MistO$_2$ Gen® Equipment Co., Oakland, Calif.)

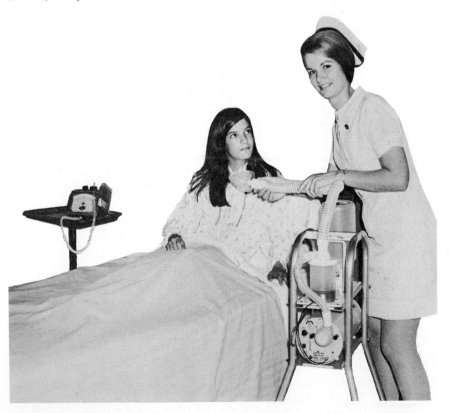

Care must be taken in the use of a nebulizer to avoid (1) bacterial contamination in the water reservoir of the equipment (such bacteria will go directly into the bronchi) and (2) excess water being added to the patient's lungs.

Colloids

The third class of liquid mixture, called colloids, consists of tiny particles suspended in a liquid. We might ask, if these particles are suspended in a liquid, why aren't they suspensions? The answer is that these colloids behave quite differently than do ordinary suspensions and have an entirely different set of properties.

Size

When a colloid (or a colloidal dispersion as it is frequently called) is poured into a funnel containing filter paper, the colloid passes through the filter paper. This indicates that the colloidal particles are smaller than the openings in a filter paper. When a colloid is placed in a membrane, the colloidal particles do not pass through. Thus, colloidal particles are larger than the openings in the membrane.

Recall that solutions pass both through filter paper and membranes, whereas suspensions pass through neither. Therefore colloidal particles must be intermediate in size between solution particles and suspension particles. Colloidal particle sizes are measured in nanometers (1 nanometer is 0.000000001 m or one millionth of a millimeter). Colloidal particles range in size from 1 to 100 nanometers (nm). Solution particles are smaller than 1 nm, whereas suspension particles are larger than 100 nm (see Figure 11–11).

Colloids have a tremendous amount of surface area because they consist of so many tiny particles. Consider a cube 1 in. on an edge. The volume of such a cube is 1 cubic inch (in.^{3}) and the surface area is 6 square inches (in.^{2}). (There are six faces to a cube and each face has an area of 1 in.^{2}).

If the cube is cut in half the total volume of both pieces is still 1 in.^{3} However, the total surface area has now increased to 8 in.^{2} (cutting introduces two new faces each having an area of 1 in.^{2}). Cutting in half again still retains the volume of 1 in.^{3} The surface area now is 10 in.^{2} Continued cutting further increases the surface area.

When the size of the individual particle reaches colloidal size—approximately 10 nm—the total amount of surface area is about 15,000,000 in.^{2} This tremendous surface area gives colloids one of their most important properties—adsorption.

Figure 11–11. Relative sizes of solution, colloid, and suspension particles.

Solution   Colloid   Suspension

Adsorption

Adsorption is defined as the property of holding substances to a surface. Colloidal charcoal will adsorb tremendous amounts of gas. It has a selective adsorption as do most colloids. Coconut charcoal is used in gas masks because it selectively adsorbs poisonous gases from the air. It does not adsorb ordinary gases from the air.

Several commonly used medications owe their use to their adsorbent properties. Charcoal tablets are administered to patients to aid indigestion. Kaolin, a finely divided aluminum silicate, is administered for the relief of diarrhea. Colloidal silver adsorbed on protein (Argyrol) is used as a germicide.

Electrical Charge

Almost all colloidal particles have an electrical charge—either positive or negative. Why? Colloids selectively adsorb ions on their surface. If a colloid selectively adsorbs negative ions, it becomes a negative colloid. If it adsorbs positive ions, it becomes a positive colloid.

Consider two negatively charged colloidal particles suspended in water. What would happen when these particles come close together? They would repel each other because of the repulsion of their like charges. Therefore, these colloidal particles will have little tendency to form large particles that would then settle out.

How can colloids be made to settle? Colloidal particles have an electrical charge which will repel all other similarly charged colloidal particles. However these charged particles *will* attract particles of opposite charge. Therefore a negative colloid may be made to coagulate (begin to settle out) by adding to it positively charged particles.

Bichloride of mercury ($HgCl_2$) is a poisonous substance. When swallowed it forms a positive colloid in the stomach. The antidote for this type of poisoning is egg white, which is a negative colloid. These two oppositely charged colloids neutralize each other and coagulate in the stomach. The stomach must then be pumped out to remove the coagulated material. If this is not done, the stomach will digest the egg white, exposing the body once again to the poisonous substance.

Protein may also be coagulated by heat. Egg white—a colloidal substance —is quickly coagulated when heated.

The charge of a colloid may be determined by placing it in a U tube containing two electrodes. When a current is passed through the U tube each electrode will attract particles of opposite charge. A negative colloid will begin to accumulate around the positive electrode and a positive colloid around the negative electrode. The movement of electrically charged suspended particles toward an oppositely charged electrode is called *electrophoresis*. Electrophoresis is a slow process because the charged particles are not soluble and are much larger than the particles in solution (see page 422).

When a strong beam of light is passed through a colloidal dispersion, the beam becomes visible because the colloidal particles reflect and scatter the light. This phenomenon is called the Tyndall effect. The Tyndall effect may also be observed when a beam of sunlight passes through a darkened room. The dust particles in the air scatter the light so that the sun's rays become visible. The blue color of the sky is due to scattered light, as is the blue color of the ocean. However, we actually do not see the colloidal particles; we merely see the light scattered by them.

When a strong beam of light is passed through a solution, no Tyndall effect is observed because the solution particles are too small to scatter the light. Thus, the Tyndall effect is a way of distinguishing between solutions and colloids.

Brownian Movement

When colloidal particles are observed with an ultramicroscope, the particles are seen to move in a haphazard irregular motion called the Brownian movement (see Figure 11–12).

What causes this irregular motion? It cannot be caused by the vibration of the slide on a microscope stand because the same motion can be observed when the microscope is mounted on a concrete pillar sunk deep into the earth. The Brownian movement can be observed during the day or the night, in the city and in the country, in warm weather and in cold weather, at high altitudes and at low altitudes.

In addition, the strangest characteristic of all is that this particular motion never ceases. It can be observed in containers which have been sealed for years. It has also been observed in liquid occlusions—in quartz samples which have been undisturbed for perhaps millions of years.

Careful investigation by scientists established the fact that the Brownian movement is not due to the colloidal particles themselves but rather to their

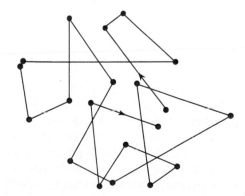

Figure 11–12. Brownian movement. (Reproduced from Sisler, H. H.; VanderWerf, C. A.; and Davidson, A. W.: *College Chemistry*, 3rd ed. Macmillan Publishing Co., Inc., New York, 1967.)

bombardment by the molecules of the suspending medium. The molecules of the suspending medium are in continuous random motion. When these rapidly moving molecules of the medium strike the colloidal particles, they cause these particles to have the random motion characteristic of the Brownian movement.

Gels and Sols

Colloidal dispersions may be subdivided into two classes. Where there is a strong attraction between the colloidal particle and the suspending liquid (water), the system is said to be hydrophilic (water loving). Systems of this type are called *gels*. Gelatin in water is an example of such a system. Gels are semisolid and semirigid; they do not flow easily.

A colloidal system where there is little attraction between the suspended particles and the suspending water is called hydrophobic (water hating). Systems of this type are called *sols*. They pour easily. A small amount of starch in water forms a sol. A hydrosol is a colloidal dispersion in water, while an aerosol is a colloidal dispersion in air.

If a gel, such as gelatin, is heated, it turns into a sol but returns to its original gel state on cooling. Protoplasm has the ability to change gel (in membranes) into sol and vice versa.

Dialysis

Dialysis is the separation of solution particles from colloidal particles by means of a membrane. Recall that solution particles can pass through a membrane but colloidal particles cannot. Suppose that a colloidal starch suspension and sodium chloride solution are placed inside a membrane that in turn is placed in a beaker of water. The starch is a colloid and cannot pass through the membrane. The salt is in solution and can pass through the membrane and does so. This is an example of diffusion. The salt will continue to pass through the membrane until the salt concentration inside the membrane is the same as that in the water surrounding the membrane.

When this happens no more salt will be removed from the mixture inside the membrane. However, if the membrane is suspended in running water, soon all the salt will be removed from inside the membrane leaving behind only the starch—the colloid. This is dialysis—the separation of a solution from a colloid by means of a membrane (see Figure 11–13). Antitoxins are prepared by this method. The impure material is placed inside a membrane suspended in running water. The soluble impurities flow out of the membrane leaving the pure antitoxin behind. The same process is used to prepare low-sodium milk. Milk is placed in a membrane suspended in running water. The soluble salt (sodium chloride) leaves the milk so that the remaining liquid is practically free from sodium compounds.

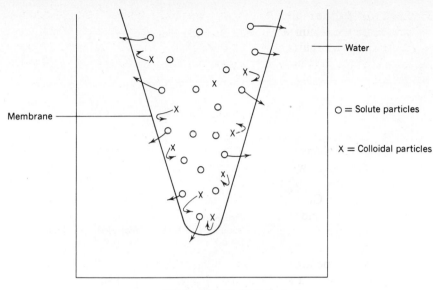

Water

O = Solute particles

X = Colloidal particles

Membrane

Figure 11–13. Dialysis.

In the body the membranes in the kidneys allow the soluble waste material to pass through. The same membranes do not allow protein to pass through since proteins are colloids.

Hemodialysis

Hemodialysis refers to the removal of soluble waste products from the bloodstream by means of a membrane. Purification of the blood can be accomplished this way because soluble particles can diffuse through a membrane whereas blood cells and plasma proteins cannot. When a patient has problems related to renal excretion, an artificial kidney machine may be used (see Figures 11–14 and 11–15). This machine applies the principles of hemodialysis.

The artificial kidney machine consists of a long cellophane tube, which is wrapped around itself to form a coil. This coil is immersed in a temperature-controlled solution whose chemical composition is carefully regulated according to the needs of the patient. The patient's blood is pumped through the coil, and the soluble end products of protein catabolism, water, and exogeneous poisons are removed from the blood. At the same time the blood cells and plasma proteins remain in the blood (recall that proteins, which are colloids, cannot pass through a membrane and here the cellophane acts as such a membrane). If the solution in which the coil is immersed has the same concentration of sodium ions as the blood, no diffusion of sodium ions will take place. The same applies to other soluble substances in the bloodstream. That is, by regulating the composition of the solution, the waste products and unwanted material can be removed from the bloodstream.

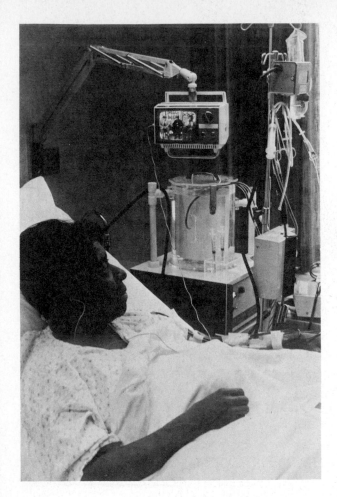

Figure 11—14. Artificial kidney machine. (Courtesy of Miami Valley Hospital, Dayton, Ohio.)

A patient may remain on the artificial kidney machine 4 to 7 hr or even more. During this time the solution must be changed at intervals to avoid accumulation of waste products. Otherwise, the waste products would no longer continue to diffuse out of the blood. After passing through the coil, the blood is returned to the patient's veins.

As we might expect, the dialysate is mostly water. The kind of water used in its preparation is of utmost importance to the patient. If the water has been softened by the zeolite process (see page 126), it may contain too high a concentration of sodium ions (and so add sodium to the blood; see pages 431–32). Improper levels of calcium and magnesium ions in the water may lead to

Figure 11—15. Hemodialysis apparatus. (Courtesy of Michael Reese Hospital and Medical Center, Chicago, Ill.)

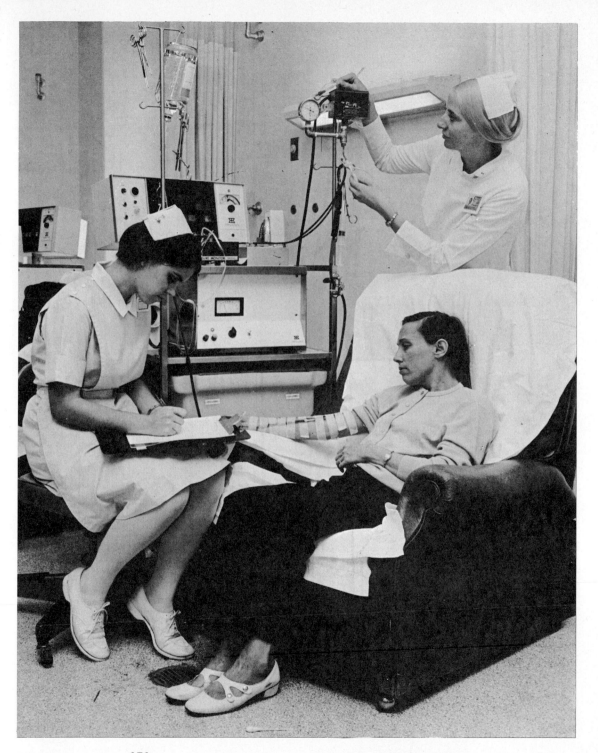

metabolic bone diseases. Traces of copper in the water (from copper tubing in the equipment, from copper pipes in the water supply, or from copper sulfate used to kill algae in the water treatment plant) have caused fatalities in dialysis patients.

Emulsions

When two liquids are mixed, either they dissolve in each other or they do not. Two liquids which are soluble in each other are called *miscible*. Two liquids which do not dissolve in each other are called *immiscible*.

Suppose two immiscible liquids such as oil and water are poured together and then shaken vigorously. The oil forms tiny drops which are suspended in the water. After a while the tiny oil drops come together to form larger drops which soon rise to the top and separate from the water. This type of a liquid mixture is called a *temporary emulsion*—emulsion because it consists of a liquid colloidally suspended in a liquid and temporary because it separates.

What are the properties of a temporary emulsion? First, as was just noted, it separates. Since it separates, it must be heterogeneous; its composition keeps on changing. Temporary emulsions, such as oil in water, are not clear. If placed in a funnel containing a piece of filter paper, they do not pass through. Neither do they pass through a membrane.

A temporary emulsion such as oil and water separates because the oil drops attract one another to form larger drops, which soon rise to the top. However, suppose that each drop of oil were given a negative charge. Then the drops would not attract one another. They would not settle out; they would form a permanent emulsion. To give the oil drops an electrical charge we need an emulsifying agent. An emulsifying agent is a protective colloid that coats the suspended oil drops and prevents them from coming together. Permanent emulsions are not clear; they are homogeneous; they do not settle; they do not pass through either filter paper or membranes.

Examples of permanent emulsions are mayonnaise, which is an emulsion of oil in vinegar with egg yolk as an emulsifying agent. Milk is an emulsion of butterfat in water with casein acting as the emulsifying agent. Soap acts as an emulsifying agent on grease and oils in water. Many medications are given in the form of emulsions, gum acacia being the most common emulsifying agent for the dispersion of fats.

Summary

Solutions consist of a solute and a solvent. Solutions are clear, homogeneous, have a variable composition, do not settle, may be separated by physical means, pass through filter paper, and pass through a membrane.

Factors affecting the solubility of a solute are temperature, pressure, surface area, agitation, and nature of the solvent.

Many chemical reactions take place in solution. Many medications are administered as solutions.

Solutions may be labeled as dilute or concentrated, as saturated or unsaturated, all of which are relative terms and do not indicate any definite amount of solute and solvent.

A percentage solution indicates the number of grams of solute per 100 ml of solution. A ratio solution indicates the number of grams of solute and the number of milliliters of solution. A molar solution indicates the number of moles of solute per liter of solution. A normal solution indicates the number of gram equivalent weights of solute per liter of solution.

When a nonvolatile solute is dissolved in a solvent, the boiling point is elevated and the freezing point depressed.

All liquids exhibit surface tension, which causes the surface molecules to form a surface film.

The movement of solute into a solvent or through a solution is called diffusion. The flow of solvent through a semipermeable membrane is called osmosis. The pressure exerted during osmosis is called the osmotic pressure.

An isotonic solution has the same salt concentration as the blood and is used for transfusions.

A hypotonic solution has a salt concentration less than that of the blood. If injected into the bloodstream, a hypotonic solution can cause hemolysis—the bursting of the red blood cells.

A hypertonic solution has a salt concentration greater than that of the blood. If injected into the bloodstream, a hypertonic solution may cause crenation or plasmolysis —the shrinking of the red blood cells.

Suspensions consist of a nonsoluble solid suspended in a liquid medium. Suspensions are not clear; they settle out; they are heterogeneous; they do not pass through a filter paper; and they do not pass through a membrane.

Colloids consist of tiny particles suspended in a liquid. Colloids do not settle; they pass through filter paper but not through membranes; they adsorb (hold) particles on their surface; they have electrical charges, owing to the adsorption of charged particles (ions); they exhibit the Tyndall effect and the Brownian movement.

Colloidal dispersions may be subdivided into two classes—sols and gels.

Dialysis is the separation of a solute from a colloid by means of a membrane.

Hemodialysis refers to the removal of soluble waste products from the bloodstream by means of a membrane. When a patient has problems related to renal excretion, an artificial kidney machine may be used.

An emulsion consists of a liquid suspended in a liquid. An emulsion that settles is called a temporary emulsion. When an emulsifying agent is added to a temporary emulsion, it becomes a permanent emulsion.

Questions and Problems

1. Give specific directions for preparing
   (a) 100 ml of a 5 per cent boric acid solution.
   (b) a 1:300 $KMnO_4$ solution.
   (c) 2 liters of a 1 M NaCl solution.

(d) 100 ml of a 0.1 N H_2SO_4 solution.

(e) 500 ml of a 0.9 per cent NaCl solution.

2. Define : (a) osmosis ; (b) dialysis ; (c) adsorption ; (d) osmotic pressure ; (e) hypertonic solution ; (f) diffusion ; (g) saturated solution.

3. Why must an isotonic solution be used during a blood transfusion?

4. What factors affect the rate of solution of a solid solute?

5. List the general properties of solutions.

6. List the general properties of suspensions.

7. List the general properties of colloids.

8. List the general properties of emulsions.

9. What is the Tyndall effect?

10. What is the Brownian movement?

11. What is surface tension?

12. How do colloids obtain their electrical charge?

13. What effect does a solute have on the boiling point and on the freezing point of a solution?

14. Explain how an artificial kidney machine functions.

15. Why must the solution in an artificial kidney machine be changed at intervals during use over a long time?

16. Why does diffusion take place more rapidly in a gas than in a liquid?

17. How may a temporary emulsion be changed into a permanent emulsion?

18. How can a colloid be made to settle? Explain.

19. Why do colloids pass through filter paper but not through membranes?

20. What is the size range of colloidal particles? Solution particles? Suspension particles?

21. What is meant by the term "tincture"?

22. Does pressure always affect the amount of a solute that will dissolve in a given solvent? Explain.

23. How can you tell whether a solution is saturated or unsaturated?

24. What is a nebulizer? For what purposes is it used in the hospital? What precautions should be taken with its use?

25. Why is the control of water important in hemodialysis? Would you expect control of temperature also to be important? Explain.

26. What is the difference between a sol and a gel? Give an example of each.

References

Ault, F. K., and Lawrence, R. M.: *Chemistry: A Conceptual Introduction.* Scott, Foresman & Co., Glenview, Ill., 1976, Chap. 6.

Baum, S. J., and Scaife, C. W.: *Chemistry: A Life Science Approach.* Macmillan Publishing Co., Inc., New York, 1975, Chap. 7.

Eastman, R. H.: *Essentials of Modern Chemistry.* Holt, Rinehart & Winston, Inc., New York, 1975, Chap. 6.

Fernandez, J. E., and Whitaker, R. D.: *An Introduction to Chemical Principles.* Macmillan Publishing Co., Inc., New York, 1975, Chap. 20.

Hein, M., and Best, L. R.: *College Chemistry.* Dickenson Publishing Co., Inc., Encino, Calif., 1976, Chaps. 14, 15.

Levenspiel, O., and de Nevers, N.: The osmotic pump. *Science*, **183**:157–160 (Jan. 18), 1974.

Longo, F. R.: *General Chemistry*. McGraw-Hill Book Co., New York, 1974, Chap. 9.

Nebergall, W. H.; Schmidt, F. C.; and Holtzclaw, H. F., Jr.: *College Chemistry*, 5th ed. D. C. Heath & Co., Lexington, Mass., 1976, Chaps. 13, 14.

Nordmann, J.: *What Is Chemistry?* Harper & Row, New York, 1974, Chap. 16.

Peters, E. I.: *Introduction to Chemical Principles*. W. B. Saunders Co., Philadelphia, 1974, Chap. 12.

Sienko, M. J., and Plane, R. A.: *Chemistry*, 5th ed. McGraw-Hill Book Co., New York, 1976, Chaps. 8, 10.

Chapter 12

Acids and Bases

Acids

Acids may be defined as compounds that yield hydrogen ions (H^+) in solution. It is this hydrogen ion that is responsible for the particular properties of acids. Most people think of acids as being liquids. However, there are many solid acids; boric acid and citric acid are two common examples.

Properties of Acids

Acids yield hydrogen ions when placed in water solution.

$$HCl \longrightarrow H^+ + Cl^-$$

hydrochloric acid   hydrogen ion   chloride ion

$$H_2SO_4 \longrightarrow 2\,H^+ + SO_4^{2-}$$

sulfuric acid   hydrogen ions   sulfate radical

$$HNO_3 \longrightarrow H^+ + NO_3^-$$

nitric acid   hydrogen ion   nitrate radical

156

TABLE 12–1

Commonly Used Acids

HCl	Hydrochloric acid
H_2SO_4	Sulfuric acid
HNO_3	Nitric acid
H_2CO_3	Carbonic acid
H_3PO_4	Phosphoric acid

Solutions of acids have a sour taste. Lemon and grapefruit juices owe their sour taste to citric acid. Vinegar owes its sour taste to acetic acid. Sour milk owes its taste partly to lactic acid.

When acids react with certain compounds, these compounds change in color. Substances which change in color in the presence of acids are called *indicators*. One of the most common indicators for acids is litmus. Blue litmus turns red in the presence of an acid (in the presence of hydrogen ions). Another common indicator—phenolphthalein—turns from red to colorless in the presence of an acid.

Acids react with metal oxides and hydroxides to form water and a salt, for example

$$2HCl + MgO \longrightarrow H_2O + MgCl_2$$
$$\text{acid} \quad \text{metal oxide} \qquad \text{water} \quad \text{salt}$$

$$H_2SO_4 + 2NaOH \longrightarrow 2H_2O + Na_2SO_4$$
$$\text{acid} \qquad \text{metal} \qquad\quad \text{water} \quad \text{salt}$$
$$\text{hydroxide}$$

The reaction of acids with certain metal hydroxides (called bases) is termed *neutralization*. That is, acids neutralize bases to form water and a salt.

The activity series of metals (Table 12–2) lists metals in order of decreasing activity. Note that hydrogen is classified with the metals. *Any metal above hydrogen in this series will replace the hydrogen from an acid.* The farther above hydrogen a metal is in the activity series, the more rapid its reaction in the replacement of hydrogen from an acid.

Acids react with any metal above hydrogen in the activity series to produce hydrogen gas and a salt, for example

$$Zn + H_2SO_4 \longrightarrow ZnSO_4 + H_2\uparrow$$
$$\text{salt}$$

$$Mg + 2HCl \longrightarrow MgCl_2 + H_2\uparrow$$
$$\text{salt}$$

These are examples of single replacement reactions—the metal replaces the

TABLE 12–2

The Activity Series of the Metals

K	Potassium
Ca	Calcium
Na	Sodium
Mg	Magnesium
Al	Aluminum
Zn	Zinc
Fe	Iron
Sn	Tin
Pb	Lead
H	Hydrogen
Cu	Copper
Hg	Mercury
Ag	Silver
Au	Gold

hydrogen in the acid. Thus acids cannot be stored in containers made of these active metals. Iron is above hydrogen in the activity series and so should replace hydrogen in an acid. Therefore, acids should not be allowed to come in contact with surgical or dental instruments. Acids are usually stored in glass or plastic containers. Reactions involving acids are carried out in glass or plastic vessels.

Acids react with carbonates and bicarbonates to form carbon dioxide, water, and salts, for example

$$2HCl + CaCO_3 \longrightarrow CaCl_2 + H_2CO_3$$

$\quad$ acid $\quad$ carbonate $\qquad\qquad$ salt $\quad$ carbonic
$\qquad\qquad\qquad\qquad\qquad\qquad\qquad\qquad$ acid

and
$$H_2CO_3 \longrightarrow CO_2\uparrow + H_2O$$

$$H_2SO_4 + 2NaHCO_3 \longrightarrow Na_2SO_4 + 2H_2CO_3$$

$\quad$ acid $\qquad$ bicarbonate $\qquad\qquad$ salt $\quad$ carbonic
$\qquad\qquad\qquad\qquad\qquad\qquad\qquad\qquad$ acid

and
$$2H_2CO_3 \longrightarrow 2H_2O + 2CO_2\uparrow$$

Calcium carbonate, $CaCO_3$, is sometimes used to remove excess acidity from the stomach. Likewise sodium bicarbonate, $NaHCO_3$, may be administered to remove excess stomach acidity. However, continued use of this latter substance may interfere with the normal digestive processes taking place in the stomach. (The prefix *bi* is sometimes used to indicate hydrogen in a compound. Sodium bicarbonate is also known as sodium hydrogen carbonate.)

Effervescent tablets or powders contain an acid and sodium bicarbonate.

When placed in water these two substances react to release carbon dioxide, thus causing the effervescence.

Strong acids will attack clothing. Vegetable fibers such as cotton and linen, animal fibers such as wool and silk, and synthetic fibers are rapidly destroyed by strong acids. All these effects are actually due to the hydrogen ions present in the acids.

Strong acids (acids with a high hydrogen ion concentration) also have an effect on tissues. Concentrated nitric acid, HNO_3, and concentrated sulfuric acid, H_2SO_4, are extremely corrosive to the skin so that great care must be exercised in handling them. If a strong acid is spilled on the skin a serious burn may result. In this case the area should be washed copiously with water. Then it should be treated with sodium bicarbonate to neutralize any remaining acid. Dilute acids, however, are not corrosive to the tissues. They may even be used internally in the body.

Uses of Acids

Acids such as HCl, hydrochloric acid (commercially known as muriatic acid), are used industrially and in laboratory work in large amounts. Several acids are also used medically; among these are hydrochloric, nitric, hypochlorous, boric, acetylsalicylic, and ascorbic acids.

Hydrochloric acid, HCl, normally found in the gastric juices, is necessary for the proper digestion of proteins in the stomach. Patients who have a lower than normal amount of hydrochloric acid in the stomach, a condition called hypoacidity, are given dilute hydrochloric acid orally before meals to overcome this deficiency.

Nitric acid, HNO_3, is used in the laboratory to test for the presence of many proteins. The protein in the skin turns yellow when it comes in contact with nitric acid. Since it will coagulate protein material, nitric acid is used to test for the presence of albumin in urine. Nitric acid has been used to remove warts, but dichloroacetic acid (bichloracetic acid) and trichloroacetic acid are now commonly used for this purpose.

Hypochlorous acid, HClO, is used as a disinfectant of floors and walls in the hospital.

Boric acid, H_3BO_3, has had extensive use as a germicide. Although boric acid has been used in eyewashes, its use in solutions or as a powder on extensive inflamed surfaces or in body cavities is now practically obsolete. Containers of boric acid should have a label reading "poison."

Acetylsalicylic acid (aspirin) is widely used as a pain killer (analgesic) and as an antipyretic (to reduce fever). Aspirin is frequently taken by people with a cold to relieve headache, muscle pain, and fever. However, the aspirin does not remove the source of infection nor effect a cure (see page 239).

Ascorbic acid (vitamin C) is normally found in citrus fruits and is used in the prevention and treatment of scurvy.

Bases

Bases may be defined as substances that yield hydroxide (OH^-) ions in solution.

If we write the ionization reactions (see Chapter 14) of the bases sodium hydroxide and potassium hydroxide, we have

$$NaOH \longrightarrow Na^+ + OH^-$$

$$KOH \longrightarrow K^+ + OH^-$$

Since hydroxide ions react with hydrogen ions to form water ($OH^- + H^+ \rightarrow H_2O$), a base may also be defined as a substance that accepts hydrogen ions. C_2H_5OH is not a base because it does not ionize; it does not yield hydroxide ions in solution.

Table 12–3 indicates several commonly used bases. Note that they consist of a metal ion ionically bonded to an OH^- radical. The only exception to this rule is the base ammonium hydroxide, NH_4OH, where the ammonium radical, NH_4^+, is considered to act as a metal ion. Bases are produced when metallic oxides are dissolved in water, for example

$$CaO + H_2O \longrightarrow Ca(OH)_2$$

metal   water   calcium hydroxide
oxide   a base

Properties of Bases

Bases yield hydroxide ions when placed in water.

Solutions of bases have a slippery, soapy feeling, and a biting, bitter taste.

Like acids, bases also react with indicators. The hydroxide ions of bases turn red litmus blue, turn methyl orange from red to yellow, and turn phenolphthalein from colorless to red.

Bases neutralize acids to form water and a salt, for example

$$Ca(OH)_2 + H_2SO_4 \longrightarrow 2H_2O + CaSO_4$$

base   acid   water   salt

TABLE 12–3

Commonly Used Bases

Name	Formula
Sodium hydroxide	$NaOH$
Ammonium hydroxide	NH_4OH
Potassium hydroxide	KOH
Calcium hydroxide	$Ca(OH)_2$
Magnesium hydroxide	$Mg(OH)_2$

Strong bases react with certain metals to produce hydrogen gas, for example

$$2Al \ + \ 6NaOH \ + \ 6H_2O \ \longrightarrow \ 3H_2\uparrow \ + \ 2Na_3Al(OH)_6$$

aluminum   sodium hydroxide   water   hydrogen   sodium aluminate
          a strong base   a soluble compound

Thus a strong base such as lye (NaOH) should never be used or stored in an aluminum container because it will rapidly react with and dissolve the container.

Strong bases have a high hydroxide ion concentration. They have a corrosive effect on tissues. If a strong base is spilled on the skin, a serious burn may result. The procedure in this case is to apply copious amounts of water followed by treatment with a weak acid such as acetic acid to neutralize any base that might be left.

Strong laundry soaps are quite basic and should not be used for washing woolen clothing because the hydroxide ion will attack the fibers and cause them to shrink. Particular care must be taken not to use a strong soap on diapers because, if it is not thoroughly removed, the basic soap may cause severe sores on the tender skin of a baby.

Strong bases will also react with proteins and fats.

Uses of Bases

Sodium hydroxide, NaOH, commonly known as lye, is used to remove fats and grease from clogged drains. It is quite caustic and care must be exercised in handling this substance.

Calcium hydroxide solution, $Ca(OH)_2$, commonly known as lime water, is used to overcome excess acidity in the stomach. It is also used medicinally as an antidote for oxalic acid poisoning because it reacts with the oxalic acid to form an insoluble compound, calcium oxalate.

Magnesium hydroxide, $Mg(OH)_2$, is commonly known as milk of magnesia. In dilute solutions it is used as an antacid for the stomach. In the form of a suspension of magnesium hydroxide in water, it is used as a laxative.

Spirits of ammonia, which contains ammonium hydroxide (NH_4OH) and ammonium carbonate ($[NH_4]_2CO_3$), is used as a heart and respiratory stimulant. Ammonium hydroxide, also known as household ammonia, is used as a water softener for washing clothes.

pH

A few drops of concentrated hydrochloric acid in water produce a dilute acid solution. A few more drops produce another solution still dilute but a little stronger than the previous one. If a piece of blue litmus paper is placed in either of these two solutions it will turn red, indicating that the solution is acidic. However, it will not tell which one is more strongly acidic. Likewise if a piece of red litmus paper turns blue when placed in a solution, it merely indicates that the solution is basic. It does not tell how strongly basic. The term *pH* is used to indicate the exact strength of an acid or a base. pH indicates the hydrogen ion concentration in a solution.

A pH of 7 indicates a neutral solution and pH values below 7, an acidic solution: pH's between 5 and 7 indicate a weakly acidic solution; values between 2 and 5, a moderate acid solution; and pH's between 0 and 2, a strongly acid solution.

Likewise, pH's above 7 indicate a basic solution: pH values between 7 and 9 indicate a weakly basic solution; those between 9 and 12, a moderate basic solution; and pH's between 12 and 14, a strongly basic solution. This may be summarized as shown in Table 12–4.

The pH of some common body fluids are listed in Table 12–5. From these values it can be seen that blood is a slightly basic liquid, the gastric juices strongly acidic, bile weakly basic, and urine and saliva both weakly acidic. As should also be expected, the pH of pure water is 7.0.

A difference of 1 in pH value represents a tenfold difference in strength. That is, an acid of pH 4.5 is ten times as strong as one of pH 5.5. Likewise, a base of pH 10.7 is ten times as strong as one of pH 9.7, and 100 times as strong (10×10) as one of pH 8.7. Therefore, a small change in pH indicates a definite change in acid or base strength.

To measure pH in a laboratory, a pH meter (Figure 12–1) may be used. This instrument is standardized by placing a solution of known pH under the electrodes to see that it is functioning and recording properly. Then a solution of unknown pH is placed under the electrodes and the pH is determined by reading the value on the scale of the pH meter. A quicker but less accurate method is to touch a drop of the liquid to a specially prepared piece of indicator paper and then determine the pH by comparison with a pH color scale.

TABLE 12–4

pH Values*

pH	0	1	2	3	4	5	6	7	8	9	10	11	12	13	14
Strength of acid (H^+ in moles/liter)	10^0	10^{-1}	10^{-2}	10^{-3}	10^{-4}	10^{-5}	10^{-6}	10^{-7}	10^{-8}	10^{-9}	10^{-10}	10^{-11}	10^{-12}	10^{-13}	10^{-14}
Strength of base (OH^- in moles/liter)	10^{-14}	10^{-13}	10^{-12}	10^{-11}	10^{-10}	10^{-9}	10^{-8}	10^{-7}	10^{-6}	10^{-5}	10^{-4}	10^{-3}	10^{-2}	10^{-1}	10^0

strong acid ◄►◄ moderate acid ►◄ weak acid ► neutral ◄ weak base ►◄ moderate base ►◄ strong base ►

* See Appendix for an explanation of scientific notation.

TABLE 12–5

pH Values of Body Fluids

Fluid	pH range
Blood	7.35–7.45
Gastric juices	1.6–1.8
Bile	7.8–8.6
Urine	5.5–7.0
Saliva	6.2–7.4

Summary

Acids are compounds that consist of hydrogen with a nonmetal or with a nonmetal radical in such a manner that the ionization of the hydrogen is possible. The general properties of acids are due to their hydrogen ions (H^+). These general properties are: presence of hydrogen ions in solution; sour taste; effect on colors of indicators; reaction with metal oxides and hydroxides (neutralization); reaction with metals to yield hydrogen gas; reaction with carbonates and bicarbonates to produce carbon dioxide gas; effect on clothing; and effect on tissues.

Figure 12–1. Digital pH meter. (Courtesy of Fisher Scientific Co., Pittsburgh, Pa.)

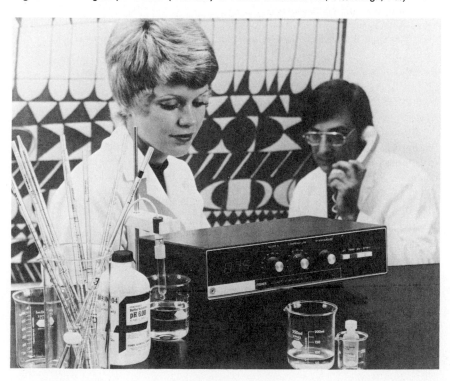

Acids and bases are used industrially, pharmaceutically, and in the hospital and laboratory.

Bases are compounds yielding hydroxide (OH^-) ions in solution. Bases have a slippery, soapy feeling; they affect the colors of indicators; neutralize acids; react with certain metals to produce hydrogen; and affect tissue and clothing.

The pH of a solution indicates numerically the acid (or base) strength of a solution in terms of its hydrogen ion concentration. A pH of 7 indicates a neutral solution; a pH below 7, an acid solution; a pH above 7, a basic solution.

Questions and Problems

1. Define: (a) acid; (b) base; (c) neutralization; (d) pH.
2. List the general properties of acids.
3. List the general properties of bases.
4. What effect does an acid have on litmus paper?
5. What effect does a base have on litmus paper?
6. Will aluminum replace the hydrogen from an acid? Explain.
7. Why are acids usually stored in a container made of glass?
8. What substances might be used to lower the acidity of the contents of the stomach?
9. What treatment should be given if a strong acid is accidentally spilled on the skin?
10. Name four acids used in the hospital and give one medical use for each.
11. Name three bases used in the hospital and give one medical use for each.
12. How may pH be measured?
13. What is the pH of blood? of urine? of saliva?
14. Are each of the substances in question 13 acidic, basic, or neutral?

References

Ault, F. K., and Lawrence, R. M.: *Chemistry: A Conceptual Introduction.* Scott, Foresman & Co., Glenview, Ill., 1976, Chap. 7.

Baum, S. J., and Scaife, C. W.: *Chemistry: A Life Science Approach.* Macmillan Publishing Co., Inc., New York, 1975, Chap. 8.

Eastman, R. H.: *Essentials of Modern Chemistry.* Holt, Rinehart & Winston, Inc., New York, 1975, Chap. 8.

Fernandez, J. E., and Whitaker, R. D.: *An Introduction to Chemical Principles.* Macmillan Publishing Co., Inc., New York, 1975, Chap. 22.

Hein, M., and Best, L. R.: *College Chemistry.* Dickenson Publishing Co., Inc., Encino, Calif., 1976, Chap. 15.

Longo, F. R.: *General Chemistry.* McGraw-Hill Book Co., New York, 1974, Chap. 13.

Nordmann, J.: *What Is Chemistry?* Harper & Row, New York, 1974, Chap. 16.

Pauling, L., and Pauling, P.: *Chemistry.* W. H. Freeman & Co., San Francisco, 1975, Chap. 12.

Peters, E. I.: *Introduction to Chemical Principles.* W. B. Saunders Co., Philadelphia, 1974, Chap. 14.

Chapter 13

Salts

CHAPTER OUTLINE

Solubility of Salts

Reactions of Salts
Hydrolysis / Reaction with Metals / Reaction with Other Salts / Reaction with Acids and Bases

Types of Salts
Normal Salts / Acid Salts / Basic Salts

Uses of Salts

Buffer Solutions

Acids have one ion in common, the hydrogen ion, H^+. Bases also have a common ion, the hydroxide ion, OH^-. Consider the ionization of the following salts

$$NaCl \longrightarrow Na^+ + Cl^-$$

$$K_2SO_4 \longrightarrow 2K^+ + SO_4^{2-}$$

$$Mg(NO_3)_2 \longrightarrow Mg^{2+} + 2NO_3^-$$

Salts have no common ion. Salts in solution yield a positive ion (a metal ion) and a negative ion (a nonmetal ion or an acid radical).

Salts are formed by the reaction of an acid and a base, for example

	Acid	Base		Salt	Water
	HCl +	KOH	$\longrightarrow$	KCl	+ H_2O

$$H_2SO_4 + Mg(OH)_2 \longrightarrow MgSO_4 + 2H_2O$$

$$2HNO_3 + Zn(OH)_2 \longrightarrow Zn(NO_3)_2 + 2H_2O$$

Solubility of Salts

Some salts are quite soluble in water. Others are classified as slightly soluble or insoluble. Table 13–1 indicates the solubility of most common salts.

The following chart indicates the solubilities in water of various salts as predicted by using Table 13–1.

Name of Salt	Formula	Solubility
Silver chloride	$AgCl$	Insoluble
Sodium sulfate	Na_2SO_4	Soluble
Zinc nitrate	$Zn(NO_3)_2$	Soluble
Barium sulfate	$BaSO_4$	Insoluble
Calcium phosphate	$Ca_3(PO_4)_2$	Insoluble
Magnesium carbonate	$MgCO_3$	Insoluble

TABLE 13–1

Solubility of Common Salts

Soluble	Insoluble
Sodium salts	Carbonates (except sodium, potassium, ammonium)
Potassium salts	
Ammonium salts	Phosphates (except sodium, potassium, ammonium)
Acetates	
Nitrates	
Chlorides (except silver, lead, and mercury +1)	Sulfides (except sodium, potassium, ammonium)
Sulfates (except calcium, barium, and lead)	Hydroxides (except sodium, potassium, ammonium)

Hydrolysis

Hydrolysis is the reaction of a compound with the hydrogen ion or the hydroxide ion derived from water. Hydrolysis of the ions of a salt produces an acid and a base:

$$\underset{\text{salt}}{NH_4Cl} + H_2O \underset{\text{neutralization}}{\overset{\text{hydrolysis}}{\rightleftharpoons}} \underset{\text{base}}{NH_4OH} + \underset{\text{acid}}{HCl}$$

The double arrow indicates an equilibrium reaction, one that proceeds in both directions simultaneously. Note that the above hydrolysis reaction may be considered as the reverse of neutralization.

The products of the hydrolysis of a salt are an acid and a base. If the hydrolysis of a salt produces a strong acid and a weak base, the resulting solution will be acidic; if the products of hydrolysis are a strong base and a weak acid, the solution will be basic; if both acid and base are strong, or if both are weak, the solution will be neutral.

For simplicity we will consider the following acids and bases as strong (Table 13–2) and assume that all other commonly used acids and bases are weak.

For a discussion of strong and weak acids and bases see page 177.

Since salts are produced by the reaction of an acid with a base, salts must contain parts of each. The positive ion of a salt is derived from a base, and the negative ion of a salt is derived from an acid.

Let us consider the hydrolysis of a salt such as sodium cyanide, NaCN. We first write the formula of the salt in ionic form, sodium ion (Na^+) and cyanide ion (CN^-). Directly below these ions we write the ionized formula for water, with the OH^- ion (here written as HO^-) below the positive ion of the salt and the H^+ ion below the negative ion of the salt, or

$$
\begin{array}{c|c}
Na^+ & CN^- \\[2mm]
HO^- & H^+
\end{array}
$$

TABLE 13–2

Strong Acids and Bases in Common Use

	Strong Acids		Strong Bases
HCl	Hydrochloric acid	NaOH	Sodium hydroxide
HNO_3	Nitric acid	KOH	Potassium hydroxide
H_2SO_4	Sulfuric acid	$Ca(OH)_2$	Calcium hydroxide

Next, if we read downward on the left side of the dividing line, we see the base NaOH (one of the strong bases listed in Table 13–2). Reading upward on the right side of the dividing line we have the acid HCN. This must be a weak acid since it is not among the strong acids listed. Therefore, since the hydrolysis of NaCN yields a strong base (NaOH) and a weak acid (HCN), the resulting solution must be basic. The hydrolysis reaction may be written as:

$$NaCN + H_2O \rightleftharpoons NaOH + HCN$$

For the hydrolysis of magnesium sulfate, $MgSO_4$, since the charge on the magnesium ion is $+2$, we must use two H_2O's for hydrolysis, or

$$
\begin{array}{c|c}
Mg^{2+} & SO_4^{2-} \\
HO^- & H^+ \\
HO^- & H^+
\end{array}
$$

Reading downward on the left side we have the base $Mg(OH)_2$, which is a weak base (it is not among the strong ones listed), and reading upward on the right side we have the acid H_2SO_4, which is a strong acid (it is listed among the strong acids). Therefore, the hydrolysis of $MgSO_4$ yields a weak base and a strong acid so that the resulting solution is acidic. The hydrolysis reaction is:

$$MgSO_4 + 2H_2O \rightleftharpoons Mg(OH)_2 + H_2SO_4$$

Reaction with Metals

Some metals react with salt solutions to form another salt and a different metal. Refer to the activity series of metals (page 158) and recall that any metal can replace a metal ion below it in the activity series. Thus zinc can replace the copper from the copper sulfate solution because the zinc is higher in the activity series than the copper. This is an example of a single replacement reaction

$$Zn + CuSO_4 \longrightarrow ZnSO_4 + Cu$$

Reaction with Other Salts

Two different salts in solution may react by double displacement. Consider the following two reactions

$$K_2SO_4 + Ba(NO_3)_2 \longrightarrow 2KNO_3 + BaSO_4\downarrow$$

$$K_2SO_4 + Zn(NO_3)_2 \longrightarrow \text{no reaction}$$

Why does the first reaction proceed and the second one not? Look at the solubility rules (page 166). Note that one of the products formed in the first reaction, $BaSO_4$, is insoluble in water. Therefore the reaction will proceed. In the second reaction the products, if formed, would be $ZnSO_4$ and KNO_3. Both of these salts are soluble and ionized so that no reaction will take place. *In order for a reaction to take place between solutions of two salts, at least one of the products must be insoluble in water, a weak electrolyte, or a nonionized compound.*

Reaction with Acids and Bases

Salts react with acids or bases to form other salts and other acids and bases. These reactions will proceed if one of the products is insoluble in water or is an insoluble gas.

$$CaCO_3 + 2HCl \longrightarrow CaCl_2 + H_2O + CO_2\uparrow \quad \text{(a gas is formed)}$$

$$AlCl_3 + 3NaOH \longrightarrow 3NaCl + Al(OH)_3\downarrow \quad \text{(a precipitate is formed)}$$

Types of Salts

Normal Salts

Normal salts contain no hydrogen ions. The following compounds are normal salts: $NaCl$, K_2CO_3, $ZnSO_4$, $Al(NO_3)_3$.

Acid Salts

Acid salts contain at least one hydrogen ion. Sodium bicarbonate, $NaHCO_3$, and potassium bisulfate, $KHSO_4$, are examples of acid salts—they both contain a hydrogen ion. When carbonic acid, H_2CO_3, reacts with a base, one or both of the hydrogen ions may be replaced

$$H_2CO_3 + NaOH \longrightarrow NaHCO_3 + H_2O$$

$$H_2CO_3 + 2NaOH \longrightarrow Na_2CO_3 + 2H_2O$$

In the first reaction, an acid salt, $NaHCO_3$, is formed. In the second reaction, Na_2CO_3, a normal salt, is formed.

Any acid that has more than one replaceable hydrogen atom can form an acid salt. Other examples of acid salts are KH_2PO_4 and Na_2HPO_4.

Basic Salts

Similarly, when only one of the hydroxyl groups (OH) of a base is replaced by reaction with an acid, a basic salt is produced. A basic salt is a salt that contains an OH$^-$ group. When bismuth hydroxide, Bi(OH)$_3$, reacts with nitric acid, two different basic salts are possible.

$$Bi(OH)_3 + HNO_3 \longrightarrow Bi(OH)_2NO_3 + H_2O$$

$$Bi(OH)_3 + 2HNO_3 \longrightarrow Bi(OH)(NO_3)_2 + 2H_2O$$

Basic salts are not very common and have limited medical use.

Uses of Salts

Salts are necessary for the proper growth and metabolism of the body. Iron salts are necessary for the formation of hemoglobin; iodine salts for the proper functioning of the thyroid gland; calcium and phosphorus salts for the formation of bones and teeth; sodium and potassium salts regulate the acid-base balance of the body. Salts regulate the irritability of nerve and muscle cells. Salts regulate the beating of the heart. Salts maintain the proper osmotic pressure of the cells.

Many salts have specific uses. Barium sulfate, BaSO$_4$, is used for x-ray work. Even though barium compounds are poisonous to the body, barium sulfate is insoluble in body fluids and so has no effect on the body. Barium sulfate is opaque to x-rays and, when swallowed, it can be used to outline the gastrointestinal (GI) system for x-ray photographs (see Figure 13–1).

Table 13–3 lists some common salts and their specific uses in medicine.

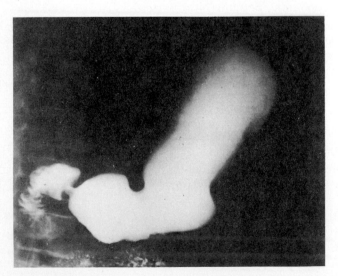

Figure 13–1. X-ray of stomach. (Courtesy of Michael Reese Hospital and Medical Center, Chicago, Ill.)

TABLE 13-3

171

Common Salts and Their Uses

Salts

Classification	Formula	Chemical Name	Common Name
Antacid			
	$CaCO_3$	Calcium carbonate	Precipitated chalk
	$NaHCO_3$	Sodium bicarbonate	Baking soda
Cathartic			
	Na_2SO_4	Sodium sulfate	Glaubers salt
	$MgSO_4$	Magnesium sulfate	Epsom salt
	$MgCO_3$	Magnesium carbonate	
	—	Magnesium citrate	Citrate of magnesia
	$KNaC_4H_4O_6 \cdot 4H_2O$	Potassium sodium tartrate	Rochelle salt
Diuretic			
	NH_4Cl	Ammonium chloride	Sal ammoniac
Expectorant			
	NH_4Cl	Ammonium chloride	Sal ammoniac
	KI	Potassium iodide	
Germicide			
	$AgNO_3$	Silver nitrate	Lunar caustic
Miscellaneous Uses			
X-ray work	$BaSO_4$	Barium sulfate	Barium
Caries reduction	NaF	Sodium fluoride	—
	SnF_2	Stannous fluoride	—
For casts	$(CaSO_4)_2 \cdot H_2O$	Calcium sulfate hydrate	Plaster of paris
Treatment of anemia	$FeSO_4$	Ferrous sulfate	—
Decrease of blood clotting time	$CaCl_2$	Calcium chloride	—
Physiologic saline solution used for irrigation and as IV replacement fluid	$NaCl$	Sodium chloride	Table salt
Thyroid treatment	KI	Potassium iodide	—
	NaI	Sodium iodide	—

Buffer Solutions

The pH of pure water (a neutral solution) is 7.0. If an acid is added to water, the pH goes down. How far below seven it goes depends on how much acid and how strong an acid is added. When a base is added to pure water, the pH rises above 7.0.

However when small amounts of acid or base are added to a buffer solution, the pH does not change. A buffer solution is defined as a solution that will not change in pH upon the addition of small amounts of either acid or base.

Buffer solutions, or buffers, are found in all body fluids and are responsible for maintaining the proper pH of those fluids. The normal pH range of the blood is 7.35–7.45. Even a slight change in pH can cause a very definite pathologic condition. When the pH falls below 7.35 the condition is known as acidosis. Alkalosis is the condition when the pH of the blood rises above 7.45.

What does a buffer solution consist of and how does it work? A blood buffer system consists of a weak acid and a salt of a weak acid. There are several buffer systems in the blood. One of these consists of carbonic acid, H_2CO_3, a weak acid, and sodium bicarbonate, $NaHCO_3$, the salt of a weak acid.

Suppose that an acid such as hydrochloric acid, HCl, enters the bloodstream. The HCl reacts with the $NaHCO_3$ part of the buffer according to the reaction

$$HCl + NaHCO_3 \longrightarrow NaCl + H_2CO_3$$

The NaCl produced is neutral; it does not hydrolyze (see page 167). The H_2CO_3 produced is part of the original buffer system and is only slightly ionized. In the body, acids (hydrogen ions) are produced by various metabolic processes. When these acids enter the bloodstream, they are removed by this reaction or a similar reaction with other buffers.

A base such as sodium hydroxide, NaOH, would react with the carbonic acid part of the buffer

$$NaOH + H_2CO_3 \longrightarrow H_2O + NaHCO_3$$

forming water, a harmless neutral normal metabolite, and sodium bicarbonate, part of the original buffer.

Thus when either acid or base is added to the buffer something neutral (NaCl or water) is formed plus more of the buffer system (H_2CO_3 or $NaHCO_3$). Thus the pH of the blood should not change.

In addition to the carbonate buffer system there are also phosphate buffers and several organic buffer systems. These will be discussed in Chapter 29.

If there is an overproduction of acid in the tissues and if these acids cannot be excreted rapidly enough, a condition known as acidosis results. Acidosis may occur in certain diseases, such as diabetes mellitus.

Prolonged vomiting may result in alkalosis because of the continued loss of the acid contents of the stomach.

Summary

Salts are formed by the reaction of an acid with a base. Salts yield ions other than hydrogen or hydroxide.

Some salts are soluble and others are insoluble in water. The solubility of most common salts can be determined by the solubility rules.

Hydrolysis is a double displacement reaction in which water is a reactant. Some salts hydrolyze to form a small amount of acid and base. If the hydrolysis produces a strong acid and a weak base, an acidic solution results. If the hydrolysis produces a strong base and a weak acid, a basic solution results.

Salts react with some metals to yield other salts and other metals. Salts may react with other salts by a double displacement reaction. Salts may react with acids or bases to form other salts and other acids or bases.

Normal salts contain no hydrogen ions. Acid salts contain at least one hydrogen ion. Basic salts contain at least one hydroxide ion.

Salts serve a definite purpose in the various metabolic processes of the body.

Buffer solutions do not change in pH upon the addition of small amounts of acid or base. Buffer solutions maintain the proper pH of the body fluids.

Questions and Problems

1. Define: (a) salt; (b) hydrolysis; (c) acid salt; (d) basic salt; (e) buffer solution.
2. Name five soluble salts.
3. Name ten insoluble salts.
4. Indicate whether the hydrolysis of the following salts will produce an acidic, a basic, or a neutral solution: (a) sodium carbonate; (b) magnesium chloride; (c) ammonium nitrate; (d) lead nitrate; (e) potassium sulfate.
5. Give an example of the reaction of a salt with a metal.
6. Give an example of the reaction of a salt with another salt.
7. Give an example of the reaction of a salt with an acid.
8. Give an example of the reaction of a salt with a base.
9. Name ten salts commonly used in the hospital and indicate what each is used for.
10. What is acidosis? alkalosis?
11. Explain how a buffer solution works.

References

Chisolm, J. J., Jr.: Lead poisoning. *Scientific American*, **224**:15–23 (Feb.), 1971.

Deevey, E. S., Jr.: The mineral cycle. *Scientific American*. **223**:148–58 (Sept.), 1970.

Hein, M., and Best, L. R.: *College Chemistry*. Dickenson Publishing Co., Inc., Encino, Calif., 1976, Chap. 15.

Hoyle, G.: How is muscle turned on and off? *Scientific American*, **222**:85–93 (Apr.), 1970.

Pauling, L., and Pauling, P.: *Chemistry*. W. H. Freeman & Co., San Francisco, 1975, Chap. 12.

Chapter 14

Ionization

Conductivity of Solutions

Figure 14–1 illustrates an apparatus designed to show whether a liquid will conduct electricity or not. When the electrodes are immersed in a liquid and the current is turned on, the bulb will light if the liquid conducts a current. If a liquid does not conduct a current the bulb will not glow at all.

It can be shown experimentally that the only water solutions which conduct electricity are those of acids, bases, and salts. The substances whose water solutions conduct electricity are called *electrolytes*. Pure water does not conduct electricity; neither does alcohol. These substances are classified as *nonelectrolytes*.

Effect of Electrolytes on
Boiling Points and Freezing Points

Electrolytes have an unusual effect on the boiling point and the freezing point of a solution. When a solid compound is dissolved in water, the

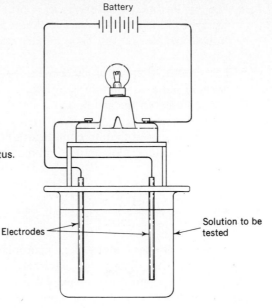

Figure 14–1. Conductivity apparatus.

resulting solution has a boiling point above 100°C. That is, solid solutes raise the boiling point of water. However, the results are considerably different if the solute is an electrolyte rather than a nonelectrolyte.

If equivalent amounts of an electrolyte such as sodium chloride and a non-electrolyte such as sugar are placed in equal volumes of water, the increase in the boiling point of the water containing the electrolyte sodium chloride will be approximately twice that for the water containing the nonelectrolyte sugar.

Likewise, the freezing point of the sodium chloride solution will be lowered approximately twice as much as that of the sugar solution.

Why do electrolytes conduct electricity? Why do they have an abnormal effect on the boiling point and freezing point of the solution?

Theory of Ionization

In 1887, a Swedish chemist, Svante Arrhenius, proposed a theory to explain the behavior of electrolytes in solution. The main points of his theory are

1. When electrolytes are placed in water, the molecules of the electrolyte break up into particles called ions. This process he called ionization.
2. Some of the ions have a positive charge, others a negative charge.
3. The sum of the positive charges is equal to the sum of the negative charges. That is, the original molecules were neutral so that the sum of the charges making up this molecule must also be neutral.

4. The conductance of electricity by solutions of electrolytes is due to the presence of ions.
5. Nonelectrolytes do not conduct electricity because of the absence of ions.
6. The abnormal effect of electrolytes on the boiling point and on the freezing point of solution is due to the increased number of particles (ions) present in the solution.

Arrhenius proposed that an equilibrium exists between the ions and the nonionized molecules. This equilibrium may be represented by the following equation

$$NaCl \rightleftharpoons Na^+ + Cl^-$$

| sodium chloride molecule | | sodium ion | chloride ion |

Arrhenius also proposed that the ionization should increase as the solution became more dilute. That is, in extremely dilute solutions, the electrolyte molecule would be almost completely dissociated (ionized).

Arrhenius' theory had to be modified as more information became available. Yet the modern theory of ionization retains most of his principles.

Arrhenius believed that the molecules of an electrolyte such as sodium chloride broke up into ions when placed in water. Later evidence proved conclusively that the sodium chloride existed as ions even in the solid state. When sodium chloride is formed from its elements, the sodium atom loses one electron to form a positively charged sodium ion. At the same time the chlorine atom gains one electron to form a negatively charged chloride ion. Thus, sodium chloride exists in an ionic state even in the solid state. When sodium chloride is dissolved in water, the ions are free to move around.

Solid sodium chloride consists of sodium ions and chloride ions bonded together by electrovalence in a definite crystalline structure. These ions are held tightly in place and cannot move about to any great extent. Thus they are not free to conduct an electrical current, and solid sodium chloride does not conduct electricity. If sodium chloride is heated until it melts (about 800°C), the resulting liquid will conduct electricity because the ions have some freedom to move around.

Conductivity of Solutions of Electrolytes

When sodium chloride crystals are placed in water, the ions present in the crystal are free to move about in the solution. These positively and negatively charged ions will be attracted to oppositely charged electrodes. Thus, if two electrodes are connected to a battery and are placed in a sodium chloride solution, the solution will conduct electricity.

The positively charged sodium ions will be attracted to the negative electrode, the cathode. Ions attracted to a cathode are called *cations*. At the same time, the negatively charged chloride ions will be attracted toward the positive electrode, the anode. Ions attracted toward an anode are called *anions*. This movement of ions through the solution consists of a flow of current. If a nonelectrolyte is placed in water, no ions are formed and no conductance takes place.

A simple method for demonstrating the migration of ions toward oppositely charged electrodes may be set up as follows:

Place a small amount of cupric dichromate solution ($CuCr_2O_7$) in a U tube. Carefully add a solution of sodium nitrate ($NaNO_3$) to both arms of the U tube, being careful not to mix the liquids. If this is done properly, there should be a colorless sodium nitrate solution above a dark solution of cupric dichromate in each arm of the U tube. Into each arm of the U tube place an electrode and connect these electrodes to a battery.

The U tube contains the following ions: sodium ion, Na^+, colorless; nitrate ion, NO_3^-, colorless; cupric ion, Cu^{2+}, blue; dichromate ion, $Cr_2O_7^{2-}$, orange.

When the current is turned on, the positively charged ions, the sodium ions and the cupric ions, will be attracted toward the negative electrode. Since the cupric ions are blue, their movement can be observed. At the same time, the migration of the orange-colored negatively charged ions, the dichromate ions, toward the positive electrode can also be observed.

Strong and Weak Electrolytes

Acids are electrolytes. Yet all acids do not behave the same when placed in water. A dilute solution of hydrochloric acid, HCl, is a strong electrolyte. When it is placed under a conductivity apparatus, the bulb glows brightly. However, when a dilute solution of acetic acid, $HC_2H_3O_2$, is placed under the conductivity apparatus, the bulb glows dimly. This indicates that the acetic acid is a weak electrolyte. Likewise, sodium hydroxide, NaOH, solution is a strong electrolyte, whereas ammonium hydroxide, NH_4OH, is a weak electrolyte.

How can one acid or base be a strong electrolyte and another one be weak? What accounts for the difference between the types of electrolytes?

When hydrochloric acid or sodium hydroxide is placed in water, it dissociates (breaks up) almost completely into ions, as indicated by the arrows pointing in one direction only.

$$HCl \longrightarrow H^+ + Cl^-$$

$$NaOH \longrightarrow Na^+ + OH^-$$

Because these substances are just about completely ionized, they are called *strong electrolytes*.

However, some acids and bases when placed in water ionize only to a limited extent. They remain primarily as nonionized molecules. These substances maintain an equilibrium between the nonionized molecule and the ions with the equilibrium being far to the left, as indicated by the arrows.

$$HC_2H_3O_2 \rightleftharpoons H^+ + C_2H_3O_2^-$$

$$NH_4OH \rightleftharpoons NH_4^+ + OH^-$$

These substances are called *weak electrolytes*.

Thus the original Arrhenius theory is true for weak electrolytes but not for strong ones.

Other Evidences of Ionization

It has already been mentioned that the conductivity of solutions of electrolytes indicates the presence of ions. Likewise, the abnormal effect of electrolytes on the boiling point and the freezing point can be explained by the presence of ions. The change in the boiling point or freezing point depends on the number of particles present in solution. Sugar, $C_6H_{12}O_6$, contributes only one particle per molecule because it is not ionized. Sodium chloride, NaCl, contributes two particles—the two ions—and so should have twice the effect on the boiling point and the freezing point.

Another factor indicating the presence of ions is the instantaneous re-action of solutions of electrolytes. When a solution of sodium chloride, NaCl, is mixed with a solution of silver nitrate, $AgNO_3$, a white precipitate of silver chloride, AgCl, is formed instantaneously.

$$Na^+ + Cl^- + Ag^+ + NO_3^- \longrightarrow AgCl\downarrow + Na^+ + NO_3^-$$

Note that the silver chloride is written as a molecule rather than as ions. This is because that substance precipitates, thus removing the ions from the solution.

When an acid reacts as a base, a salt and water are formed. If potassium hydroxide, KOH, is reacted with nitric acid, HNO_3, potassium nitrate and water are formed. The ionic reaction is as follows

$$K^+ + OH^- + H^+ + NO_3^- \longrightarrow K^+ + NO_3^- + H_2O$$

Potassium nitrate is a soluble salt, an electrolyte, which is ionized. Water is a nonelectrolyte; it is not ionized and is therefore written as a molecule.

Hydrolysis of salts (see page 167) may also be explained on the basis of ionization. When sodium acetate, $NaC_2H_3O_2$, is placed in water the acetate ion reacts with the water according to the following equation

$$Na^+ + \underset{\substack{\text{acetate} \\ \text{ion}}}{C_2H_3O_2^-} + H_2O \rightleftharpoons \underset{\substack{\text{acetic} \\ \text{acid}}}{HC_2H_3O_2} + Na^+ + OH^-$$

The acetic acid, being a weak acid, does not ionize appreciably to furnish any hydrogen ions and so is written as a molecule in the above reaction. If the ions common to both sides of the above equation (the sodium ions) are eliminated the net reaction becomes

$$C_2H_3O_2^- + H_2O \longrightarrow HC_2H_3O_2 + OH^-$$

That is, the acetate ion (which came from a weak acid) hydrolyzes (reacts with water) to form nonionized acetic acid, a weak electrolyte, leaving hydroxide ions in solution. Therefore, when sodium acetate is placed in water, hydroxide ions are formed causing the solution to be basic. Note that ions from a strong acid or a strong base do not hydrolyze. Thus, the sodium ion above does not react with water and so can be eliminated from both sides of the equation.

Importance of Ions in Body Chemistry

Ions play the chief role in the various processes that take place in the body. Many of the body's vital processes take place in the ionic state within the cell. The following list indicates some of the more important ions that are found in the body.

Calcium ion	Ca^{2+}	Necessary for clotting of the blood; for formation of milk curd during digestion in the stomach; for formation of bones and teeth.
Iron ion	Fe^{2+}	Necessary for formation of hemoglobin.
Sodium ion	Na^+	Present in body fluids.
Potassium ion	K^+	Present inside body cells.
Chloride ion	Cl^-	Regulates acidity in the gastric juices.
Bicarbonate ion	HCO_3^-	Helps regulate pH of the blood.
Fluoride ion	F^-	Helps prevent tooth decay.
Iodide ion	I^-	Present in thyroid hormones.
Ammonium ion	NH_4^+	Plays a major role in maintaining body's acid-base balance.
Phosphate ion	PO_4^{3-}	Plays an important role, along with calcium ions, in the formation of bones and teeth.
Magnesium ion	Mg^{2+}	An important activator for many enzyme systems. (See Table 2–1, page 19, for other ions necessary for enzymes.)

In addition, ions are necessary as part of the blood buffer system. They cause osmotic pressure in the cells and are necessary to control the contracting and relaxing of muscles. Ions are necessary to carry nerve impulses and help regulate the digestive processes (see pages 427, 433).

Summary

Substances whose water solutions conduct electricity are called electrolytes. Soluble acids, bases, and salts are electrolytes. Solutions that do not conduct electricity are nonelectrolytes.

Electrolytes have an abnormal effect on the boiling and freezing point of a solution. This abnormal effect is due to the presence of ions.

Ionization (the formation of ions) accounts for the electrical conductivity of a solution. When a substance ionizes, the sum of the positive charges equals the sum of the negative charges. The positive ions, cations, are attracted toward the cathode (negative electrode) while the negative ions, the anions, are attracted toward the anode (positive electrode).

The modern theory of ionization considers that ions are already present in a crystalline salt and that these ions are released to move about when that substance is placed in solution.

Salts are strong electrolytes because they are completely (strongly) ionized. Acids and bases that are strong electrolytes are highly ionized. Acids and bases that are poor electrolytes are weakly ionized.

The presence of ions is of great importance in the body in the maintenance of the electrolyte balance of the body fluids.

Questions and Problems

1. Describe a laboratory experiment to determine whether a substance is an electrolyte or not.
2. What effect does an electrolyte have on the boiling point of a solution? Why?
3. What effect does an electrolyte have on the freezing point of a solution? Why?
4. What effect does a nonelectrolyte have on the freezing point of a solution? Why?
5. State the main ideas of Arrhenius' theory of ionization.
6. How does the modern theory of ionization differ from that of Arrhenius?
7. How does an electrolyte conduct a current?
8. What is an anion? a cation?
9. What is a strong electrolyte? Give an example.
10. What is a weak electrolyte? Give an example.
11. How may hydrolysis be explained in terms of weak electrolytes?
12. List several ions necessary for the proper functioning of the body. What purpose does each ion serve?

References

Baum, S. J., and Scaife, C. W.: *Chemistry: A Life Science Approach.* Macmillan Publishing Co., Inc., New York, 1975, Chap. 10.

Hein, M.: *Foundations of College Chemistry*, 3rd ed. Dickenson Publishing Co., Inc., Encino, Calif., 1973, Chap. 15.

Longo, F. R.: *General Chemistry.* McGraw-Hill Book Co., New York, 1974, Chaps. 13, 14.

Pauling, L., and Pauling, P.: *Chemistry.* W. H. Freeman & Co., San Francisco, 1975, Chap. 6.

Peters, E. I.: *Introduction to Chemical Principles.* W. B. Saunders Co., Philadelphia, 1974, Chap. 12.

unit two

Organic Chemistry

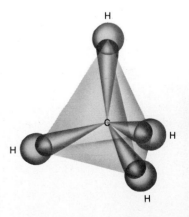

Chapter 15

Introduction to Organic Chemistry

CHAPTER OUTLINE

Importance of Organic Chemistry

Comparison of Organic and Inorganic Compounds

Bonding

Structural Formulas

Isomers

Three-Dimensional Arrangement of the Bonds in the Carbon Atom

Bonding Ability of Carbon

Carbon Compounds with Other Elements

In the eighteenth century it was believed that a "vital force" was needed to make the compounds produced by living cells. However, this belief was overthrown by a German chemist, Fredrick Wöhler, in 1828. He prepared urea, a compound normally found in the blood and urine, from simple inorganic reagents.

After Wöhler's work many other organic compounds were produced in the laboratory. This led to the subdivision of chemistry into two parts — inorganic and organic. *Organic chemistry* was defined as being the chemistry of carbon compounds. Why make one category for the element carbon and have all of the other elements in the other category? The answer is that although there are tens of thousands of inorganic compounds known today, *millions* of organic compounds are known.

183

Importance of Organic Chemistry

Organic chemistry is important in that it is the chemistry associated with all living matter in both plants and animals. Carbohydrates, fats, proteins, vitamins, hormones, enzymes, and many drugs are organic compounds. Wool, silk, cotton, linen, and such synthetic fibers as nylon, rayon, and dacron contain organic compounds. So do perfumes, dyes, flavors, soaps, detergents, plastics, gasolines, and oils.

Comparison of Organic and Inorganic Compounds

Organic compounds differ from inorganic compounds in many ways. The most important of these are listed below.

1. Most organic compounds are combustible. Most inorganic compounds are not combustible.
2. Organic compounds generally react much more slowly than do inorganic compounds because most organic compounds are nonelectrolytes.
3. Most organic compounds have a low melting point. Most inorganic compounds have a high melting point.
4. Most organic compounds are insoluble in water. Many inorganic compounds are soluble in water.
5. Organic reactions usually take place between molecules. Inorganic reactions usually take place between ions.
6. Organic compounds generally contain many atoms. Inorganic compounds contain relatively few atoms.
7. Organic compounds have a complex structure. Inorganic compounds have a simpler structure.

Bonding

Organic compounds—compounds of carbon—are held together by covalent bonds. Recall that covalent bonds are formed by shared electrons. In organic chemistry the term *bond* is used to designate a shared pair of electrons. Thus, the statement is made that carbon forms four bonds; it has an oxidation number of -4. Bonds are usually represented by a short, straight line connecting the atoms. The four bonds of the carbon atom may be represented as follows

$$-\overset{|}{\underset{|}{C}}-$$

The bonds are symmetrically arranged here. However, they need not be symmetrically arranged. They can also be arranged as follows

$$-\overset{|}{C}= \quad \text{or} \quad -C\equiv \quad \text{or} \quad C\equiv$$

Note that in each example the carbon atom has four bonds connected to it.

Likewise, since the oxygen atom has an oxidation number of -2 it must have two bonds attached to it. These bonds may be separated or placed together as follows

$$-O- \quad \text{or} \quad O=$$

The hydrogen atom with an oxidation number of $+1$ must have only one bond.

$$H-$$

The halogens, all with an oxidation number of -1, have only one bond.

$$F-, \ Cl-, \ Br-, \ I-$$

Structural Formulas

As will be seen later, organic compounds are often written using a structural rather than a molecular formula. What is the difference between a structural formula and a molecular formula? And why use the former and not the latter?

Consider an organic compound with a molecular formula C_2H_6O. In inorganic chemistry a formula of this type designates a specific compound. However, in organic chemistry this is not always true; the same formula may designate more than one compound. Let us see how this is possible.

If the carbons, hydrogens, and oxygens in C_2H_6O are arranged in such a manner that each carbon atom has four bonds attached to it, each hydrogen atom has one bond, and the oxygen has two bonds, there will be two possible structures, both having the formula C_2H_6O.

These two structural formulas are

$$
\begin{array}{ccc}
\text{H} & \text{H} & \\
| & | & \\
\text{H}-\text{C}-\text{C}-\text{O}-\text{H} & \quad \text{and} \quad & \text{H}-\overset{\text{H}}{\underset{\text{H}}{\text{C}}}-\text{O}-\overset{\text{H}}{\underset{\text{H}}{\text{C}}}-\text{H} \\
| & | & \\
\text{H} & \text{H} &
\end{array}
$$

Note that each compound contains two carbons, six hydrogens, and one oxygen. Note also that each satisfies all of the bond requirements. These two compounds have different structures, different properties, and actually represent different compounds. The first compound is called ethyl alcohol, the second dimethyl ether. This difference in the structure of compounds having the same molecular formula illustrates the importance of using structural rather than molecular formulas for organic compounds.

In the previous diagrams the lines represent bonds or shared electrons. If the same structures were written using electron dots, they would be

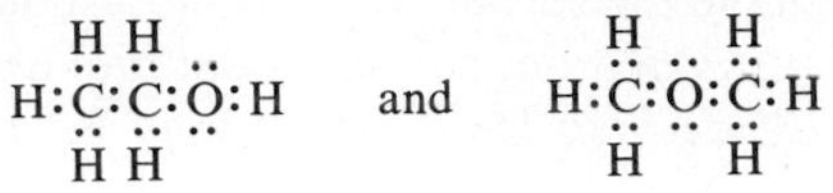

$$\begin{array}{cc} \text{H H} & \text{H} \quad \text{H} \\ \text{H:C:C:O:H} & \quad \text{and} \quad \text{H:C:O:C:H} \\ \text{H H} & \text{H} \quad \text{H} \end{array}$$

Note that it is simpler to write a structural formula using the "bond" notation.

Isomers

Isomers are defined as compounds having the same molecular formula but different structural formula. Thus, the compound C_2H_6O has two isomers. The compound C_6H_{14} has five isomers, as illustrated in Figure 15–1.

$C_6H_{12}O_6$ is usually called sugar. Actually this molecular formula represents 16 different compounds or isomers, each of which is a different sugar. (see page 265).

Figure 15–1. Isomers of C_6H_{14}.

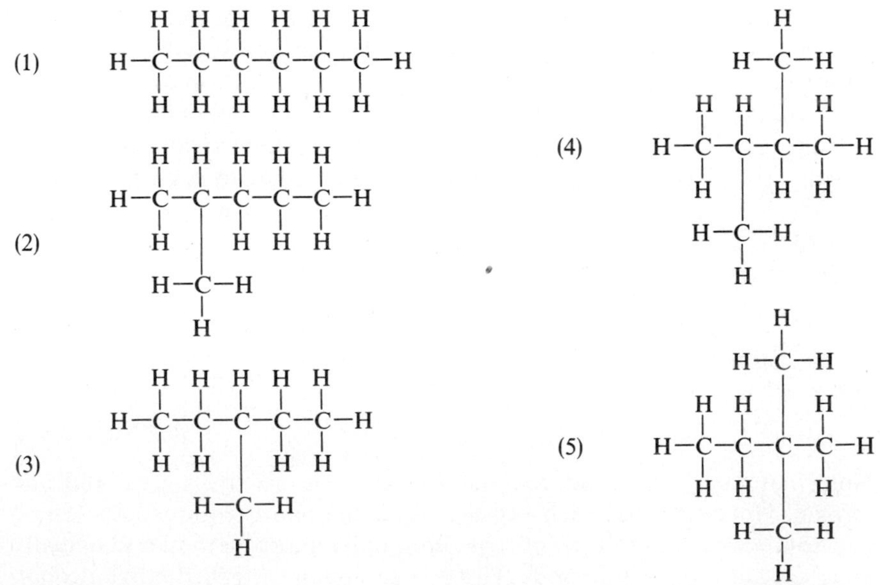

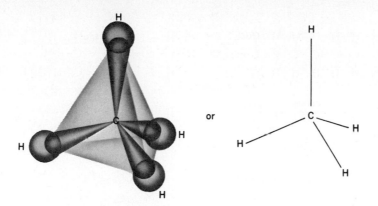

Figure 15–2. Tetrahedral structure of methane, CH_4.

Three-Dimensional Arrangement of the Bonds in the Carbon Atom

Care should be taken to realize that the compounds represented by structural formulas are three dimensional and not planar.

For example, each carbon atom has four bonds attached to it. If these bonds are symmetrically arranged then the structure, in a planar representation, is as follows

$$-\overset{|}{\underset{|}{C}}-$$

What does it actually look like in three dimensions? The four bonds of the carbon atom are arranged in a tetrahedral shape. The carbon atom lies at the center of the tetrahedron; the angle between the bonds is 109.5°. Therefore the compound CH_4, methane, may be represented as shown in Figure 15–2.

Bonding Ability of Carbon

What is so unique about the element carbon that it forms so many compounds? The other elements form relatively few compounds. The answer is that the carbon atom has the ability to bond other carbon atoms to itself to form very large and complex molecules. Carbon atoms may join together to form continuous or branched chains of carbon atoms. Compounds of this type are called aliphatic or chain compounds (Figure 15–3*a*). Carbon compounds may also bond together in the shape of rings to form ring or cyclic compounds (Figure 15–3*b*). A third type of organic compound is the heterocyclic compounds, which also have a ring structure. However, this ring

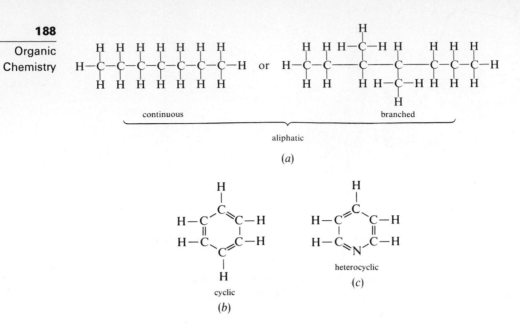

Figure 15–3 region:

continuous ⎫ branched

aliphatic

(a)

cyclic

(b)

heterocyclic

(c)

Figure 15–3. Types of carbon compounds.

structure contains some element other than carbon in the ring (Figure 15–3c). In this particular case the other element is nitrogen, which has three bonds attached to it.

Carbon Compounds with Other Elements

Carbon also forms compounds with other elements. If a chlorine atom is bonded to a carbon atom, the compound chloromethane or methyl chloride is formed.

Again note how much simpler it is to write the structure as above, rather than as the electron-dot structure.

Also a carbon atom may combine with chlorine and hydrogen atoms to form other compounds of similar structures.

$$\underset{\text{dichloromethane}}{H-\overset{\displaystyle Cl}{\underset{\displaystyle H}{C}}-Cl} \qquad \underset{\substack{\text{trichloromethane} \\ \text{(chloroform)}[1]}}{H-\overset{\displaystyle Cl}{\underset{\displaystyle Cl}{C}}-Cl} \qquad \underset{\substack{\text{tetrachloromethane} \\ \text{(carbon tetrachloride)}}}{Cl-\overset{\displaystyle Cl}{\underset{\displaystyle Cl}{C}}-Cl}$$

Similar compounds can be formed with bromine and iodine. $CHCl_3$ is called chloroform. Likewise, $CHBr_3$ is called bromoform, and CHI_3 is called iodoform.

$$\underset{\text{chloroform}}{H-\overset{\displaystyle Cl}{\underset{\displaystyle Cl}{C}}-Cl} \qquad \underset{\text{bromoform}}{H-\overset{\displaystyle Br}{\underset{\displaystyle Br}{C}}-Br} \qquad \underset{\text{iodoform}}{H-\overset{\displaystyle I}{\underset{\displaystyle I}{C}}-I}$$

Summary

Organic chemistry is defined as the chemistry of carbon compounds. Most organic compounds differ from inorganic compounds as follows: they are combustible; they react more slowly; they have lower melting points; they are insoluble in water; reaction takes place between molecules rather than between ions; the molecules contain many atoms; the molecules have a complex structure.

In organic chemistry the term *bond* is used rather than oxidation number. A bond is indicated by a short line and represents a pair of shared electrons. The carbon atom always has four bonds associated with it; the oxygen atom, two; the hydrogen atom, one.

Structural formulas are used for organic compounds rather than molecular formulas because the same molecular formula may represent more than one structural formula.

Isomers are compounds having the same molecular formula but different structural formulas.

The four bonds of the carbon atom are arranged in three dimensions in a tetrahedral structure with the carbon atom at the center and each bond pointing to a corner.

Organic compounds may be divided into three categories: aliphatic, or chain, compounds; ring, or cyclic, compounds; and heterocyclic compounds. Heterocyclic compounds contain elements other than carbon in the ring.

Carbon also forms compounds with other elements.

Questions and Problems

1. What was Wöhler's best-known contribution to organic chemistry?
2. Compare the properties of organic and inorganic compounds.
3. What is a "bond" in an organic compound?

[1] Because it is carcinogenic in rats and mice, the Food and Drug Administration has banned the use of chloroform in drug, cosmetic, and food packaging products.

4. How are the bonds arranged around the carbon atom in methane?
5. Why are structural formulas rather than molecular formulas used for organic compounds?
6. What are isomers? Do all organic compounds have isomers?
7. The compound C_5H_{12} has three isomers. Draw their structures.
8. What are the three major types of organic compounds?
9. Draw the structure of chloromethane; of chloroform.
10. Draw the structure of carbon tetrachloride; dichloromethane; iodoform; bromoform.

References

Baum, S. J., and Scaife, C. W.: *Chemistry: A Life Science Approach.* Macmillan Publishing Co., Inc., New York, 1975, Chap. 12.

Brown, R. F.: *Organic Chemistry.* Wadsworth Publishing Co., Inc., Belmont, Calif., 1975, Chap. 1.

Gutsche, D. D., and Pasto, D. C.: *Fundamentals of Organic Chemistry.* Prentice-Hall, Inc., Englewood Cliffs, N.J., 1975, Chap. 1.

Hein, M., and Best, L. R.: *College Chemistry.* Dickenson Publishing Co., Inc., Encino, Calif, 1976, Chap. 22.

Morrison, R. T., and Boyd, R. N.: *Organic Chemistry*, 3rd ed. Allyn & Bacon, Inc., Boston, 1973, Chap. 1.

Stacy, G. W.: *Organic Chemistry.* Harper & Row, New York, 1975, Chap. 1.

Hydrocarbons

Alkanes

As the name implies, *hydrocarbons* are compounds that contain carbon and hydrogen only.

Consider the hydrocarbon of only one carbon atom. Since a carbon atom must have four bonds, four hydrogen atoms may be attached to that carbon atom. The hydrocarbon thus formed is called *methane* and has the structural formula shown in Figure 16–1.

If two carbon atoms are bonded together, six hydrogen atoms may be joined to them. This hydrocarbon is called *ethane*. The molecular formula is C_2H_6, and the structure is shown in Figure 16–2.

The hydrocarbon of three carbon atoms needs eight hydrogens to satisfy all the bonds. This compound, C_3H_8, is called *propane* (see Figure 16–3).

These compounds are called *alkanes*, and they are said to be *saturated*; that is, they have single covalent bonds between carbon atoms.

Table 16–1 lists several hydrocarbons. Note that the names of all alkanes end in *-ane*. The names of the first four compounds have to be memorized;

H—C—H   or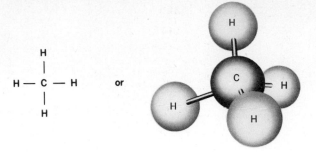

Figure 16–1. Methane.

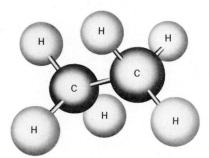

or

Figure 16–2. Ethane.

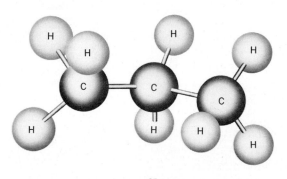

Figure 16–3. Propane.

or

192

TABLE 16–1

Alkanes

No. of Carbon Atoms	Name	Molecular Formula	Structural Formula	Condensed Structural Formula
1	Methane	CH_4	H | H—C—H | H	CH_4
2	Ethane	C_2H_6	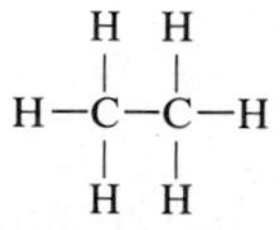 H H | | H—C—C—H | | H H	$CH_3—CH_3$
3	Propane	C_3H_8	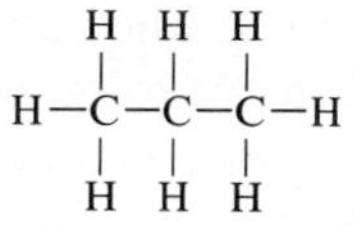 H H H | | | H—C—C—C—H | | | H H H	$CH_3—CH_2—CH_3$

4	Butane	C_4H_{10}	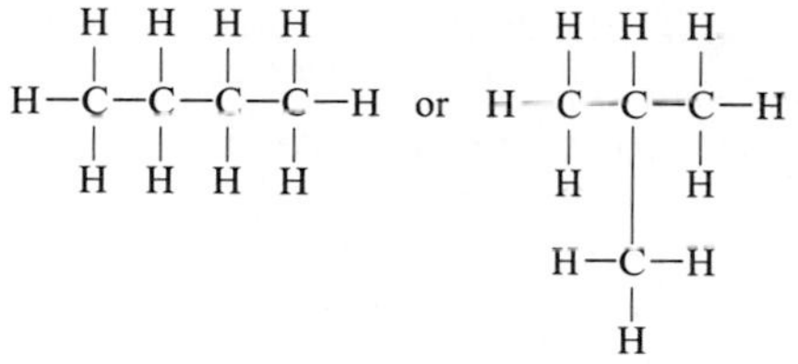 H H H H H H H
			| | | | | | |
			H—C—C—C—C—H or H—C—C—C—H
			| | | | | | |
			H H H H H H
			                H—C—H
			                  |
			                  H
				$CH_3—CH_2—CH_2—CH_3$
				or
				$CH_3—CH—CH_3$
				         |
				       CH_3

5	Pentane	C_5H_{12}	3 isomers*	
6	Hexane	C_6H_{14}	5 isomers	
7	Heptane	C_7H_{16}	9 isomers	
8	Octane	C_8H_{18}	18 isomers	
9	Nonane	C_9H_{20}	35 isomers	
10	Decane	$C_{10}H_{22}$	75 isomers	

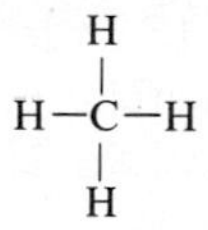

* Note that the number of isomers increases rapidly as the compound contains more carbon atoms. $C_{40}H_{82}$ has 62,491,178,805,831 possible isomers.

they were named before a system of nomenclature was devised. Beginning with the hydrocarbon containing five carbon atoms, however, the names follow a definite pattern. The name *pentane* was derived from the prefix *penta-*, meaning five; pentane contains five carbon atoms. The name *hexane* is derived from the prefix *hexa-*, which means six; and so on through the series.

Structural formulas may be condensed, as indicated in Table 16 1.

General Formula

The general formula for alkanes is

$$C_nH_{2n+2}$$

where n represents the number of carbon atoms. That is, if a compound contains n carbon atoms the number of hydrogen atoms is twice n plus two more.

Butane contains four carbon atoms. The number of hydrogens should be $(2 \times 4) + 2$ or 10, which agrees with the molecular formula for butane listed in Table 16–1. Likewise, octane has eight carbon atoms and so should have $(2 \times 8) + 2$ or 18 hydrogen atoms, which it does. .

Continuing with a larger number of carbon atoms, the alkane containing 16 carbon atoms would have $(2 \times 16) + 2$ or 34 hydrogen atoms, giving it the formula $C_{16}H_{34}$.

Radicals

When a hydrogen atom is removed from a hydrocarbon, an *organic radical* is formed. The names of the radicals are obtained by changing the ending of the name from *-ane* to *-yl*. The radical of one carbon atom formed from the hydrocarbon methane is called the methyl radical.

The radical of two carbon atoms formed from ethane is called the ethyl radical. The radical of three carbon atoms formed from propane is called the propyl radical (see Table 16–2).

The compound CH_3Cl, or

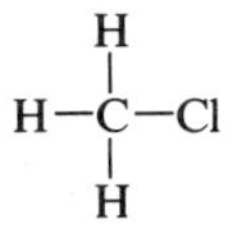

$$H-\overset{\overset{\displaystyle H}{|}}{\underset{\underset{\displaystyle H}{|}}{C}}-Cl$$

is made up of a methyl radical (CH_3-) attached to a chlorine atom. It is

TABLE 16–2

Radicals of Simple Alkanes

Name of Radical	Condensed Structural Formula
Methyl	CH_3-
Ethyl	CH_3-CH_2- or C_2H_5-
Propyl	$CH_3-CH_2-CH_2-$ or C_3H_7-
Butyl	$CH_3-CH_2-CH_2-CH_2-$ or C_4H_9-

called methyl chloride. The compound C_2H_5I consists of an ethyl radical
(C_2H_5-) bonded to an iodine. It is called ethyl iodide.

$$H-\overset{\displaystyle H}{\underset{\displaystyle H}{\overset{|}{\underset{|}{C}}}}-\overset{\displaystyle H}{\underset{\displaystyle H}{\overset{|}{\underset{|}{C}}}}-I$$

Naming Hydrocarbon Isomers

An international system of nomenclature for organic compounds has been devised and is recognized and used by chemists all over the world. This system was devised and approved by the International Union of Pure and Applied Chemistry and is frequently designated by the initials IUPAC. The rules of the IUPAC system are:

1. Pick out the longest continuous chain of carbon atoms.
2. Identify that chain as an alkane.
3. Pick out the radicals attached to that chain.
4. Number the carbons in the chain, starting at whichever end of the chain that will give the smallest number to the carbon to which the radical is attached. Continue the numbering of this carbon chain in the same direction from one end to the other.
5. List the numbers and the names of the radicals.

Identify the following compound

$$CH_3-\underset{\displaystyle CH_3}{\overset{\displaystyle |}{CH}}-CH_2-CH_2-CH_2-CH_3$$

The longest chain contains six carbon atoms; therefore, this compound is some kind of a *hexane*. To identify the radical attached to the chain and also to identify the chain itself, it is sometimes easier to draw a box around the chain. Then whatever is attached to the chain, the radicals, will be outside the box and can be easily noticed. Thus the compound can be written with the six-carbon chain inside the box.

$$\boxed{CH_3-\overset{\displaystyle |}{CH}-CH_2-CH_2-CH_2-CH_3}$$
$$CH_3$$

Attached to the chain (sticking out of the box) is a radical of one carbon atom, the CH_3- or methyl radical. Thus this compound is a methylhexane.

The next step calls for numbering the carbons in the chain. They can be numbered in either direction, from left to right or from right to left. It should be observed that the methyl radical is on the second carbon atom from the left end or on the fifth carbon atom from the right end. The rule states that the numbering should be such that the carbon to which the radical is attached has the smallest number. Therefore the correct name of this compound is 2-methylhexane, which indicates that there is a methyl radical on the second carbon from the end in a chain consisting of six carbon atoms. If the methyl radical had been pointing upward instead of down, the compound would still be the same and so would the name.

Identify the following compound

$$CH_3-CH_2-\overset{\displaystyle CH_3}{\underset{\displaystyle CH_3}{\overset{|}{\underset{|}{CH}}}-\overset{}{CH}}-CH_3$$

The longest chain contains five carbon atoms. This compound is some kind of a *pentane*. There are two methyl radicals attached to the chain, one on the second carbon and one on the third carbon. This time the numbering is from right to left in order to obtain the lowest numbers for the radicals.

$$\overset{5}{CH_3}-\overset{4}{CH_2}-\overset{3}{\underset{\underset{\displaystyle CH_3}{|}}{CH}}-\overset{\overset{\displaystyle CH_3}{|}}{\overset{2}{CH}}-\overset{1}{CH_3}$$

This compound is called 2,3-dimethylpentane, where the prefix *di-* indicates that there are two identical radicals. Dimethyl means two methyl radicals and the numbers tell us on which carbon atoms they are located.

Whenever a radical appears more than once in a compound, a prefix is used to designate how many of these radicals are present in that compound. The most commonly used prefixes are:

di- which means two
tri- which means three
tetra- which means four
penta- which means five

Identify the following compound

$$CH_3-\overset{}{\underset{\underset{\displaystyle CH_3}{|}}{CH}}-\overset{\overset{\displaystyle CH_3}{|}}{CH}-CH_2-\overset{}{\underset{\underset{\displaystyle C_2H_5}{|}}{CH}}-\overset{\overset{\displaystyle CH_3}{|}}{CH}-CH_2-CH_3$$

We first draw a box around the longest chain. This chain contains eight carbon atoms so the compound is some kind of octane.

$$CH_3-\overset{\displaystyle CH_3}{\underset{\displaystyle CH_3}{CH}}-\overset{\displaystyle CH_3}{CH}-CH_2-\overset{\displaystyle CH_3}{CH}-\underset{\displaystyle C_2H_5}{CH}-CH_2-CH_3$$

Attached to the chain are three methyl radicals (radicals of one carbon atom, or CH_3- radicals) and one ethyl radical (a radical of two carbon atoms, or a C_2H_5- radical).

The chain should be numbered from left to right in order to obtain radicals of the lowest numbers. There are methyl radicals on carbons numbered 2, 3, and 6 and an ethyl radical on carbon number 5 of the eight carbon chain.

The correct name of this compound is 2,3,6-trimethyl-5-ethyloctane. Note that the methyl radicals are named first because they occur first in the chain. Also note that the prefix *tri-* is used to indicate three radicals of the same type.

Identify the following compound

$$CH_3-CH-\overset{\displaystyle CH_3}{\underset{\displaystyle \underset{\displaystyle CH_3}{CH_2}}{CH}}-CH_2-\overset{\displaystyle}{\underset{\displaystyle CH_3}{CH}}-CH_3$$

First draw a box around the longest chain. Note that this chain contains seven carbon atoms so that this compound is a heptane.

$$CH_3-\overset{\displaystyle CH_3}{\underset{\displaystyle \underset{\displaystyle 7CH_3}{6CH_2}}{\underset{5}{CH}}}-\underset{4}{CH}-\underset{3}{CH_2}-\overset{\displaystyle}{\underset{\displaystyle CH_3}{\underset{2}{CH}}}-\underset{1}{CH_3}$$

Attached to the chain are three methyl groups at carbons numbered 2, 4, and 5. Thus the name of this compound is 2,4,5-trimethylheptane. Recall rule number 1 (page 195) which says "pick out the longest continuous chain of carbon atoms. If you had picked a chain of six carbon atoms (straight across) this would not have been the longest continuous chain of carbon atoms.

Alkenes

Alkanes have a single bond between the carbon atoms. Alkenes have a double bond (two bonds) between two of the carbon atoms.

Consider two carbon atoms connected by a double bond or $C=C$. Attached to these two carbon atoms there can be placed a total of only four hydrogen atoms in order to satisfy all of the bonds. Recall that a single bond represents a pair of shared electrons; a double bond represents two pairs of shared electrons.

This compound thus becomes

$$\begin{matrix} & H & H & \\ & | & | & \\ H- & C & = C & -H \end{matrix} \qquad \left(\begin{matrix} & H & H \\ H: & \overset{..}{C}::\overset{..}{C} & :H \end{matrix} \right)$$

and has the molecular formula of C_2H_4. It is called *ethene*.

When three carbon atoms are arranged in a chain with a double bond between two of the carbon atoms, $C=C-C$, how many hydrogen atoms must be connected to these carbon atoms in order to satisfy all of the bond requirements? The answer is six, so that the structure becomes

$$\begin{matrix} & H & H & H & \\ & | & | & | & \\ H- & C & = C & - C & -H \\ & & & | & \\ & & & H & \end{matrix}$$

where the molecular formula is C_3H_6. The name of this compound is *propene*.

Note that the names of these compounds end in *-ene*. This is true of all alkenes. It should be noted also that the names of these compounds are similar to the alkanes except for the ending, which is *-ene* instead of *-ane*.

Compare the structures of ethane (C_2H_6) and ethene (C_2H_4).

$$\begin{matrix} & H & H & & & & H & H & \\ & | & | & & & & | & | & \\ H- & C & - C & -H & \qquad & H- & C & = C & -H \\ & | & | & & & & & & \\ & H & H & & & & & & \end{matrix}$$

ethane ethene

The three-carbon alkane is propane, and the three-carbon alkene is propene. Likewise the four-carbon alkane is called butane while the corresponding alkene is called butene.

The names and formulas of some of the alkenes are listed in Table 16–3. The general formula for alkenes is

$$C_nH_{2n}$$

There are twice as many hydrogen atoms as carbon atoms in every alkene. Thus, octene has 8 carbon atoms and 16 hydrogen atoms, and the formula of an alkene of 15 carbon atoms would be $C_{15}H_{30}$.

TABLE 16-3

Some Simple Alkenes

No. of Carbon Atoms	Name	Molecular Formula	Structural Formula	Condensed Structural Formula
2	Ethene	C_2H_4	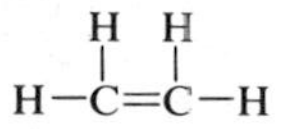 H H | | H—C=C—H	$CH_2{=}CH_2$
3	Propene	C_3H_6	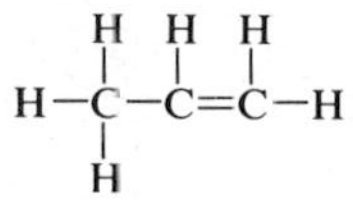 H H H | | | H—C—C=C—H | H	$CH_3{-}CH{=}CH_2$
4	Butene	C_4H_8	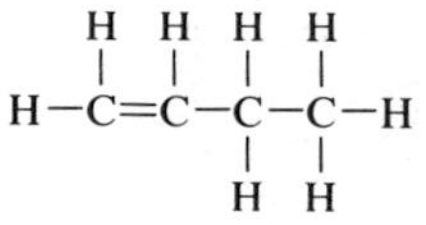 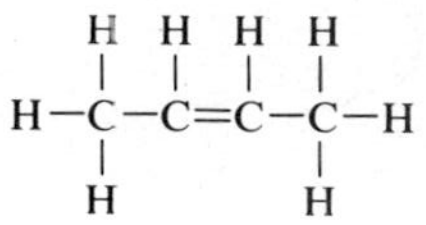 H H H H | | | | H—C=C—C—C—H _or_ H—C—C=C—C—H | | | | H H H H	$CH_2{=}CH{-}CH_2{-}CH_3$ _or_ $CH_3{-}CH{=}CH{-}CH_3$
5	Pentene	C_5H_{10}	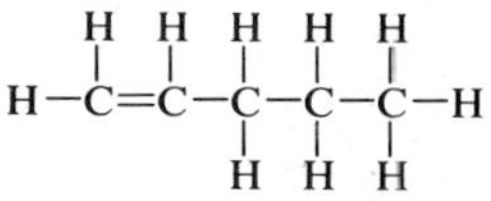 H H H H H | | | | | H—C=C—C—C—C—H _or_ 5 other isomers | | | | H H H H	$CH_2{=}CH{-}CH_2{-}CH_2{-}CH_3$
6	Hexene	C_6H_{12}	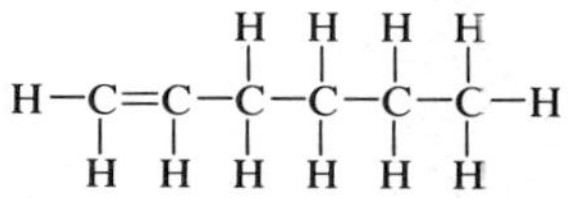 H H H H H H | | | | | | H—C=C—C—C—C—C—H _or_ 14 other isomers | | | | | H H H H H	$CH_2{=}CH{-}CH_2{-}CH_2{-}CH_2{-}CH_3$
7 8	Heptene Octene	C_7H_{14} C_8H_{16}	30 isomers 66 isomers	

Alkynes

Consider two carbon atoms connected by a triple bond.

$$C \equiv C$$

How many hydrogen atoms must be connected to these two carbon atoms in order to satisfy all the bond requirements? The answer is two, so that the molecular formula of this compound is C_2H_2. This compound is called *ethyne*. Its structure is

$$H : C \equiv C : H \qquad (H:C:::C:H)$$

Ethyne is also known by the old-fashioned name acetylene. However, this name is not preferred because the ending *-ene* denotes a double bond, whereas this compound actually has a triple bond between the carbon atoms.

If three carbon atoms are placed in a chain with a triple bond between two of these carbon atoms

$$C \equiv C - C$$

only four hydrogen atoms can be placed around these carbons in order to satisfy all the bonds. The compound then becomes

$$H - C \equiv C - \overset{\displaystyle H}{\underset{\displaystyle H}{\overset{|}{\underset{|}{C}}}} - H$$

with the molecular formula C_3H_4. This compound is called *propyne*.

These two compounds, ethyne and propyne, are alkynes. They have a triple bond between two of the carbon atoms. All their names end in *-yne*. The general formula for alkynes is

$$C_nH_{2n-2}$$

Thus hexyne, which has six carbon atoms, has the formula $C_6H_{(2 \times 6)-2}$ or C_6H_{10}. Likewise, octyne has the molecular formula C_8H_{14}.

It should be noted that there can be no hydrocarbon with four bonds between the carbon atoms because then there would be no bonds available for any hydrogen atoms. (Recall that hydrocarbons must contain both carbon and hydrogen.)

A summary of the names and formulas of some hydrocarbons will be found in Table 16–4.

Saturated and Unsaturated Hydrocarbons

Saturated hydrocarbons are those that have only single bonds between the carbon atoms. Alkanes are saturated compounds.

TABLE 16–4

201

Hydrocarbons

Hydrocarbons

No. of Carbon Atoms	Alkanes	Alkenes	Alkynes
1	Methane CH_4	—	—
2	Ethane C_2H_6	Ethene C_2H_4	Ethyne C_2H_2
3	Propane C_3H_8	Propene C_3H_6	Propyne C_3H_4
4	Butane C_4H_{10}	Butene C_4H_8	Butyne C_4H_6
General formula	C_nH_{2n+2}	C_nH_{2n}	C_nH_{2n-2}

Unsaturated hydrocarbons contain at least one double bond or triple bond. Both alkenes and alkynes are unsaturated compounds.

Reactions of Saturated Hydrocarbons

Saturated hydrocarbons react by a process known as substitution. When ethane reacts with chlorine, Cl_2 or $Cl-Cl$, one of the chlorine atoms substitutes for one of the hydrogen atoms in the saturated compound.

$$H-\overset{\overset{\displaystyle H}{|}}{\underset{\underset{\displaystyle H}{|}}{C}}-\overset{\overset{\displaystyle H}{|}}{\underset{\underset{\displaystyle H}{|}}{C}}\boxed{-H + Cl-}Cl \longrightarrow H-\overset{\overset{\displaystyle H}{|}}{\underset{\underset{\displaystyle H}{|}}{C}}-\overset{\overset{\displaystyle H}{|}}{\underset{\underset{\displaystyle H}{|}}{C}}-Cl + HCl$$

ethane + chlorine ⟶ ethyl chloride (chloroethane) + hydrogen chloride

In general the reactions of the saturated hydrocarbons are very slow compared to those of the unsaturated hydrocarbons.

Reactions of Unsaturated Hydrocarbons

When ethene reacts with hydrogen, H_2, it reacts by a process known as addition. That is, the hydrogen atoms add to the double bond making a single bond out of it. Chlorine, Cl_2, reacts similarly with propene.

$$H-\overset{\overset{\displaystyle H}{|}}{C}=\overset{\overset{\displaystyle H}{|}}{C}-H + H_2 \longrightarrow H-\overset{\overset{\displaystyle H}{|}}{\underset{\underset{\displaystyle H}{|}}{C}}-\overset{\overset{\displaystyle H}{|}}{\underset{\underset{\displaystyle H}{|}}{C}}-H$$

ethene   + hydrogen ⟶   ethane

$$H-\underset{\underset{\displaystyle H}{|}}{\overset{\overset{\displaystyle H}{|}}{C}}-\overset{\overset{\displaystyle H}{|}}{C}=\overset{\overset{\displaystyle H}{|}}{C}-H \quad + \quad Cl-Cl \quad \longrightarrow \quad H-\overset{\overset{\displaystyle H}{|}}{C}-\overset{\overset{\displaystyle H}{|}}{\underset{\underset{\displaystyle Cl}{|}}{C}}-\overset{\overset{\displaystyle H}{|}}{\underset{\underset{\displaystyle Cl}{|}}{C}}-H$$

propene   + chlorine $\longrightarrow$ 1,2-dichloropropane

Alkynes behave in the same manner but at a much more rapid rate. Ethyne (acetylene) reacts with chlorine by a process of addition to form a single bond between the carbon atoms.

$$H-C\equiv C-H \quad + \quad 2Cl-Cl \quad \longrightarrow \quad H-\overset{\overset{\displaystyle Cl}{|}}{\underset{\underset{\displaystyle Cl}{|}}{C}}-\overset{\overset{\displaystyle Cl}{|}}{\underset{\underset{\displaystyle Cl}{|}}{C}}-H$$

ethyne   + chlorine $\longrightarrow$ 1,1,2,2-tetrachloroethane

Sources of Hydrocarbons

The chief sources of hydrocarbons are petroleum and natural gas. Petroleum is a very complex mixture of solid, liquid, and gaseous hydrocarbons plus a few compounds of other elements. Natural gas is primarily a mixture of alkanes of one to four carbon atoms. In general, alkanes are gaseous if the compounds contain between 1 and 4 carbon atoms; liquid if they contain between 5 and 16 carbon atoms; and solid if they contain over 16 carbon atoms.

The various hydrocarbons present in petroleum are isolated by a process known as fractional distillation.

Uses of Hydrocarbons

Methane is used for heating and cooking purposes both in the laboratory and in the home. Mineral oil is a mixture of saturated hydrocarbons and is used extensively in the hospital for lubricating purposes and also as a laxative. Rubber is also a hydrocarbon. Ethene (ethylene) and propene (propylene) have been used extensively to make plastics called polyethylene and polypropylene, respectively. Ethylene has been used as an anesthetic, but its popularity has declined since the introduction of neuromuscular blocking drugs. Acetylene has been employed by European surgeons as an anesthetic; however, the main use of acetylene (ethyne) in North America is for welding.

A hydrocarbon derivative used as an anesthetic is halothane, 2-bromo-2-chloro-1,1,1-trifluoroethane

$$H-\overset{\overset{\displaystyle Cl}{|}}{\underset{\underset{\displaystyle Br}{|}}{C}}-\overset{\overset{\displaystyle F}{|}}{\underset{\underset{\displaystyle F}{|}}{C}}-F$$

Its main advantage over other anesthetics formerly used is that it is non-flammable and nonirritating to the respiratory passages. Halothane is usually used in conjunction with nitrous oxide (see page 106) and with muscle relaxants to provide general anesthesia for surgery of all types, but it can also be administered alone.

Summary

Hydrocarbons are organic compounds that contain the elements carbon and hydrogen only. The simplest hydrocarbon, the hydrocarbon containing only one carbon atom, is methane (CH_4). The hydrocarbon of two carbon atoms is called ethane; that of three carbon atoms, propane. These compounds are called alkanes. The general formula for alkanes is C_nH_{2n+2} where n is the number of carbon atoms. The names of all alkanes end in -*ane*. Beginning with pentane, the prefixes indicate the number of carbon atoms present.

When an alkane loses a hydrogen atom, it forms an organic radical whose name ends in -*yl*. The radical of one carbon atom, derived from methane, is called the methyl radical (CH_3—). The radical of two carbon atoms, derived from ethane, is called the ethyl radical (C_2H_5—).

To identify and name a hydrocarbon compound, pick out the longest continuous chain of carbon atoms and name that chain. Then pick out the radicals attached to that chain. Number the carbons, starting at whichever end of the chain will give radicals with the smallest numbers. List the numbers and names of the radicals, using prefixes to designate radicals occurring more than once.

Alkenes have a double bond between two of the carbon atoms. The names of all alkenes end in -*ene*. The general formula for alkenes is C_nH_{2n}.

Alkynes have a triple bond between two of the carbon atoms and have the general formula C_nH_{2n-2}. The names of all alkynes end in -*yne*.

A saturated hydrocarbon has only single bonds between the carbon atoms. An unsaturated hydrocarbon has double or triple bonds between its carbon atoms. Saturated hydrocarbons react by a process known as substitution. Unsaturated hydrocarbons react by a process known as addition. The addition reaction is usually much more rapid than the substitution reaction.

The chief sources of hydrocarbons are petroleum and natural gas.

Questions and Problems

1. Name the following compounds
    (a) Alkane of five carbon atoms; two carbon atoms.
    (b) Alkyne of four carbon atoms; of three carbon atoms.
    (c) Alkene of eight carbon atoms; of two carbon atoms.

2. What type of hydrocarbon is each of the following?
    (a) C_3H_6
    (b) C_9H_{20}
    (c) $C_{14}H_{26}$
    (d) C_3H_8
    (e) $C_{22}H_{44}$
    (f) $C_{45}H_{88}$
    (g) C_5H_8
    (h) C_5H_{10}

3. What is an organic radical? Name and draw the structure of the radical derived from ethane.

4. Name the following compounds

(a) $CH_3-CH_2-\underset{\underset{\displaystyle CH_3}{|}}{CH}-CH_2-CH_3$

(b) $CH_3-\underset{\underset{\displaystyle CH_3}{|}}{CH}-\overset{\overset{\displaystyle CH_3}{|}}{CH}-CH_2-CH_2-CH_3$

(c) $CH_3-\underset{\underset{\displaystyle CH_3}{|}}{CH}-CH_2-\underset{\underset{\displaystyle CH_3}{|}}{CH}-CH_2-\overset{\overset{\displaystyle CH_3}{|}}{CH}-CH_2-CH_3$

5. What does the term saturated refer to in hydrocarbons?

6. What does the term unsaturated mean in terms of hydrocarbons?

7. How do saturated hydrocarbons react? Give an example of such a reaction.

8. How do unsaturated hydrocarbons react? Give an example of such a reaction.

9. What are the chief sources of hydrocarbons?

10. List several uses of hydrocarbons.

11. What is halothane? For what purpose is it used?

References

Baum, S. J., and Scaife, C. W.: *Chemistry: A Life Science Approach*. Macmillan Publishing Co., Inc., New York, 1975, Chaps. 13, 14.

Brown, R. F.: *Organic Chemistry*. Wadsworth Publishing Co., Inc., Belmont, Calif., 1975, Chap. 13.

Gutsche, D. D., and Pasto, D. C.: *Fundamentals of Organic Chemistry*. Prentice-Hall, Inc., Englewood Cliffs, N.J., 1975, Chaps. 3, 4.

Hein, M., and Best, L. R.: *College Chemistry*. Dickenson Publishing Co., Inc., Encino, Calif., 1976, Chap. 22.

Morrison, R. T., and Boyd, R. N.: *Organic Chemistry*, 3rd ed. Allyn & Bacon, Inc., Boston, 1973, Chaps. 2, 3, 5, 6, 8.

O'Leary, M. H.: *Contemporary Organic Chemistry*. McGraw-Hill Book Co., New York, 1976, Chaps. 3, 4.

Stacy, G. W.: *Organic Chemistry*. Harper & Row, New York, 1975, Chaps. 4, 5.

Alcohols and Ethers

Alcohols

Alcohols are derivatives of hydrocarbons in which one or more of the hydrogen atoms has been replaced by a hydroxyl (—OH) functional group. The hydroxyl group is called a *functional group* because it imparts certain properties to the radical to which it is attached. Other functional groups, which will be discussed in Chapter 18, are the aldehyde, ketone, acid, ester, amine, and amide groups (see Appendix B).

Alcohols are named according to the radical attached to the —OH group. CH_3OH is methyl alcohol; it consists of a methyl radical, CH_3—, attached to an —OH group.

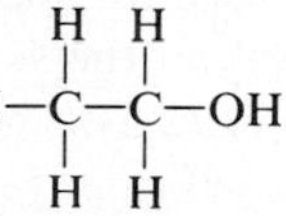

$$
\begin{array}{c}
\text{H} \\
| \\
\text{H}-\text{C}-\text{OH} \\
| \\
\text{H}
\end{array}
$$

methyl alcohol

C_2H_5OH is ethyl alcohol; it contains an ethyl radical, C_2H_5— attached to an —OH group. Its structural formula is

$$
\begin{array}{c}
\text{H} \quad \text{H} \\
| \quad\ | \\
\text{H}-\text{C}-\text{C}-\text{OH} \\
| \quad\ | \\
\text{H} \quad \text{H}
\end{array}
$$

ethyl alcohol

In the IUPAC system (see page 195), alcohols are named according to the longest continuous chain of carbon atoms with the ending -ol, to indicate the functional group. Thus, methyl alcohol is known as methanol and ethyl alcohol as ethanol.

The general formula for an alcohol is ROH, where the R signifies a hydrocarbon radical attached to an —OH functional group.

Alcohols contain a hydroxyl (—OH) group that does not ionize. Because alcohols do not ionize, their reactions are much slower than those of inorganic bases, which contain a hydroxide ion (OH⁻). Solutions of alcohols are non-electrolytes; they are not bases. However, alcohols do react with acids to form a type of organic salt called an ester, which will be discussed in Chapter 18.

Uses

Methyl Alcohol. Methyl alcohol (methanol), CH_3OH, is commonly known as wood alcohol. It is used as a solvent in many industrial reactions. Methyl alcohol should never be used medicinally. If taken internally even small amounts can cause blindness and paralysis, and large amounts may be fatal.

Ethyl Alcohol. Ethyl alcohol (ethanol), C_2H_5OH, is known commonly as grain alcohol. In the hospital it is referred to simply as "alcohol." When no type is specified the word "alcohol" always means ethyl alcohol.

One important property of ethyl alcohol is its ability to coagulate protein. Because of this property, ethyl alcohol is widely used as an antiseptic. As an antiseptic, 70 per cent alcohol is preferred to a stronger solution. It would seem that if a 70 per cent alcohol is a good antiseptic then 100 per cent alcohol would be even better; however, the reverse is actually true. The 70 per cent alcohol is actually a better antiseptic than the 100 per cent alcohol. Pure alcohol coagulates protein on contact. Suppose that pure alcohol is poured over a single-celled organism. The alcohol would penetrate the cell wall of that organism in all directions, coagulating the protein just inside of the cell wall, as shown in Figure 17–1. This ring of coagulated protein would then prevent the alcohol from penetrating farther into the cell so that no more coagulation would take place. At this time the cell would become dormant, but it is not dead. Under the proper conditions the organism can again begin to function. If 70 per cent alcohol is poured over a single-celled organism, the diluted alcohol also coagulates the protein, but at a slower rate, so that it penetrates all the way through the cell before coagulation takes place. Then the entire cell protein is coagulated and the organism dies (see Figure 17–2).

Alcohol (ethyl) may also be used for sponge baths to reduce the fever of a patient. When alcohol is placed on the skin it evaporates rapidly. In order to evaporate, alcohol requires heat. This heat comes from the patient's skin. Thus an alcohol sponge bath will remove heat from the patient's skin and so lower the body temperature. A water sponge bath will do the same thing,

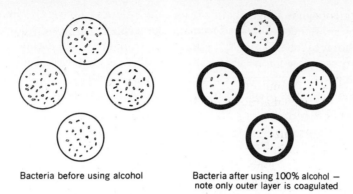

Bacteria before using alcohol

Bacteria after using 100% alcohol —
note only outer layer is coagulated

Figure 17–1. Effect of 100% alcohol on bacteria.

but water evaporates more slowly than alcohol and the heat removal is slower. However water sponge baths are in common use in many hospitals. Note that since alcohol is flammable it may not be used in a room where oxygen is in use.

Alcohol is used as a solvent for many substances. Alcohol solutions are called tinctures. Tincture of benzidine consists of benzidine dissolved in alcohol. Tincture of iodine consists of iodine dissolved in alcohol.

Ethyl alcohol is also used for drinking purposes. The concentration of alcohol in alcoholic beverages is expressed as a "proof." The proof is twice the percentage of alcohol in the solution. Thus a beverage marked "100 proof" contains 50 per cent alcohol. Alcohol slows metabolic processes so that driving under the influence of alcohol can be very dangerous. Alcohol is not a stimulant. It actually depresses the nervous system and removes the inhibitions that a person normally has.

Figure 17–2. Effect of 70% alcohol on bacteria.

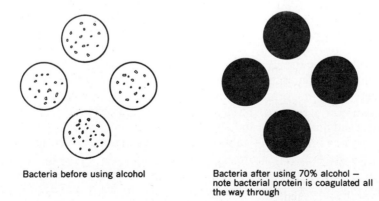

Bacteria before using alcohol

Bacteria after using 70% alcohol —
note bacterial protein is coagulated all
the way through

Pure alcohol has a high tax rate because of its use as a beverage. For use in industry, substances are added to make the alcohol unfit for drinking. Such an alcohol is called *denatured*. It is not fit to drink but is still useful for industrial purposes. Common denaturants are wood alcohol, formaldehyde, and gasoline.

Alcohol (ethyl) is prepared by the fermentation of sugar with the use of certain enzymes found in yeast.

$$C_6H_{12}O_6 \xrightarrow{\text{enzymes}} 2C_2H_5OH + 2CO_2\uparrow$$

glucose   ethyl alcohol

Ethyl alcohol may also be prepared synthetically.

Isopropyl Alcohol. Isopropyl alcohol has the structural formula

$$\begin{array}{ccc} H & H & H \\ | & | & | \\ H-C- & C- & C-H \\ | & | & | \\ H & OH & H \end{array}$$

isopropyl alcohol

The IUPAC name for this compound is 2-propanol, indicating that the —OH functional group is on the second carbon of a three-carbon chain. Isopropyl alcohol is commonly used as rubbing alcohol and as an astringent. An astringent is a medication used externally to contract tissue and decrease the size of blood vessels. It should never be taken internally because of its toxic effects.

Glycerol or Glycerin. Glycerol or glycerin, as it is commonly known, is a trihydric alcohol; it contains three —OH groups. Recall that an alcohol must contain at least one —OH group. A monohydric alcohol contains one —OH group, a dihydric alcohol two. The structural formula for glycerol is

$$\begin{array}{ccc} H & H & H \\ | & | & | \\ H-C- & C- & C-H \\ | & | & | \\ OH & OH & OH \end{array}$$

glycerol

The IUPAC name for glycerol is 1,2,3-propanetriol, since there is an —OH functional group on each carbon atom of the three-carbon chain.

Glycerol is an important alcohol in terms of body chemistry, especially as a constituent of fats (see Chapter 22). Glycerol is used in the preparation of cosmetics and hand lotions and also in suppositories. Glycerin is used in the laboratory as a lubricant for rubber tubing and stoppers. When treated with nitric acid glycerin forms nitroglycerin, an explosive. Medicinally, nitroglycerin is used as a heart stimulant. It causes a dilation of the coronary arteries, thus increasing the supply of blood to the heart muscles.

Primary Alcohols. A primary alcohol is one that contains an —OH functional group attached to a carbon that has one or no carbon atoms attached to it.

$$H-\underset{\underset{H}{|}}{\overset{\overset{H}{|}}{C}}-OH \quad \text{(a primary alcohol)} \qquad H-\underset{\underset{H}{|}}{\overset{\overset{H}{|}}{C}}-\underset{\underset{H}{|}}{\overset{\overset{H}{|}}{C}}-OH \quad \text{(a primary alcohol)}$$

methyl alcohol　　　　　　　　　　　ethyl alcohol

$$H-\underset{\underset{H}{|}}{\overset{\overset{H}{|}}{C}}-\underset{\underset{H}{|}}{\overset{\overset{H}{|}}{C}}-\underset{\underset{H}{|}}{\overset{\overset{H}{|}}{C}}-OH \quad \text{(a primary alcohol)}$$

propyl alcohol

Methyl alcohol (methanol) is an example of a primary alcohol in which the —OH functional group is attached to a carbon atom with no other carbons attached to it. Ethyl alcohol (ethanol) and propyl alcohol (propanol) are examples of primary alcohols in which the —OH functional group is attached to a carbon atom with one carbon atom attached to it. Note that in a primary alcohol the functional group (—OH) is at the end of the chain.

Secondary Alcohols. A secondary alcohol is one in which the —OH is attached to a carbon atom having two other carbon atoms attached to it.

$$CH_3-\underset{\underset{OH}{|}}{CH}-CH_3 \quad \text{(a secondary alcohol)}$$

isopropyl alcohol, 2-propanol

Isopropyl alcohol is an example of a secondary alcohol. Note that in a secondary alcohol the functional group is not at the end of the chain.

Tertiary Alcohols. A tertiary alcohol is one in which the —OH is attached to a carbon atom having three carbon atoms attached to it.

$$CH_3-\underset{\underset{OH}{|}}{\overset{\overset{CH_3}{|}}{C}}-CH_3 \quad \text{(a tertiary alcohol)}$$

2-methyl-2-propanol,
commonly known as
tertiary butyl alcohol

Ethers

An ether is formed by the reaction of an alcohol with sulfuric acid. In this reaction the sulfuric acid may be considered as a dehydrating agent that removes water from two molecules of alcohol. Consider the following reaction, where R indicates a radical and hence ROH indicates an alcohol.

$$\underset{\text{alcohol}}{R-OH} + \underset{\text{alcohol}}{HO-R} \xrightarrow{\text{H}_2\text{SO}_4} \underset{\text{ether}}{R-O-R} + H_2O$$

This equation indicates that two molecules of alcohol (with the second molecular formula written backward) react in the presence of sulfuric acid to form water and an ether. When methyl alcohol is reacted with sulfuric acid, methyl ether (also called dimethyl ether) is formed.

$$\underset{\substack{\text{methyl} \\ \text{alcohol}}}{CH_3-O\boxed{H + HO}CH_3} \xrightarrow{\text{H}_2\text{SO}_4} \underset{\text{methyl ether}}{CH_3-O-CH_3} + H_2O$$

When ethyl alcohol is treated with sulfuric acid, ethyl ether (also called diethyl ether) is formed.

$$\underset{\text{ethyl alcohol}}{CH_3-CH_2\boxed{OH + H}O-CH_2-CH_3} \xrightarrow{\text{H}_2\text{SO}_4} \underset{\text{ethyl ether}}{CH_3-CH_2-O-CH_2-CH_3} + H_2O$$

Note that the ether is named for the alcohol from which it is made: ethyl ether from ethyl alcohol and methyl ether from methyl alcohol. Note also that the general formula for an ether is ROR.

Under the IUPAC system, the $-OCH_3$ group is called methoxy and the $-OCH_2CH_3$ group is called ethoxy. Thus, methyl ether, CH_3-O-CH_3, is called methoxymethane and $CH_3-CH_2-O-CH_2-CH_3$, ethyl ether, is called ethoxyethane.

If two different alcohols are reacted, a mixed ether is formed. Methyl ethyl ether is an example of a mixed ether.

$$CH_3-O-CH_2-CH_3$$
methyl ethyl ether
or methoxyethane

Ether as an Anesthetic

Ethyl ether, commonly known as ether, has been used quite extensively as a general anesthetic. It is very easy to administer, is an excellent muscular relaxant, and has very little effect upon the rate of respiration, blood pressure, or pulse rate. However, the disadvantages of ether outweigh its advantages. The disadvantages are that it is very flammable and combustible; it is irritating to the membranes of the respiratory tract; and it has an aftereffect of nausea. Today ether is infrequently employed as a general anesthetic except when used in laboratory work. It has been replaced by such nonflammable anesthetics as nitrous oxide (see page 106) and halothane (see page 202).

Recent research has indicated that the direct application of ether to a common cold sore produces alleviation of symptoms.

Methoxyflurane, 2,2-dichloro-1,1-difluoroethylmethyl ether, has also been used as a general anesthetic, but its use has been discontinued in several hospitals because of its effects on renal function.

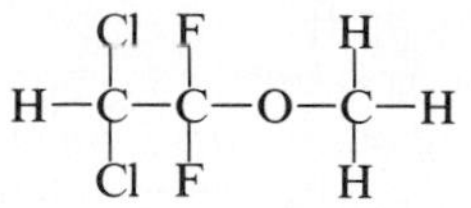

$$\begin{array}{c} \text{Cl} \quad \text{F} \qquad\quad \text{H} \\ | \quad\ \ | \qquad\quad\ | \\ \text{H}-\text{C}-\text{C}-\text{O}-\text{C}-\text{H} \\ | \quad\ \ | \qquad\quad\ | \\ \text{Cl} \quad \text{F} \qquad\quad \text{H} \end{array}$$

methoxyflurane

Divinyl ether has also been used as a surgical anesthetic but is now obsolete for this purpose.

The vinyl group has the following structure

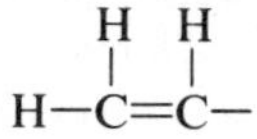

$$\begin{array}{c} \text{H} \ \ \text{H} \\ | \quad | \\ \text{H}-\text{C}=\text{C}- \end{array}$$

The structure of divinyl ether is

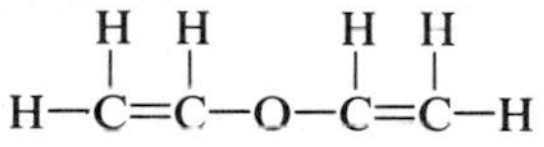

$$\begin{array}{c} \text{H} \ \ \text{H} \qquad\ \ \text{H} \ \ \text{H} \\ | \quad | \qquad\ \ | \quad | \\ \text{H}-\text{C}=\text{C}-\text{O}-\text{C}=\text{C}-\text{H} \end{array}$$

divinyl   ether

Vinyl chloride is used in the manufacture of such consumer products as floor tile, raincoats, phonograph records, fabrics, and furniture coverings. However, evidence has shown that several workers exposed to vinyl chloride during their work have died from a very rare form of liver cancer. In addition, exposure to vinyl chloride is suspected as being responsible for certain types of birth defects.

The structure of vinyl chloride is

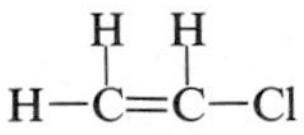

$$\begin{array}{c} \text{H} \ \ \text{H} \\ | \quad | \\ \text{H}-\text{C}=\text{C}-\text{Cl} \end{array}$$

Summary

Alcohols are derivatives of hydrocarbons with one or more of the hydrogen atoms replaced by an —OH group. The —OH (hydroxyl) group is a functional group that imparts to alcohol its particular properties.

The general formula for an alcohol is ROH, where the R represents a radical attached to the —OH group.

Alcohols do not ionize; they are not bases; their reactions are slower than those of inorganic hydroxides.

The simplest alcohol is methyl alcohol, CH_3OH, also known as methanol. This alcohol is poisonous and should never be used internally.

Ethyl alcohol, C_2H_5OH, is also known as ethanol. It is used as a disinfectant because it has the property of coagulating protein. It may also be used for sponge baths to reduce body temperature. Solutions of medications in alcohol are called tinctures.

Isopropyl alcohol is used primarily as rubbing alcohol. It is toxic and should never be used internally.

Glycerol or glycerin is a trihydric alcohol; it has three —OH groups in its molecule. Glycerol is a constituent of fats.

Alcohols may be divided into three categories: primary, where the —OH group is attached to a carbon atom having one or no carbon atoms attached to it; secondary, where the —OH group is attached to a carbon atom having two carbon atoms attached to it; and tertiary, where the —OH is attached to a carbon atom having three carbon atoms attached to it.

Ethers are produced by the dehydration of an alcohol. Ethers are named according to the alcohol or alcohols from which they were produced.

Ethyl ether has been used as a general anesthetic. It is an excellent muscular relaxant and has little effect on the rate of respiration or pulse rate. However, ether is irritating to the membranes of the respiratory tract, it may cause nausea, and it is very flammable. It has been replaced by nonflammable, nonirritating anesthetics such as halothane and nitrous oxide.

Questions and Problems

1. What is the general formula for an alcohol?
2. Indicate several general properties of alcohols.
3. Draw the structure of (a) methyl alcohol; (b) ethyl alcohol; (c) isopropyl alcohol; (d) propyl alcohol; (e) glycerol. Give the IUPAC name for each.
4. Why should methyl alcohol never be used medicinally?
5. Explain why 70% alcohol is a better disinfectant than 100% alcohol.
6. Why is alcohol frequently used for sponge baths?
7. Why may water be substituted for alcohol in a sponge bath?
8. What is a tincture?
9. What is denatured alcohol? Where is it used?
10. What does the word "proof" mean in terms of alcohol?
11. How may alcohol be prepared commercially?
12. What is a common use for isopropyl alcohol?
13. Why is glycerol important in the body?
14. What is nitroglycerin used for in the body? How may it be prepared?
15. List the three types of alcohols and give an example of each.
16. How may an ether be prepared? What reagents are needed?
17. Write the equation for the formation of ethyl ether from ethyl alcohol.
18. What is the structural formula for vinyl ether? Vinyl chloride?
19. What is a mixed ether? How may it be prepared?
20. List disadvantages of ether as a general anesthetic.

References

Baum, S. J., and Scaife, C. W.: *Chemistry: A Life Science Approach.* Macmillan Publishing Co., Inc., New York, 1975, Chap. 17.

Brown, R. F.: *Organic Chemistry*. Wadsworth Publishing Co., Inc., Belmont, Calif., 1975, Chap. 13.

Gutsche, D. D., and Pasto, D. C.: *Fundamentals of Organic Chemistry*. Prentice-Hall, Inc., Englewood Cliffs, N.J., 1975, Chaps. 10, 11.

Hein, M., and Best, L. R.: *College Chemistry*. Dickenson Publishing Co., Inc., Encino, Calif., 1976, Chap. 23.

Morrison, R. T., and Boyd, R. N.: *Organic Chemistry*, 3rd ed. Allyn & Bacon, Inc., Boston, 1973, Chaps. 15, 16, 17.

O'Leary, M. H.: *Contemporary Organic Chemistry*. McGraw-Hill Book Co., New York, 1976, Chap. 5.

Stacy, G. W.: *Organic Chemistry*. Harper & Row, New York, 1975, Chap. 10.

Chapter 18

Other Organic Compounds

The oxidation of primary, secondary, and tertiary alcohols gives different types of products. If we consider oxidation as the removal of hydrogens (see page 109), then the oxidation of an alcohol may be said to involve the removal of one hydrogen from the —OH group of the alcohol and the removal of a second hydrogen from the carbon atom to which the —OH group is attached. This reaction may be written as

$$R-\overset{\overset{\displaystyle H}{|}}{\underset{\underset{\displaystyle H}{|}}{C}}-OH + [O] \longrightarrow R-\overset{\overset{\displaystyle H}{|}}{C}=O + H_2O$$

The oxidation of an alcohol involves the use of some oxidizing agent such as $KMnO_4$, $K_2Cr_2O_7$, O_2, or CuO. However, for the sake of simplicity

214

the oxidizing agent in the preceding and in the following reactions is simply listed as [O], which stands for any substance that will yield the oxygen needed for the reaction.

Aldehydes

Oxidation of a Primary Alcohol

Recall that a primary alcohol has the —OH functional group bonded to a carbon with one or no other carbon atom attached to it. The following equation indicates the oxidation of methanol, CH_3OH, a primary alcohol.

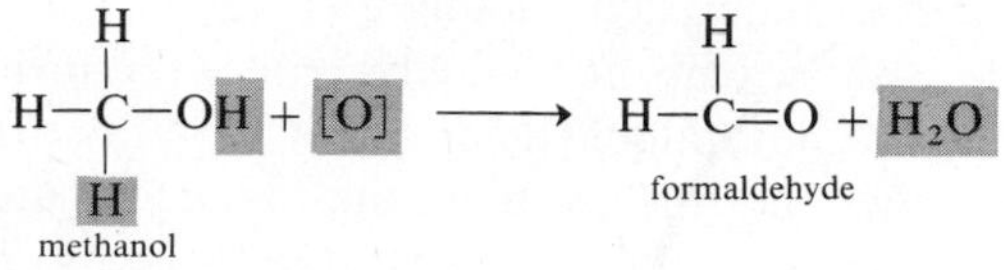

$$\begin{array}{c} H \\ | \\ H-C-OH \\ | \\ H \\ \text{methanol} \end{array} + [O] \longrightarrow \begin{array}{c} H \\ | \\ H-C=O \\ \text{formaldehyde} \end{array} + H_2O$$

Observe that during the oxidation, one H was removed from the —OH group and another H from the carbon to which the —OH group was attached (the only carbon in this compound). Water is one product of this reaction; the other product is a new kind of compound called an aldehyde. In this example, the product is called formaldehyde. The formula for formaldehyde may also be written as HCHO.

The following reaction indicates the oxidation of ethyl alcohol (also a primary alcohol).

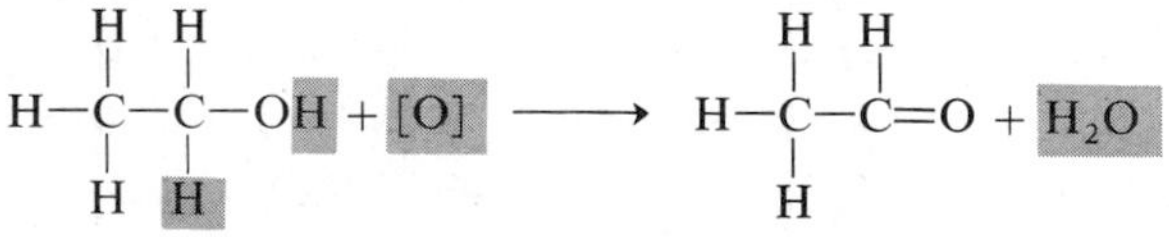

$$\begin{array}{c} H \quad H \\ | \quad | \\ H-C-C-OH \\ | \quad | \\ H \quad H \end{array} + [O] \longrightarrow \begin{array}{c} H \quad H \\ | \quad | \\ H-C-C=O \\ | \\ H \end{array} + H_2O$$

The oxidation of ethyl alcohol, a primary alcohol, yields acetaldehyde, whose formula may also be written as CH_3CHO. In general,

$$\text{primary alcohol} \xrightarrow{\text{oxidation}} \text{aldehyde}$$

Aldehydes all contain the —CHO group. The general formula for an aldehyde is RCHO, which indicates that some radical (R) is attached to a CHO group.

As we have seen, the oxidation of methyl alcohol, a primary alcohol of one carbon atom, yields an aldehyde of one carbon atom, HCHO, formaldehyde. The oxidation of ethyl alcohol, a primary alcohol of two carbon atoms, yields an aldehyde of two carbon atoms, CH_3CHO, acetaldehyde. Note that the term *aldehyde* comes from the words *al*cohol *dehyd*rogenation.

The IUPAC names for aldehydes end in -al. Thus, formaldehyde is known as methanal and acetaldehyde as ethanal.

Uses of Aldehydes

Formaldehyde is a colorless gas with a very sharp odor. It is used in the laboratory as a water solution containing about 40 per cent formaldehyde. A 4 per cent formaldehyde solution is used in histology to harden tissue. The 40 per cent solution, commonly known as formalin, is an effective germicide for the disinfection of excreta, rooms, and clothing. Formaldehyde solutions should not be used directly on the patient or even in the room with the patient because of irritating fumes.

Formaldehyde and its oxidation product, formic acid, are primarily responsible for the systemic toxicity of methyl alcohol.

Glutaraldehyde is superior to formaldehyde as a sterilizing agent and is gradually replacing it. Glutaraldehyde does not have the disagreeable odor that formaldehyde does and it is less irritating to the eyes and skin.

Paraldehyde is formed by the polymerization (joining) of molecules of acetaldehyde. Paraldehyde depresses the central nervous system. It is used as an hypnotic—a sleep producer. In therapeutic dosages it is nontoxic; it does not depress heart action or respiration. Its disadvantages are its disagreeable taste and its unpleasant odor.

Test for Aldehydes

In general aldehydes are good reducing agents. Laboratory tests for the presence of aldehyde are based on their ability to reduce copper (II) or cupric ions to form copper (I) or cuprous oxide. When an aldehyde is heated with Benedict's or Fehling's solution or treated with a Clinitest tablet [all of which contain $Cu(OH)_2$], a red precipitate of copper (I) or cuprous oxide (Cu_2O) is formed. This is actually the test for glucose (sugar) in urine since glucose is an aldehyde (see page 265).

Ketones

Oxidation of a Secondary Alcohol

Recall that a secondary alcohol is one in which the —OH group is bonded to a carbon atom having two carbon atoms attached to it. Isopropyl alcohol is an example of a secondary alcohol.

$$CH_3-\overset{\overset{\displaystyle H}{|}}{\underset{\underset{\displaystyle OH}{|}}{C}}-CH_3$$

isopropyl alcohol (2-propanol)

The oxidation of isopropyl alcohol is indicated by the following equation

$$CH_3-\underset{\underset{OH}{|}}{\overset{\overset{H}{|}}{C}}-CH_3 + [O] \longrightarrow CH_3-\underset{\overset{||}{O}}{C}-CH_3 + H_2O$$

isopropyl alcohol or   acetone or
2-propanol   propanone

As before, the oxygen atom from the oxidizing agent reacts with the H from the —OH group and with the H attached to the same carbon as the —OH group, forming water and a new class of compounds called ketones. The ketone produced by the oxidation of isopropyl alcohol is called acetone. In general, names of ketones end in *-one*. The IUPAC name for acetone is propanone.

Note that the ketone has the same number of carbon atoms as does the alcohol from which it was made. Acetone is a good solvent for fats and oils. It is also frequently used in fingernail polish and in polish remover. Acetone is normally present in small amounts in the blood and urine. In diabetes mellitus it is present in larger amounts in the blood and urine and even in the expired air.

The oxidation of a secondary alcohol yields a ketone, of the general formula RCOR. That is, a ketone has two radicals attached to a $>$C=O group which is called a *carbonyl group*. This carbonyl group is present in both aldehydes and ketones. However, the carbonyl group is at the end of the chain in an aldehyde and not at the end in a ketone.

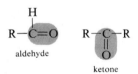

$$R-\underset{}{\overset{\overset{H}{|}}{C}}=O \qquad R-\underset{\overset{||}{O}}{C}-R$$

aldehyde   ketone

Oxidation of a Ketone

What would happen when a ketone is oxidized? Consider the formula for a ketone, acetone, indicated below:

$$CH_3-\underset{\overset{||}{O}}{C}-CH_3$$

acetone

There are no hydrogen atoms on the carbon atom to which the carbonyl group is attached. Therefore, there would be no simple reaction. In general, ketones are not easily oxidized.

The following alcohol, tertiary butyl alcohol, is not easily oxidized. Why?

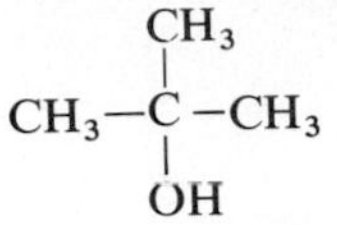

$$CH_3-\underset{\underset{OH}{|}}{\overset{\overset{CH_3}{|}}{C}}-CH_3$$

tertiary butyl alcohol
(2-methyl-2-propanol)

Note that there is no hydrogen atom on the carbon that has the —OH group attached to it. Therefore, oxidation of a tertiary alcohol does not easily take place.

Organic Acids

Oxidation of an Aldehyde

The oxidation of a primary alcohol yields an aldehyde. Aldehydes in turn can be easily oxidized.

When an aldehyde is oxidized, the reaction is as follows

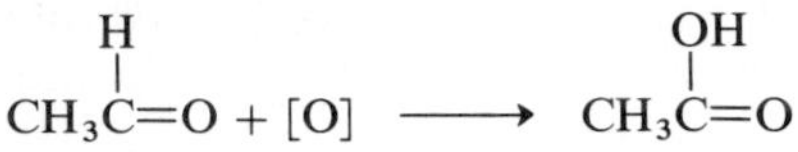

$$CH_3\overset{\overset{H}{|}}{C}=O + [O] \longrightarrow CH_3\overset{\overset{OH}{|}}{C}=O \quad \text{(also written as } CH_3COOH)$$

acetaldehyde   acetic acid

The resulting compound is acidic because it yields hydrogen ions in solution. (Note that this reaction involved oxidation because of a gain in oxygen—see page 109.)

The functional group of an organic acid is —COOH, so that the oxidation of an aldehyde to an acid may be written functionally as

$$R-CHO \xrightarrow{[O]} R-COOH$$

aldehyde   acid

The oxidation of methyl alcohol, a primary alcohol, to an aldehyde and then to an acid is illustrated in the following equation

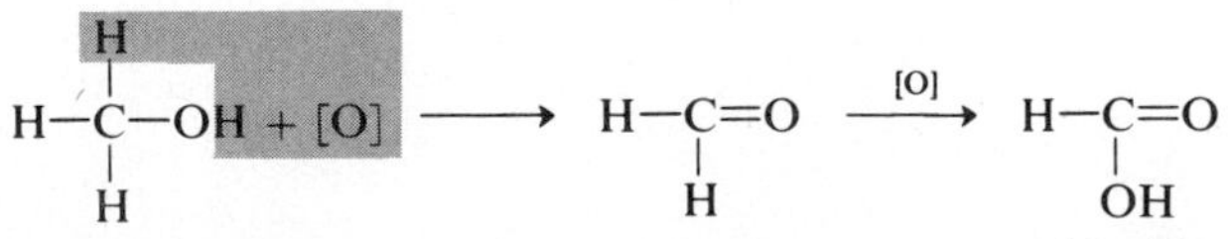

$$H-\underset{\underset{H}{|}}{\overset{\overset{H}{|}}{C}}-OH + [O] \longrightarrow H-\overset{\overset{}{|}}{C}=O \xrightarrow{[O]} H-\overset{}{C}=O$$

methyl alcohol   formaldehyde   formic acid
(methanol)   (methanal)   (methanoic acid)

(also written as HCOOH)

Note that the name of the acid is derived from the name of the aldehyde. Formaldehyde yields formic acid; acetaldehyde yields acetic acid.

The IUPAC names for acids end in -oic acid. Thus,

$$\text{methanol} \xrightarrow{[O]} \text{methanal} \xrightarrow{[O]} \text{methanoic acid (formic acid)}$$

$$\text{ethanol} \xrightarrow{[O]} \text{ethanal} \xrightarrow{[O]} \text{ethanoic acid (acetic acid)}$$

The general formula for an acid is RCOOH. All organic acids contain at least one —COOH group. This group is called the *carboxyl group* and it is this group that yields hydrogen ions.

Properties and Reactions of Organic Acids

Most organic acids are relatively weak acids since they ionize very slightly in water.

$$\underset{\text{acetic acid}}{CH_3COOH} \rightleftharpoons \underset{\substack{\text{hydrogen} \\ \text{ion}}}{H^+} + \underset{\text{acetate ion}}{CH_3COO^-}$$

Organic acids react with bases to form salts and water. The general formula for the reaction of an organic acid with a base to form a salt and water may be written as follows

$$\underset{\text{organic acid}}{RCOOH} + \underset{\text{base}}{NaOH} \rightleftharpoons \underset{\text{organic salt}}{RCOONa} + \underset{\text{water}}{H_2O}$$

$$\underset{\text{acetic acid}}{CH_3COOH} + \underset{\substack{\text{sodium} \\ \text{hydroxide}}}{NaOH} \rightleftharpoons \underset{\substack{\text{sodium acetate} \\ \text{(a salt)}}}{CH_3COONa} + \underset{\text{water}}{H_2O}$$

Organic acids also react with alcohols to form a class of compounds called esters, which will be dealt with later in this chapter.

Medically Important Organic Acids

Formic acid, HCOOH, is a colorless liquid with a sharp irritating odor. Formic acid is found in the sting of bees and causes the characteristic pain and swelling when it is injected into the tissues. It is one of the strongest organic acids.

Acetic acid, CH_3COOH, is one of the components of vinegar where it is usually found as a 4 to 5 per cent solution. Anhydrous acetic acid freezes at 17°C and then looks like a chunk of ice. For this reason it is called *glacial acetic acid.*

Citric acid is found in citrus fruits. Its formula indicates that it is an alcohol as well as an acid. Note that citric acid contains one alcohol (—OH) group and three acid (—COOH) groups.

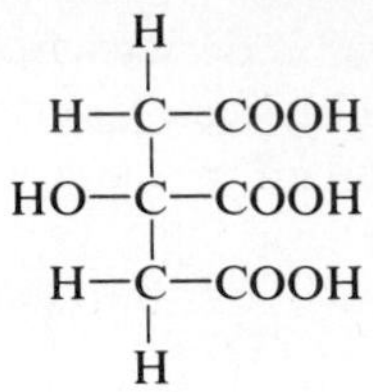

$$
\begin{array}{c}
H \\
| \\
H-C-COOH \\
| \\
HO-C-COOH \\
| \\
H-C-COOH \\
| \\
H
\end{array}
$$

citric acid

Magnesium citrate, a salt of citric acid, is used as a cathartic (a medication for stimulating the evacuation of bowels). Sodium citrate, another salt of citric acid, is used as a blood anticoagulant.

Lactic acid is found in sour milk. It is formed from the fermentation of milk sugar, lactose. Its formula is

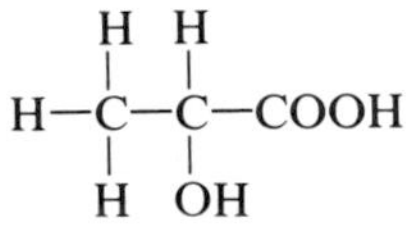

$$
\begin{array}{c}
H \quad H \\
| \quad | \\
H-C-C-COOH \\
| \quad | \\
H \quad OH
\end{array}
$$

lactic acid

Note that lactic acid is both an acid and an alcohol but it reacts primarily as an acid because of the greater reactivity of the acid functional group. Lactic acid is formed in the muscle during exercise.

Oxalic acid is another one of the strong, naturally occurring organic acids. Its formula is

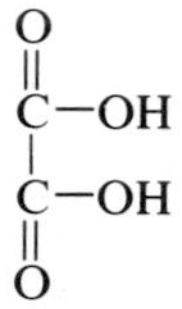

$$
\begin{array}{c}
O \\
\| \\
C-OH \\
| \\
C-OH \\
\| \\
O
\end{array}
$$

oxalic acid

Oxalic acid is used to remove stains, particularly rust and potassium permanganate stains, from clothing. It is poisonous when taken internally. Oxalate salts are added to blood samples to prevent clotting. However, this may be done only for blood samples that are to be analyzed in the laboratory. Oxalate salts cannot be added directly to the bloodstream because of their poisonous nature.

Tartaric acid is another one of the organic acids that is both an acid and an alcohol. Its formula is written as

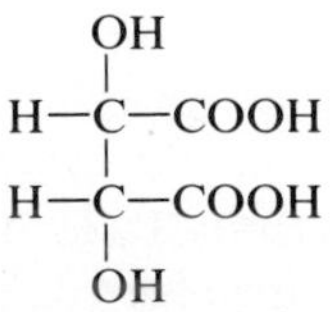

$$
\begin{array}{c}
OH \\
| \\
H-C-COOH \\
| \\
H-C-COOH \\
| \\
OH
\end{array}
$$

tartaric acid

Tartaric acid is found in several fruits, particularly grapes. Potassium hydrogen tartrate, an acid salt called cream of tartar, is used in making baking powders. Rochelle salts, or potassium sodium tartrate, is used as a mild cathartic. Antimony potassium tartrate, known as tartar emetic, is used in the treatment of schistosomiasis (a disease caused by a type of parasitic flatworm), but its use is inadvisable in the presence of severe hepatic, renal, or cardiac insufficiency.

Stearic acid is a solid greaselike acid which is insoluble in water. Its formula is $C_{17}H_{35}COOH$. The sodium salt of stearic acid, sodium stearate, is a commonly used soap.

$$C_{17}H_{35}COOH + NaOH \longrightarrow C_{17}H_{35}COONa + H_2O$$

stearic acid   sodium stearate

There are two other groups of acids that are of great importance in the body: the fatty acids and the amino acids. Fatty acids will be discussed in Chapter 22 and amino acids in Chapter 23.

Esters

Esters are classified as organic salts. Esters are produced by the reaction of an organic acid with an alcohol and have the general formula

$$RCOOR' \quad \text{or} \quad \overset{\displaystyle O}{\overset{\displaystyle \|}{R}}COR'$$

The general reaction of an alcohol with an acid is illustrated by the following equation (where R and R' may be the same or different radicals)

$$RCOOH + R'OH \rightleftharpoons RCOOR' + H_2O$$

acid   alcohol   ester   water

Note that the reaction is written with a double arrow, indicating that it is an equilibrium reaction; that is, the reverse reaction also takes place. Thus, esters hydrolize to form organic acids and alcohols.

Although esters are organic salts, they do not readily ionize in water solution.

When an organic acid is reacted with an alcohol, the name of the ester formed is determined as follows

1. Write the name of the alcohol.
2. Write the name of the acid minus the ending -ic.
3. Add the ending -ate to the name of the acid.

In the reaction of acetic acid with methyl alcohol

$$CH_3-\overset{\displaystyle O}{\overset{\displaystyle \|}{C}}-OH + HOCH_3 \rightleftharpoons CH_3\overset{\displaystyle O}{\overset{\displaystyle \|}{C}}-OCH_3 + H_2O$$

acetic acid   methyl alcohol   methyl acetate   water

TABLE 18–1

Esters and Synthetic Flavors

Ester	Flavor
Amyl acetate	Banana
Ethyl butyrate	Pineapple
Amyl butyrate	Apricot
Isoamyl acetate	Pear
Octyl acetate	Orange

the name of the ester is methyl (from methyl alcohol) acetate (from acetic minus -ic plus -ate). Likewise, the ester formed by the reaction of stearic acid and isopropyl alcohol is called isopropyl stearate. The IUPAC name for the above ester formed by the reaction of methanol and ethanoic acid is methyl ethanoate.

Esters are important as solvents, perfumes, and flavoring agents. Table 18–1 indicates some of the esters used in preparing synthetic flavors. Many esters are also used medicinally.

Medical Uses of Esters

A few of the many esters used medicinally are listed in Table 18–2.

Amines

Amines are organic compounds derived from ammonia. There are three classes of amines: primary, secondary, and tertiary. Note that in relation to amines, the terms primary, secondary, and tertiary refer directly to the number

TABLE 18–2

Some Medical Uses of Esters

Esters	Use
Ethyl aminobenzoate (benzocaine)	Local anesthetic
Glyceryl trinitrate (nitroglycerin)	Used to dilate coronary arteries and lower blood pressure
Methyl salicylate (oil of wintergreen)	Used as a flavoring agent and also as a counterirritant in many liniments
Phenyl mercuric acetate	Used to disinfect instruments and as an antiseptic on cutaneous and mucosal surfaces

of hydrogen atoms of ammonia that have been replaced by organic radicals. In the case of alcohols, the terms primary, secondary, and tertiary refer to the number of carbon atoms attached to the carbon having the —OH group on it.

Primary amines are those in which one of the hydrogen atoms of ammonia (NH_3) has been replaced by an organic radical. Primary amines have the general formula RNH_2.

Secondary amines are those in which two of the hydrogen atoms of the ammonia have been replaced by organic radicals. Secondary amines have the general formula R_2NH.

Tertiary amines are those in which all three hydrogen atoms of the ammonia have been replaced by organic radicals. Tertiary amines have the general formula R_3N.

Table 18–3 illustrates the various types of amines and compares their structures with that of ammonia.

Among the amines that are found in the body or are used medicinally are histamines (pages 396, 498), barbiturates (page 248), nucleic acids (page 319), and niacin (page 453).

TABLE 18–3

Some Common Amines

Name	Formula	Example	
Ammonia	NH_3	$\begin{array}{c} H \\ \mid \\ N-H \\ \mid \\ H \end{array}$	
Primary amine	RNH_2	$\begin{array}{c} H\ \ H \\ \mid\ \ \mid \\ H-C-N \\ \mid\ \ \mid \\ H\ \ H \end{array}$	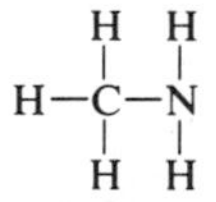 or CH_3NH_2 (methyl amine)
Secondary amine	R_2NH	(dimethyl amine structure) N—H	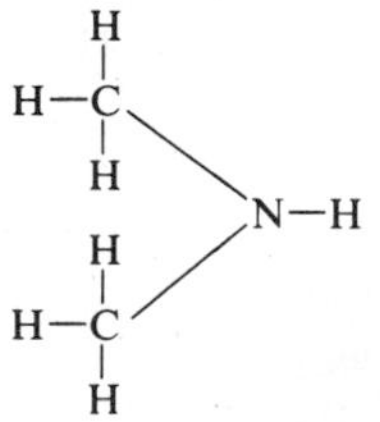 or $(CH_3)_2NH$ (dimethyl amine)
Tertiary amine	R_3N	(trimethyl amine structure)	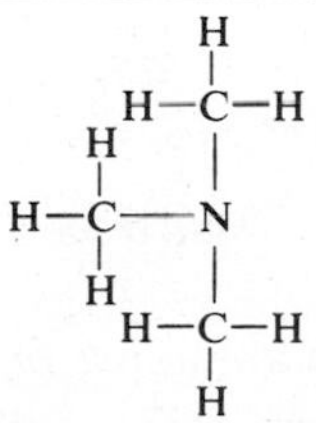 or $(CH_3)_3N$ (trimethyl amine)

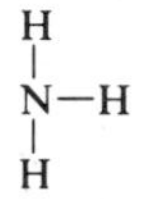

Amines in general are basic compounds and easily form salts. Quaternary ammonium salts are formed by the action of tertiary amines with organic halogen compounds. An example of this type of reaction is that between trimethyl amine and methyl iodide which forms tetramethyl ammonium iodide is

$$
\begin{array}{ccc}
\underset{\substack{\text{trimethyl}\\\text{amine}}}{\overset{\displaystyle CH_3}{\underset{\displaystyle CH_3}{\overset{|}{\underset{|}{CH_3-N}}}}}
& + \underset{\substack{\text{methyl}\\\text{iodide}}}{CH_3I}
\longrightarrow
& \underset{\substack{\text{tetramethyl}\\\text{ammonium}\\\text{iodide}}}{\overset{\displaystyle CH_3^{+}}{\underset{\displaystyle CH_3}{\overset{|}{\underset{|}{CH_3-N-CH_3}}}}} + I^{-}
\end{array}
$$

Some quaternary ammonium salts have both a detergent action and antibacterial activity. They are used medicinally as antiseptics. Benzalkonium chloride (zephiran chloride) is used in 0.1 per cent solution for storage of sterilized instruments. A 0.01 to 0.02 per cent solution is applied as a wet dressing to denuded areas. A 0.005 per cent solution is used for irrigations of the bladder and the urethra.

Amino Acids

Amino acids are organic acids that contain an amine group. Examples of amino acids are

$$
\underset{\substack{\text{glycine, an}\\\alpha\text{-amino acid}}}{\overset{\displaystyle H}{\underset{\displaystyle NH_2}{\overset{|}{\underset{|}{H-C-COOH}}}}}
\qquad
\underset{\substack{\text{an }\alpha\text{-amino}\\\text{acid}}}{\overset{\displaystyle H\ \ H}{\underset{\displaystyle H\ \ NH_2}{\overset{|\ \ |}{\underset{|\ \ |}{H-C-C-COOH}}}}}
\qquad
\underset{\substack{\text{a }\beta\text{-amino}\\\text{acid}}}{\overset{\displaystyle H\ \ \ \ H}{\underset{\displaystyle NH_2\ H}{\overset{|\ \ \ \ |}{\underset{|\ \ \ \ |}{H-C-\!-\!-C-COOH}}}}}
$$

Note that Greek letters are used to designate the position of the amino group in the chain. The carbon atom next to the acid group, the —COOH group, is called the alpha (α) carbon. Then next in order comes the beta (β), the gamma (γ), and the delta (δ) carbon atoms.

Amino acids contain an acid group, —COOH, which naturally is acidic. Amino acids also contain an —NH_2 group, which is basic. Thus, amino acids exhibit both acidic and basic properties; amino acids can react with either acids or bases. Compounds that can act as or react with either acids or bases are called *amphoteric compounds.*

Amino acids are the building blocks of proteins. That is, proteins are polymers of amino acids (see Chapter 23).

Amides

Organic acids can react with ammonia or with amines to form a class of compounds called *amides.*

$$R-\overset{\overset{\text{O}}{\|}}{C}-OH + H-\overset{\overset{H}{|}}{N}-H \xrightarrow{heat} R-\overset{\overset{\text{O}}{\|}}{C}-NH_2 + H_2O$$

acid   ammonia   amide

Amides are named by dropping the ending -*ic* and the word *acid* from the name of that acid and adding *amide*. Thus, when acetic acid reacts with ammonia, the product is called acetamide.

$$CH_3-\overset{\overset{\text{O}}{\|}}{C}-OH + H-\overset{\overset{H}{|}}{N}-H \xrightarrow{heat} CH_3-\overset{\overset{\text{O}}{\|}}{C}-NH_2 + H_2O$$

acetic acid   ammonia   acetamide

If an organic acid reacts with a primary amine, a substituted amide is formed.

$$R-\overset{\overset{\text{O}}{\|}}{C}-OH + H-\overset{\overset{H}{|}}{N}-R' \longrightarrow R-\overset{\overset{\text{O}}{\|}}{C}-\overset{\overset{H}{|}}{N}-R' + H_2O$$

acid   primary   a substituted
        amine   amide

Note that amides have a bond between a carbonyl group ($-\overset{\overset{\text{O}}{\|}}{C}-$), and a nitrogen.

If two amino acids combine with the acid part of one reacting with the amine part of the other, the following type of reaction may occur.

$$R-\underset{\underset{NH_2}{|}}{CH}-\overset{\overset{\text{O}}{\|}}{C}-OH + H-\overset{\overset{H}{|}}{N}-\underset{\underset{R}{|}}{CH}-COOH$$

amino acid   amino acid

$$R-\underset{\underset{NH_2}{|}}{C}-\overset{\overset{\text{O}}{\|}}{C}-\overset{\overset{H}{|}}{N}-\underset{\underset{R}{|}}{CH}-COOH + H_2O$$

a peptide

These two amino acids are said to be linked by a peptide (amide) bond. Compounds of this type will be discussed in Chapter 23.

Niacin, one of the **B** vitamins, is administered as an amide, niacinamide. Urea, one of the metabolic products of protein metabolism, may be considered as the diamide of carbonic acid.

$$HO-\overset{\overset{\text{O}}{\|}}{C}-OH \qquad H_2N-\overset{\overset{\text{O}}{\|}}{C}-NH_2$$

carbonic   urea
acid

Summary

During the oxidation of an alcohol, the oxygen atom reacts with the H from the —OH group and with a hydrogen attached to the same carbon that has the —OH group on it.

The oxidation of a primary alcohol yields an aldehyde. Aldehydes contain the —CHO group and have the general formula RCHO. The aldehyde of one carbon atom is known as formaldehyde or methanal; that of two carbon atoms, acetaldehyde or ethanal. A water solution of formaldehyde, known as formalin, is commonly used as a germicide. Paraldehyde, formed by polymerizing molecules of acetaldehyde, depresses the central nervous system.

Aldehydes are good reducing agents. When an aldehyde is heated with $Cu(OH)_2$, a red precipitate of Cu_2O is formed.

The oxidation of a secondary alcohol yields a ketone. A ketone may not be further oxidized without decomposing it. Likewise, tertiary alcohols are not easily oxidized.

The oxidation of an aldehyde yields an acid. The general formula for an acid is RCOOH. The —COOH group, called the carboxyl group, furnishes hydrogen ions and so causes the acidic properties. Organic acids react with bases to form organic salts.

Formic acid, HCOOH, is found in the sting of bees. Acetic acid, CH_3COOH, is one of the components of vinegar. Citric acid is found in citrus fruits, lactic acid in milk.

When an organic acid reacts with an alcohol, a compound called an ester is produced. Esters have the general formula RCOOR′. Esters are named according to the alcohol and the acid from which they were made (with the -ic ending of the acid removed and replaced by -ate).

Amines are organic compounds in which an organic radical has replaced one or more of the hydrogen atoms of ammonia, NH_3. Amines are basic compounds and readily form salts.

Amino acids are organic acids that contain an amine group. Amino acids are the building blocks of protein.

Amides are formed by the reaction of an organic acid with ammonia or with an amine. Amino acids are held together by peptide (amide) bonds.

Questions and Problems

1. Write the equation for the oxidation of methyl alcohol. What product is formed?
2. Write the equation for the formation of acetaldehyde from ethyl alcohol.
3. What is the general formula for an aldehyde?
4. What does the suffix -al indicate?
5. What does the suffix -one indicate?
6. Indicate some uses for formaldehyde.
7. What is paraldehyde? What is it used for?
8. Describe the test for aldehydes.
9. Write the reaction for the oxidation of isopropyl alcohol.
10. The oxidation of a primary alcohol yields what type of compound?
11. The oxidation of a secondary alcohol yields what type of product?
12. Describe the oxidation of a tertiary alcohol; of a ketone.
13. Write the equation for the oxidation of formaldehyde and name the product.
14. Write the equation for the oxidation of acetaldehyde and name the product.

15. What is the general formula for an acid?
16. Write the equation for the reaction between formic acid and potassium hydroxide.
17. Name several organic acids and indicate where each may be found.
18. What is an ester? Indicate its general formula.
19. Write the equation for the reaction between formic acid and ethyl alcohol. Name the product.
20. Name and give the medical use of two esters.
21. What is an amine? Compare the structures of primary, secondary, and tertiary amines.
22. How do the terms *primary, secondary,* and *tertiary* for amines compare with the usage of these same terms for alcohols?
23. What is an amino acid? What properties does it have?
24. For what purposes does the body use amino acids?
25. Why are amino acids amphoteric?
26. What is the type structure for an amide?
27. Write the reaction when formic acid and ammonia are heated. What is the name of the product?
28. What is a peptide bond? Where is it found?
29. Why is urea called a diamide?

References

Baum, S. J., and Scaife, C. W.: *Chemistry: A Life Science Approach.* Macmillan Publishing Co., Inc., New York, 1975, Chaps. 18, 19, 20.

Brown, R. F.: *Organic Chemistry.* Wadsworth Publishing Co., Inc., Belmont, Calif., 1975, Chaps. 16, 17.

Gutsche, D. D., and Pasto, D. C.: *Fundamentals of Organic Chemistry.* Prentice-Hall, Inc., Englewood Cliffs, N.J., 1975, Chaps. 13, 14.

Hein, M., and Best, L. R.: *College Chemistry.* Dickenson Publishing Co., Inc., Encino, Calif., 1976, Chap. 24.

Morrison, R. T., and Boyd, R. N.: *Organic Chemistry,* 3rd ed. Allyn & Bacon, Inc., Boston, 1973, Chap. 12.

O'Leary, M. H.: *Contemporary Organic Chemistry.* McGraw-Hill Book Co., New York, 1976, Chaps. 6, 7, 9, 10.

Stacy, G. W.: *Organic Chemistry.* Harper & Row, New York, 1975, Chap. 12.

Chapter 19

Cyclic Compounds

Types of Cyclic Compounds

Saturated Cyclic Compounds

A cyclic compound is a ring compound. The simplest cyclic alkane is cyclopropane.

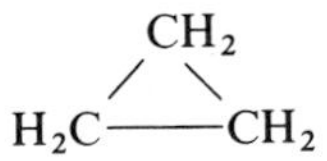

$$
\begin{array}{c}
CH_2 \\
\diagup \quad \diagdown \\
H_2C \text{——} CH_2
\end{array}
$$

Cyclopropane is used medicinally as a general anesthetic. Both induction and recovery time are short. Muscular relaxation is greater than with nitrous oxide but less than with ether. As with ether the danger of explosion is great, so care must be taken in its use. Note that cyclopropane is a saturated hydrocarbon since it contains only single bonds.

Other saturated cyclic hydrocarbons, such as cyclobutane and cyclohexane, are used commercially in gasolines.

The term *aromatic* originally referred to certain compounds that had a pleasant odor and similar chemical and physical properties. Further studies of these compounds showed that they all had a ring-shaped structure. However, many compounds have an aromatic odor and do not have this ring-shaped structure. The term *aromatic* is now used to designate compounds whose structure is based upon that of benzene, C_6H_6.

Benzene

Structure

Benzene has the formula C_6H_6. Since there is one hydrogen atom for every carbon atom, we might expect benzene, because it is unsaturated, to be very reactive. On the contrary, however, benzene is quite stable. The structure for the benzene molecule was first deduced by Kekulé in 1865. He stated that the six carbon atoms were arranged in a ring with alternate single and double bonds, each carbon atom having one hydrogen atom attached to it. Kekulé also suggested that the position of the double and single bonds could change, producing two structures both of which represent benzene. This shifting of the bonds is called *resonance*, and is indicated by a double-headed arrow.

resonance structures of benzene

The electron-cloud picture of the benzene molecule shows two continuous doughnut-shaped electron clouds, one above and the other below the plane of the atoms (see Figure 19–1).

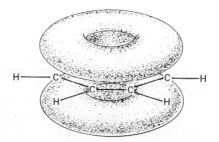

Figure 19–1. Electron-cloud picture of benzene.

The benzene structure is the basis of many thousands of organic compounds. The chemist has devised an abbreviated version of the benzene formula.

Abbreviated resonance structures for benzene found in older texts are shown below.

Note that in these representations it is assumed that a hydrogen atom is present on each carbon or at each corner unless otherwise indicated.

If one of the hydrogen atoms has been replaced by a methyl group, CH_3-, it may be indicated as follows

Properties

Benzene is a colorless liquid with a distinct gasolinelike odor. It is insoluble in water but is soluble in alcohol and ether. Benzene is toxic when taken internally. Contact with the skin is harmful and continued inhalation of benzene vapors decreases red and white blood cell count.

Halogen Derivatives

When benzene is treated with chlorine, chlorobenzene is produced.

The structure of chlorobenzene indicates that one of the hydrogen atoms in the benzene ring has been replaced by a chlorine atom. Since all six carbon atoms and all six hydrogen atoms are equivalent in the benzene ring, there is only one possible *mono*substitution product. That is, all of the structures indicated below are identical—they represent the same compound, chlorobenzene.

However, when two of the hydrogen atoms in the benzene ring are replaced, more than one possible *di*substitution product is possible. One of the disubstitution products obtained upon the reaction of chlorine with chlorobenzene is indicated as follows.

To name this compound, the benzene ring is numbered from 1 to 6. The numbers must be such that the substituents attached to the ring have the lowest possible numbers. Therefore, the ring may be numbered as indicated below. The name of this compound is 1,2-dichlorobenzene, which indicates two chlorines on a benzene ring in positions 1 and 2.

Actually, when chlorine reacts with chlorobenzene, three different disubstitution products are obtained. Their structures and names are as follows:

1,2-dichlorobenzene   1,3-dichlorobenzene   1,4-dichlorobenzene

If the disubstitution products on the benzene ring are different, the same system of naming may be used. Consider the compound indicated below.

It may be numbered with either the bromine atom at position 1 and the chlorine atom at position 2 or vice versa, as shown in the following

Thus the compound may be named either 1-bromo-2-chlorobenzene or 1-chloro-2-bromobenzene. The preferred name lists the substituents in alphabetic order, but frequently the pharmaceutical companies use several different systems for naming their drugs.

Another system for naming disubstituted benzene compounds is based upon the use of prefixes rather than numbers to designate positions in the benzene ring. The prefix *ortho* indicates substances on the benzene ring in positions next to each other. The compound shown below is called *ortho*-dichlorobenzene, or simply *o*-dichlorobenzene.

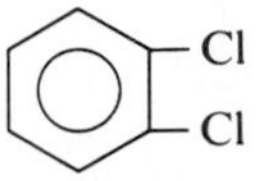

The compound shown below may be called either *o*-chlorobromobenzene or *o*-bromochlorobenzene; the alphabetic sequence is preferred.

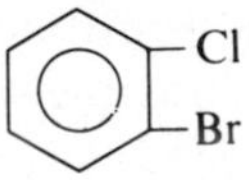

When substituents on the benzene ring are separated by one carbon atom (in positions 1 and 3), the prefix used is *meta*. The compound shown below is called *meta*dichlorobenzene or *m*-dichlorobenzene.

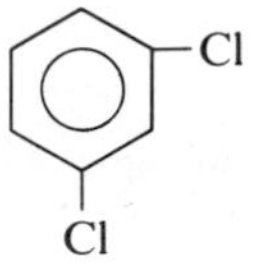

And the next compound is called *m*-bromochlorobenzene.

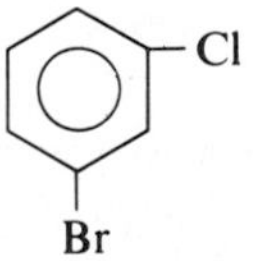

When two substituents on the benzene ring are separated by two carbon atoms (in positions 1 and 4) the prefix used is *para*. The compound indicated below is called *para*dichlorobenzene or *p*-dichlorobenzene.

The following compound is *p*-bromochlorobenzene.

Other Derivatives

Toluene. The methyl derivative of benzene is commonly called toluene. Its structure is

$$\text{—CH}_3$$

Toluene is a colorless liquid with a benzenelike odor. It is insoluble in water and soluble in alcohol and ether. Toluene is used as a preservative for urine specimens and in the preparation of dyes and explosives.

Xylene. Xylene is a dimethylbenzene. Again, with two substitutions on the benzene ring, there are three possible structures of xylene.

o-xylene   *m*-xylene   *p*-xylene

Xylenes are good solvents for oils and are used in cleaning lenses in microscopes.

Naphthalene. Naphthalene, $C_{10}H_8$, is an aromatic compound containing two benzene rings. These two rings are attached to each other in such a manner that they share two carbon atoms.

or

Naphthalene is a white crystalline solid obtained from coal tar. Naphthalene crystals are frequently used in the home under the name mothballs. Functional groups may be attached to naphthalene in either of two positions, called alpha and beta. These positions in alpha- and beta-naphthol are

alpha-naphthol
or α-naphthol

beta-naphthol
or β-naphthol

These compounds are carcinogenic and should only be used under strictly controlled conditions. Note that naphthalene has four alpha positions and also four beta positions.

Anthracene and Phenanthrene. Anthracene and phenanthrene are aromatic compounds containing three benzene rings joined together. Their structures are indicated below.

anthracene

phenanthrene

Anthracene is used commercially in the manufacture of dyes. Phenanthrene is an isomer of anthracene. It also contains three benzene rings but in a different structural arrangement. Phenanthrene has the basic structure of many biologically and medically important compounds. Among these are the male and female sex hormones, vitamin D, cholesterol, bile acids, and some alkaloids.

Other Aromatic Compounds

Phenols

When an —OH group is attached to a ring system, a class of compounds known as phenols is formed. The simplest member of this class is phenol, a compound in which an —OH group is attached to a benzene ring. Generally, phenols are like alcohols but have been placed in a class by themselves since phenols are weak acids, but alcohols are not.

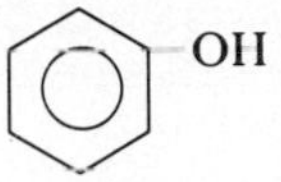

phenol

Pure phenol is a white crystalline solid with a low melting point, 41°C. However, on exposure to light and air phenol turns reddish. Phenol is poisonous if taken internally. Externally it causes deep burns and blisters on the skin. If phenol should accidentally be spilled on the skin, it should be removed as quickly as possible with 50 per cent alcohol, glycerin, or sodium bicarbonate solution, or water.

Phenol is used as a disinfectant for surgical instruments and utensils, clothing and bed linens, floors, toilets, and sinks. Phenol is used commercially in the manufacture of dyes and plastics.

Phenol is the standard of reference for germicidal activity of disinfectants; that is, their activity is compared to that of phenol.

Phenol Derivatives

The methyl derivatives of phenol are called cresols. There are three different cresols—*ortho*, *meta*, and *para*.

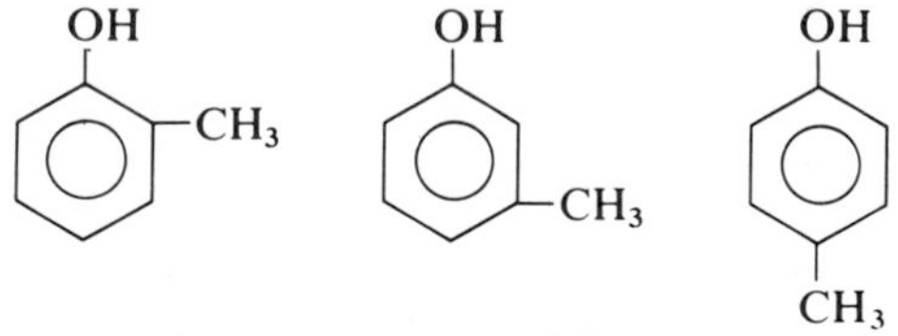

o-cresol   m-cresol   p-cresol

Commonly, cresol is a mixture of all three of these isomers. Cresol is a better antiseptic than phenol and also is less toxic. One commonly used disinfectant, lysol, is a mixture of the three cresols in water with soap added as an emulsifying agent. Even though cresols are less toxic than phenol, they are still poisonous and should be used for external purposes only.

Resorcinol is *m*-dihydroxybenzene. It is also an antiseptic but is not as good as phenol. Its structure is

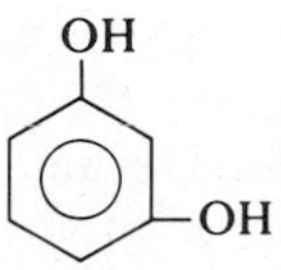

resorcinol

A resorcinol derivative, hexylresorcinol, is a much better antiseptic and germicide. It is commonly used in mouthwashes. Its structure is

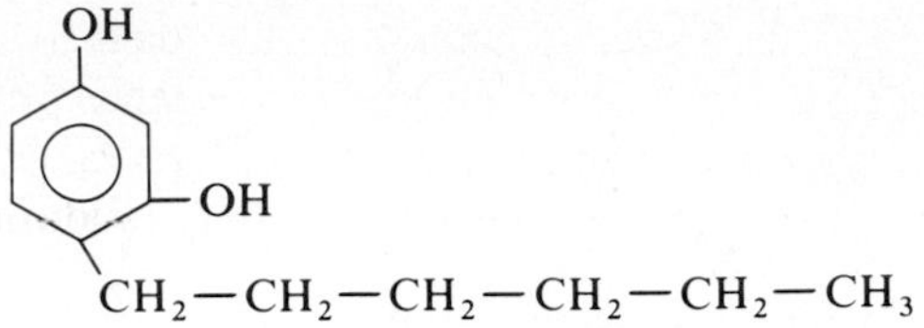

OH

OH

$CH_2-CH_2-CH_2-CH_2-CH_2-CH_3$

hexylresorcinol

Hexachlorophene has marked antibacterial activity. Its activity is retained in the presence of soaps so that soaps containing hexachlorophene are used for surgical scrubs and also by food handlers and dentists. Hexachlorophene has also been employed in deodorants and cleansing creams, but its use for all these purposes has largely been discontinued because of the possibility of adverse effects on brain tissue, particularly in premature infants. The structure of hexachlorophene is

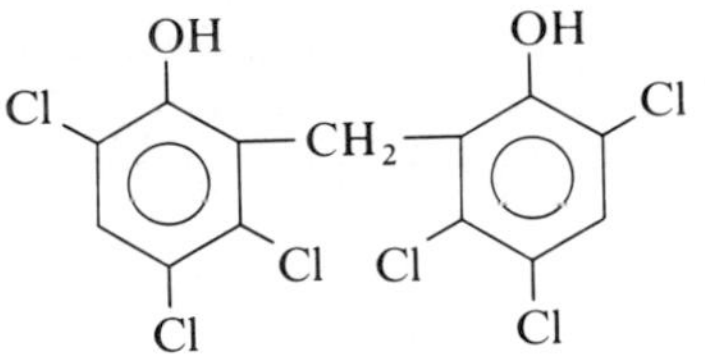

OH   OH

Cl   Cl

CH_2

Cl   Cl

Cl   Cl

hexachlorophene

Picric Acid

Picric acid, 2,4,6-trinitrophenol, is prepared by heating phenol with a mixture of nitric and sulfuric acids, as indicated below. Note that this compound is named numerically, with the —OH group being understood to be in position 1:

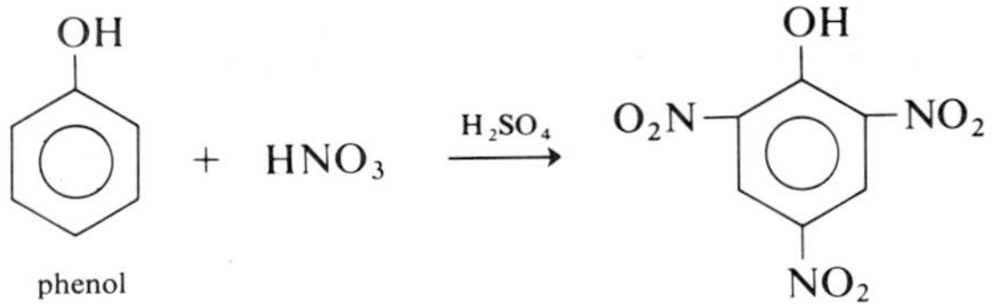

OH   OH

$+ \ HNO_3 \quad \xrightarrow{H_2SO_4} \quad O_2N$ NO_2

phenol NO_2

picric acid

Picric acid is a bright yellow solid that has been used as an antiseptic for the treatment of burns. It coagulates the protein on the surface of the burn and thus prevents the loss of blood serum. Picric acid has been replaced for this purpose by newer and more effective drugs such as silver sulfadiazine. Commercially, picric acid is used as an explosive.

Aldehydes

Aromatic aldehydes have a general formula ArCHO, where Ar stands for an aromatic ring. The simplest aromatic aldehyde is benzaldehyde, which consists of an aldehyde group attached to a benzene ring. Benzaldehyde is prepared by the mild oxidation of toluene, as is shown below. Note that the side chain, the methyl group, is more susceptible to oxidation than the fairly stable benzene ring.

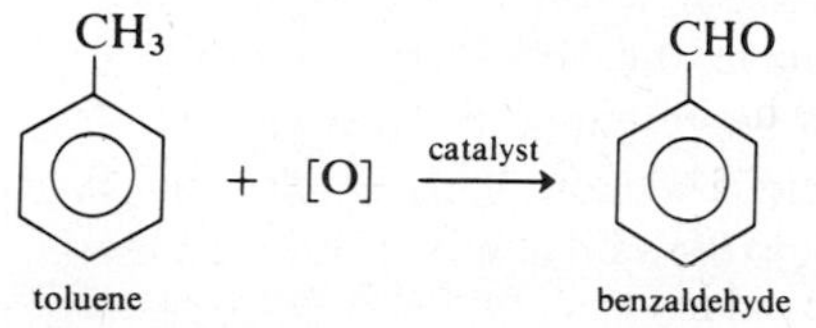

toluene   benzaldehyde

Benzaldehyde is a colorless oily liquid with an almondlike odor. It is used in the preparation of flavoring agents, perfumes, drugs, and dyes.

Vanillin occurs in vanilla beans and gives the particular taste and odor to vanilla extract. It also has an aldehyde structure. Cinnamic aldehyde is present in oil of cinnamon and oil found in cinnamon bark. Both vanillin and cinnamic aldehyde may be prepared synthetically, and both are used as flavoring agents.

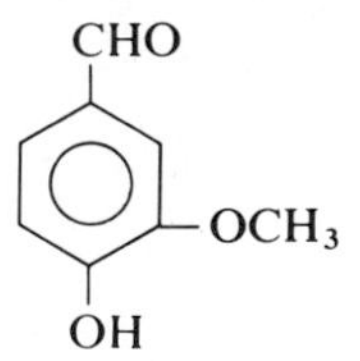

vanillin 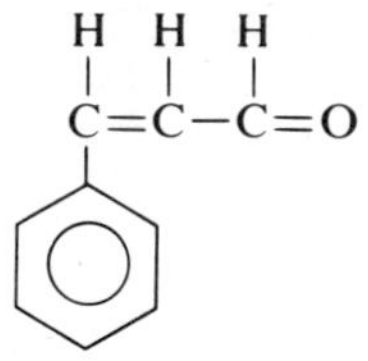   cinnamic aldehyde

Ketones

Aromatic ketones have the general formula ArCOAr′ or ArCOR. The simplest aromatic ketone is acetophenone.

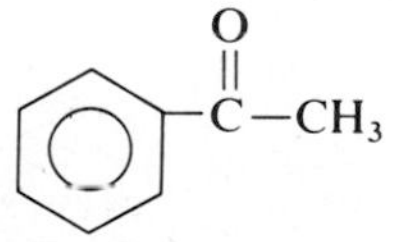

acetophenone

Acetophenone has been used as a hypnotic but has been supplanted for this purpose by newer and safer drugs.

Chloracetophenone is a lacrimator and is used as a tear gas.

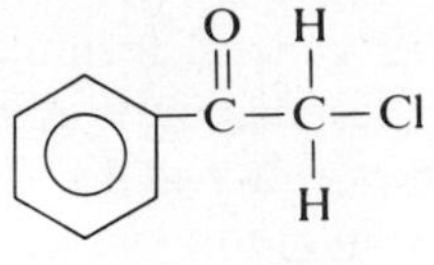

chloracetophenone

Acids

Aromatic acids have the —COOH group just as do aliphatic acids. Aromatic acids are represented by the general formula ArCOOH. The simplest aromatic acid is benzoic acid, which consists of a —COOH attached to a benzene ring. Benzoic acid can be produced by the oxidation of toluene, as is shown below.

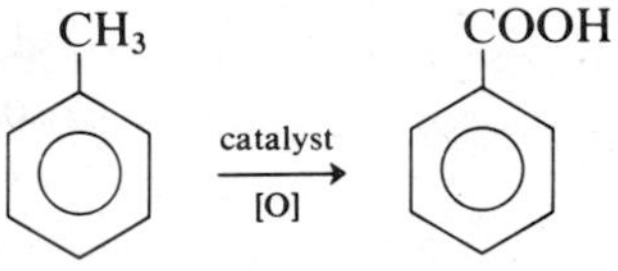

toluene   benzoic acid

Benzoic acid is a white crystalline compound slightly soluble in cold water and more soluble in hot water. Benzoic acid is used medicinally as an antifungal agent. The sodium salt of benzoic acid, sodium benzoate, is used as a food preservative.

Salicylic acid is both an alcohol and an acid, as can be seen from its structure.

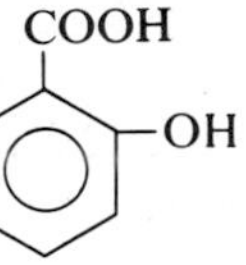

salicylic acid

Salicylic acid is a white crystalline compound with properties similar to those of benzoic acid. It is used in the treatment of fungal infections and also for the removal of warts and corns.

The commonly used salts of salicylic acid are sodium salicylate and methyl salicylate.

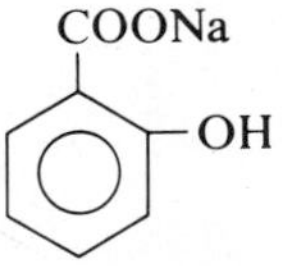

sodium salicylate 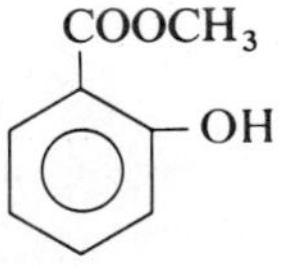   methyl salicylate

Sodium salicylate is used as an antipyretic (to reduce fever) and also to relieve pain of arthritis, bursitis, and headache. Methyl salicylate is a liquid with a pleasant odor, that of wintergreen. It is used topically to relieve pain in muscles and joints.

The acetyl derivative of salicylic acid (the acetyl group is $CH_3CO—$) is acetylsalicylic acid, more commonly known as aspirin.

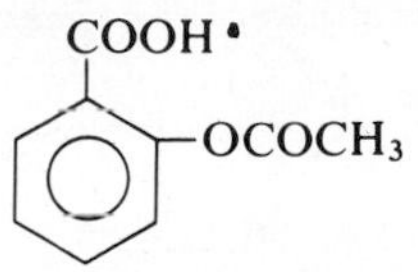

acetylsalicylic acid, aspirin

Aspirin is used as an analgesic (to relieve pain), as an antipyretic, and for the treatment of colds, headaches, minor aches, and pains. Aspirin is also used in the treatment of rheumatic fever. Over 43 tons of aspirin are used daily in the United States. Aspirin is contraindicated after surgery because it interferes with the normal clotting of the blood and so may induce hemorrhaging.

Aspirin can cause bleeding of the stomach and, therefore, should not be taken on an "empty" stomach. About one person in 10,000 is allergic to aspirin.

The action of aspirin is related directly to that of the prostaglandins (see page 285).

Recent evidence appears to indicate that aspirin can prevent blood clots from forming by interfering with the action of the blood platelets. Evidence has also shown that aspirin seems to help prevent heart attacks. While aspirin can prevent sickling of red blood cells (see page 285) in test tubes, clinical evidence of this effect on humans is far from complete.

Amines

The simplest aromatic amine is aniline, which consists of an amine group attached to the benzene ring. It is prepared by the reduction of nitrobenzene.

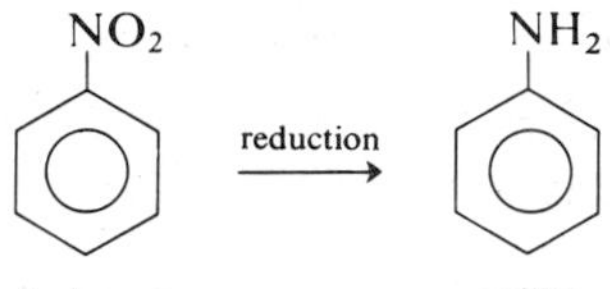

nitrobenzene   aniline

Aniline is used commercially in the preparation of many dyes and drugs. When aniline is reacted with acetic acid, acetanilide is produced.

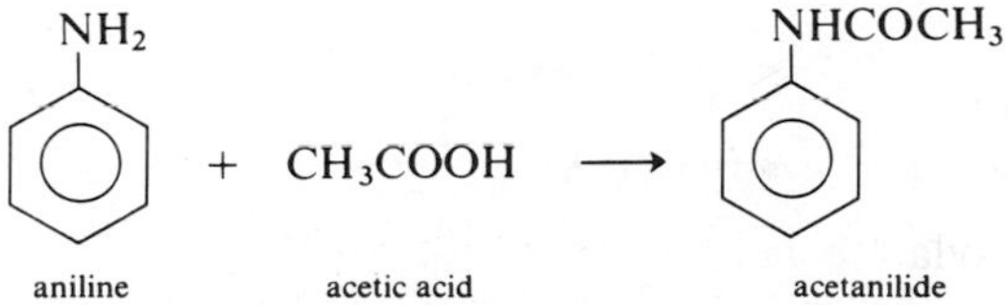

aniline   acetic acid   acetanilide

Acetanilide has been used as an antipyretic and as an analgesic, but it has largely been supplanted by a related compound, phenacetin, which is less toxic.

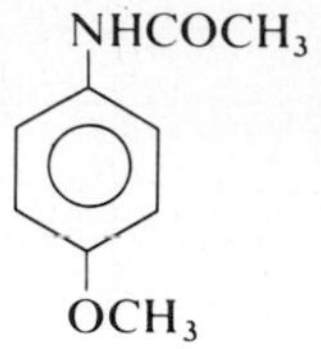

NHCOCH$_3$

OCH$_3$

phenacetin

When aniline is heated with concentrated sulfuric acid, it forms *p*-aminobenzenesulfonic acid, commonly known as sulfanilic acid.

NH$_2$

SO$_3$H

sulfanilic acid

In 1936 a derivative of sulfanilic acid, sulfanilamide, was discovered to have definite therapeutic effects against such diseases as pneumonia, diarrhea, and streptococcal infections. However, further investigation showed that sulfanilamide had several disadvantages. It caused nausea, dizziness, anemia, and other toxic reactions in the body. Pharmaceutical chemists then developed a group of related sulfanilamide compounds with equal or better therapeutic value and with greatly decreased toxic effects.

All of the sulfa drugs contain the *p*-aminobenzenesulfonamide group which gives them their characteristic properties.

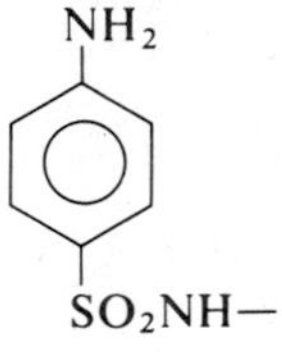

NH$_2$

SO$_2$NH—

p-aminobenzenesulfonamide group

Examples of some sulfa drugs are shown below.

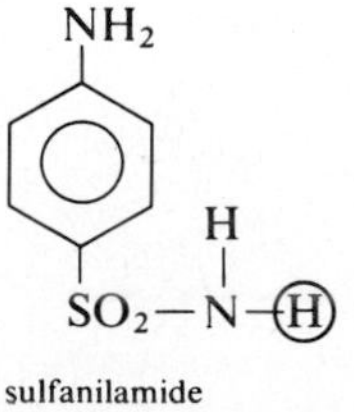

NH$_2$

SO$_2$—N—H

sulfanilamide

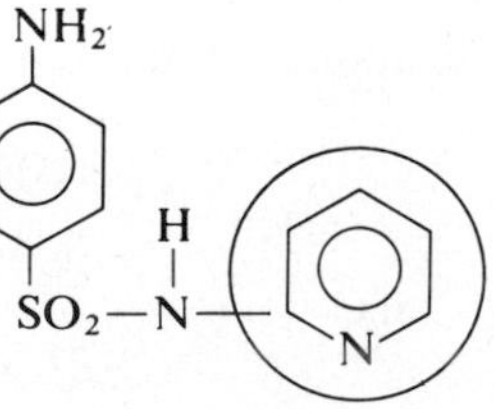

NH$_2$

SO$_2$—N

sulfapyridine

NH$_2$

SO$_2$—N

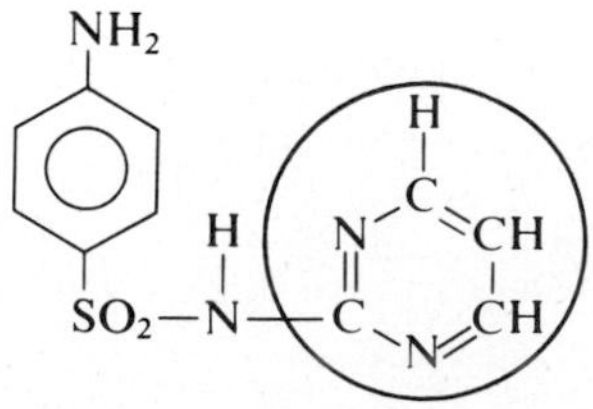

sulfadiazine

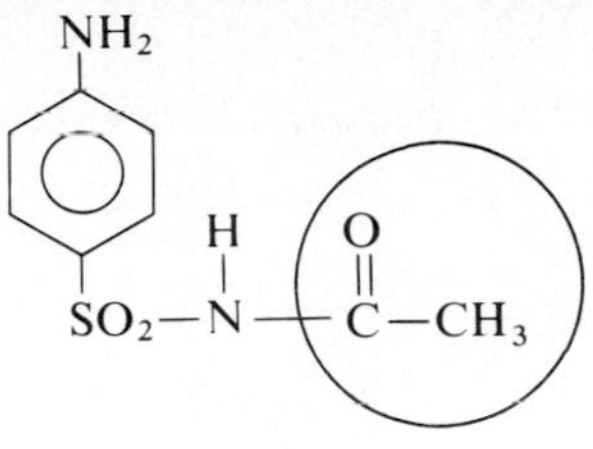

NH$_2$

SO$_2$—N—C—CH$_3$

sulfacetamide

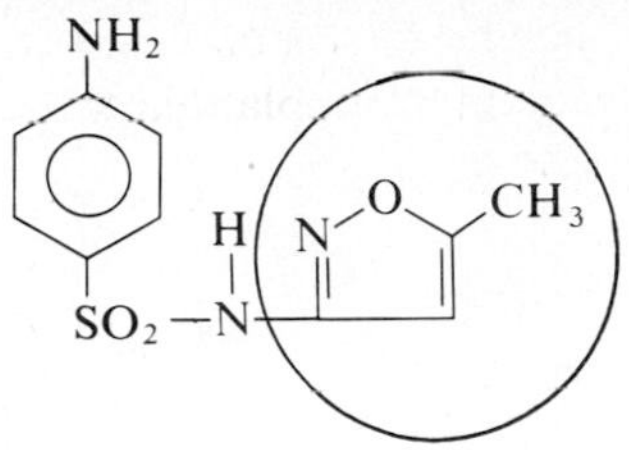

NH$_2$

SO$_2$—N

sulfamethoxazole

Since 1935, when the antibacterial action of Prontosil was discovered by Domagk, over 5000 sulfonamide (sulfa-type) compounds have been prepared and tested. Of these only a few have clinical use. The different groups attached to the parent substance determine the potency and toxicity. Most of the sulfa drugs are white crystalline substances that are insoluble in water. Their sodium compounds, however, are readily soluble.

The sulfa drugs are classified as bacteriostatic and not bactericidal. They appear to weaken or inhibit the growth of susceptible bacteria and make them more vulnerable to the action of phagocytes in the bloodstream of the host.

It is believed that the sulfa drugs interfere and compete with the reaction of p-aminobenzoic acid (PABA) on a necessary group of enzymes. Note the similarity of structures of the sulfa drugs and PABA (see page 339).

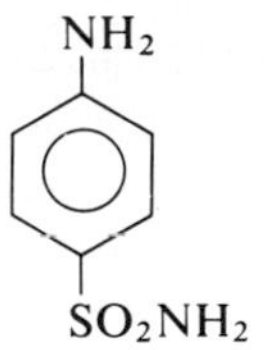

NH$_2$

SO$_2$NH$_2$

sulfanilamide, a sulfa drug

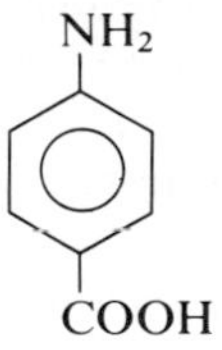

NH$_2$

COOH

PABA

Amides

The simplest aromatic amide is benzamide, which is formed by the reaction of benzoic acid with ammonia.

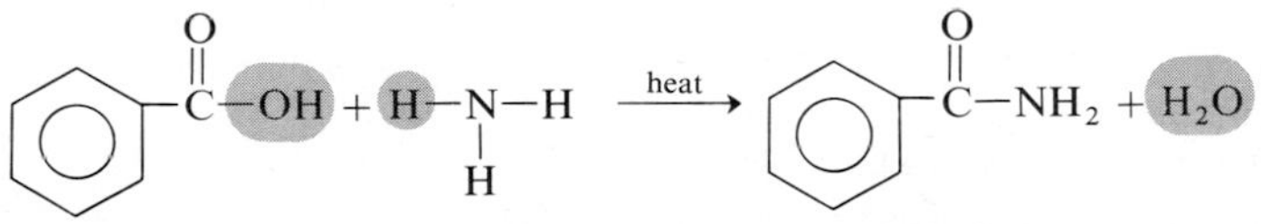

C—OH + H—N—H $\xrightarrow{\text{heat}}$ C—NH$_2$ + H$_2$O
             |
             H

benzoic acid   ammonia   benzamide

Aromatic amides have the general formula ArCONH$_2$. The amide of niacin, niacinamide, is one of the B vitamins (see page 453).

niacin

niacinamide

Summary

A cyclic hydrocarbon is a ring compound. This ring compound may be saturated, as in the case of cyclopropane, or unsaturated, as in the case of benzene. Benzene, C_6H_6, is a symmetrical six-sided ring compound with a hydrogen at each corner. One or more of the hydrogen atoms of benzene may be replaced by a halogen yielding halogen derivatives.

There is only one monosubstitution product of benzene because all six positions on the ring are equivalent.

There are three possible disubstitution products of benzene. These compounds may be named numerically. If both the substituents are chlorine, the disubstitution products are 1,2-dichlorobenzene, 1,3-dichlorobenzene, and 1,4-dichlorobenzene.

Prefixes are also used to designate positions on the benzene ring. The prefix *ortho* corresponds to positions 1,2; *meta-* to 1,3; and *para-* to 1,4.

Naphthalene contains two benzene rings joined together. Anthracene and phenanthrene contain three benzene rings joined together.

The same general formulas are used for aromatic compounds; ArOH can indicate an alcohol, with Ar representing an aromatic ring.

The simplest ring alcohol is phenol, a benzene ring with an —OH group attached. Phenol is used as a disinfectant of surgical instruments.

Cresols are methyl derivatives of phenol. Cresol is a better antiseptic than phenol and is also less toxic.

Benzoic acid consists of a benzene ring with a carboxyl (—COOH) group attached. Benzoic acid may be produced by the oxidation of toluene.

Acetylsalicylic acid, commonly known as aspirin, is an analgesic, an antipyretic, and is used in the treatment of colds, headaches, minor aches, and pains. Aspirin is contraindicated after surgery.

Aniline consists of an amine group attached to a benzene ring. Aniline is the basic compound for many dyes and drugs. Among these are the sulfa drugs.

Questions and Problems

1. Draw the structure of cyclopropane. What is this compound used for?
2. Draw the structure of cyclobutane.
3. What are aromatic compounds?
4. Indicate the resonance structures for benzene.
5. What is the more recent abbreviation for the structure of benzene?
6. What effects do benzene vapors have on the body?
7. Draw the structure of chlorobenzene.
8. Why is there only one monosubstitution product for benzene?

9. Why are there three and only three disubstitution products for benzene?
10. Draw the structures for the following compounds: (a) 1,3-dibromobenzene; (b) 1,4-diiodobenzene; (c) 1-chloro-2-iodobenzene.
11. Name the following compounds numerically:

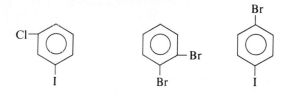

12. Name the compounds in question 11 using the prefix system.
13. Draw the structure of each of the following compounds: (a) *o*-dichlorobenzene; (b) *m*-bromochlorobenzene; (c) *p*-xylene.
14. Draw the structure of toluene and indicate one medical use for it.
15. What are xylenes used for?
16. Draw the structure of naphthalene; of alpha-naphthol.
17. Why does naphthalene have only two possible monosubstitution products?
18. Draw the structures of phenanthrene and anthracene. Why is the phenanthrene structure important biologically?
19. Draw the structure of phenol. For what is this compound used?
20. What are cresols? How many cresols are there? For what are they used?
21. Draw the structures of resorcinol and hexylresorcinol.
22. What is the structure of picric acid? From what compound may it be made?
23. How may benzaldehyde be prepared? Write the equation.
24. What kind of compound is acetophenone?
25. How may benzoic acid be prepared? Write the equation.
26. Write the formula for sodium benzoate. What is it used for?
27. What is methyl salicylate used for? Sodium salicylate?
28. Draw the structure of aspirin. What is its chemical name?
29. List several uses for aspirin.
30. Why is the use of aspirin after surgery contraindicated?
31. How may acetanilide be prepared?
32. Draw the structure of sulfanilic acid. Of what medical importance is this compound?
33. To what do sulfa drugs owe their activity?
34. What is the general structure for the following types of aromatic compounds: acids, aldehydes, amines, amides?
35. Write the reaction between benzoic acid and ammonia; between benzoic acid and methyl amine.

References

Baum, S. J., and Scaife, C. W.: *Chemistry: A Life Science Approach.* Macmillan Publishing Co., Inc., New York, 1975, Chap. 15.
Collier, H. O.: Aspirin. *Scientific American*, **209**:97–108 (Nov.), 1963.

Morrison, R. T., and Boyd, R. N.: *Organic Chemistry*, 3rd ed. Allyn & Bacon, Inc., Boston, 1973, Chap. 10.

O'Leary, M. H.: *Contemporary Organic Chemistry*. McGraw-Hill Book Co., New York, 1976, Chap. 11.

Stacy, G. W.: *Organic Chemistry*. Harper & Row, New York, 1975, Chap. 8.

Heterocyclic Compounds

Heterocyclic compounds are ring compounds that contain some element other than carbon in the ring. The elements most commonly found in the ring other than carbon are nitrogen, sulfur, and oxygen. Heterocyclics are primarily five- and six-membered rings, sometimes joined to one another.

Some Common Heterocyclic Compounds

Listed below are a few of the common rings in structural and abbreviated form; also indicated are the biologically and medically important compounds based on these structures. The compound after which the group is named is given in italics.

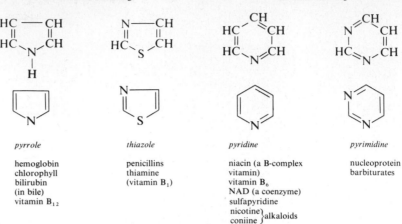

five-membered rings   six-membered rings

pyrrole   *thiazole*   *pyridine*   *pyrimidine*

hemoglobin   penicillins   niacin (a B-complex   nucleoprotein
chlorophyll   thiamine   vitamin)   barbiturates
bilirubin   (vitamin B₁)   vitamin B₆
(in bile)   NAD (a coenzyme)
vitamin B₁₂   sulfapyridine
                                    nicotine ⎫
                                    coniine ⎭ alkaloids

fused five- and six-membered rings

purine

caffeine
dramamine
theobromine
theophylline
nucleic acids
ATP, ADP,
AMP, C-AMP
coenzyme A

The DNA (deoxyribonucleic acid) structure, which is the basis of genetic information in the cell, contains four heterocyclic rings; two of these have a purine base and two have a pyrimidine base. The purines are adenine and guanine; the pyrimidines are cytosine and thymine. The cell messenger, ribonucleic acid (RNA) (see Chapter 33), contains the pyrimidine uracil in place of thymine. Their structures are

adenine   guanine

cytosine   uracil   thymine

Mind Drugs

Mind drugs affect the central nervous system. There are several categories of mind drugs, depending upon their action and the responses they elicit. Among them are the opiates, barbiturates, amphetamines, psychedelics, and cannabinoids (marihuana). Most of these drugs are heterocyclic compounds containing one or more nitrogen molecules in their rings.

Opiates

Opiates are morphine narcotics that are obtained from the opium poppy. The dried unripened pod of this plant is the source of opium, which is a complex mixture of many narcotics. Among them are morphine, codeine. and papaverine. Heroin can be prepared from morphine by a process called acetylation. Synthetic opiates, compounds with morphinelike properties, include meperidine (Demerol) and methadone.

The opiates depress the central nervous system, although the mechanism by which they act is unknown. The principal effects are analgesia, sedation, drowsiness, lethargy, and euphoria. The major reason for the abuse of opiates is the relief of apprehension and elevation of mood. However, as time passes, the euphoria wears off and the user becomes apathetic and suddenly falls asleep. Opiates depress the respiratory center, and death from an overdose results from respiratory failure.

Morphine is used as an analgesic for the relief of postoperative pain, for cardiac pain, and for conditions where other narcotics are ineffective. However, the chief disadvantage of the continued use of morphine is that it leads to addiction.

Codeine (methylmorphine) depresses the cough center of the brain and so is used in cough medicines. However, it also is addictive.

Papaverine, another opiate, is an antispasmodic; it is a nonspecific smooth muscle relaxant. The drug has been used to dilate blood vessels but is now practically obsolete.

The most commonly abused opiate is heroin, which is a bitter-tasting, water-soluble compound. Heroin is two and one-half times as powerful as morphine and has a correspondingly greater possibility of addiction. Its use in the United States has been outlawed except for research purposes. A user of hereoin develops a dependence upon the drug and also builds up a tolerance to it. As a tolerance develops, increased amounts of heroin are required, so that the addict may be using an amount that would be fatal to a nonuser.

Rapid intravenous injection of heroin produces a warm flushing of the skin and a sensation in the lower abdomen described by addicts as a "thrill." Morphine behaves similarly. Actually, heroin is converted into morphine by the body.

If an addict is without the drug for 10 to 12 hours, the following symptoms appear: vomiting, diarrhea, tremors, pains, restlessness, and mental disturbances. A pregnant heroin addict can pass the dependency on to the child, who must then spend the first few days of life withdrawing from the drug. If not treated properly, the infant may die within a week.

Heroin and other opiates reduce aggression and sexual drive and, therefore, are unlikely to induce crime. However, many individuals commit crimes to pay for their habit.

The currently preferred treatment for heroin addiction is the use of methadone. Methadone is itself an addictive narcotic, but it does not induce euphoria. It eliminates the desire for heroin and also reduces the withdrawal symptoms that normally accompany abstention from heroin. The amount of methadone used is gradually reduced as withdrawal symptoms lessen.

Barbiturates

The barbiturates are a class of synthetic drugs that have sedative and hypnotic (sleep-producing) properties when administered in therapeutic dosages. The structure of a barbiturate is

Table 20–1 indicates some of the various barbiturates that can be prepared by varying the constituents R_1 and R_2.

In general, the more rapid the onset of action of a barbiturate, the shorter the duration of its physiologic effect. Phenobarbital is a long-acting barbiturate (10 to 12 hours), while secobarbital (Seconal) and pentobarbital (Nembutal) are short-acting drugs (3 to 4 hours).

Phenobarbital is used as an anticonvulsant agent in the treatment of epilepsy, but its mechanism of action is unknown.

In general, barbiturates depress the central nervous system. A person who takes repeated doses of barbiturates develops a tolerance for the drug and soon requires many times the original amount to produce the desired effect. Small amounts of barbiturates produce the symptoms of chronic alcoholism—the user is sociable, relaxed, and good humored, but his ability

TABLE 20–1

Composition and Names of Some Barbiturates

Name	R_1	R_2	Trade Name	"Street" Name
Barbital	ethyl, C_2H_5-	ethyl, C_2H_5-	Veronal	
Amobarbital	ethyl, C_2H_5-	isopentyl, $CH_3CH(CH_3)CH_2CH_2-$	Amytal	Blue heaven
Phenobarbital	ethyl, C_2H_5-	phenyl, C_6H_5-	Luminal	
Probarbital	ethyl, C_2H_5-	isopropyl, $(CH_3)_2CH-$	Ipral	
Pentobarbital	ethyl, C_2H_5-	1-methylbutyl, $CH_3CH_2CH_2CH(CH_3)-$	Nembutal	Yellow jackets
Secobarbital	allyl, $CH_2{=}CHCH_2-$	1-methylbutyl, $CH_3CH_2CH_2CH(CH_3)-$	Seconal	Red devils

to react and his alertness are decreased. Large amounts of the drug result in blurred speech, confusion, mental sluggishness, and loss of emotional control and may induce a coma.

Barbiturates are a definite hazard to the user in that, even though the tolerance builds up with increased use, the lethal dose remains about the same, thus reducing an addict's margin of safety.

In the United States the names of all barbiturates end in -al, even though the compounds are ketones rather than aldehydes. In the British Commonwealth, the ending -one is used.

Amphetamines

The amphetamines, known as "pep pills" or uppers, are one type of the synthetic drugs that stimulate the central nervous system. These drugs are useful in the treatment of such conditions as narcolepsy (an overwhelming compulsion to sleep), mild depression, and obesity. However, their use for the latter condition is not recommended because as soon as the drug is withdrawn the appetite returns. Frequent misusers of amphetamines are overtired truckdrivers and businessmen, students cramming for exams, and athletes who need a "pickup" before a game. Prolonged use or misuse leads to serious effects because the amphetamines tend to concentrate in the brain and cerebrospinal fluid. A tolerance to the drug develops rapidly, and increasing amounts lead to long periods of sleeplessness, loss of weight, and severe paranoia.

Amphetamine (Benzedrine) was synthesized to simulate the action of epinephrine (Adrenaline) (see page 479). Note the similarity of their structures.

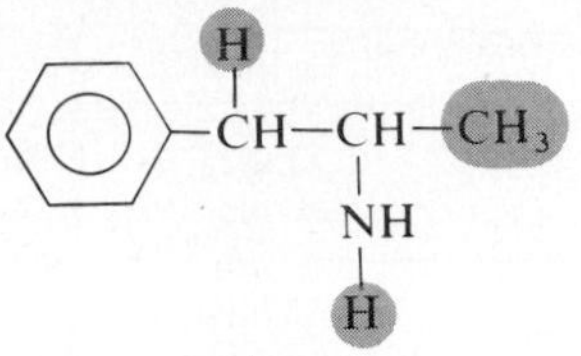

amphetamine
(Benzedrine)

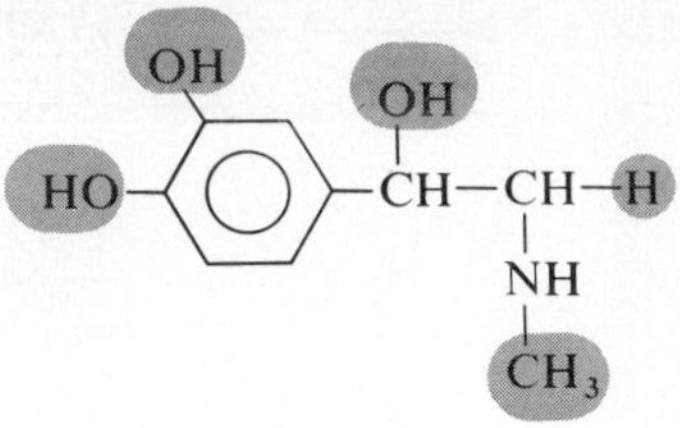

epinephrine
(Adrenaline)

Drug users refer to Benzedrine as "bennies" and dextroamphetamine as "dexies." Amphetamine and methamphetamine, used intravenously, are known as "speed."

Psychedelics

The psychedelic drugs produce "visions." Among these drugs are the *hallucinogens*, drugs that stimulate sensory perceptions, and the *psychomimetics*, drugs that seem to mimic psychoses.

The "psychedelic state" includes several major effects. There is a heightened awareness of sensory input. The individual has an enhanced sense of clarity but a reduced control over what is experienced. One part of the individual seems to be observing what the other part is seeing and doing. The environment may be perceived as harmonious and beautiful with interplays of light and color in minute detail. Colors may be heard and sounds may be seen. There is a loss of boundary between the real and the unreal.

The most abused of these drugs are the hallucinogens, such as LSD (lysergic acid diethylamide), mescaline (3,4,5-trimethoxyphenylethylamine), and STP (2,5-dimethoxy-4-methylamphetamine) (the initials STP in drug culture refer to serenity, tranquility, and peace).

The structures of LSD, mescaline, and STP are indicated below.

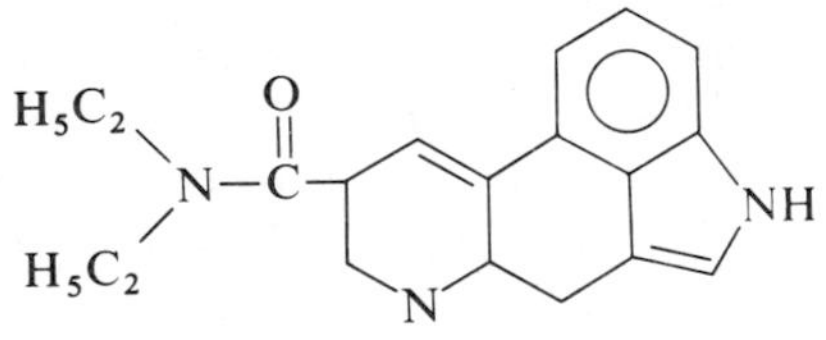

LSD

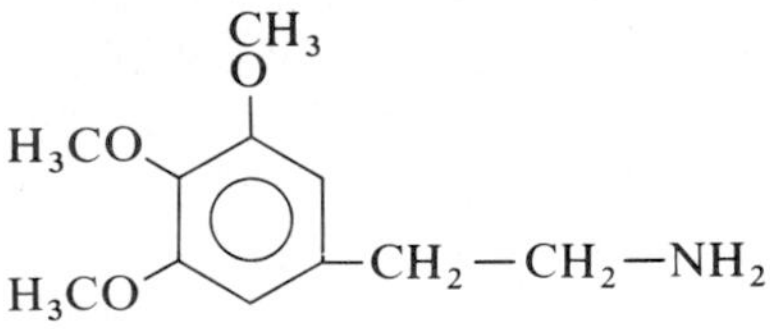

mescaline

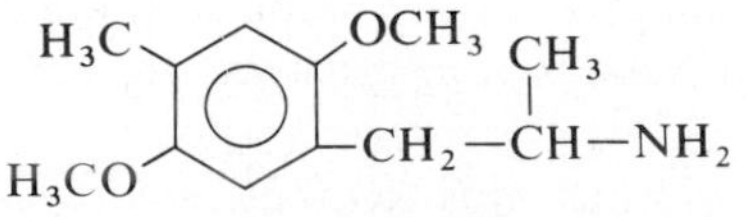

STP

Of all the hallucinogens, LSD ("acid") is the most powerful. An average "trip" dose of LSD is 0.1 mg. To obtain the same effect from STP requires 5 mg, and from mescaline 400 mg. Certain susceptible individuals feel the psychologic effects of LSD in doses as low as 20–25 μg (0.020 to 0.025 mg). The effects on the central nervous system begin within an hour and may last 8 to 12 hours.

In addition to the hallucinogenic effects, LSD causes dilation of the pupils of the eyes, an increased blood pressure, tachycardia, nausea, and increased body temperature.

It is believed that LSD exerts its effects by interfering with serotonin, a hormone found in the brain, which plays an important part in the thought process. Note the similarity in structure between LSD (above) and serotonin (below).

serotonin

LSD has been used by psychiatrists and psychologists in the treatment of psychoneuroses and social delinquency. It has also been used for the treatment of terminal cancer patients because it lessens their pain and awareness.

Cannabinoids (Marihuana)

Cannabis, also known as marihuana, hashish, "pot," "grass," or "Mary Jane," is obtained from the flowery tops of hemp plants. In the Middle East and North Africa the dried resinous extract of the tops is called hashish; in the Far East, charas. In the United States the term *marihuana*, or *marijuana*, is used to describe any part of the plant that induces psychic changes in man.

Marihuana acts on both the central nervous system and the cardiovascular system. When it is smoked, the effects can be felt within a few minutes. If eaten with food, the effects are delayed about an hour. The physiologic effects include a voracious appetite, nausea, bloodshot eyes, increased pulse rate, dry mouth and throat, and dilation of the pupils. In addition, there is a sense of euphoria, intensified visual images, a keener sense of hearing, and a distortion of space and time.

Large doses of marihuana can induce hallucinations, delusions, and paranoiac feelings. Anxiety may replace euphoria.

Marihuana is not a narcotic in that it is not a derivative of opium or synthetic substitutes, but it is so labeled by law. It is not addictive, and there is no medical evidence of the effects of its long-term use.

Other Medically Used Heterocyclics

Atropine. Obtained from the belladonna plant, atropine is administered as a sulfate salt. It is used extensively to dilate the pupil of the eye. It is also used as an antispasmodic (that is, to prevent spasms), to check excessive secretion of saliva and mucous, and in the treatment of Parkinson's disease.

Caffeine. Caffeine occurs in coffee and tea. It stimulates the central nervous system and also acts as a diuretic. Caffeine is used for various headache remedies in conjunction with analgesic drugs.

Cocaine. Obtained from the leaves of a coca shrub in South America, cocaine is administered medicinally in the form of a hydrochloride. Because it blocks nerve conduction, cocaine is widely used as a local anesthetic. However, it has largely been supplanted by synthetic drugs which are less habit forming and less toxic.

Nicotine. Found in tobacco leaves, nicotine is one of the few alkaloids soluble in water. It has no therapeutic use but is used as an insecticide. Nicotine is found in the body after cigarette smoking. It increases the blood pressure and pulse rate and constricts the blood vessels.

Quinine. Obtained from the bark of the cinchona tree in South America, quinine was formerly used primarily in the treatment of malaria. It has been replaced by less toxic and more effective synthetic antimalarial drugs. Quinine has been used as an antipyretic and analgesic for muscular pain and headaches.

Emetine. Emetine is found in the roots of the *Cephaelis ipecacuanha* shrub which grows in South America. The dried root is known as ipecac. Emetine hydrochloride is used to control acute amebic dysentery and to treat amebic hepatitis. Emetine is usually given subcutaneously. It may be administered orally but, if so given, it usually induces nausea and vomiting.

Ephedrine. Obtained from twigs of a Chinese plant ma-huang, ephedrine is used to elevate blood pressure during spinal anesthesia. It is used in the treatment of asthma, coughs, and colds. It is also used as a nasal decongestant.

Reserpine. Obtained from the shrub of *Rauwolfia serpentina*, in India and the Malay peninsula, reserpine is used primarily for the treatment of hypertension. It is also used as a sedative and as a relaxant.

Scopolamine. Scopolamine is obtained from the scopolar plant. It depresses the central nervous system and is used as a sedative and hypnotic.

Scopolamine is used in conjunction with morphine as a preanesthetic combination. This mixture quiets the patient, reduces apprehension, and also reduces shock to the nervous system.

Summary

Heterocyclics are compounds containing some element other than carbon in the ring, the most common being nitrogen, sulfur, and oxygen. Heterocyclics are usually five- or six-membered rings sometimes joined together.

The parent compounds of some of the more important heterocyclic compounds are: pyrrole, a five-membered ring with a nitrogen in it; pyridine, a six-membered ring containing a nitrogen; pyrimidine, a six-membered ring containing two nitrogens; and purine, a six-membered ring containing two nitrogens joined to a five-membered ring containing two nitrogens.

The purines and the pyrimidines are the parent compounds for the base of the nucleic acids DNA and RNA.

Mind drugs affect the central nervous system. Among the categories of mind drugs are the opiates, barbiturates, amphetamines, psychedelics, and cannabinoids.

The following heterocyclics are but a few of the many used medically: atropine, to dilate the pupil of the eye; caffeine, as a stimulant; cocaine, as a local anesthetic; quinine, for treatment of malaria; emetine, to control dysentry; ephedrine, as a nasal decongestant; reserpine, for treatment of hypertension; scopolamine, as a sedative.

Questions and Problems

1. Define the term *heterocyclic*.
2. Which heterocyclic bases are used in the DNA and RNA molecules?
3. What are opiates?
4. Name two naturally occurring opiates; one semisynthetic opiate; two synthetic opiates.
5. What are the effects of opiates on the body?
6. For what purposes is morphine used? What is its chief disadvantage?
7. What is codeine used for? How is it related structurally to morphine?
8. What are the dangers of use of heroin? What are the symptoms of heroin withdrawal?
9. How is heroin addition treated?
10. What properties do barbiturates have? How do they differ structurally?
11. What is one use for phenobarbital?
12. Why is there greater danger of a fatal overdose with barbiturates than with opiates?
13. What is narcolepsy? How is it treated?
14. Who are frequent misusers of amphetamines?
15. What are the effects of the use of psychedelic drugs?
16. What is the most powerful psychedelic drug?
17. What is believed to be the method of action of LSD?

18. Cannabis is known by what other names? What are the effects of small amounts of cannabis? Large amounts?
19. Indicate one use for each of the following alkaloids: (a) caffeine; (b) quinine; (c) reserpine; (d) cocaine; (e) atropine.
20. Why is emetine not given orally? How is it usually given?
21. Why is scopolamine used in conjunction with morphine?

References

Baum, S. J., and Scaife, C. W.: *Chemistry: A Life Science Approach.* Macmillan Publishing Co., Inc., New York, 1975, Chap. 20.

Brown, R. F.: *Organic Chemistry.* Wadsworth Publishing Co., Inc., Belmont, Calif., 1975, Chap. 28.

Goodman, L. S., and Gilman, A. (eds): *The Pharmacological Basis of Therapeutics*, 5th ed. Macmillan Publishing Co., Inc., New York, 1975, Chap. 16.

Morrison, R. T., and Boyd, R. N.: *Organic Chemistry*, 3rd ed. Allyn & Bacon, Inc., Boston, 1973, Chap. 31.

unit three

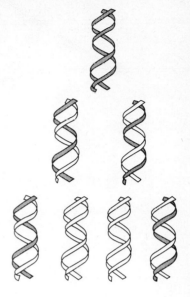

Biochemistry

BIOCHEMISTRY IS THE STUDY OF LIFE ITSELF—

the interrelationships of the metabolism of carbohydrates, fats, and proteins

the need for enzymes, vitamins, and hormones in various body processes

the formation and excretion of waste products

the properties and functions of the body's fluids, including blood and urine

the cause and cure of certain pathologic conditions

the study of human reproduction and its molecular basis as evidenced by the study of DNA

Chapter 21

Carbohydrates

C H A P T E R   O U T L I N E

Classification

Origin
Oxygen–Carbon Dioxide Cycle in Nature

Optical Activity
Structural Isomers / Functional Isomers / Geometric Isomers / Optical Isomers

Monosaccharides
Trioses / Pentoses / Hexoses / Reactions of the Hexoses

Disaccharides
Reducing Properties / Fermentation / Structure of the Disaccharides / Sucrose / Maltose / Lactose

Polysaccharides
Starch / Cellulose / Glycogen / Dextrin / Heparin / Structure of the Polysaccharides

Test for Carbohydrates

Carbohydrates are a class of organic compounds that includes sugars, starches, and cellulose. Originally all known carbohydrates were considered to be hydrates of carbon because they contain hydrogen and oxygen in the ratio of two to one just as in water. The formula for glucose, $C_6H_{12}O_6$, was written as $C_6(H_2O)_6$. Likewise sucrose, $C_{12}H_{22}O_{11}$, was written as $C_{12}(H_2O)_{11}$. However, later investigation showed that rhamnose, another carbohydrate, has the formula $C_6H_{12}O_5$. It did not fit the general formula for a hydrate of carbon yet it was a carbohydrate. Also such compounds as acetic acid, $C_2H_4O_2$, and pyrogallol, $C_6H_6O_3$, did fit such a system but were not carbohydrates.

Carbohydrates are now defined as polyhydroxyaldehydes or poly-hydroxyketones or substances that yield these compounds on hydrolysis. *Polyhydroxy* means "containing several alcohol groups." Thus simple carbohydrates are alcohols and also are either aldehydes or ketones (they contain a carbonyl group, $>C=O$; see Chapter 18).

Classification

Carbohydrates are divided into three major categories: monosaccharides, disaccharides, and polysaccharides.

Monosaccharides (*mono-* means one) are simple sugars. They cannot be changed into simpler sugars upon hydrolysis (reaction with water).

Disaccharides (*di-* means two) are double sugars. On hydrolysis, they yield two simple sugars.

$$\text{disaccharides} \xrightarrow{\text{hydrolysis}} \text{2 monosaccharides}$$

Polysaccharides (*poly-* means many) are complex sugars. On hydrolysis, they yield many simple sugars.

$$\text{polysaccharides} \xrightarrow{\text{hydrolysis}} \text{many simple sugars}$$

Monosaccharides, or simple sugars, are called either *aldoses* or *ketoses*, depending upon whether they contain an aldehyde ($-CHO$) or a ketone ($>C=O$) group. Aldoses and ketoses are further classified according to the number of carbon atoms they contain. An aldopentose is a five-carbon simple sugar containing an aldehyde group. A ketohexose is a six-carbon simple sugar containing a ketone group.

Although there are simple sugars with three carbon atoms (trioses), four carbon atoms (tetroses), and five carbon atoms (pentoses), the hexoses (six-carbon simple sugars) are the most common in terms of the human body because they are the body's main energy-producing compounds.

Polysaccharides are sometimes called hexosans or pentosans, depending upon the type of monosaccharide they yield on hydrolysis. That is, hexosans on hydrolysis yield hexoses and pentosans yield pentoses.

Origin

Plants pick up carbon dioxide from the air and water from the oil and combine them to form carbohydrates in a process called *photosynthesis.* Enzymes, chlorophyll, and sunlight are necessary. The overall reaction can be represented as follows

$$6CO_2 + 6H_2O \xrightarrow[\substack{\text{chlorophyll} \\ \text{enzymes}}]{\text{sunlight}} \underset{\text{sugar}}{C_6H_{12}O_6} + 6O_2\uparrow$$

However, it should be understood that even though this reaction appears to be simple, it is very complex with many intermediate steps between the original reactants and the final products.

During photosynthesis oxygen is given off into the air, thus renewing our vital supply of this element.

The carbohydrate produced in the previous reaction, $C_6H_{12}O_6$, is a monosaccharide. Plant cells also have the ability to combine two molecules of a monosaccharide into one of a disaccharide.

$$2C_6H_{12}O_6 \longrightarrow C_{12}H_{22}O_{11} + H_2O$$

monosaccharide   disaccharide

Note that this reaction is the reverse of hydrolysis—water is removed when two molecules of a monosaccharide combine.

Plant (and animal) cells can also combine many molecules of monosaccharide into large polysaccharide molecules. The n in the following equation represents a number larger than 2:

$$nC_6H_{12}O_6 \longrightarrow (C_6H_{10}O_5)_n + nH_2O$$

monosaccharide   polysaccharide

Polysaccharides occur in plants as cellulose in the stalks and stems and as starches in the roots and seeds. Monosaccharides and disaccharides are generally found in plants in their fruits.

Plants as well as animals are able to convert carbohydrates into fats and proteins.

Oxygen–Carbon Dioxide Cycle in Nature

Although plants have the ability to pick up carbon dioxide from the air and water from the ground to form carbohydrates, animals are unable to do this and must rely on plants for their carbohydrates.

Animals oxidize carbohydrates in their bodies to yield carbon dioxide, water, and energy:

$$C_6H_{12}O_6 + 6O_2 \longrightarrow 6CO_2 + 6H_2O + energy$$

Again, this reaction is not as simple as it appears; many steps are involved between the reactants and the products, and many different enzymes are required.

It should be noted that this reaction during metabolism is the reverse of that taking place during photosynthesis. Both reactions can be summarized as follows

$$\text{energy} + 6CO_2 + 6H_2O \underset{\substack{\text{animal} \\ \text{metabolism}}}{\overset{\substack{\text{plant} \\ \text{photosynthesis}}}{\rightleftharpoons}} C_6H_{12}O_6 + 6O_2$$

Thus, there is a cycle in nature. During photosynthesis, plants pick up carbon dioxide from the air and give off oxygen; both plants and animals pick up oxygen from the air and give off carbon dioxide.

During photosynthesis, the energy from the sun is needed for the reaction (that is, the reaction is endothermic). During metabolism of these carbohydrates in animals, this same amount of energy is liberated (the reaction is exothermic). Thus, all the energy from the burning of carbohydrates by animals comes originally from the sun. Plants store solar energy in carbohydrates, and this energy is utilized by all living organisms during the metabolic process. It has been estimated that only about 1 per cent of the total solar energy falling on plants is converted into useful stored energy.

Optical Activity

In Chapters 15 and 16 the topic of isomers was mentioned. However, isomers may be subdivided into four classifications. They are structural, functional, geometric, and optical.

Structural Isomers

Structural isomers are compounds with the same molecular formula, same functional groups, but different structural formulas. Examples are propyl alcohol and isopropyl alcohol, o-dichlorobenzene and m-dichlorobenzene, and 2-methylhexane and 3-methylhexane.

Functional Isomers

Functional isomers are compounds having the same molecular formula but different functional groups. One example is ethyl alcohol and dimethyl ether (see page 185).

Geometric Isomers

Geometric isomers are compounds with the same molecular formula but different structural formulas owing to a restricted rotation because of either a double bond or a ring system.

If two carbon atoms are connected by a double bond, those two carbon atoms and the four groups attached to them all lie in a single plane. There are two possible isomeric structures, one having similar groups on the same "side" of the double bond and one having similar groups on opposite "sides" of the double bond. An example of geometric isomerism is

cis-1,2-dichloroethene   *trans*-1,2-dichloroethene

The prefix *cis-* means on the same side and *trans-* means across or on opposite sides.

If two groups attached to one of the doubly bonded carbon atoms are identical, only one structure is possible. The two groups attached to each carbon must be different for geometric isomerism to occur.

Cis–trans isomerism is also possible for ring structures. In cyclopropane, the three carbon atoms all lie in one plane. Substituents on adjacent carbons may both be on the same side of the plane (*cis*) or on opposite sides of the plane (*trans*), such as

cis-1,2-dimethylcyclopropane   *trans*-1,2-dimethylcyclopropane

Cis–trans isomerism occurs in fatty acids (see Chapter 22) and is very important in the rhodopsin–vitamin A cycle (see page 445).

Optical Isomers

Optical isomers are compounds with the same molecular formula but with structural formulas that are the mirror image of one another. Such isomers rotate the plane of polarized light equally but in opposite directions. Let us see what this means and why.

Polarized light vibrates in one plane only as opposed to ordinary light, which vibrates in all planes. When polarized light is passed through a solution of an optically active substance, the plane of polarized light is rotated (see Figure 21–1).

What causes such a rotation of the plane of polarized light? According to the theory of van't Hoff, such an effect upon the plane of polarized light is

Figure 21–1. Schematic representation of the production of polarized light and its rotation by an optically active substance.

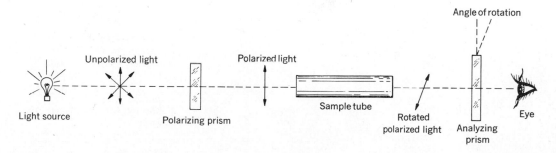

due to the presence of one (or more) *asymmetric* carbon atoms. An asymmetric carbon atom is one having four different groups attached to it.

The simplest carbohydrate is glyceraldehyde, whose structural formula is

$$
\begin{array}{c}
\text{CHO} \\
| \\
\text{H—C—OH} \\
| \\
\text{CH}_2\text{OH}
\end{array}
$$

glyceraldehyde

Glyceraldehyde has an aldehyde group at one end of the molecule, a primary alcohol at the other end, and a secondary alcohol in the middle. The central carbon atom in glyceraldehyde is asymmetric—it has four different groups attached to it. They are:

$$-\text{CHO} \qquad -\text{OH} \qquad -\text{CH}_2\text{OH} \qquad -\text{H}$$

This compound can exist in two optically active forms. These two structures are mirror images of one another (see Figure 21–2); they are *not* superimposable. Your right and left hands are mirror images; they are not superimposable. You wear a right and a left shoe; they too are mirror images and are not superimposable.

The Fischer projection formula is a two-dimensional representation of the above structures. In this system, the two isomers of glyceraldehyde are represented as

$$
\begin{array}{c}
\text{CHO} \\
| \\
\text{H—C—OH} \\
| \\
\text{CH}_2\text{OH}
\end{array}
\qquad\qquad
\begin{array}{c}
\text{CHO} \\
| \\
\text{HO—C—H} \\
| \\
\text{CH}_2\text{OH}
\end{array}
$$

D-glyceraldehyde   L-glyceraldehyde

The horizontal lines indicate bonds extending forward from the plane, and the vertical lines indicate bonds extending backward from the plane. Compare these two diagrams with those above. The Fischer projection formulas are *always* written with the aldehyde (or ketone) group—the most highly oxidized—at the top. Thus, in the above Fischer formulas, the —H and —OH

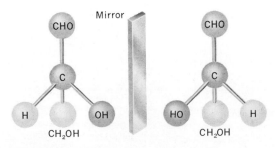

Figure 21–2. Ball-and-stick models of the two forms of glyceraldehyde. (Reproduced from Baum, S. J., and Scaife, C. W.: *Chemistry: A Life Science Approach*. Macmillan Publishing Co., Inc., New York, 1975.)

groups project forward from the plane and the —CHO and —CH$_2$OH

groups project backward from the plane.

Fischer called the glyceraldehyde with the —OH group on the right side of the asymmetric carbon atom D-glyceraldehyde. Likewise, the one with the —OH group on the left side of the asymmetric carbon he called L-glyceraldehyde.

Glyceraldehyde is considered the parent compound from which more complex sugars may be derived. Those whose terminal structure (the primary alcohol and the asymmetric carbon) is similar to that of D-glyceraldehyde belong to the D-series; those whose terminal end is similar to that of L-glyceraldehyde belong to the L-series. Note below that D-erythrose and D-threose have structures similar to that of D-glyceraldehyde, and the corresponding L-structures are also similar. Note also that D and L refer to structure only, not to the direction of rotation of polarized light.

The number of optical isomers depends upon the number of asymmetric carbon atoms present in a compound and may be calculated by using the formula 2^n where n is the number of asymmetric carbons. Thus, glyceraldehyde, which has one asymmetric carbon, has 2^1 or 2 optical isomers, as was shown. Glucose, which will be discussed later in this chapter, has four asymmetric carbons and so has 2^4 or 16 optical isomers. Of these 16 isomers, 8 belong to the D-series and 8 to the L-series (one set of 8 is the mirror image of the other set).

In the human body, the D-series is the primary configuration for carbohydrates.

Monosaccharides

Trioses

A triose is a three-carbon simple sugar. Trioses are formed during the metabolic breakdown of hexoses in muscle metabolism.

An example of a triose is glyceraldehyde (glycerose), whose optical isomers are shown on page 262.

Tetroses

Tetroses are four-carbon sugars, Four examples are shown here.

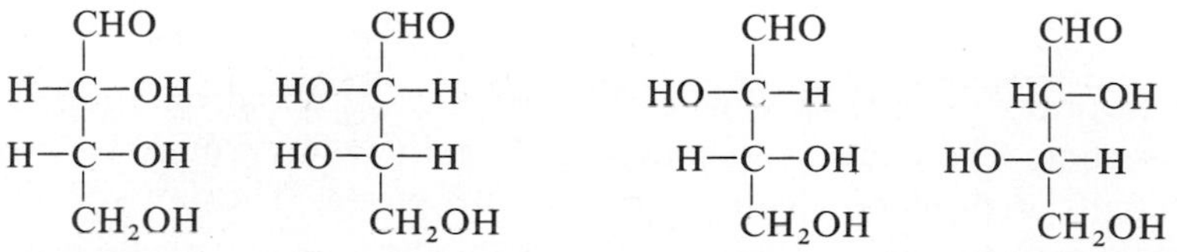

mirror images   mirror images

Erythrose is an intermediate in the hexose-monophosphate shunt for the oxidation of glucose (see page 366).

Pentoses

Pentoses are five-carbon sugar molecules. The most important of these are ribose and deoxyribose which are found in nucleic acids. Ribose forms part of ribonucleic acid (RNA), and deoxyribose forms part of deoxyribonucleic acid (DNA). Both DNA and RNA are components of every cell nucleus and cytoplasm. The prefix *de* means without, so *deoxy* means without oxygen. Note that deoxyribose has one less oxygen atom than does ribose.

$$
\begin{array}{c}
\text{CHO} \\
\text{H}-\overset{|}{\text{C}}-\text{OH} \\
\text{H}-\overset{|}{\text{C}}-\text{OH} \\
\text{H}-\overset{|}{\text{C}}-\text{OH} \\
\text{CH}_2\text{OH}
\end{array}
\qquad
\begin{array}{c}
\text{CHO} \\
\text{H}-\overset{|}{\text{C}}-\text{H} \\
\text{H}-\overset{|}{\text{C}}-\text{OH} \\
\text{H}-\overset{|}{\text{C}}-\text{OH} \\
\text{CH}_2\text{OH}
\end{array}
$$

D-ribose   D-deoxyribose

Ribose and deoxyribose may also be represented as ring structures with a carbon atom being understood at each corner except where some other element is indicated; the H's and OH's are assumed to be above and below the plane of the ring.

ring structures of D-ribose and D-deoxyribose, respectively

D-Ribose may also be represented as a "stick" figure.

The vertical line represents the carbon atoms, with a carbon atom at each intersection. The circle represents a carbonyl (C=O) group; the triangle represents a primary alcohol, or CH_2OH group. The horizontal lines indicate —OH groups.

Other pentoses are ribulose, which is formed in the metabolic breakdown of glucose; lyxose, which is found in heart muscle; arabinose, which is found in gum arabic and the gum of the cherry tree; and xylose, which is obtained from the hydrolysis of wood gums.

Hexoses

The hexoses, the six-carbon sugars, are the most common of all the carbohydrates. Of the several hexoses, the most important as far as the human body is concerned are *glucose, fructose,* and *galactose.* All three of these hexoses have the same molecular formula, $C_6H_{12}O_6$, but different structural formulas; they are isomers.

Glucose. Glucose, $C_6H_{12}O_6$, is an aldohexose and may be represented structurally as follows.

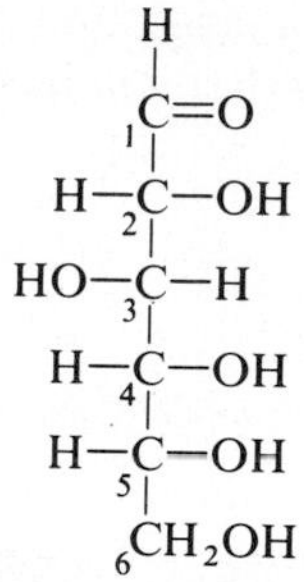

D-glucose   or   D-glucose

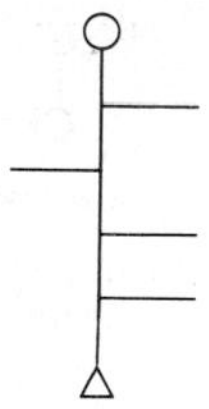

Note that glucose contains four asymmetric carbon atoms (numbers 2, 3, 4, 5) and so has 2^4 or 16 optical isomers.

Glucose may also be represented as a ring compound. Actually, solutions of glucose are composed primarily of ring-shaped molecules. If the H atom of the —OH group on carbon atom number 5 in the previous structure becomes attached to the oxygen atom on the aldehyde group (carbon atom number 1) then the O atom of the OH group on carbon atom number 5 can bond to carbon atom number 1, giving the following structures

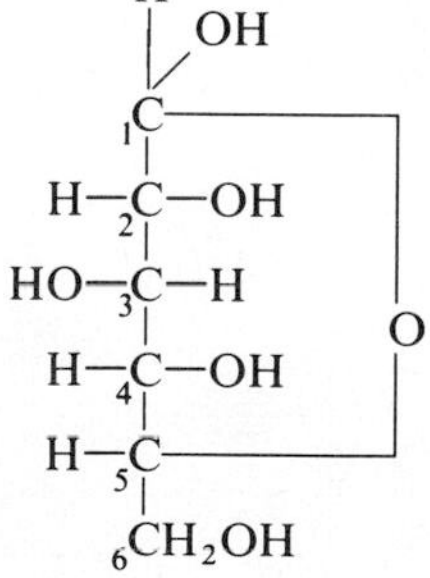

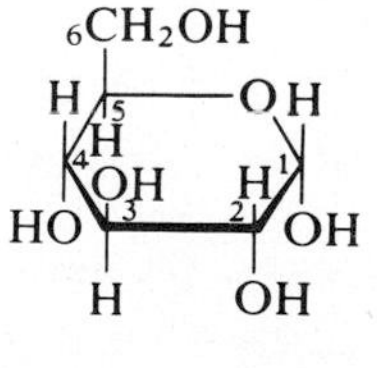

ring structures for glucose[1]

[1] The chemical name for this ring compound is alpha-D-glucopyranose, but this system of naming is of interest primarily to organic chemists.

Glucose is known commonly as dextrose or grape sugar. It is a white crystalline solid, soluble in water and insoluble in most organic liquids. It is found, along with fructose, in many fruit juices. It may be prepared by the hydrolysis of sucrose, a disaccharide, or by the hydrolysis of starch, a polysaccharide.

Glucose is the most important of all monosaccharides. It is normally found in the bloodstream and in the tissue fluids. As will be discussed in Chapter 26 (Carbohydrate Metabolism), glucose requires no digestion and can be given intravenously to patients who are unable to take food by mouth (see Figure 21–3).

Glucose shows up in the urine of patients suffering from diabetes mellitus and is an indication of this disease. The presence of glucose in the urine is called glycosuria. Glucose may also show up in the urine during extreme excitement (emotional glycosuria), after ingestion of large amounts of sugar (alimentary glycosuria), or because of other factors that will be discussed in the chapter on carbohydrate metabolism.

Figure 21–3. Intravenous infusion of glucose. (Courtesy of Fort Sanders Presbyterian Hospital, Knoxville, Tenn.)

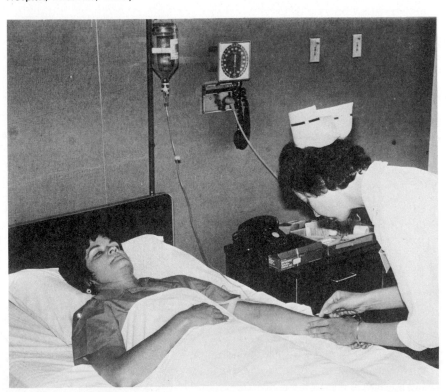

Galactose. Galactose, an isomer of glucose, is also an aldohexose. Different representations for the structures of galactose are illustrated below.

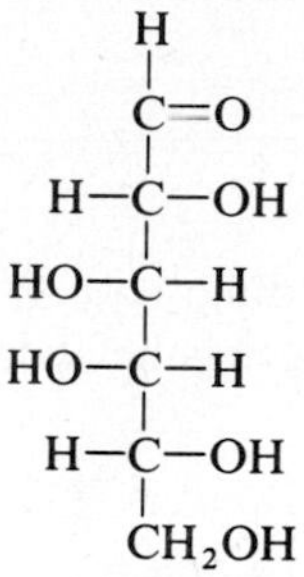

D-galactose   D-galactose   D-galactose

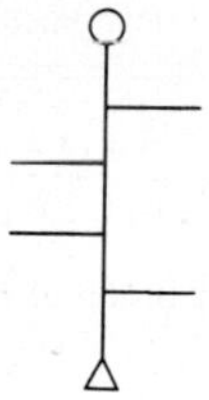

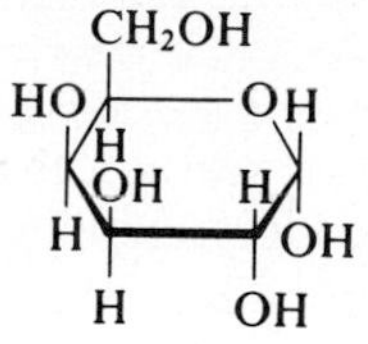

Glucose and galactose differ from each other only in the configuration of the H and OH about a single carbon atom. Two sugars that differ only in the configuration about a single carbon atom are called *epimers.* D-Galactose is converted to D-glucose in the liver by a specific enzyme called an epimerase.

Fructose. Fructose is a ketohexose. Its molecular formula, like that of glucose and galactose, is $C_6H_{12}O_6$. It too may be represented as a straight-chain or a ring compound (see below).

The ring structure is predominant.

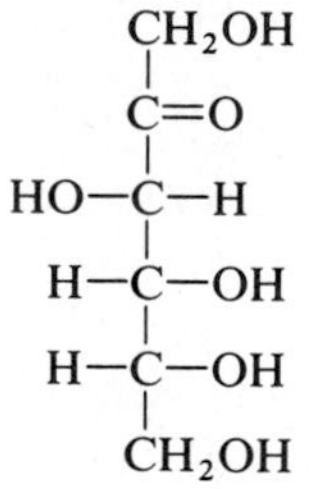

fructose   fructose   ring structures for D-fructose

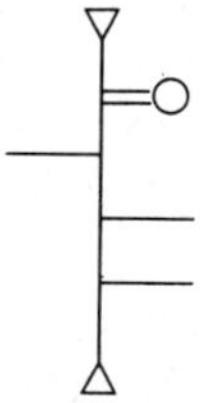

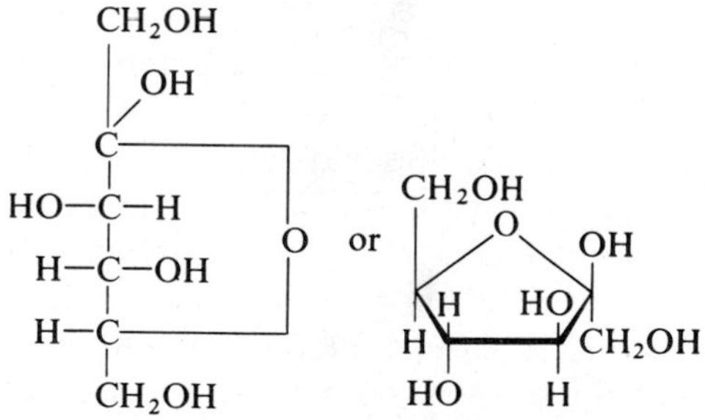

Fructose is often called levulose or fruit sugar. It occurs naturally in fruit juices and in honey. It may be prepared by the hydrolysis of sucrose, a disaccharide, and also by the hydrolysis of inulin, a polysaccharide found in Jerusalem artichokes. Fructose is the most soluble sugar and also the sweetest of all sugars, being approximately twice as sweet as glucose.

Reactions of the Hexoses

Reduction. Hexoses, which are either aldoses or ketoses, show reducing properties. This reducing property is the basis of the test for sugar in the urine and in the blood. When a reducing agent is treated with an oxidizing agent such as copper (II) hydroxide, $Cu(OH)_2$, a red orange precipitate of copper (I) oxide, Cu_2O, is formed. The unbalanced equation for the reaction of an aldehyde with copper (II) hydroxide may be written as follows

$$\text{aldehyde} + \underset{\substack{\text{deep blue}\\\text{solution}}}{Cu(OH)_2} \xrightarrow{\text{heat}} \text{acid} + \underset{\substack{\text{red-orange}\\\text{precipitate}}}{Cu_2O\downarrow} + \text{water}$$

In this reaction the aldehyde is oxidized to the corresponding acid. When glucose is treated with copper (II) hydroxide and the mixture is heated, the reaction is as shown below.

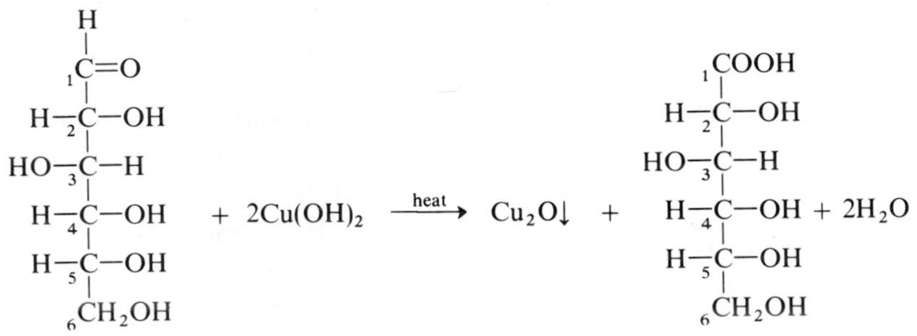

D-glucose; copper (II) hydroxide (deep blue color); copper (I) oxide (red-orange precipitate); D-gluconic acid

Laboratory tests for the presence of glucose in urine use Benedict's solution, Fehling's solution, or Clinitest tablets, all of which contain copper (II) hydroxide. If sugar (glucose) is present in the urine, the color of the copper (II) hydroxide will change from blue to green, yellow, orange, or red-orange, depending on the amount of glucose present.

Glucose does not normally appear in the urine for any extended period of time. Its persistent presence usually indicates something wrong with the metabolism of carbohydrates—such as diabetes mellitus.

Another laboratory test for the presence of a reducing sugar uses Tollen's reagent, silver hydroxide, AgOH. In this reaction glucose is oxidized to gluconic acid as before and the silver hydroxide is reduced to free silver, which appears as a bright shiny mirror on the inside of the test tube.

$$\text{glucose} + \underset{\substack{\text{Tollen's}\\\text{reagent}}}{AgOH} \xrightarrow{\text{heat}} \text{gluconic acid} + \underset{\substack{\text{silver}\\\text{mirror}}}{Ag\downarrow} + \text{water}$$

The Molisch test is a general qualitative test for the presence of carbohydrates. When a solution of a carbohydrate is mixed with alpha-naphthol and then concentrated sulfuric acid is carefully added so as to form a layer below the mixture, a red-violet ring is formed at the interface of the two liquids.

Fermentation. Glucose ferments in the presence of yeast, forming ethyl alcohol and carbon dioxide. This reaction will not readily occur in the absence of yeast. Yeast contains enzymes that catalyze this particular reaction.

$$C_6H_{12}O_6 \xrightarrow{\text{enzymes}} 2C_2H_5OH + 2CO_2\uparrow$$

glucose ethyl alcohol

Fructose will also ferment; galactose will not readily ferment. Pentoses do not ferment in the presence of yeast.

Oxidation. An aldose contains an aldehyde group as well as several —OH groups. If the aldehyde end of the molecule is oxidized, the product is called an *onic acid*. When the aldehyde end of glucose is oxidized, the product is called gluconic acid (see page 268). When the aldehyde end of galactose is oxidized, the product is called galactonic acid. If the alcohol at the end opposite the aldehyde (the other end of the molecule) is oxidized, the product is called a *uronic acid*. The oxidation of the alcohol end of glucose yields glucuronic acid.

glucose glucuronic acid

Likewise, if the alcohol end of galactose is oxidized, galacturonic acid is formed.

If both ends of the glucose molecule are oxidized at the same time, the product is called saccharic acid.

glucose saccharic acid

Reduction. The aldohexoses may be reduced to alcohols. When glucose is reduced, sorbitol is formed.

$$
\begin{array}{ccc}
\text{CHO} & & \text{CH}_2\text{OH} \\
\text{H---C---OH} & & \text{H---C---OH} \\
\text{HO---C---H} & \xrightarrow[\text{catalyst}]{\text{H}_2} & \text{HO---C---H} \\
\text{H---C---OH} & & \text{H---C---OH} \\
\text{H---C---OH} & & \text{H---C---OH} \\
\text{CH}_2\text{OH} & & \text{CH}_2\text{OH}
\end{array}
$$

glucose   sorbitol

Reduction of galactose yields dulcitol and reduction of fructose yields a mixture of mannitol and sorbitol.

Sorbitol is used as a sweetening agent for diabetics because half of it is excreted unchanged in the urine. In addition, the rest is slowly absorbed and does not contribute very much to the blood sugar level. When administered intravenously, sorbitol withdraws fluid from the body into the circulatory system, thereby causing a dehydrating action. Thus, sorbitol is used to relieve edema, to lower cerebrospinal fluid pressure, and to reduce intraocular tension due to glaucoma.

Mannitol behaves similarly to sorbitol. Most of it is eliminated unchanged in the urine. The rest is minimally absorbed. Mannitol is used intravenously, as is sorbitol, because of its dehydrating action. In addition, mannitol is used as a diagnostic agent to measure kidney function (see page 410).

Disaccharides

There are three common disaccharides; all of these are isomers with the molecular formula $C_{12}H_{22}O_{11}$. They are sucrose, maltose, and lactose. On hydrolysis these disaccharides yield two monosaccharides. The general reaction may be written as follows

$$
C_{12}H_{22}O_{11} + H_2O \xrightarrow{\text{hydrolysis}} C_6H_{12}O_6 + C_6H_{12}O_6
$$

a disaccharide	⟶	a monosaccharide + a monosaccharide
sucrose	⟶	glucose + fructose
maltose	⟶	glucose + glucose
lactose	⟶	glucose + galactose

The disaccharides, just like the monosaccharides, are white, crystalline, sweet solids. Sucrose is very soluble in water; maltose is fairly soluble; and lactose is only slightly soluble. The disaccharides are also optically active; they rotate the plane of polarized light. However, even though they are soluble in water, they are too large to pass through cell membranes.

Reducing Properties

When sucrose is treated with warm copper (II) hydroxide, no reaction is observed. This lack of reducing power indicates that sucrose does not have a free aldehyde or ketone group. This may be explained by considering that, when sucrose is formed from glucose and fructose, the aldehyde group of glucose combines with the ketone group of fructose, thus destroying the reducing power of both carbonyl groups (see the formula given below).

When maltose or lactose is treated with warm copper (II) hydroxide, a positive test for a reducing sugar is observed. In maltose and lactose there must be a free aldehyde (carbonyl) group to cause this reducing property. In maltose the aldehyde group of one glucose reacts with an —OH group of the other glucose, thus leaving an aldehyde (reducing) group free. Likewise, in lactose one aldehyde group is free (see formulas given). Thus, sucrose is not a reducing sugar, whereas lactose and maltose are reducing sugars.

The following structural formulas show both the condensed chain and the ring structures of the three disaccharides.

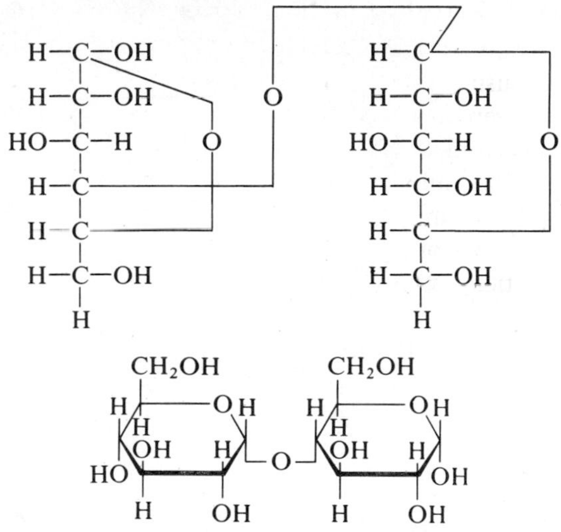

maltose

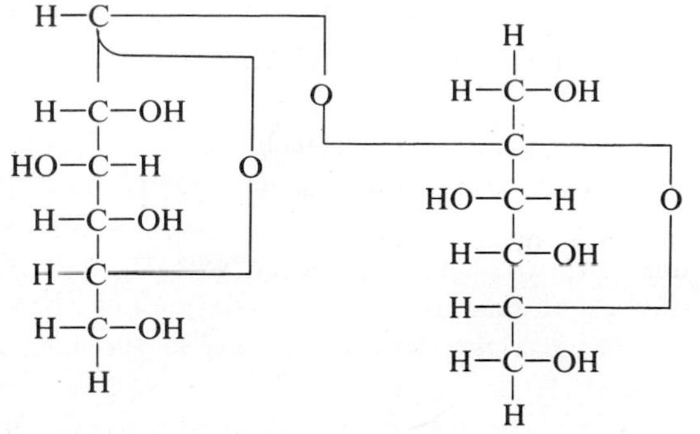

sucrose

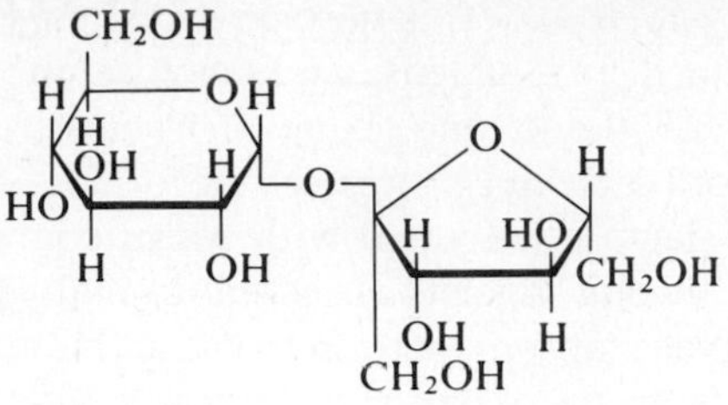

CH₂OH

sucrose

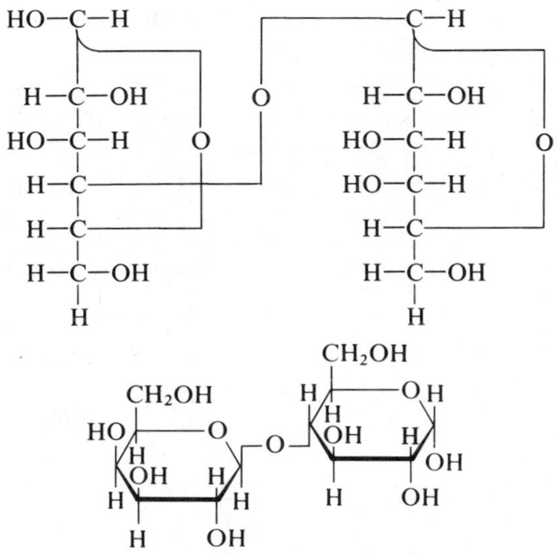

lactose

Fermentation

Sucrose and maltose will ferment when yeast is added because yeast contains the enzymes sucrase and maltase; lactose will not ferment when yeast is added because yeast does not contain lactase. The identity of a disaccharide can be deduced on the basis of its fermentation reaction and its reducing properties.

Suppose that a test tube contains a disaccharide, $C_{12}H_{22}O_{11}$. Is it sucrose, lactose, or maltose? The identity can be determined by the following method.

1. Mix the unknown disaccharide with copper (II) hydroxide and warm gently. If there is no reaction, the disaccharide must be sucrose. In this case, no further test is necessary to prove the identity of the disaccharide.

2. If the unknown disaccharide gives a positive test with copper (II) hydroxide, it must be either maltose or lactose. In this case, another

sample of the disaccharide is mixed with yeast and allowed to stand to observe whether fermentation takes place or not. If the disaccharide does ferment, then it must be maltose. If it does not ferment, then it must be lactose.

The same two laboratory tests may be performed in reverse order with the same results.

1. Mix the unknown disaccharide with yeast and allow to stand, to observe whether fermentation takes place. If no fermentation is observed, the disaccharide is lactose, and no further test is necessary.
2. If fermentation does occur, the disaccharide is either sucrose or maltose. In this case take another sample of the unknown disaccharide, mix with copper (II) hydroxide, and warm gently. If the color remains blue, the unknown disaccharide must be sucrose. If the unknown gives a positive test with copper (II) hydroxide, it must be maltose.

Structure of the Disaccharides

The three common disaccharides—sucrose, lactose, and maltose—all have the molecular formula $C_{12}H_{22}O_{11}$. Their structures are different, as we have seen. Note that these structures can be drawn either as chain or ring compounds. The ring structures are the correct and preferred ones, although the chain structures are frequently used simply because they are easier to work with.

Sucrose

Sucrose is the sugar used ordinarily in the home. It is also known as cane sugar. Sucrose is produced commercially from sugar cane and sugar beets. It also occurs in sorghum, pineapple, and carrot roots.

When sucrose is hydrolyzed, it forms a mixture of glucose and fructose. This 50:50 mixture of glucose and fructose is called *invert sugar*. Honey contains a high percentage of invert sugar.

Maltose

Maltose, commonly known as malt sugar, is present in germinating grain. It is produced commercially by the hydrolysis of starch.

Lactose

Lactose, commonly known as milk sugar, is present in milk. It differs from the above sugars in that it has an animal origin. Certain bacteria cause lactose to ferment, forming lactic acid. When this reaction occurs, the milk is said to be sour. Lactose is used in high-calcium diets and in infant foods. Lactose can be used for increasing calorie intake without adding much sweetness. Lactose is found in the urine of pregnant women and, since it is a reducing sugar, it gives a positive test with $Cu(OH)_2$.

Polysaccharides

Polysaccharides are polymers of monosaccharides. Complete hydrolysis of polysaccharides produces many molecules of monosaccharides. The polysaccharides differ from monosaccharides and disaccharides in many ways, as is indicated in Table 21–1.

Polysaccharides can be formed from pentoses (five-carbon sugars) or from hexoses (six-carbon sugars). Polysaccharides formed from pentoses are called *pentosans*. Those formed from hexoses are called *hexosans* (or sometimes glucosans).

The hexosans (or glucosans) are the most important in terms of physiology. The hexosans have the general formula $(C_6H_{10}O_5)_x$ where x is some large number. Some of the common hexosans are starch, cellulose, glycogen, and dextrin.

Starch

Plants store their foods in the form of starch granules. Starch is actually a mixture of the polysaccharides amylopectin and amylose. Amylopectin is a branched polysaccharide present in starch to a large extent (80 to 85 per cent). It is usually present in the covering of the starch granules. Amylose is a nonbranched polysaccharide present in starch to an extent of 15 to 20 per cent.

Starch is insoluble in water. When starch is placed in boiling water, the granules rupture forming a paste which gels on cooling. When a small

TABLE 21–1

Comparison of Polysaccharides with Monosaccharides and Disaccharides

Property	Monosaccharides and Disaccharides	Polysaccharides
Molecular weight	Low	Very high
Taste	Sweet	Tasteless
Solubility in water	Soluble	Insoluble
Size of particles	Pass through a membrane	Do not pass through a membrane
Test with $Cu(OH)_2$ (an oxidizing agent)	Positive (except for sucrose)	Negative

amount of starch is added to a large amount of boiling water, a colloidal
dispersion of starch in water is formed.

Starch gives a characteristic deep blue color with iodine. This test is used
to detect the presence of starch because it is conclusive even when only a
small amount of starch is present. That is, if iodine is added to an unknown
and a blue color is produced, starch is present. This test may also be used to
check for the presence of iodine. If starch is added to an unknown and a
blue color is produced, iodine must be present.

When starch is hydrolyzed, it forms dextrins (amylodextrin, erythrodextrin,
achroodextrin), then maltose, and finally glucose. Erythrodextrins turn red
in the presence of iodine. Both maltose and glucose produce no color in the
presence of iodine. Thus, it is possible to follow the hydrolysis of the starch
by observing the changing colors when iodine is added.

starch $\longrightarrow$ erythrodextrins $\longrightarrow$ maltose $\longrightarrow$ glucose

blue   red   colorless   colorless

Cellulose

Wood, cotton, and paper are composed primarily of cellulose. Cellulose
is the supporting and structural substance of plants. Like starch, cellulose is a
polysaccharide composed of many glucose units. However, cellulose is not
affected by any of the enzymes present in the human digestive system and
so cannot be digested. However, it does serve a purpose when eaten with
other foods: it gives bulk to the feces and prevents constipation.

Cellulose does not dissolve in water nor in most ordinary solvents. It
gives no color test with iodine and gives a negative test with copper (II)
hydroxide.

Cotton is nearly pure cellulose. When cotton fibers are treated with a
strong solution of sodium hydroxide, then stretched and dried, the fibers
take on a high luster. Such cotton is called *mercerized* cotton.

Cellulose is also used to make rayon. In this process, purified wood pulp
(nearly pure cellulose) is converted into a viscous liquid called viscose by
treatment with sodium hydroxide and carbon disulfide. When the viscose is
forced through small openings in a block suspended in an acid solution, the
cellulose is regenerated into fibers which can then be formed into threads.

Glycogen

Glycogen is present in the body and is stored in the liver and the muscles,
where it serves as a reserve supply of carbohydrates. Glycogen has an
animal origin, as opposed to the plant origin of starch.

Glycogen forms a colloidal dispersion in water and gives a red color with
iodine. It gives no test with copper (II) hydroxide. Glycogen is formed in the
body cells from molecules of glucose. This process is called *glycogenesis* (see

page 358). When glycogen is hydrolyzed into glucose, the process is called *glycogenolysis*.

$$\text{glucose} \underset{\text{glycogenolysis}}{\overset{\text{glycogenesis}}{\rightleftharpoons}} \text{glycogen}$$

Dextrin

Dextrin is produced during the hydrolysis of starch. Dextrin is an intermediate between starch and maltose. It forms sticky colloidal suspensions with water and is used in the preparation of adhesives. The glue on the back of postage stamps is a dextrin.

Heparin

Heparin is a polysaccharide used as a blood anticoagulant. It inhibits the conversion of prothrombin to thrombin, which in turn acts as a catalyst for the formation of the clot (see page 423). The structure of heparin consists of repeating units of glucuronic acid and glucosamine with some sulfate groups on the amino and hydroxyl groups. Heparin is the strongest organic acid present in the body.

heparin

Structure of the Polysaccharides

The ring structures of some of the polysaccharides are indicated opposite.

Tests for Carbohydrates

Table 21–2 lists the chief laboratory tests for carbohydrates.

Summary

Carbohydrates are polyhydroxyaldehydes or polyhydroxyketones or substances that yield these compounds on hydrolysis. Carbohydrates are divided into three categories: monosaccharides, disaccharides, and polysaccharides—based upon hydrolytic possibilities.

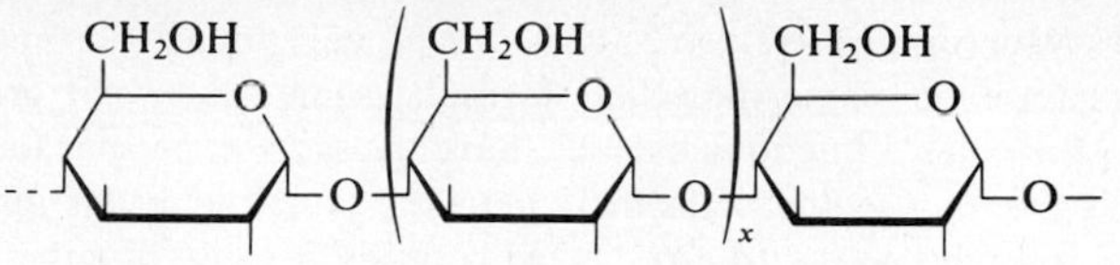

amylose

$x = 100$ to 400

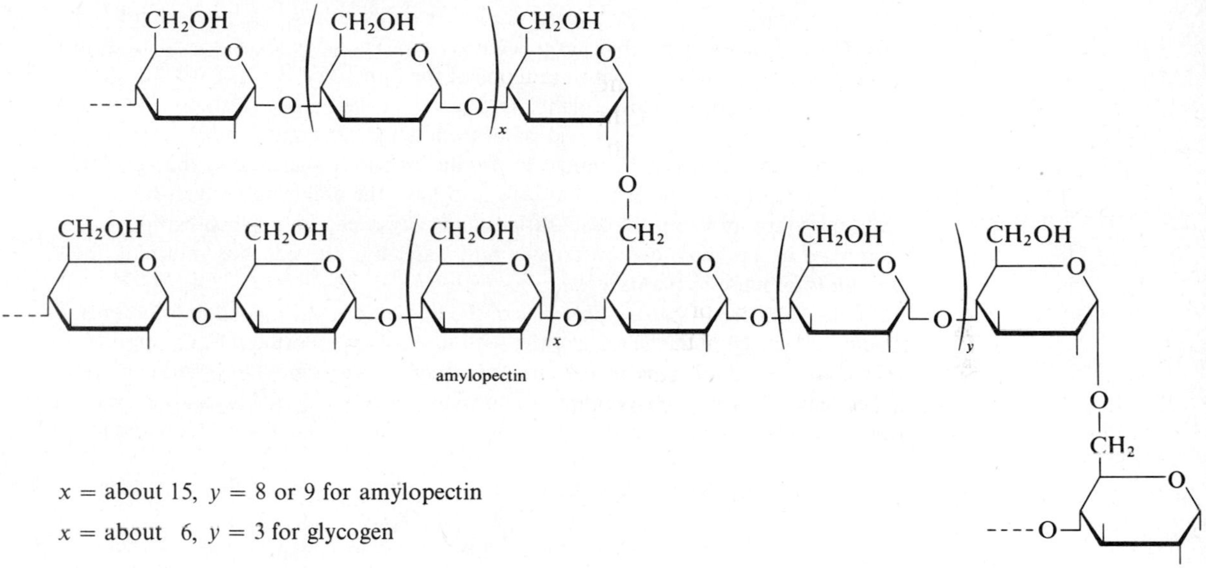

amylopectin

x = about 15, y = 8 or 9 for amylopectin
x = about 6, y = 3 for glycogen

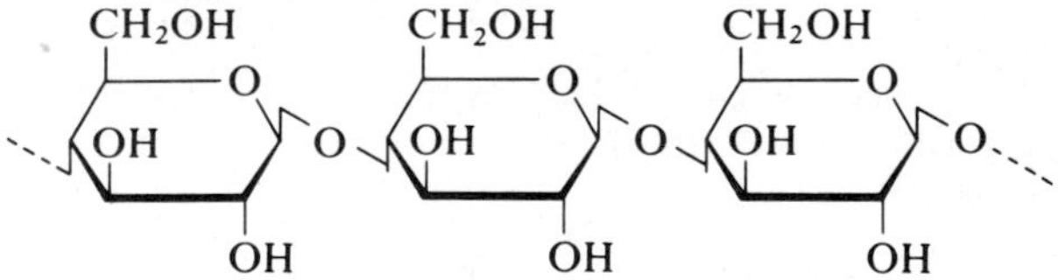

cellulose

TABLE 21–2

Laboratory Tests for Carbohydrates

Carbohydrate	Molisch	Fermentation	Benedict's
Glucose	+	+	+
Galactose	+	−	+
Fructose	+	+	+
Lactose	+	−	+
Sucrose	+	+	−
Maltose	+	+	+
Starch	+	+	−
Glycogen	+	+	−

There are four types of isomers: structural, functional, geometric, and optical. Structural isomers have the same molecular formula, same functional groups, but different structural formulas. Functional isomers have the same molecular formula but different functional groups. Geometric isomers have the same molecular formula but different structural formulas owing to a restricted rotation because of either a double bond or a ring system. Optical isomers have the same molecular formula but have structural formulas that are mirror images of each other.

Monosaccharides, or simple sugars, are either aldoses or ketoses, depending upon whether they contain an aldehyde or a ketone group. The six-carbon monosaccharides, the hexoses, are the most common in terms of the human body.

Carbohydrates are formed in plants by a process called photosynthesis. Plant cells take carbon dioxide from the air and water from the ground and combine them in the presence of sunlight and chlorophyll to produce monosaccharides, at the same time giving off oxygen into the air. Plant cells also have the ability to convert the monosaccharides thus formed into disaccharides and polysaccharides. When carbohydrates are burned in the body, carbon dioxide and water are formed, thus returning these substances for reuse by plants.

The most important hexoses in terms of the human body are glucose, fructose, and galactose. These compounds are isomers with the molecular formula $C_6H_{12}O_6$.

Glucose is an aldohexose whose structure may be represented as a linear or ring-shaped molecule. Glucose is commonly known as dextrose or grape sugar. It is the most important of all monosaccharides and is normally found in the bloodstream and in the tissue fluids.

Galactose is also an aldohexose. It occurs in nature as one of the constituents of lactose.

Fructose, a ketohexose, is commonly known as levulose or fruit sugar. Fructose is the sweetest of all sugars. It occurs in nature as one of the constituents of sucrose and is found free in fruit juices and in honey.

The hexoses are either aldehydes or ketones and can act as reducing agents. When a hexose is treated with $Cu(OH)_2$ (Fehling's solution, Benedict's solution, or Clinitest), a red-orange precipitate of Cu_2O is formed. This reaction is the basis for the test for sugar (hexoses) in the urine. Hexoses will also reduce Tollen's reagent (AgOH) to free silver.

Hexoses will ferment in the presence of enzymes found in yeast.

When the aldehyde end of a monosaccharide is oxidized, an *onic* acid is formed.

When the alcohol end of a monosaccharide is oxidized, a *uronic* acid is formed.

The three common disaccharides—sucrose, maltose, and lactose—are isomers with the molecular formula $C_{12}H_{22}O_{11}$. On hydrolysis a disaccharide yields two monosaccharides.

Of the three disaccharides only maltose and lactose show reducing properties with $Cu(OH)_2$. Sucrose is not a reducing sugar.

Sucrose and maltose will ferment with yeast owing to the presence of the enzymes sucrase and maltase. Lactose will not ferment with yeast because of the absence of the enzyme lactase.

Polysaccharides are polymers of monosaccharides and yield monosaccharides upon hydrolysis.

Polysaccharides have a high molecular weight, are insoluble in water, are tasteless, and give negative tests for reducing sugars. These properties are the opposite of those for monosaccharides and disaccharides.

Three common polysaccharides are starch, cellulose, and glycogen. Plants store their food as starch; plants use cellulose as supporting and as structural parts; animals use glycogen as a reserve supply of carbohydrate.

Questions and Problems

1. What is a carbohydrate?
2. What are aldoses? Aldopentoses? Ketohexoses?
3. What are structural isomers? Give an example.
4. What are geometric isomers? Give an example.
5. What are functional isomers? Give an example.
6. What are optical isomers? Give an example.
7. Draw the structure for (a) *cis*-1,2-dibromocyclopentane, (b) *trans*-1,2-dibromo-propene.
8. What is meant by the term *asymmetric carbon atom*? Give an example.
9. How can the number of asymetric carbon atoms be used to predict the number of optical isomers?
10. If a compound has three asymmetric carbons, how many optical isomers are possible?
11. What is the difference between ordinary light and polarized light?
12. What effect does a solution of an optical isomer have upon the plane of polarized light?
13. What do the letters D and L refer to in terms of optical isomers?
14. What are epimers? Give an example.
15. Is the D or the L form the predominant one in terms of carbohydrates that the body uses?
16. What is the reference compound for the configurations of carbohydrates?
17. What is a hexosan? A pentosan?
18. What are the three types of carbohydrates?
19. How are carbohydrates formed in nature? Are they formed directly as poly-saccharides?
20. Are animals able to synthesize carbohydrates from raw materials? Explain.
21. What are trioses? Pentoses? Hexoses?
22. Name the three important hexoses and draw their linear and stick structural formulas.
23. How do the three important hexoses differ in molecular formula? In structure?
24. Why can glucose be given intravenously whereas sucrose cannot?
25. Where is glucose normally found in nature?
26. Where are fructose and galactose normally found in nature?
27. Where is glucose found in the body?
28. How do the hexoses affect copper (II) hydroxide? What use is made of this reaction?
29. What are the trade names given to copper (II) hydroxide used in testing for hexoses?
30. What is Tollen's reagent? What is it used for?
31. Do all hexoses ferment in the presence of yeast? Why or why not?
32. What product is formed when the aldehyde end of glucose is oxidized?
33. What product is formed when the alcohol end of glucose is oxidized?

34. What product is formed when both ends of glucose are oxidized?
35. What product is formed when the aldehyde end of glucose is reduced?
36. Do all three disaccharides act as reducing agents? Why?
37. Do all three disaccharides ferment in the presence of yeast? Why
38. Explain how you could identify a disaccharide on the basis of its reducing action and its fermentation.
39. Compare the ring structures of the three disaccharides and use these structures to explain their reaction toward $Cu(OH)_2$.
40. Where is sucrose found in nature?
41. Where is maltose found in nature?
42. Where is lactose found in nature?
43. Compare the properties of the polysaccharides with those of monosaccharides and disaccharides.
44. What is starch?
45. What is the test for the presence of starch? Of iodine?
46. What products are formed when starch is slowly hydrolyzed? How may the presence of these products be detected?
47. For what purpose do plants use cellulose?
48. What is mercerized cotton?
49. For what purpose does the body use glycogen?
50. What is glycogenesis? Glycogenolysis?
51. What are dextrins? How are they used commercially?
52. What type of liquid mixture does starch form in boiling water?

References

Barker, R.: *Organic Chemistry of Biological Compounds*. Prentice-Hall, Inc., Englewood Cliffs, N.J., 1971, Chap. 5.

Baum, S. J., and Scaife, C. W.: *Chemistry: A Life Science Approach*. Macmillan Publishing Co., Inc., New York, 1975, Chap. 23.

Bronk, J. R.: *Chemical Biology*. Macmillan Publishing Co., Inc., New York, 1973, Chap. 4.

Harper, H. A.: *Review of Physiological Chemistry*, 15th ed. Lange Medical Publications, Los Altos, Calif., 1975, Chap. 1.

Lehninger, A. H.: *Biochemistry*, 2nd ed. Worth Publishers, Inc., New York, 1975, Chap. 10.

Morrison, R. T., and Boyd, R. N.: *Organic Chemistry*, 3rd ed. Allyn & Bacon, Inc., Boston, 1973, Chaps. 34, 35.

Rafelson, M. E.; Binkley, S. B.; and Hayashi, J. A.: *Basic Biochemistry*, 3rd ed. Macmillan Publishing Co., Inc., New York, 1971, Chap. 5.

Stacy, G. W.: *Organic Chemistry*. Harper & Row, New York, 1975, Chap. 13.

Lipids

General Properties

A second group of organic compounds that serve as food for the body are called lipids. Lipids have the following general properties.

1. Insoluble in water.
2. Soluble in organic solvents such as alcohol, ether, acetone, and carbon tetrachloride.
3. Contain carbon, hydrogen, and oxygen; sometimes contain nitrogen and phosphorus.

4. Yield fatty acids on hydrolysis or combine with fatty acids to form an ester.
5. Take part in plant and animal metabolism.

Classification

Lipids are divided into three main categories: simple, compound, and derived.

Simple Lipids

Simple lipids are esters of fatty acids. The hydrolysis of a simple lipid may be expressed as

$$\text{simple lipid} + H_2O \xrightarrow{\text{hydrolysis}} \text{fatty acid} + \text{alcohol}$$

If the hydrolysis of a simple lipid yields a fatty acid and *glycerol*, the simple lipid is called a *fat*.

If the hydrolysis of a simple lipid yields a fatty acid and a high-molecular-weight alcohol, the simple lipid is called a *wax*.

Compound Lipids

Compound lipids on hydrolysis yield a fatty acid(s), an alcohol, and some other type of compound. In this category are phospholipids and glycolipids (also called cerebrosides because they are found in the cerebrum of the brain). These compound lipids undergo hydrolysis as follows

$$\text{phospholipids} + H_2O \xrightarrow{\text{hydrolysis}} \text{fatty acid} + \text{alcohol} + \text{phosphoric acid} + \text{a nitrogen compound}$$

$$\text{glycolipids} + H_2O \xrightarrow{\text{hydrolysis}} \text{fatty acid} + \text{an alcohol} + \text{a carbohydrate} + \text{a nitrogen compound}$$

Derived Lipids

Derived lipids are compounds derived from simple and compound lipids on hydrolysis. Derived lipids include such substances as fatty acids, glycerol, other alcohols, and sterols, which are solid alcohols having a high molecular-weight.

Fats

Fatty Acids

Both simple and compound lipids yield fatty acids on hydrolysis. Fatty acids are straight-chain organic acids. The fatty acids that are found in natural

TABLE 22–1

Common Fatty Acids

Name	Formula		Source
		Saturated Fatty Acids	
Butyric	C_3H_7COOH		Butter fat
Caproic	$C_5H_{11}COOH$		Butter fat
Caprylic	$C_7H_{15}COOH$		Coconut oil
Capric	$C_9H_{19}COOH$		Palm oil
Lauric	$C_{11}H_{23}COOH$		Laurel
Myristic	$C_{13}H_{27}COOH$		Nutmeg oil, coconut oil
Palmitic	$C_{15}H_{31}COOH$		Palm oil, lard, cottonseed oil
Stearic	$C_{17}H_{35}COOH$		Plant and animal fats such as lard, peanut oil
Arachidic	$C_{19}H_{39}COOH$		Peanut oil
		Unsaturated Fatty Acids	
Oleic	$C_{17}H_{33}COOH$	(contains 1 double bond)	Olive oil
Linoleic	$C_{17}H_{31}COOH$	(contains 2 double bonds)	Linseed oil
Linolenic	$C_{17}H_{29}COOH$	(contains 3 double bonds)	Linseed oil
Arachidonic	$C_{19}H_{31}COOH$	(contains 4 double bonds)	Animal tissues, corn oil, linseed oil

fats usually contain an even number of carbon atoms. Fatty acids may be of two types—saturated and unsaturated. Saturated fatty acids contain only single bonds between carbon atoms. Unsaturated fatty acids contain one or more double bonds between carbon atoms.

Table 22–1 lists some of the common fatty acids and indicates where they are found in nature. Note that they all contain an even number of carbon atoms.

Linoleic and linolenic acids are called *essential fatty acids*—they are essential for the complete nutrition of the human body. They cannot be synthesized in the body and must be supplied from food we eat. Arachidonic acid, which was formerly also designated as an essential fatty acid, can be synthesized in the body from linoleic acid. Linoleic acid is found in large concentrations in oils such as corn, cottonseed, peanut, and soybean but *not* in cocoanut or olive oils. The essential fatty acids are necessary for the synthesis of the prostaglandins (see page 285).

The absence of these essential fatty acids from the diet of an infant causes loss of weight and also eczema. Such conditions may be cured by administering corn oil or boiled linseed oil.

Note that commercially available boiled linseed oil should never be used for this purpose because it contains litharge, a lead compound that is poisonous to the body.

The percentages of fatty acids in corn oil, linseed oil, butter, and lard are listed in Table 22–2.

TABLE 22–2

Average Percentage of Fatty Acids in Fats and Oils

		Saturated				Unsaturated			
		Myristic Acid	Palmitic Acid	Stearic Acid	Other Acids	Oleic Acid	Linoleic Acid	Other Acids	Iodine Number
Vegetable oils	cotton- seed oil	0–3	17–23	1–3	—	23–44	34–55	0–1	103–115
	corn oil	0–2	8–10	1–4	—	36–50	34–56	0–3	116–130
Animal fats	butter	8–13	25–32	8–13	4–11	22–29	3	3–9	26–45
	lard	1	25–30	12–16	—	41–51	3–8	5–8	46–66

Note that the percentages listed in Table 22–2 are given as averages. This is because the percentage composition of a fat or oil can vary considerably owing to weather conditions and/or the type of food eaten by the animal.

In addition to these straight-chain fatty acids, there are also cyclic fatty acids. An example of this is chaulmoogric acid, whose formula is indicated below.

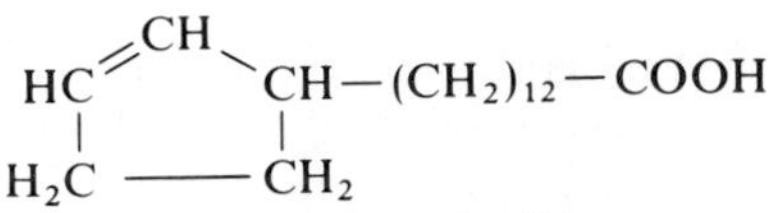

$$HC \overset{\text{CH}}{\diagup} \diagdown CH-(CH_2)_{12}-COOH$$
$$H_2C \text{———} CH_2$$

chaulmoogric acid

Chaulmoogric acid occurs in chaulmoogra oil and has been used in the treatment of leprosy, although it has been supplanted in this use by newer and more effective drugs.

Oleic acid, $C_{17}H_{33}COOH$, occurs in nature as the *cis* configuration (see page 260), as do most naturally occurring unsaturated fatty acids. The *trans* form is called elaidic acid.

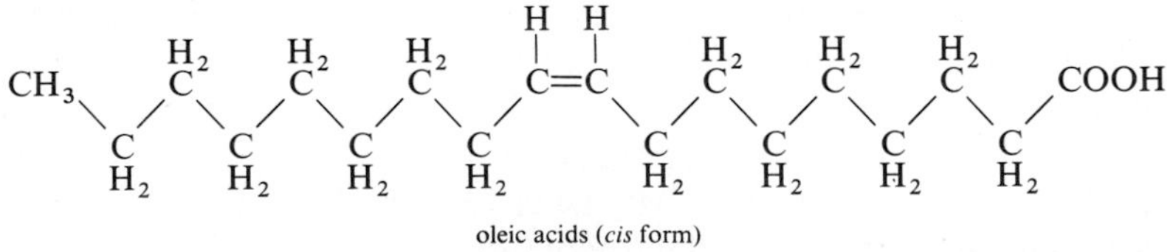

oleic acids (*cis* form)

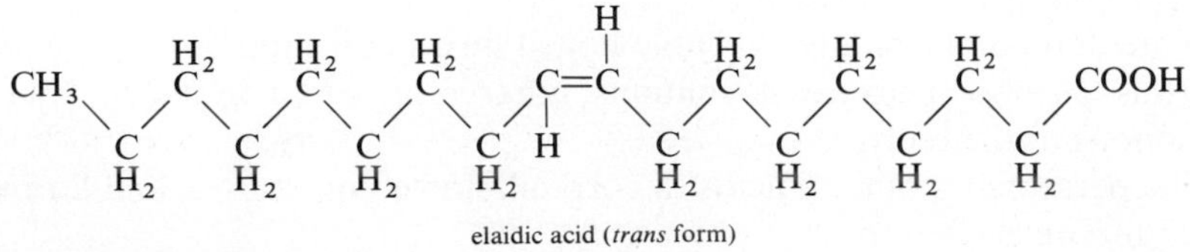

elaidic acid (*trans* form)

The prostaglandins consist of 20-carbon unsaturated fatty acids containing a five-membered ring and two side chains. One side chain has seven carbon atoms and ends with an acid group (COOH). The other chain contains eight carbon atoms with an —OH group on the third carbon from the ring (see structure below). The E series of prostaglandins has, in addition to four asymmetric carbon atoms, a *trans* configuration.

Prostaglandins are derived from arachidonic acid, which is formed from the essential fatty acid linoleic acid. The structures of arachidonic acid and prostaglandin E_1 are indicated below.

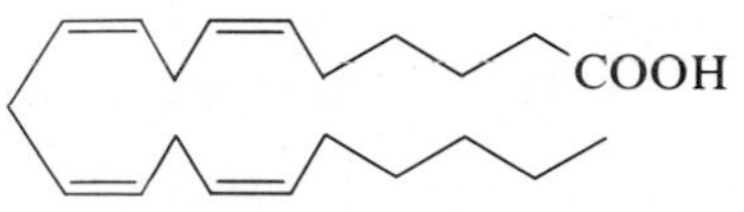

arachidonic acid 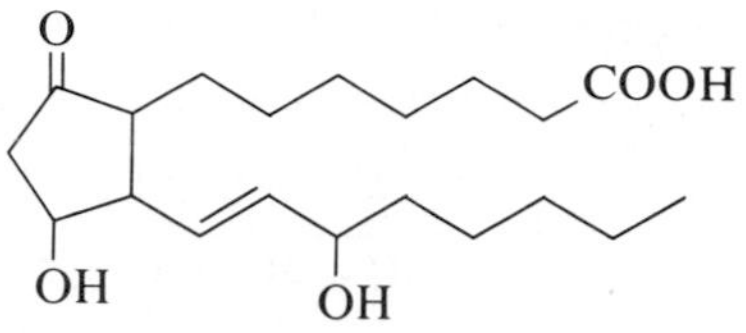   prostaglandin E_1

Prostaglandins have been isolated from most mammalian tissues, including male and female reproductive systems, liver, kidneys, pancreas, heart, lungs, brain, and intestines. The richest source of prostaglandins is the human seminal fluid.

Prostaglandins have a wide range of physiologic effects. They seem to be involved in the body's natural defenses against all forms of change including those induced by chemical, mechanical, physiologic, and pathologic stimuli. Aspirin and other anti-inflamatory drugs appear to operate by inhibiting prostaglandin synthesis. Prostaglandins are involved at the cellular level in regulating many body functions, including gastric acid secretion, contraction and relaxation of smooth muscles, inflammation and vascular permeability, body temperature, and blood platelet aggregation. In addition, the interaction of prostaglandin with the membranes of the red blood cells seems to be involved in sickle-cell anemia (see page 505). It is thought that inhibitors of prostaglandin synthesis might alleviate the sufferings of patients with this disease. Prostaglandins have also been used clinically to induce abortion or to induce labor in a term pregnancy, to treat hypertension, to relieve bronchial asthma, and to heal peptic ulcers.

Prostaglandins also appear to be essential intermediaries in the hormone–cyclic AMP process in that the E prostaglandins regulate the cellular concentration of cyclic-AMP, and the F prostaglandins regulate the cellular concentration of cyclic-GMP (see page 324).

Iodine Number

Unsaturated fats and oils will readily combine with iodine, whereas saturated fats and oils will not do so very readily. The more unsaturated the fat or oil, the more iodine it will combine with.

The iodine number of a fat or oil is the number of grams of iodine that will react with the double bonds present in 100 g of that fat or oil. The higher the iodine number, the greater the degree of unsaturation of the fat or oil. The iodine numbers of some fats and oils are listed in Table 22–2.

In general, animal fats have a lower iodine number than do vegetable oils. This indicates that vegetable oils are more unsaturated. This increasing unsaturation is also accompanied by a change of state—animal fats are solid, vegetable oils are liquid.

Fats and Oils

Structure. Fats are esters formed by the combination of a fatty acid with one particular alcohol, glycerol. If one molecule of glycerol reacts with one molecule of stearic acid (a fatty acid), glyceryl monostearate is formed. The

$$
\begin{array}{ccc}
 & \text{H} & \text{H} \\
 & | & | \\
C_{17}H_{35}CO\boxed{OH + H}O-C-H & & C_{17}H_{35}COO-C-H \\
 & | & | \\
 & HO-C-H \longrightarrow & HO-C-H + H_2O \\
 & | & | \\
 & HO-C-H & HO-C-H \\
 & | & | \\
 & \text{H} & \text{H}
\end{array}
$$

stearic acid   glycerol   glyceryl monostearate

product of this reaction may react with a second molecule and then with a third molecule of stearic acid.

$$
\begin{array}{ccc}
 & \text{H} & \text{H} \\
 & | & | \\
 & C_{17}H_{35}COO-C-H & C_{17}H_{35}COO-C-H \\
 & | & | \\
C_{17}H_{35}CO\boxed{OH + H}O-C-H \longrightarrow & C_{17}H_{35}COO-C-H + H_2O \\
 & | & | \\
 & HO-C-H & HO-C-H \\
 & | & | \\
 & \text{H} & \text{H}
\end{array}
$$

glyceryl distearate

$$
\begin{array}{ccc}
 & \text{H} & \text{H} \\
 & | & | \\
 & C_{17}H_{35}COO-C-H & C_{17}H_{35}COO-C-H \\
 & | & | \\
 & C_{17}H_{35}COO-C-H \longrightarrow & C_{17}H_{35}COO-C-H + H_2O \\
 & | & | \\
C_{17}H_{35}CO\boxed{OH + H}O-C-H & C_{17}H_{35}COO-C-H \\
 & | & | \\
 & \text{H} & \text{H}
\end{array}
$$

glyceryl tristearate
a fat

Glyceryl tristearate (also called tristearin) is formed by the reaction of one molecule of glycerol with three molecules of stearic acid. Since stearic acid is a saturated fatty acid, the product is a fat. If the fatty acid had been unsaturated, the product would have been an unsaturated fat, also called an oil.

The glycerol molecule contains three —OH groups and so combines with three fatty acid molecules. However, these fatty acid molecules do not have to be the same. Fats and oils can contain three different fatty acid molecules, either saturated, unsaturated, or some combination of these.

An example of a mixed triglyceride[1] formed from the reaction of glycerol with three different fatty acid molecules is indicated below. The fatty acids are oleic, stearic, and linoleic.

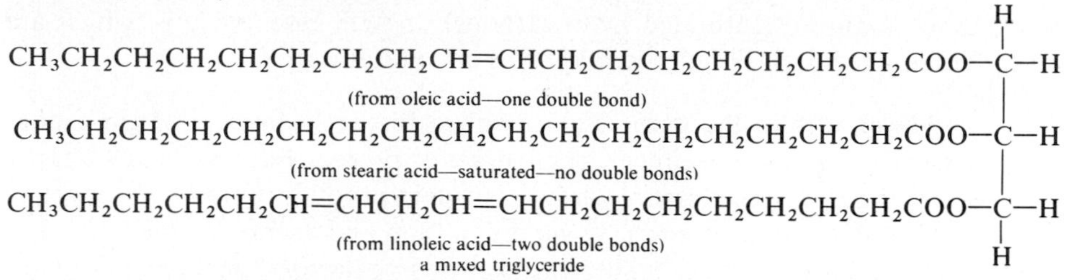

$$CH_3CH_2CH_2CH_2CH_2CH_2CH_2CH_2CH{=}CHCH_2CH_2CH_2CH_2CH_2CH_2CH_2COO{-}\overset{\displaystyle H}{\underset{\displaystyle |}{C}}{-}H$$

(from oleic acid—one double bond)

$$CH_3CH_2CH_2CH_2CH_2CH_2CH_2CH_2CH_2CH_2CH_2CH_2CH_2CH_2CH_2CH_2CH_2COO{-}\overset{|}{C}{-}H$$

(from stearic acid—saturated—no double bonds)

$$CH_3CH_2CH_2CH_2CH_2CH{=}CHCH_2CH{=}CHCH_2CH_2CH_2CH_2CH_2CH_2COO{-}\underset{\displaystyle |}{\underset{\displaystyle H}{\overset{|}{C}}}{-}H$$

(from linoleic acid—two double bonds)
a mixed triglyceride

Oleic acid has a *cis* configuration around its double bond; linoleic acid has a *cis-cis* configuration.

The preceding formula for a mixed triglyceride may be written in condensed form as shown here.

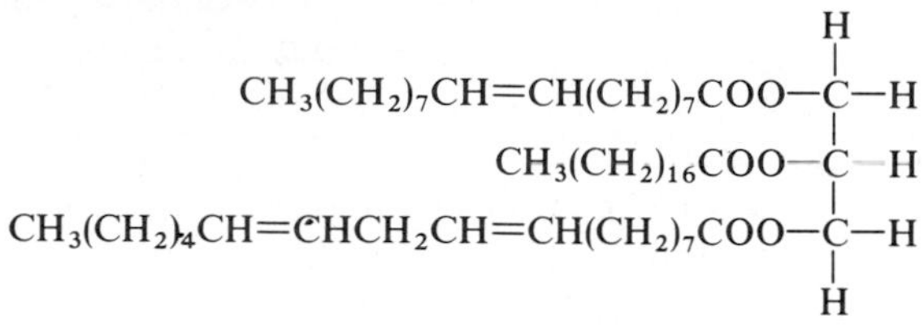

$$CH_3(CH_2)_7CH{=}CH(CH_2)_7COO{-}\overset{\displaystyle H}{\underset{\displaystyle |}{C}}{-}H$$

$$CH_3(CH_2)_{16}COO{-}\overset{|}{C}{-}H$$

$$CH_3(CH_2)_4CH{=}CHCH_2CH{=}CH(CH_2)_7COO{-}\underset{\displaystyle |}{\underset{\displaystyle H}{\overset{|}{C}}}{-}H$$

or simply as:

$$C_{17}H_{33}COO{-}\overset{\displaystyle H}{\underset{\displaystyle |}{C}}{-}H$$

$$C_{17}H_{35}COO{-}\overset{|}{C}{-}H$$

$$C_{17}H_{31}COO{-}\underset{\displaystyle |}{\underset{\displaystyle H}{\overset{|}{C}}}{-}H$$

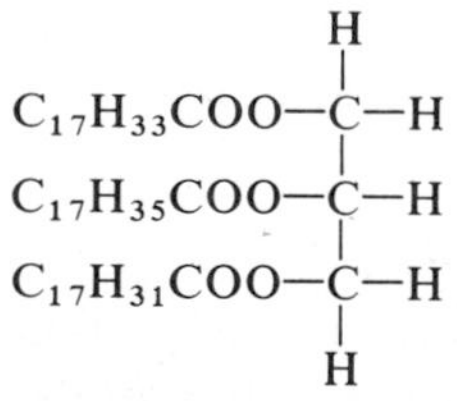

A more correct representation of the structure of the above triglyceride is

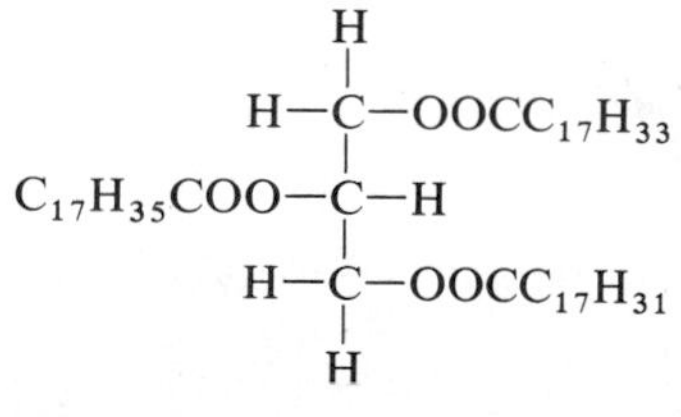

$$H{-}\overset{\displaystyle H}{\underset{\displaystyle |}{C}}{-}OOCC_{17}H_{33}$$
$$C_{17}H_{35}COO{-}\overset{|}{C}{-}H$$
$$H{-}\underset{\displaystyle |}{\underset{\displaystyle H}{\overset{|}{C}}}{-}OOCC_{17}H_{31}$$

but for simplicity we will use the former representation.

[1] According to the International Union of Pure and Applied Chemistry (IUPAC) and the International Union of Biochemistry (IUB), monoglycerides are to be designated as monoacylglycerols, diglycerides as diacylglycerols, and triglycerides as triacylglycerols. Throughout this text, both systems may be used interchangeably.

In general, fats have an iodine number below 70, whereas oils have iodine numbers above 70. Since the iodine number of the compound shown above is 88, it will be an oil similar to peanut oil, which has an iodine number between 83 and 98.

Animal and vegetable oils should not be confused with mineral oil, which is a mixture of saturated hydrocarbons, or with essential oils, which are volatile aromatic liquids used as flavors and perfumes.

Use of Fats in the Body.   Fats serve as a fuel in the body, producing more energy per gram than either carbohydrate or protein. Fat produces 9 Cal per gram, whereas either carbohydrate or protein produces only 4 Cal per gram.

Fats also serve as a reserve supply of food and energy for the body. Fat is stored in the adipose tissue. Fats serve as protectors for the vital organs. That is, fats surround the vital organs to keep them in place and also act as shock absorbers. Fats in the outer layers of the body act as heat insulators, helping to keep the body warm in cold weather.

Physical Properties.   Pure fats and oils are generally white or yellow solids and liquids, respectively. Pure fats and oils are also odorless and tasteless. However, over a period of time fats become rancid; they develop an unpleasant odor and taste.

Fats and oils are insoluble in water but are soluble in such organic liquids as benzene, acetone, and ether. Fats do not diffuse through a membrane. Fats are lighter than water and have a greasy feeling. Fats and oils form a temporary emulsion when shaken with water. The emulsion may be made permanent by the addition of an emulsifying agent such as soap. Fats and oils must be emulsified by bile in the body before they can be digested.

Chemical Properties. Hydrolysis.   When fats are treated with enzymes or heated with steam in the presence of certain catalysts or enzymes, they hydrolyze to form fatty acids and glycerol. When tripalmitin (glyceryl tripalmitate) is hydrolyzed, it forms palmitic acid and glycerol and requires three molecules of water. Recall that during the formation of a fat, water is a product.

$$
\begin{array}{c}
\mathrm{C_{15}H_{31}COO-\overset{\displaystyle H}{\overset{|}{C}}-H} \\[2pt]
\mathrm{C_{15}H_{31}COO-\overset{|}{C}-H} + 3H_2O \xrightarrow[\substack{\text{catalyst}\\\text{or}\\\text{enzyme}}]{\text{heat}} 3C_{15}H_{31}COOH + \begin{array}{c}\mathrm{HO-\overset{\displaystyle H}{\overset{|}{C}}-H}\\[2pt]\mathrm{HO-\overset{|}{C}-H}\\[2pt]\mathrm{HO-\overset{|}{\underset{|}{C}}-H}\\ \mathrm{H}\end{array} \\[2pt]
\mathrm{C_{15}H_{31}COO-\overset{|}{\underset{|}{C}}-H} \\
\mathrm{H}
\end{array}
$$

tripalmitin   palmitic acid   glycerol

When fats are hydrolyzed to fatty acids and glycerol, the glycerol separates from the fatty acids and can be drawn off and purified. Glycerol is used both medicinally and industrially.

Saponification. Saponification is the reaction of a fat with a strong base such as sodium hydroxide to produce glycerol and the salt of a fatty acid.

$$
\begin{array}{c}
\text{H} \\
| \\
C_{17}H_{35}COO{-}\overset{\displaystyle}{C}{-}H \\
| \\
C_{17}H_{35}COO{-}\overset{\displaystyle}{C}{-}H \; + \; 3NaOH \\
| \\
C_{17}H_{35}COO{-}\overset{\displaystyle}{C}{-}H \\
| \\
\text{H}
\end{array}
\quad\longrightarrow\quad
3C_{17}H_{35}COONa \; + \;
\begin{array}{c}
\text{H} \\
| \\
HO{-}\overset{\displaystyle}{C}{-}H \\
| \\
HO{-}\overset{\displaystyle}{C}{-}H \\
| \\
HO{-}\overset{\displaystyle}{C}{-}H \\
| \\
\text{H}
\end{array}
$$

tristearin   sodium stearate, a soap   glycerol

The sodium salt (or potassium salt) of a fatty acid is called a soap. Reactions and properties of soaps will be discussed later in this chapter.

Hydrogenation. Fats and oils are similar compounds except that fats are saturated compounds whereas oils are unsaturated; that is, oils contain double bonds. These double bonds may be changed to single bonds upon the addition of hydrogen. Vegetable oils may be converted to fats by the addition of hydrogen in the presence of a catalyst. This process is called hydrogenation. Hydrogenation is used to produce the so-called vegetable shortenings such as those used in the home. Oleomargarine is prepared by the hydrogenation of certain fats and oils with the addition of flavoring and coloring agents, plus vitamins A and D. Compounds that give butter its characteristic flavor are sometimes added.

$$
\begin{array}{c}
\text{H} \\
| \\
C_{17}H_{33}COO{-}\overset{\displaystyle}{C}{-}H \\
| \\
C_{17}H_{33}COO{-}\overset{\displaystyle}{C}{-}H \; + \; 3H_2 \\
| \\
C_{17}H_{33}COO{-}\overset{\displaystyle}{C}{-}H \\
| \\
\text{H}
\end{array}
\quad\xrightarrow{\text{catalyst}}\quad
\begin{array}{c}
\text{H} \\
| \\
C_{17}H_{35}COO{-}\overset{\displaystyle}{C}{-}H \\
| \\
C_{17}H_{35}COO{-}\overset{\displaystyle}{C}{-}H \\
| \\
C_{17}H_{35}COO{-}\overset{\displaystyle}{C}{-}H \\
| \\
\text{H}
\end{array}
$$

triolein, an oil   tristearin, a fat

In actual practice, vegetable oils are not completely hydrogenated. Enough hydrogen is added to produce a solid at room temperature. If the oil were completely hydrogenated, the solid fat would be hard and brittle and unsuitable for cooking purposes.

As should be expected, hydrogenation lowers the iodine number to a value within the range of fats.

Acrolein Test. The acrolein test, which is a test for the presence of glycerol, is sometimes used as a test for fats and oils, since all fats and oils contain glycerol.

When glycerol is heated to a high temperature, especially in the presence of a dehydrating agent such as potassium bisulfate ($KHSO_4$), a product called acrolein is produced.

$$
\begin{array}{ccc}
& H & \\
& | & \\
H\!-\!\overset{\displaystyle H}{\underset{\displaystyle |}{C}}\!-\!OH & & H\!-\!C\!=\!O \\
H\!-\!\overset{|}{C}\!-\!OH & \xrightarrow[\text{KHSO}_4]{\text{heat}} & H\!-\!C \quad +2\,H_2O \\
H\!-\!\overset{|}{C}\!-\!OH & & H\!-\!C \\
& | & | \\
& H & H \\
\text{glycerol} & & \text{acrolein}
\end{array}
$$

This substance is easily recognized by its strong, pungent odor. When fats or oils are heated to a high temperature or are burned, the disagreeable odor is that of acrolein.

Rancidity. Fats develop an unpleasant odor and taste when allowed to stand at room temperature for a short period of time. That is, they become rancid. Rancidity is due to two types of reactions—hydrolysis and oxidation.

When butter is allowed to stand at room temperature, hydrolysis takes place between the fats and the water present in the butter. The products of this hydrolysis are fatty acids and glycerol. One of the fatty acids produced, butyric acid, has the disagreeable odor that causes one to say that the butter is rancid. The catalysts necessary for the hydrolysis reaction are produced by the action of microorganisms present in the air acting upon the butter. At room temperature this reaction proceeds rapidly so that the butter soon turns rancid. This effect may be overcome by keeping the butter refrigerated and covered.

Oxygen present in the air can oxidize some unsaturated parts of fats and oils. If this oxidation reaction produces short chain acids or aldehydes, the fat turns rancid as evidenced by a disagreeable odor and taste. Since oxidation, as well as hydrolysis, takes place more rapidly at higher temperatures, fats and foods containing a high percentage of fats should be stored in a cool place. Oxidation of fats, especially in hydrogenated vegetable compounds, can be inhibited by the addition of antioxidants, substances which prevent oxidation. Two naturally occurring antioxidants are vitamin C and vitamin E.

Soaps

Soaps are produced by the saponification of fats. Soaps are salts of fatty acids. When the saponifying agent used is sodium hydroxide, a sodium soap is produced. Sodium soaps are bar soaps. When the saponifying agent used is potassium hydroxide, a potassium soap is produced. Potassium soaps are liquid soaps.

$$C_{17}H_{35}COOH + NaOH \longrightarrow C_{17}H_{35}COONa + H_2O$$

stearic acid   sodium   sodium stearate
          hydroxide   a soap

Soaps may also be produced by the reaction of a fatty acid with an inorganic base, although this method is much too expensive to be of commercial value.

Various substances may be added to soaps to give them a pleasant color and odor. Floating soaps contain air bubbles. Germicidal soaps contain a germicide. Scouring soaps contain some abrasive. Tincture of green soap is a solution of a potassium soap in alcohol.

Calcium and magnesium ions present in hard water react with soap to form insoluble calcium and magnesium soaps.

$$Na\ soap + Ca^{2+} \longrightarrow Ca\ soap\downarrow + Na^+$$

$$Na\ soap + Mg^{2+} \longrightarrow Mg\ soap\downarrow + Na^+$$

This precipitated soap is seen as "the ring around the bathtub." The harder the water is, the more soap is required to produce a lather.

Soap emulsifies grease and oil. Washing the hands with soap thus emulsifies the oil and grease that hold the dirt on the skin so that they may be easily washed away. Soap has little effect as an antibacterial agent. Nurses and surgeons in the operating room scrub for at least 10 min to remove most of the debris, such as keratin and natural fats, from the skin. A germicidal soap, one that contains a germ-killing compound, usually is used.

Zinc stearate is an insoluble soap used as a dusting powder for infants. It has antiseptic properties but is irritating to mucous membranes. Zinc undecylenate is used in the treatment of athlete's foot.

Children who suffer from celiac disease cannot absorb fatty acids from the small intestine. The unabsorbed fatty acids combine with calcium ions to form insoluble calcium compounds, or soaps. These calcium compounds are eliminated from the body, and the body will become deficient in calcium unless additional amounts of this necessary element are given to the child.

Detergents

Detergents (syndets) are synthetic compounds used as cleansing agents. They work like soaps but do not have several of the disadvantages that soaps have.

Detergents work as well in hard water as they do in soft water. That is, calcium and magnesium salts of detergents are soluble and do not precipitate out of solution (as do calcium and magnesium soaps). Recall that soaps do not work as well in hard water because insoluble calcium and magnesium salts precipitate out of solution. Detergents are generally neutral compounds when compared to soaps, which are usually alkaline or basic substances. Therefore, detergents may be used on silks and woolens but soaps may not. Detergents are used in washing clothes and also as cleansing agents in toothpastes and toothpowders.

Detergents are sodium salts of long chain alcohol sulfates. For example, sodium lauryl sulfate may be prepared by treating lauryl alcohol, a 12-carbon alcohol, with sulfuric acid and then neutralizing with sodium hydroxide. The reactions are

$$C_{11}H_{23}CH_2OH + H_2SO_4 \longrightarrow C_{11}H_{23}CH_2OSO_3H + H_2O$$

lauryl alcohol   lauryl hydrogen sulfate

$$C_{11}H_{23}CH_2OSO_3H + NaOH \longrightarrow C_{11}H_{23}CH_2OSO_3Na + H_2O$$

lauryl hydrogen   sodium lauryl sulfate
sulfate   a detergent

Detergents containing straight chains are *biodegradable* and do not cause water pollution, whereas those containing branched chains are nonbiodegradable and cause pollution.

Waxes

A wax is a compound produced by the reaction of a fatty acid with a high molecular weight alcohol such as myricyl alcohol ($C_{30}H_{61}OH$) and ceryl alcohol ($C_{26}H_{53}OH$). Carnauba wax is largely $C_{25}H_{51}COOC_{30}H_{61}$, an ester of myricyl alcohol. Beeswax is largely $C_{15}H_{31}COOC_{30}H_{61}$, also an ester of myricyl alcohol.

Note that waxes are primarily esters of long-chain fatty acids with an even number of carbon atoms and long-chain alcohols, also with an even number of carbon atoms.

Some of the common waxes are listed in Table 22–3.

Paraffin wax is different from these waxes because it is merely a mixture of hydrocarbons and is not an ester.

TABLE 22–3

Common Waxes

Name	Source	Use
Beeswax	Honeycomb of bee	Polishes and pharmaceutical products
Spermaceti	Sperm whale	Cosmetics and candles
Carnauba	Carnauba palm	Floor waxes and polishes
Lanolin	Wool	Skin ointments

Phospholipids are phosphate esters and may be divided into two categories —phosphoglycerides and phosphosphingosides—depending on whether the alcohol is glycerol or sphingosine. As indicated on page 282, phospholipids also contain a nitrogen compound. Phospholipids are found in all tissues in the human body, particularly in brain, liver, and spinal tissue and in the membranes of most cells.

Phosphoglycerides

Phosphoglycerides may be subdivided into several types, depending upon the nitrogen compound present. Among these are the lecithins and the cephalins.

Lecithins. Lecithins, now called phosphatidyl cholines, are compounds that are particularly important in the metabolism of fats by the liver. Lecithins are similar to fats in that they are esters of glycerol (they are phosphoglycerides). The differences between a fat and a lecithin are indicated below.

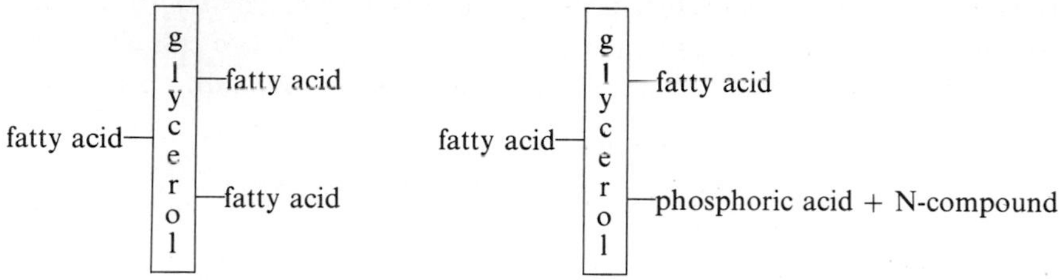

a fat   a lecithin

In lecithins, the nitrogen compound is choline, an alcohol.

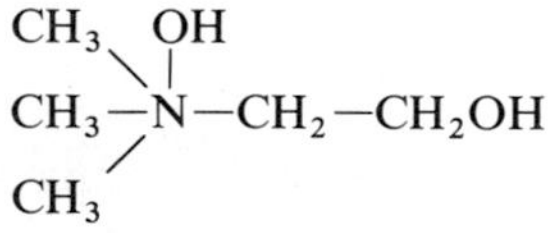

$$CH_3-\underset{\underset{CH_3}{|}}{\overset{\overset{CH_3}{\diagdown}\quad OH}{N}}-CH_2-CH_2OH$$

choline

A typical formula for a lecithin (phosphatidyl choline) is shown below.

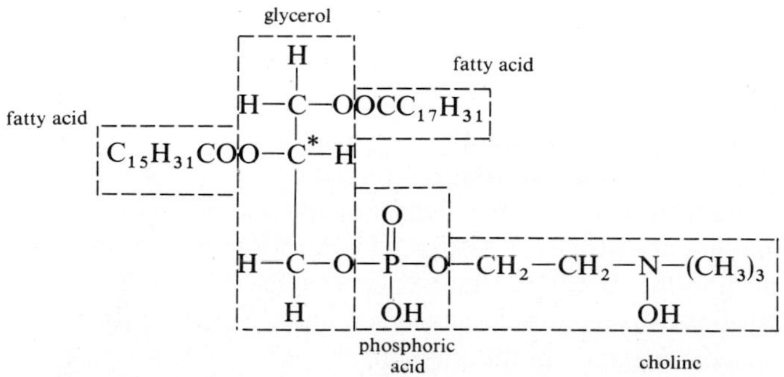

The carbon marked with an asterisk is asymmetric, indicating optical activity. Naturally occurring lecithins have the L-form (see page 261).

Lecithins are insoluble in water but are good emulsifying agents. Lecithins are also a good source of phosphoric acid, which is needed for the synthesis of new tissue.

Fats are partly converted to lecithins in the body and are transported as lecithins from one part of the body to another.

Dipalmityl lecithin (lecithin where the two fatty acids are palmitic acid) is a very good surface active agent (see page 138). It prevents adherence of the inner surfaces of the lungs.

Removal of one molecule of fatty acid from lecithin produces lysolecithin. The removal of this molecule of fatty acid is catalyzed by the enzyme lecithinase A, which is found in the venom of poisonous snakes. This venom is poisonous because it produces lysolecithin, which in turn causes hemolysis—the destruction of the red blood cells.

Cephalins. Cephalins are similar to lecithins except that another nitrogen compound is present instead of choline. In cephalins the nitrogen compound may be ethanolamine, serine, or inositol. The newer names for these compounds, indicating the nitrogen compound present, are phosphatidyl ethanolamine, phosphatidyl serine, and phosphatidyl inositol, respectively.

Cephalins are important in the clotting of the blood and also are sources of phosphoric acid for the formation of new tissue.

Phosphosphingosides

Phosphosphingosides, also called sphingolipids, differ from phosphoglycerides in that they contain the alcohol sphingosine in place of glycerol. One particular type of sphingolipid, called sphingomyelin, is present in large amounts in brain and nerve tissue. The general formula for a sphingolipid and the structural formula for sphingomyelin are given below.

In Niemann-Pick disease, a disease of infancy or early childhood, sphingomyelins accumulate in the brain, liver, and spleen. The accumulation of the sphingomyelins causes mental retardation and early death. It is caused by the lack of a specific enzyme, sphingomyelinase.

$$\text{choline} + \text{phosphoric acid} - \boxed{\begin{array}{c} s \\ p \\ h \\ i \\ n \\ g \\ o \\ s \\ i \\ n \\ e \end{array}} - \text{fatty acid}$$

a sphingolipid

$$\begin{array}{c} \fbox{$CH_3 - (CH_2)_{12}$} \quad \text{sphingosine} \\ H - C \\ \parallel \\ C - H \\ H - C - OH \\ H - C - NH - \overset{\displaystyle O}{\underset{\displaystyle \parallel}{C}} - C_{17}H_{31} \quad \text{fatty acid} \\ (CH_3)_3 - \overset{+}{N} \quad CH_2 - CH_2 - O - \overset{\displaystyle O}{\underset{\displaystyle \parallel}{P}} - O - C - H \\ \underset{\text{choline}}{} \qquad \underset{\text{phosphoric acid}}{OH} \quad H \end{array}$$

sphingomyelin

The ratio of lecithins to sphingomyelins sloughed off by an infant into the amniotic fluid may be used as a test for hyaline membrane disease. A ratio of less than 1.5:1 indicates that the infant's lungs have not developed sufficiently.

Glycolipids

Glycolipids are similar to sphingomyelins except that they contain a carbohydrate, often galactose, in place of the choline and phosphoric acid. The general structure of a glycolipid is represented as follows

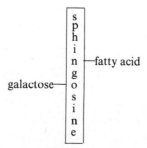

$$\text{galactose} - \boxed{\begin{array}{c} s \\ p \\ h \\ i \\ n \\ g \\ o \\ s \\ i \\ n \\ e \end{array}} - \text{fatty acid}$$

Glycolipids produce no phosphoric acid on hydrolysis because they do not contain this compound. Glycolipids are also called cerebrosides because they are found in large amounts in the brain tissue.

Among the glycolipids are kerasin, cerebron, nervon, oxynervon, and the gangliosides.

In Gaucher's disease glycolipids accumulate in the brain and cause severe mental retardation and death by age three. Juvenile and adult forms of this disease are characterized by enlarged spleen and kidneys, hemorrhaging, mild anemia, and fragile bones. This disease is caused by the lack of a specific enzyme, beta-glucosidase.

In the absence of a particular enzyme, hexosaminidase A, glycolipids accumulate in the tissues of the brain and eyes. This effect, called Tay-Sachs disease, is usually fatal to infants before they reach age two.

Phospholipids are found in the membranes of all cells. Their peculiar properties are responsible for passage of various substances into and out of the cells. What gives phospholipids their strange properties? Consider the phosphoglycerides, whose structure may be represented as shown below.

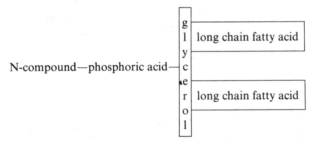

The fatty acid chains are *hydrophobic*—they point away from water. The other end of the molecule, the one containing the nitrogen compound and phosphoric acid, is *hydrophilic* and dissolves in water. Molecules of this type, with a hydrophobic and a hydrophilic end, are said to be *amphipathic*.

Because phospholipids are amphipathic, they may form a bilayer at a position between two aqueous layers. In the following diagram, the long lines represent the fatty acid chains, the hydrophobic ends, while the circles represent the hydrophilic ends which dissolve in water.

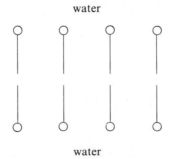

Nonpolar molecules can pass through the membrane, which is mostly hydrocarbon and also nonpolar, but polar molecules and ions find this layer a barrier. How do they pass through? Embedded in this type of bilayer and extending through it are proteins that consist of long chains of amino acids bonded together. These proteins, which are looped and coiled in many different ways, transport polar molecules and ions through the membrane. Selectivity is given by different types of protein molecules.

Steroids

Steroids are high molecular weight tetracyclic (four-ring) compounds. Those containing one or more —OH groups and no C=O groups are called sterols. The most common sterol is cholesterol, which is found in animal fats but not in plant fats. Cholesterol is found in all animal tissues, particularly in brain and nervous tissue, in the bloodstream, and as gallstones. Cholesterol aids in the absorption of fatty acids from the small intestine. Several theories have been proposed relating cholesterol to coronary malfunction, but no definite evidence has yet been found linking the two together.

Most of the body's cholesterol is derived or synthesized from other substances, such as carbohydrates and proteins as well as from fats.

Atherosclerosis, a form of arteriosclerosis, results from the deposition of excess lipids, primarily triglycerides and cholesterol, from the bloodstream. Of these two, cholesterol poses a greater threat to the well-being of a person, although excess triglycerides also present a significant risk. One way of combating heart disease and atherosclerosis is to reduce the concentration of lipids in the bloodstream—either by reducing lipid intake or by the use of antihyperlipidemic drugs, those that tend to reduce blood lipid levels. It has been found that certain unsaturated fish and vegetable oils, when substituted for saturated fats, lower the serum cholesterol level.

Ergosterol is a sterol similar to cholesterol. When ergosterol is irradiated (exposed to radiation) with ultraviolet light, one of the products formed is calciferol (vitamin D_2) (see page 446).

Other steroids include bile salts, the sex hormones, and the hormones of the adrenal cortex. The similarities of the structure of some of these steroids are indicated below.

cholesterol

estradiol

testosterone

ergosterol

Summary

Lipids yield fatty acids on hydrolysis or combine with fatty acids to form esters. Lipids are insoluble in water but are soluble in organic solvents such as ether, acetone, and carbon tetrachloride.

Lipids may be classified into three types: simple, compound, and derived. Simple lipids are esters of fatty acids. A simple lipid that yields fatty acids and glycerol upon hydrolysis is called a fat or oil. A simple lipid that, on hydrolysis, yields fatty acids and a high molecular weight alcohol is called a wax.

Compound lipids yield fatty acids, alcohol, and some other type of compound on hydrolysis.

Derived lipids are compounds derived from simple or compound lipids on hydrolysis.

Fatty acids are straight-chain organic acids. Those found in nature usually contain an even number of carbon atoms. Saturated fatty acids have only single bonds between carbon atoms. Unsaturated fatty acids have one or more double bonds in the molecule and occur in nature in the *cis* form.

The essential fatty acids, so called because they are necessary in the diet of an infant, are linoleic and linolenic acids.

Prostaglandins consist of 20-carbon fatty acids containing a five-membered ring. Prostaglandins have a wide range of physiologic effects.

Unsaturated fats and oils react with iodine, whereas saturated ones do not. The iodine number of a fat or oil is the number of grams of iodine that will react with (the double bonds present in) 100 g of that fat or oil.

Fats serve as fuel for the body—1 g of fat producing 9 Cal as compared to only 4 Cal per gram of carbohydrate. Fats protect nerve endings and also act as insulators to keep the body warm in cold weather.

Fats and oils are odorless and tasteless when pure. They are insoluble in water but are soluble in organic solvents. Fats and oils must be emulsified before being digested.

When fats are hydrolyzed, they form fatty acids and glycerol. When fats are saponified, they form salts of fatty acids (soaps) and glycerol.

When oils are hydrogenated, the double bonds are changed to single bonds and the (liquid) oil becomes a (solid) fat.

When fats or oils are heated to a high temperature, especially in the presence of a dehydrating agent, a product known as acrolein is produced. The odor of burning fat or oil is due to the presence of acrolein.

When fats are allowed to stand at room temperature, they become rancid because of hydrolysis and also because of oxidation. Keeping fats cool prevents their becoming rancid.

Soaps are salts of fatty acids—sodium soaps being solid or bar soaps, and potassium soaps being liquid soaps.

Calcium and magnesium soaps are insoluble in water and are formed when sodium or potassium soaps are used in hard water. The precipitated calcium and magnesium soaps are seen as the "ring around the bathtub."

Detergents are similar to soaps in their cleansing properties. However, detergents do not precipitate in hard water because their calcium and magnesium compounds are soluble.

A wax is a compound produced by the reaction of a fatty acid with a high molecular weight alcohol.

Phospholipids contain fatty acids, an alcohol, a nitrogen compound, and phosphoric acid. Two types of phospholipids are phosphoglycerides and phosphosphingosides.

Phosphoglycerides include phosphatidyl cholines (lecithins) and phosphatidyl ethanolamines, serines, and inositols (cephalins). Lecithins are important in the metabolism of fats by the liver and also are a source of phosphoric acid, which is needed for the formation of new tissue. Cephalins are important in clotting of the blood and also are a source of phosphoric acid for the formation of new tissue.

Sphingomyelins are an example of phosphosphingosides and are present in large amounts in brain and nerve tissue.

Glycolipids, also called cerebrosides, are found in large amounts in brain tissue.

Phospholipids are found in all cell membranes as bilayers and are responsible for the passage of various substances into and out of the cells.

Steroids are high-molecular-weight, four-ring compounds. Those containing —OH groups are called sterols. The most common sterol in the body is cholesterol. Other steroids are vitamin D, bile salts, sex hormones, and hormones of the adrenal cortex.

Questions and Problems

1. State the general properties of lipids.
2. What are simple lipids? *ester of FA*
3. What is the difference between a fat and a wax? A fat and an oil?
4. What is a compound lipid? Give examples of several.
5. What is a derived lipid? Give examples of several.
6. What is a fatty acid?
7. What is the difference between a saturated and an unsaturated fatty acid?
8. What is unusual about the number of carbon atoms in naturally occurring fatty acids?
9. What are the essential fatty acids? Why are they important?

10. Why should commercial boiled linseed oil never be used to overcome a deficiency of the essential fatty acids?
11. What fatty acid has been used in the treatment of leprosy?
12. What is the iodine number of a fat or oil? What determines this number?
13. Which have higher iodine numbers, animal or vegetable fats?
14. Draw the structure of glyceryl tripalmitate.
15. Draw the structure of the fat formed from the reaction of glycerol with palmitic, stearic, and oleic acids.
16. What is the function of fat in the body?
17. What are the general physical properties of fats and oils?
18. What products are formed when a fat is hydrolyzed?
19. What products are formed when a fat is saponified?
20. How may an unsaturated oil be changed to a saturated fat? Is the fat thus formed completely saturated? Why?
21. What is the test for the presence of a fat or oil?
22. What causes rancidity in a fat or oil?
23. Compare soaps with detergents on the basis of structure.
24. Compare soaps with detergents on the basis of their reaction with hard water.
25. What causes a soap to float?
26. What is a germicidal soap? How is it made?
27. Why do surgeons scrub so vigorously before surgery?
28. What causes celiac disease?
29. What is a phospholipid?
30. What are the types of phospholipids?
31. What is a lysolecithin? What effect does it have on the body? How is it produced?
32. Compare the structures of the phosphoglycerides.
33. How do phospholipids help control the passage of materials into or out of cells?
34. What is meant by the term *amphipathic*?
35. What causes atherosclerosis? What can be done to prevent it?
36. What causes Niemann-Pick disease? Tay-Sachs disease? Gaucher's disease?
37. What is the most common sterol in the body?
38. Give the names of several steroids. Where is each found?
39. Why does the elimination of cholesterol-containing foods in the diet have little effect on the body's cholesterol? What can be done to reduce serum cholesterol?

References

Barker, R.: *Organic Chemistry of Biological Compounds*. Prentice-Hall, Inc., Englewood Cliffs, N. J., 1971, Chap. 7.

Baum, S. J., and Scaife, C. W.: *Chemistry: A Life Science Approach*. Macmillan Publishing Co., Inc., New York, 1975, Chap. 24.

Bronk, J. R.: *Chemical Biology*. Macmillan Publishing Co., Inc., New York, 1973, Chap. 4.

Capaldi, R. A.: A dynamic model of cell membranes. *Scientific American*, **230**:26–33 (March), 1974.

Cullitun, B. J.: Prostaglandins—something for everyone. *Science News*, **98**:306–307 (Oct. 10), 1970.

Green, D. E.: Synthesis of fat. *Scientific American*, **202**:46–51 (Feb.), 1960.

Harper, H. A.: *Review of Physiological Chemistry*, 15th ed. Lange Medical Publications, Los Altos, Calif., 1975, Chap. 2.

Lehninger, A. H.: *Biochemistry*, 2nd ed. Worth Publishers, Inc., New York, 1975, Chap. 11.

Morrison, R. T., and Boyd, R. N.: *Organic Chemistry*, 3rd ed. Allyn & Bacon, Inc., Boston, 1973, Chap. 33.

Rafelson, M. E.; Binkley, S. B.; and Hayashi, J. A.: *Basic Biochemistry*, 3rd ed. Macmillan Publishing Co., Inc., New York, 1971, Chap. 6.

Russell, P. M. G.: Obstetric endocrinology. *Nursing Mirror*, **133**:33–34 (Nov. 26), 1971.

Stacy, G. W.: *Organic Chemistry*. Harper & Row, New York, 1975, Chap. 16.

Chapter 23

Proteins

Other than water, proteins are the chief constituents of all cells of the body.
Proteins are much more complex than either carbohydrates or fats. All
proteins contain the elements carbon, hydrogen, oxygen, and nitrogen. Most
proteins also contain sulfur; some contain phosphorus; and a few, such as
hemoglobin, contain some other element.

Plants synthesize proteins from material present in the air and in the soil. Animals cannot synthesize proteins from such starting materials. Animals must obtain proteins from plants, or from other animals which in turn have obtained it from plants.

Animals excrete waste materials containing many nitrogen compounds. These nitrogen compounds along with decaying animal and plant matter are converted into soluble nitrogen compounds by soil bacteria. Plants in turn use these soluble nitrogen compounds to manufacture more protein, thus completing the cycle in nature. A simplified version of the nitrogen cycle is shown in Figure 23–1.

Functions

Proteins function in the body in the building of new cells, the maintenance of existing cells, and the replacement of old cells. Thus, proteins are the most important type of compound in the body. The word *protein* is derived from

Figure 23–1. The nitrogen cycle.

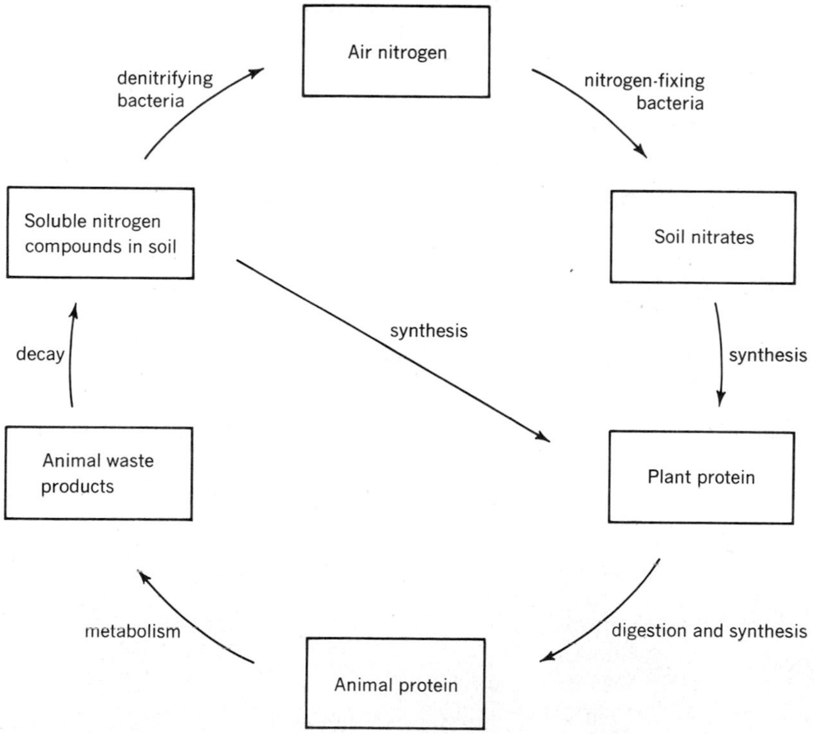

the Greek word *proteios*, which means "of first importance." Proteins are also a valuable source of energy in the body. The oxidation of 1 g protein yields 4 Cal—just as does the oxidation of 1 g carbohydrate. Proteins are also necessary for the formation of the various enzymes and hormones found in the body. Antibodies are proteins, as are parts of viruses. Keratins are the protein found in hair; collagen is a protein in connective tissue and muscle.

Molecular Weight

Proteins have very high molecular weights. A comparison of the molecular weight of proteins with carbohydrates and fats can be seen in Table 23–1.

Amino Acids

Proteins are built up from simple units called amino acids. Hydrolysis of proteins yields amino acids. There are 20 known amino acids that can be produced by the hydrolysis of protein. All these amino acids, except glycine, which has no asymmetric carbon, have the L configuration.

TABLE 23–1

Molecular Weights of Various Proteins, Carbohydrates, and Fats

Type of Compound	Molecular Weight
Inorganic compounds	
Water	18
Sodium chloride	58.5
Plaster of paris	290
Organic compounds	
Benzene	78
Ether	46
Carbohydrates	
Glucose	180
Sucrose	342
Lipids	
Tristearin	891
Cholesterol	384
Proteins	
Insulin	6,300
Lactalbumin	17,500
Zein	50,000
Hemoglobin	68,000
Serum globulin	180,000
Fibrinogen	450,000
Thyroglobin	630,000
Hemocyanin	9,000,000
Tobacco mosaic virus	40,000,000

Amino acids are organic acids having an amine ($-NH_2$) group attached to a chain containing an acid group. Although the amine group can be anywhere on the chain, amino acids found in nature all have the amine group on the alpha carbon—that is, the carbon atom next to the acid group. These alpha amino acids may be represented by the following general formula

$$
\begin{array}{c}
\text{COOH} \\
| \\
\text{H}_2\text{N}-\text{C}-\text{H} \\
| \\
\text{R}
\end{array}
$$

where R can be many different radicals. The structures of a few of the alpha L-amino acids are shown

glycine   alanine   leucine   threonine

valine   methionine   glutamic acid   serine

phenylalanine   tryptophan

The body can synthesize some, but not all, of the amino acids that it needs. Those that it cannot synthesize must be supplied from the food consumed. These are called the *essential amino acids* and are listed below.

1. Arginine (can be synthesized by the body but too slowly to be of practical value)
2. Histidine (required only during childhood)
3. Isoleucine
4. Leucine
5. Lysine
6. Methionine
7. Phenylalanine
8. Threonine
9. Tryptophan
10. Valine

Amphoteric Nature

Amino acids are amphoteric compounds; that is, they can react with either acids or bases. How is this possible? Recall that amino acids contain a —COOH group, which is acidic, and an —NH₂ group, which is basic.

The amphoteric nature of glycine, the simplest amino acid, can be illustrated by its reaction with a base, NaOH, and with an acid, HCl.

$$\underset{\underset{NH_2}{|}}{CH_2}-COOH + NaOH \longrightarrow \underset{\underset{NH_2}{|}}{CH_2}COONa + H_2O$$

$$\underset{\underset{NH_2}{|}}{CH_2}-COOH + HCl \longrightarrow \underset{\underset{NH_3Cl}{|}}{CH_2}-COOH$$

Since amino acids are amphoteric, proteins, which are made up of amino acids, are also amphoteric. This amphoteric nature of proteins accounts for their ability to act as buffers in the blood; they can react with either acids or bases to prevent an excess of either.

Isoelectric Point

When an amino acid is placed in an acid solution, it forms a positive ion which is attracted to a negatively charged electrode.

$$H-\underset{\underset{NH_2}{|}}{\overset{\overset{H}{|}}{C}}-\overset{\overset{O}{||}}{C}-OH + H^+Cl^- \longrightarrow H-\underset{\underset{NH_3^+}{|}}{\overset{\overset{H}{|}}{C}}-\overset{\overset{O}{||}}{C}-OH + Cl^-$$

positively charged ion

TABLE 23–2

Isoelectric Points of Some Proteins

Protein	Isoelectric Point (pH)
Egg albumin	4.7
Casein	4.6
Hemoglobin	6.7
Insulin	5.3
Serum globulin in blood	5.4
Fibrinogen in blood	5.6

When an amino acid is placed in a basic solution, it forms a negatively charged ion which is attracted toward a positively charged electrode.

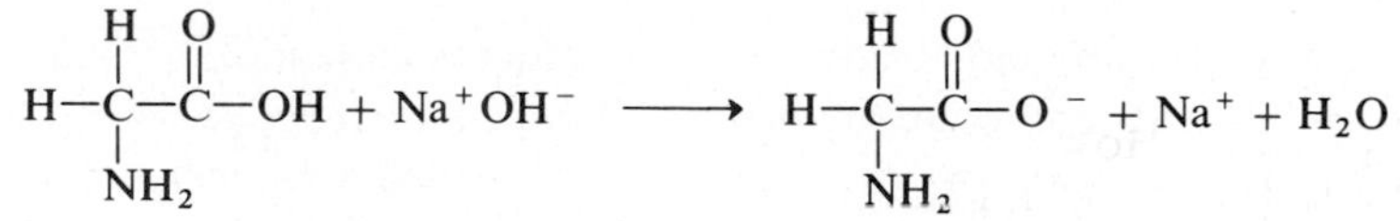

$$
\begin{array}{c}
\text{H} \quad \text{O} \\
| \quad \| \\
\text{H} - \text{C} - \text{C} - \text{OH} + \text{Na}^+\text{OH}^- \\
| \\
\text{NH}_2
\end{array}
\longrightarrow
\begin{array}{c}
\text{H} \quad \text{O} \\
| \quad \| \\
\text{H} - \text{C} - \text{C} - \text{O}^- + \text{Na}^+ + \text{H}_2\text{O} \\
| \\
\text{NH}_2
\end{array}
$$

negatively charged ion

At a certain pH (that is, a certain hydrogen ion concentration) amino acids will not migrate toward either the positive or the negative electrode. At this pH amino acids will be neutral; there will be an equal number of positive and negative ions. This point is called the *isoelectric point*.

Proteins, which are composed of amino acids, also have an isoelectric point which is different for each protein. At their isoelectric point proteins have a minimum solubility, a minimum viscosity, and also a minimum osmotic pressure. At a pH above the isoelectric point, a protein has more negative than positive charges. At a pH below the isoelectric point a protein has more positive than negative charges. The isoelectric points of a few proteins are listed in Table 23–2.

Dipeptides

Proteins consist of many amino acids joined together by what is called a *peptide linkage* or a *peptide bond* (see page 225). Suppose that a glycine molecule reacts with an alanine molecule. This reaction may occur in two different ways. That is, the amine part of the glycine may react with the acid part of the alanine, or the acid part of the glycine may react with the amine part of the alanine. Both of these reactions are illustrated below. (Note that amino acid combinations are usually abbreviated by using the first three letters of each amino acid in the order in which it appears in the peptide.)

$$CH_3-CH-\underset{\underset{NH_2}{|}}{\overset{\overset{O}{\|}}{C}}-\boxed{OH + H}-NH-CH_2-COOH \longrightarrow$$

alanine   glycine

$$CH_3-CH-\underset{\underset{NH_2}{|}}{\overset{\overset{O}{\|}}{C}}-NH-CH_2-COOH + H_2O$$

alanyl-glycine (ala-gly)

$$NH_2-CH_2-\overset{\overset{O}{\|}}{C}-OH + CH_3-CH-COOH \longrightarrow$$

glycine   H NH   alanine

$$NH_2-CH_2-\overset{\overset{O}{\|}}{C}-NH-\underset{\underset{CH_3}{|}}{CH}-COOH + H_2O$$

glycyl-alanine (gly-ala)

When two amino acids combine, the product is called a *dipeptide*. When three amino acids combine, the product is called a *tripeptide*. When four or more amino acids join together, the product is called a *polypeptide*. For just two amino acids, glycine and alanine, two different combinations have already been indicated—glycyl-alanine and alanyl-glycine, where the first member of each group acts as the one furnishing the OH from the acid group. For three different amino acids—such as glycine, alanine, and valine—there are six possible combinations (or tripeptide linkages).

1. Glycyl-alanyl-valine (gly-ala-val)
2. Glycyl-valyl-alanine (gly-val-ala)
3. Alanyl-glycyl-valine (ala-gly-val)
4. Alanyl-valyl-glycine (ala-val-gly)
5. Valyl-glycyl-alanine (val-gly-ala)
6. Valyl-alanyl-glycine (val-ala-gly)

Proteins contain a large number of peptide linkages, and the number of possible combinations of the many amino acids in the formation of a protein

Figure 23-2. Structure of human insulin. Note that insulin contains disulfide bridges between cys groups. Breaking these disulfide bridges inactivates insulin.

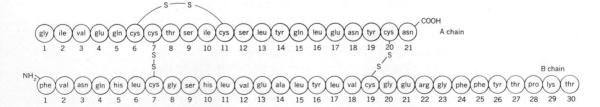

is beyond all comprehension. Insulin (see Figure 23–2) illustrates the peptide linkages. It has an A chain, containing 21 amino acids, and a B chain, which contains 30 amino acids. Note that both chains are connected by two sulfide bridges.

Structure

When a protein is hydrolyzed (by acids, bases, or certain enzymes), it breaks down into smaller and smaller units, eventually forming amino acids. Likewise, when amino acids combine (under the influence of certain enzymes), they first form dipeptides, then tripeptides, then polypeptides, and so on, until they eventually form a protein, a molecule with a large molecular weight because it is formed from so many amino acid groups.

$$\text{protein} \underset{H_2O}{\overset{H_2O}{\rightleftharpoons}} \text{proteoses} \underset{H_2O}{\overset{H_2O}{\rightleftharpoons}} \text{peptones} \underset{H_2O}{\overset{H_2O}{\rightleftharpoons}} \text{polypeptides} \underset{H_2O}{\overset{H_2O}{\rightleftharpoons}}$$

$$\text{tripeptides} \underset{H_2O}{\overset{H_2O}{\rightleftharpoons}} \text{dipeptides} \underset{H_2O}{\overset{H_2O}{\rightleftharpoons}} \text{amino acids}$$

Proteins have a three-dimensional structure that can be considered as being composed of simpler structures.

The *primary structure* of a protein refers to the number and sequence of the amino acids in the protein. These amino acids are held together by peptide bonds. The primary structures of human insulin and of ribonuclease are indicated in Figures 23–2 and 23–3, respectively.

The *secondary structure* of a protein refers to the regular recurring arrangement of the amino acid chain, such as in a coiled shape (see Figure 23–4). These structures are held together by hydrogen bonds (bonds between the —H of the —NH$_2$ of one amino acid and the O of the C=O of the acid part of another amino acid, as indicated above) and by disulfide bonds (bonds between sulfur atoms of two adjacent sulfur-containing amino acids, as indicated in Figures 23–2 and 23–3).

The *tertiary structure* of a protein refers to the specific folding and bending of the coils into specific layers or fibers (see Figure 23–5). It is the tertiary structure that gives proteins their specific biologic activity (see enzymes, page 331).

Some proteins have a *quaternary structure*, which occurs when several protein units, each with its own primary, secondary, and tertiary structure, combine to form a more complex unit (see Figure 23–6).

Percentage Composition

The average percentage of nitrogen present in protein is 16 per cent; that is, about one sixth of protein is nitrogen. Because protein is the major

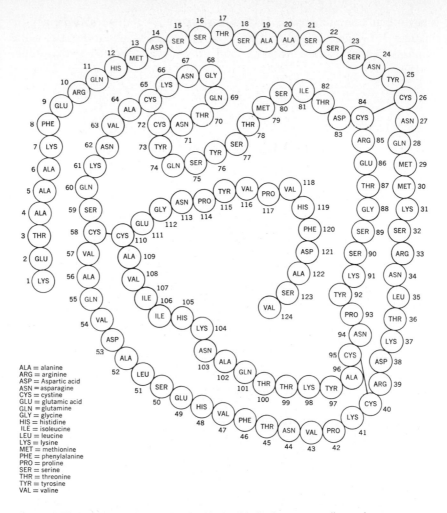

Figure 23–3. Sequence of amino acids in the enzyme ribonuclease.

ALA = alanine
ARG = arginine
ASP = Aspartic acid
ASN = asparagine
CYS = cystine
GLU = glutamic acid
GLN = glutamine
GLY = glycine
HIS = histidine
ILE = isoleucine
LEU = leucine
LYS = lysine
MET = methionine
PHE = phenylalanine
PRO = proline
SER = serine
THR = threonine
TYR = tyrosine
VAL = valine

food that contains nitrogen, the chemist can determine the amount of protein present in a food substance by determining the amount of nitrogen present. This amount is about one sixth of the amount of protein present. Therefore, the amount of protein in the food can be calculated by multiplying the weight of nitrogen by six and converting this to a percentage of the total. For example, suppose that a 100-g sample of food yielded 4 g nitrogen upon chemical analysis. Since the amount of nitrogen in protein is one sixth of the total amount of protein present, the amount of protein present is 6 × 4 g, or 24 g. Then the percentage of protein present in the original 100-g sample is 24 per cent.

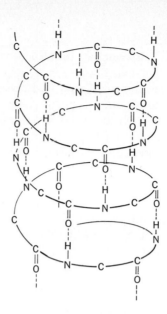

Figure 23-4. Secondary structure of a protein. (Reproduced from Rafelson, M. E.; Binkley, S. B.; and Hayashi, J. A.: *Basic Biochemistry*, 3rd ed. Macmillan Publishing Co., Inc., New York, 1971.)

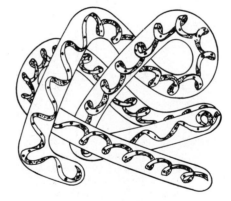

Figure 23-5. Tertiary structure of a protein. Note that the coiled helix represents the secondary structure, which in turn is made up of the various amino acids in the sequence specified by the primary structure.

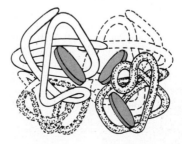

Figure 23-6. Quaternary structure of hemoglobin.

Classification

Proteins are divided into three categories; simple, conjugated, and derived. On hydrolysis, simple proteins yield only amino acids or derivatives of amino acids. On hydrolysis, conjugated proteins yield amino acids plus some other type of compound. Conjugated protein consists of a simple protein combined with a nonprotein compound. Derived proteins are protein derivatives—produced by the action of chemical, enzymes, and physical forces on the other two classes of protein.

Simple Proteins

Simple proteins are further classified according to their solubility in various reagents and also as to whether they are coagulated by heat (see Table 23–3).

Conjugated Proteins

Conjugated proteins are also subdivided into several classes. These are listed in Table 23–4.

TABLE 23–3

Properties of Simple Proteins

Type of Simple Protein	Solubility	Coagulated by Heat	Examples
Albumins	Soluble in water, precipitated by saturated salt solution	Yes	Egg albumin; serum albumin; lactalbumin
Globulins	Soluble in dilute salt solution, insoluble in water and moderately concentrated salt solution	Yes	Serum globulin; lactoglobulin; vegetable globulin
Glutenins	Soluble in dilute acid and alkali, insoluble in neutral solvents	Yes	Glutenin from wheat
Prolamins	Soluble in 70–80 per cent alcohol; insoluble in pure alcohol and other neutral solvents	No	Zein from corn; gliadin from wheat
Albuminoids	Insoluble in all neutral solvents and in dilute acid and alkali	No	Keratin in hair, nails; feathers; collagen
Histones	Soluble in water and very dilute acid; insoluble in very dilute NH_4OH	No	Nucleohistone in thymus gland; globin in hemoglobin
Protamines	Soluble in water and NH_4OH	No	Salmine from salmon; sturine from sturgeon

TABLE 23-4

313

Proteins

Conjugated Proteins

Type	Nonprotein Portion of the Combination	Examples
Nucleoproteins	Nucleic acid	Found in all cells and in glandular tissue
Glycoproteins	Carbohydrates	Mucin in saliva
Phosphoproteins	Phosphate	Casein in milk
Chromoproteins	Chromophore group (color-producing group)	Hemoglobin, hemocyanin, flavoproteins, cytochrome
Lipoproteins	Lipids	Fibrin in blood
Metalloproteins	Metals	Ceruloplasmin (containing Cu) and siderophilin (containing Fe) in blood plasma

Derived Proteins

Derived proteins are products of hydrolysis somewhere between a protein and an amino acid in structure. Derived proteins include proteoses, peptones, polypeptides, tripeptides, and dipeptides.

Properties

Colloidal Nature

Proteins form colloidal dispersions in water. Protein, being colloidal, will pass through a filter paper but not through a membrane. The inability of protein to pass through a membrane is of great importance in the body. Proteins present in the bloodstream cannot pass through the cell membranes and should remain in the bloodstream. Since proteins cannot pass through membranes, there should be no protein material present in the urine. The presence of protein in the urine indicates damage to the membranes in the kidneys—possibly nephritis.

Denaturation

Denaturation of a protein refers to the unfolding and rearrangement of the secondary and tertiary structures of a protein (see Figure 23-7). When a protein is denatured, it loses its biologic activity; some become insoluble and coagulate or precipitate from solution. Proteins may be denatured by a variety of agents, as indicated in the following section.

Figure 23–7. Protein structure before and after denaturation.

Precipitation

Alcohol. Alcohol coagulates (precipitates) all types of protein except prolamines. Alcohol (70%) is used as a disinfectant because of its ability to coagulate the protein present in bacteria (see page 206).

Salting Out. Most proteins are insoluble in saturated salt solutions and precipitate out unchanged. To separate a protein from a mixture of other substances, the mixture is placed in a saturated salt solution [such as NaCl, Na_2SO_4, or $(NH_4)_2SO_4$]. The protein precipitates out and is removed by filtration. The protein is then purified from the remaining salt by the process of dialysis (see page 148).

Salts of Heavy Metals. Heavy metal salts such as mercuric chloride (bichloride of mercury) or silver nitrate (lunar caustic) precipitate protein. They are very poisonous if taken internally because they coagulate and destroy protein present in the body. The antidote for mercuric chloride or silver nitrate when these poisons are taken internally is egg white. The heavy metal salts react with the egg white and precipitate out. (The egg white colloid has a charge opposite to that of the heavy metal ion and so attracts it.) The precipitate thus formed must be removed from the stomach by an emetic or else the stomach will digest the egg white and return the poisonous material to the system.

Dilute silver nitrate solution is used as a disinfectant in the eyes of newborn infants. Stronger solutions of silver nitrate are used to cauterize fissures and destroy excessive granulation tissues.

Heat. Heat coagulates almost all protein. Egg white, a substance containing a high percentage of protein, coagulates on heating. Heat coagulates and destroys protein present in bacteria. Hence sterilization of instruments and clothing for use in operating rooms requires the use of high heats. The presence of protein in the urine may be determined by heating a sample of urine, which will cause the coagulation of any protein material that might be present.

Alkaloidal Reagents. Alkaloidal reagents such as tannic acid and picric acid form insoluble compounds with proteins. Tannic acid has been used extensively in the treatment of burns. When this substance is applied to a burn area, it causes the protein to precipitate as a tough covering, thus reducing the amount of water loss from the area. It also reduces exposure to air.

Newer drugs have taken the place of tannic acid for burns, but an old-fashioned remedy still in use for emergencies involves the use of wet tea bags (which contain tannic acid).

Concentrated Inorganic Acids. Proteins are coagulated by such strong acids as concentrated hydrochloric, sulfuric, and nitric acids. Casein is precipitated from milk as a curd when in contact with the hydrochloric acid of the stomach. Heller's ring test is used to detect the presence of albumin in urine. A layer of concentrated nitric acid is carefully placed under a sample of urine in a test tube. If albumin is present, it will precipitate out as a white ring at the interface of the two liquids.

Radiation. Ultraviolet or x-rays can cause protein to coagulate. In the human body the skin absorbs and stops ultraviolet rays from the sun so that they do not reach the inner cells. Proteins in cancer cells are more susceptible to radiation than those present in normal cells, so x-radiation is used to destroy cancerous tissue (see page 50).

Color Tests

Color tests for the presence of proteins depend upon the presence of certain amino acids in that protein. It may be necessary to try several tests before deciding whether a substance is a protein or not.

Xanthoproteic Test

The word xanthoproteic means yellow protein. The test consists of adding concentrated nitric acid to a protein. The protein will then turn yellow and precipitate. Anyone who has spilled nitric acid on his hands will recall the yellow color produced by the reaction of the nitric acid with the protein of the skin. The xanthoproteic test works only for a protein that consists of amino acids containing a benzene ring, such as tyrosine or phenylalanine.

Biuret Test

If a protein suspension is made alkaline with sodium hydroxide solution and copper (II) sulfate solution is added, a violet color is produced. This test is positive for substances that contain two or more peptide linkages—that is, such substances as proteins or polypeptides. It is negative for amino acids, which do not contain a peptide linkage, or for a dipeptide, which contains only one peptide linkage.

Millon's Test

Millon's reagent consists of mercury dissolved in nitric acid (forming a mixture of mercuric and mercurous nitrates). When Millon's reagent is added to a protein, a white precipitate forms. This white precipitate on heating turns to a brick-red color. Millon's test is specific for the amino acid tyrosine.

Hopkins-Cole Test

In the Hopkins-Cole test, a protein is mixed with glyoxalic acid, CHOCOOH, and then carefully placed over a layer of concentrated sulfuric acid in a test tube. If the amino acid tryptophan is present, a purple color will appear at the area of contact of the two liquids.

Ninhydrin Test

When proteins are boiled with ninhydrin (a benzene-type compound) a blue-purple color and CO_2 are produced. This test indicates the presence of α-amino acids or peptide groups.

Chromatography

Chromatography is a method used to separate very complex mixtures of amino acids, proteins, or lipids. It may be used for very small samples or for components present in extremely low concentrations. There are several types of chromatography. Among them are column, paper, thin layer, and vapor phase.

Column Chromatography. In this method a glass tube is packed with a chemically inert material that tends to adsorb solids on its surface. A solution containing a sample is poured into the column, where it is held by packing material at the top (see Figure 23–8, *A*). A solvent in which the components of the mixture have different solubilities is poured into the column, gradually flushing the components of the mixture down the column. The component of the mixture that is most soluble (or least strongly adsorbed by the

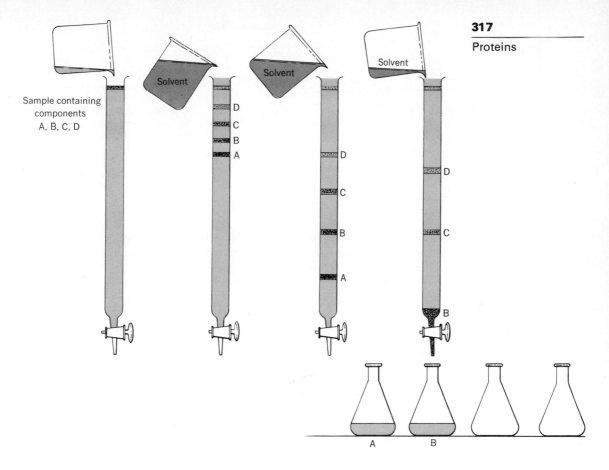

Figure 23–8. Column chromatography for a mixture of four components.

inert material) moves down the column most rapidly, while the component that is least soluble in the solvent (or most strongly adsorbed by the inert material) moves downward the slowest. As more solvent is added, the components move downward with greater and greater separation, until they may be collected individually.

Paper Chromatography. In this method, a piece of filter paper is used in place of the adsorbent. The paper is suspended vertically in a stoppered container. A drop of unknown is placed near the bottom of the paper and solvent added until its level is near that of the unknown (see Figure 23–9). As the solvent rises through the paper by capillary action, the components of the mixture are gradually separated. Different amino acids move at different rates and can be identified by calculations of ratios of distances moved. If, after the separation of a mixture into its components, the paper

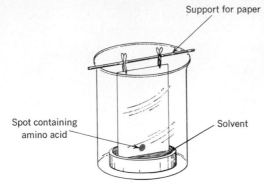

Figure 23–9. Paper chromatography. (Reproduced from Rafelson, M. E.; Binkley, S. B.; and Hayashi, J. A.: *Basic Biochemistry*, 3rd ed. Macmillan Publishing Co., Inc., New York, 1971.)

is dried, turned at right angles, and then placed in another solvent, a two-dimensional chromatogram is obtained. In this method, there is a further separation of the amino acids (see Figure 23–10).

Thin-Layer Chromatography. This method is very similar to that of paper chromatography, except that a glass plate covered with a thin layer of adsorbent is used in place of the filter paper. The advantages of this method over paper chromatography are the rapidity of separation and the choice of adsorbents, which permits separations not possible with filter paper.

Vapor-Phase Chromatography. This is probably the most widely used of all chromatographic methods. In this system, a column is packed with an inert material coated with a thin layer of a high-boiling, nonvolatile liquid. A very small sample (0.001 to 0.010 ml) is injected into the top of the column. The column is heated, and an inert carrier such as helium is passed through. The mixture separates gradually as the more volatile components vaporize first. These more volatile components move faster and are carried out of the

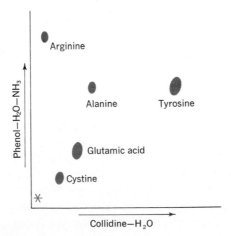

Figure 23–10. Two-dimensional paper chromatography. (Reproduced from Rafelson, M. E.; Binkley, S. B.; and Hayashi, J. A.: *Basic Biochemistry*, 3rd ed. Macmillan Publishing Co., Inc., New York, 1971.)

column first. As each component of the mixture is eluted from the column, a detector produces an electrical signal which is proportional to the amount of the substance being detected. This electrical signal is used to produce a tracing of concentration versus time. Such a trace is known as a chromatogram.

Nucleoproteins

Nucleoproteins are conjugated proteins; they contain a protein part and a nonprotein part, a nucleic acid. During digestion, nucleoproteins are hydrolyzed into nucleic acids and proteins. The proteins are hydrolyzed into amino acids and follow the normal metabolic processes for amino acids.

The nucleic acids are hydrolyzed in several steps as indicated in the following sequence

$$\text{nucleic acid} \longrightarrow \text{nucleotides}$$

$$\text{nucleotides} \longrightarrow \text{nucleosides} + \text{phosphoric acid}$$

$$\text{nucleosides} \longrightarrow \text{purines or pyrimidines} + \text{pentose}$$

Two kinds of nucleic acids are deoxyribonucleic acid (DNA) and ribonucleic acid (RNA). Both DNA and RNA consist of chains of four nucleotides (see Chapter 33). The hydrolysis of a nucleic acid and the products of hydrolysis of DNA and RNA are indicated as follows

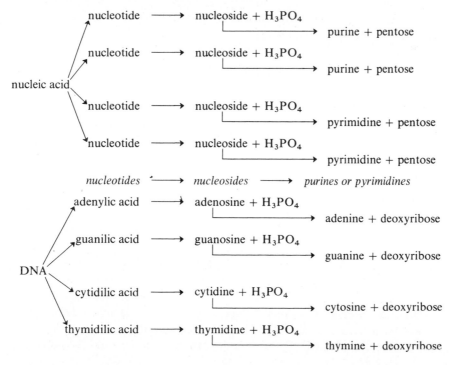

The hydrolysis products of DNA are phosphoric acid, deoxyribose (a pentose), two purines (adenine and guanine), and two pyrimidines (cytosine and thymine).

The hydrolysis products of RNA are the same, except that the pentose is ribose and the pyrimidines are cytosine and uracil rather than cytosine and thymine. The corresponding nucleotide and nucleoside produced from uracil are uridilic acid and uridine. Note that both ribose and deoxyribose have the D-configuration (see page 264).

The Pentoses

RNA contains the pentose ribose whereas DNA is made from deoxyribose, a ribose molecule from which one oxygen atom has been removed.

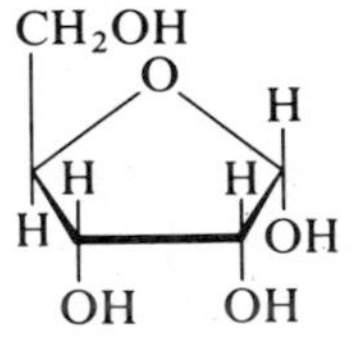

ribose 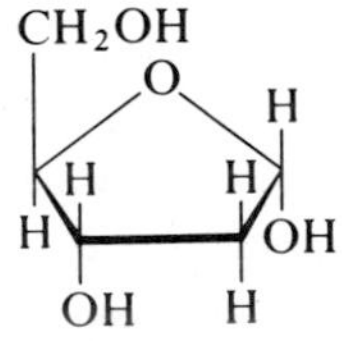   deoxyribose

The Purines

The purines, which are found in nucleic acids, are derivatives of a substance, purine, that does not occur naturally.

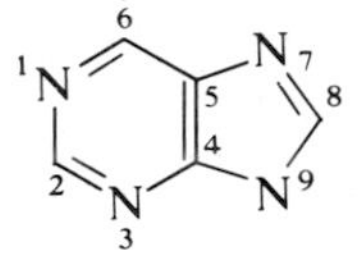

purine

As indicated by their structures, adenine is 6-aminopurine and guanine is 2-amino-6-oxypurine.

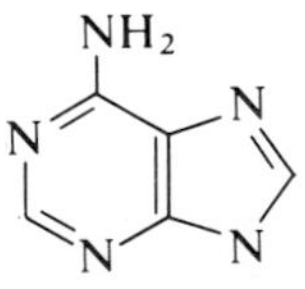

adenine 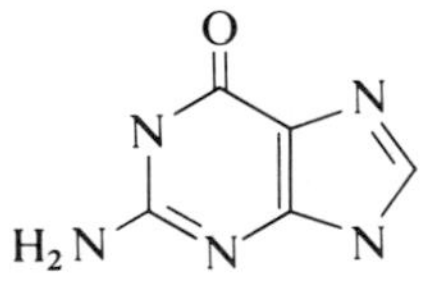   guanine

The end product of the metabolism of purines is uric acid. The uric acid concentration in the blood is increased during the disease called *gout*. In this disease, uric acid crystals are deposited in and around the joints and

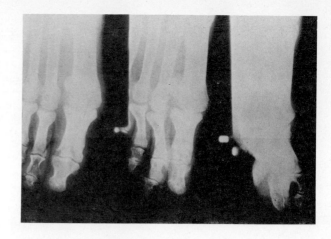

Figure 23–11. X-ray of gouty large toe. (Courtesy of X-ray Department, Michael Reese Hospital and Medical Center, Chicago, Ill.)

in the cartilage, chiefly in the large toe (see Figure 23–11) and in the ear cartilage.

Caffeine is a stimulant for the central nervous system and also a diuretic. Caffeine is found in coffee and tea. Its chemical name is 1,3,7-trimethyl-2,6-dioxypurine. Theophylline, 1,3-dimethyl-2,6-dioxypurine, is found in tea and is used medicinally as a diuretic and for bronchial asthma.

uric acid   caffeine   theophylline

The Pyrimidines

Pyrimidine is a six-membered heterocyclic ring containing two nitrogen atoms. Three important pyrimidines found in nucleic acids are thymine (2,6-dioxy-5-methylpyrimidine), cytosine (2-oxy-6-aminopyrimidine), and uracil (2,6-dioxypyrimidine). Their structures are as follows

pyrimidine   thymine   cytosine   uracil

Other important compounds containing pyrimidines are thiamin (vitamin B_1) and the barbiturates.

321

TABLE 23–5

Names of Nucleosides

Purines	Nucleosides Combined with Ribose	Nucleosides Combined with Deoxyribose
Adenine	Adenosine	Deoxyadenosine
Guanine	Guanosine	Deoxyguanosine
Pyrimidines		
Thymine	Thymidine	Deoxythymidine
Cytosine	Cytidine	Deoxycytidine
Uracil	Uridine	

The Nucleosides

The nucleosides are purines or pyrimidines combined with a pentose, either ribose or deoxyribose. The nucleosides derived from the purines have names ending in *-osine*, whereas those derived from the pyrimidines have names ending in *-idine* (see Table 23–5).

The structures of two nucleosides, adenosine and deoxycytidine, are indicated below.

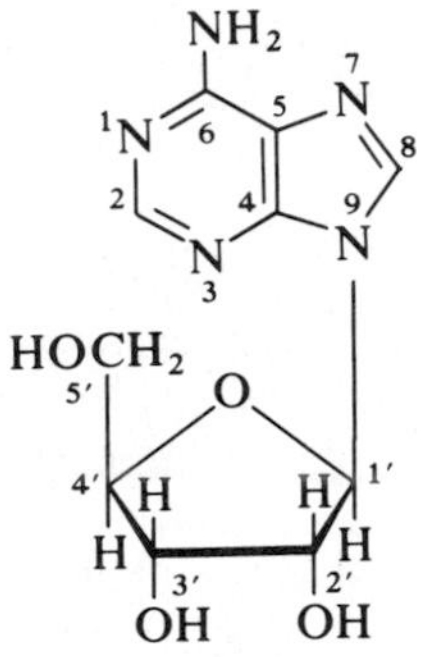

adenosine
(adenine riboside)

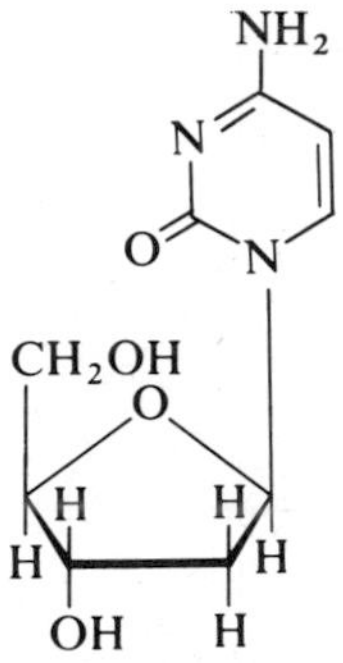

deoxycytidine
(cytosine deoxyriboside)

The Nucleotides

The nucleotides are the phosphate esters of the nucleosides. Because the phosphoric acid contains ionizable hydrogens, the nucleotides are named as acids.

nucleoside + phosphoric acid $\longrightarrow$	*nucleotide*
adenosine	adenylic acid
guanosine	guanilic acid
cytidine	cytidilic acid
uridine	uridilic acid
thymidine	thymidilic acid

Adenylic acid, which is formed by the reaction of one molecule of adenosine with one molecule of phosphoric acid, is also known as adenosine monophosphate, AMP. If two phosphate groups are involved in the reaction, adenosine diphosphate (ADP) is formed, and adenosine triphosphate (ATP) is formed during the reaction with three phosphate groups.

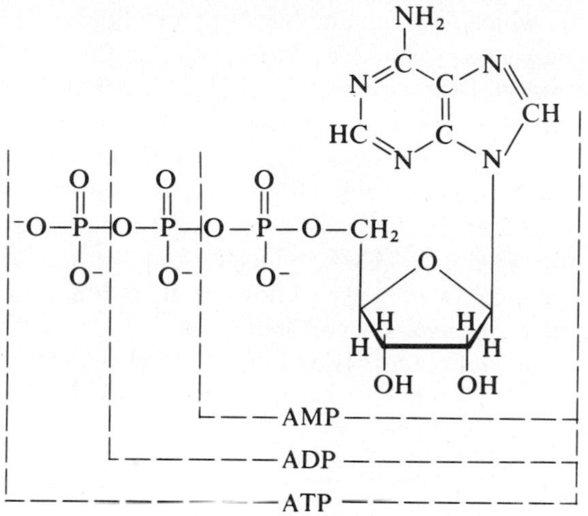

AMP—adenosine monophosphate
ADP—adenosine diphosphate
ATP—adenosine triphosphate

ATP, ADP, and AMP are involved in various metabolic processes involving the storage and release of energy from their phosphate bonds.

Cyclic-AMP

Cyclic-AMP (c-AMP) acts as a chemical messenger to regulate enzyme activity within the cells that store carbohydrate and fat. Without cyclic-AMP the activity of all the enzymes working at maximum speed within the cells would soon create chaos. It also appears that an inadequate supply of cyclic-AMP can lead to one type of uncontrolled cell growth which we call cancer.

Cyclic-AMP is present in almost every type of body cell to the extent of about one part per million. It is produced through the activity of the enzyme adenylate cyclase (found in the membranes of the cell walls) on ATP.

When a hormone, a chemical messenger, reaches its target cell, it interacts with the cell membrane and triggers the production or release of the E prostaglandins, which in turn activate the enzyme adenylate cyclase, causing the production of more cyclic-AMP. This cyclic-AMP, diffusing into the cell, acts as a messenger and instructs the cell to respond to the hormone in a certain characteristic manner.

The concentration of cyclic-AMP in the cells is controlled primarily by two different methods: (1) by the regulation of its rate of synthesis and (2) by changing it into an inactive form, AMP, through the action of certain enzymes.

Abnormalities in the metabolism of cyclic-AMP may explain the effects caused by certain diseases. For example, the bacteria that cause cholera produce a toxin, which in turn stimulates the intestinal cells to accumulate cyclic-AMP. The excess cyclic-AMP instructs the cells to secrete a salty fluid. The accumulation and subsequent loss of large amounts of this salty fluid and the resulting dehydration, if unchecked, can cause cholera to be fatal.

Cyclic-AMP activates a protein kinase that catalyzes phosphorylation and thus activates RNA polymerase. RNA polymerase, in turn, stimulates RNA production, which then functions as a messenger for protein synthesis (see page 499). In this way cyclic-AMP influences the synthesis of protein.

The role of cyclic-AMP in the breakdown of liver and muscle glycogen to glucose is indicated on pages 359 and 360.

The structure of cyclic-AMP is as follows. Cyclic-AMP is not found in plants.

cyclic-AMP

A similar compound, cyclic guanosine monophosphate (cyclic-GMP), is involved in the actions of the F prostaglandins, whose effects run counter to those of the E prostaglandins. It has been suggested that cyclic-AMP and cyclic-GMP function reciprocally in regulating cellular activity, but much more research needs to be done in this area.

Summary

Proteins are high molecular weight compounds containing the elements carbon, hydrogen, oxygen, and nitrogen. Some proteins also contain other elements. Proteins are the chief constituents of all cells of the body.

Animals cannot synthesize protein from raw materials. They must obtain their protein from plants or from other animals, which in turn have obtained the protein from plants.

Protein serves to build new cells, to maintain existing cells, and to replace old cells in the body. Protein is necessary for the formation of the various enzymes and hormones in the body. The oxidation of protein yields 4 Cal per gram.

Proteins are composed of amino acids. Hydrolysis of protein yields amino acids. Amino acids are organic acids with an amine group attached to the alpha carbon. The body can synthesize very few amino acids. Those that it needs and cannot synthesize are called essential amino acids. All the amino acids in the body, except glycine, have the L configuration.

Amino acids are amphoteric—they act as either acids or bases because they contain an acid group ($-COOH$) and a basic group ($-NH_2$).

When an amino acid is placed in an acid solution, it forms a positive ion and migrates toward the negative electrode. When an amino acid is placed in a basic solution it forms a negative ion and migrates toward the positive electrode. At a certain pH, the isoelectric point, the amino acid will be neutral; it will not migrate toward either electrode.

When two amino acids combine, a dipeptide is formed. When four or more amino acids combine, a polypeptide is formed. Polypeptides in turn form peptones, then proteoses, and finally protein. The hydrolysis of protein proceeds through the same types of compounds in reverse order, forming proteoses, peptones, polypeptides, dipeptides, and amino acids.

The primary structure of a protein refers to the number and sequence of the amino acids in the protein chain. The secondary structure refers to the regular recurring arrangement of the amino acid chain into a coil or a pleated sheet. The tertiary structure refers to the specific folding and bending of the coils into specific layers or fibers. The quaternary structure of a protein occurs when several protein units combine to form a more complex unit.

In general, the nitrogen content of protein is 16 per cent.

Proteins are classified into three types: simple, conjugated, and derived. On hydrolysis, simple proteins yield only amino acids. On hydrolysis, conjugated proteins yield amino acids plus some other type of compound. Derived proteins are actually protein derivatives. Simple proteins are classified according to their solubility in various solvents and also according to their ability to be coagulated by heat.

Denaturation of a protein refers to the unfolding and rearrangement of the secondary and tertiary structures of a protein.

Proteins do not dissolve in water; rather they form colloidal dispersions. Proteins, being colloids, cannot pass through membranes and should not normally be present in the urine.

Proteins may be coagulated (precipitated) by means of alcohol, concentrated salt solutions, salts of heavy metals, by heating, by the use of alkaloidal reagents, by concentrated inorganic acids, and by radiation.

Proteins give certain color tests based upon the presence of certain amino acids. Among these tests are the xanthoproteic test (for benzene-ring proteins), the biuret test (for substances containing two or more peptide linkages), the Millon test (for the amino acid tyrosine), the Hopkins-Cole test (for the amino acid tryptophan), and the ninhydrin test (for α-amino acids).

Mixtures of amino acids, proteins, and lipids may be separated by chromatographic methods. Among the systems used are column, paper, thin layer, and vapor phase chromatography.

Nucleoproteins are conjugated proteins which contain a protein part and a nonprotein part, the nucleic acid. Hydrolysis of nucleic acids yields four nucleotides.

Hydrolysis of nucleotides yields nucleosides and phosphoric acid. Hydrolysis of nucleosides yields a pentose and either a purine or a pyrimidine.

The pentose present in RNA is ribose, whereas deoxyribose is present in DNA.

The nucleotides are the phosphate esters of the nucleosides. Among the important nucleotides are AMP (adenosine monophosphate), ADP (adenosine diphosphate), ATP (adenosine triphosphate), and c-AMP (cyclic adenosine monophosphate).

Questions and Problems

1. All proteins contain which elements?
2. What additional elements are present in most proteins?
3. Where do plants obtain their protein?
4. Where do animals obtain their protein?
5. Describe the nitrogen cycle in nature.
6. Where did the word *protein* originate?
7. How does the energy value of protein compare with that of carbohydrate? With fat?
8. How does the molecular weight of a protein compare with that of other organic compounds?
9. What are amino acids? State the names and write the structure of three amino acids.
10. Name the essential amino acids.
11. Why are amino acids amphoteric? Why are they optically active? What configuration do they have?
12. What is meant by the term *isoelectric point* in relation to proteins?
13. What happens to the solubility of a protein at its isoelectric point?
14. What is a peptide linkage?
15. Write the reaction of glycine with valine in two different ways.
16. What is a polypeptide?
17. What amino acids are present in ribonuclease?
18. Describe the types of structure of proteins.
19. When proteins are slowly hydrolyzed, what products are formed?
20. What percentage of protein is usually nitrogen? What use is made of this fact?
21. How are proteins classified?
22. List several different types of simple protein.
23. List several different types of conjugated protein.
24. Why should protein not normally be found in the urine?
25. Why is 70% alcohol a better disinfectant than 100% alcohol?
26. What is meant by the term *salting out* of a protein?
27. What is the antidote for $HgCl_2$ poisoning? Why must the stomach be pumped afterward?
28. What use is made of the fact that heat coagulates protein?
29. Name five color tests for protein. What reagent does each use? What is each test for?
30. What is a nucleoprotein?
31. Indicate the hydrolysis products of nucleic acids.
32. How many nucleotides are present in DNA? in RNA?
33. Compare the hydrolysis products of DNA and RNA.

34. Indicate the structural difference between ribose and deoxyribose.
35. What are the purines? What is the end product of the metabolism of purines?
36. What is gout and what causes it?
37. Name several important pyrimidines.
38. What is a nucleoside?
39. What is a nucleotide?
40. What role does cyclic-AMP play in the cell? Cyclic-GMP?
41. Describe the separation of amino acids by column chromatography.
42. How does paper chromatography work? What is meant by the term two-dimensional paper chromatography?
43. What is thin-layer chromatography? What are its advantages over paper chromatography?
44. Describe vapor-phase chromatography.

References

Baum, S. J., and Scaife, C. W.: *Chemistry: A Life Science Approach*. Macmillan Publishing Co., Inc., New York, 1975, Chap. 25.

Bronk, J. R.: *Chemical Biology*. Macmillan Publishing Co., Inc., New York, 1973, Chap. 3.

Clark, B. F. C., and Marcker, K. A.: How proteins start. *Scientific American*, **218**: 36–42 (Jan.), 1968.

Dayhoff, M. A.: Computer analysis of protein evolution. *Scientific American*, **221**: 86–96 (July), 1969.

Dickerson, R. E.: The structure and history of an ancient protein. *Scientific American*, **226**: 58-72 (April), 1972.

Engelman, D. M., and Moore, P. B.: Neutron-scattering studies of the ribosome. *Scientific American*, **235**: 44–54 (Oct.), 1976.

Fox, C. F.: The structure of cell membranes. *Scientific American*, **226**: 30–38 (Feb.), 1972.

Harper, H. A.: *Review of Physiological Chemistry*, 15th ed. Lange Medical Publications, Los Altos, Calif., 1975, Chap. 1.

Lehninger, A. H.: *Biochemistry*, 2nd ed. Worth Publishers, Inc., New York, 1975, Chaps. 4, 5, 6.

Morrison, R. T., and Boyd, R. N.: *Organic Chemistry*, 3rd Ed. Allyn & Bacon, Inc., Boston, 1973, Chaps. 34, 35.

Murray, J. M., and Weber, A.: The cooperative action of muscle protein. *Scientific American*, **236**: 58–71 (Feb.), 1974.

Old, L. J., *et al.*: L-Asparagine and leukemia. *Scientific American*, **219**: 34–40 (Aug.), 1972.

Pastan, I.: Cyclic AMP. *Scientific American*, **227**: 97–105 (Aug.), 1972.

Reisfield, R. A., and Kahan, B. D.: Markers of biological individuality. *Scientific American*, **226**: 28–37 (June), 1972.

Sharon, N.: Glyco proteins. *Scientific American*, **230**: 78–86 (May), 1974).

Stacy, G. W.: *Organic Chemistry*. Harper & Row, New York, 1975, Chap. 17.

Stroud, R. M.: A family of protein cutting proteins. *Scientific American*, **231**: 74–88 (July), 1974.

Chapter 24

Enzymes

Enzymes are biologic catalysts. Catalysts are substances that alter the speed of a chemical reaction. Some catalysts increase the speed of a reaction, others decrease the speed. Although a catalyst influences a chemical reaction, it is not itself changed chemically nor does it cause the reaction to occur; that is, a catalyst can affect the speed of a reaction but cannot cause that reaction if it would not normally occur in the absence of that catalyst. Catalysts are nonspecific; they can affect the speed of many different reactions. Since catalysts are not used up, they may be used over and over again.

Enzymes are organic catalysts produced by living organisms. Enzymes differ from nonbiologic catalysts in that enzymes are highly specific. Each enzyme will affect only one specific substance (called the *substrate*). An enzyme is like a key that will unlock only one door, whereas most catalysts are like master keys—they can unlock many doors.

Properties of Enzymes

Composition

Enzymes are proteins and, as such, undergo all the reactions that proteins do. That is, enzymes may be coagulated by heat, alcohol, strong acids, and alkaloidal reagents. Many different enzymes have now been prepared in crystalline form.

Effects of Temperature

The speed of all chemical reactions is affected by the temperature: the higher the temperature the faster the rate of the reaction (see Figure 24–1). This is also true for reactions involving enzymes. However, if the temperature is raised too much, the enzyme (protein) will be inactivated and be unable to function.

The best temperature for enzyme function—the temperature at which the rate of a reaction involving an enzyme is the greatest—is called the *optimum temperature* for that particular enzyme. At higher temperatures, the enzyme will coagulate and be unable to function. At temperatures below the optimum temperature, the rate of reaction will be decreased.

Figure 24–1. Effect of temperature on enzyme activity.

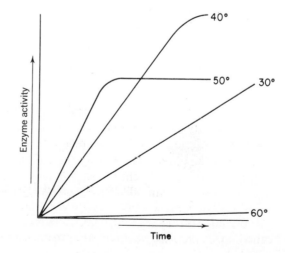

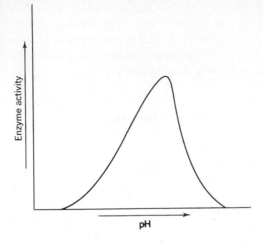

Figure 24–2. Effect of pH on enzyme activity.

Many enzymes have an optimum temperature near 40°C, or close to that of body temperature, so that they function at maximum efficiency in the body.

pH

Each enzyme has a pH range at which it can best function (see Figure 24–2). This is called the *optimum pH range* for that particular enzyme. For example, the optimum pH of pepsin, an enzyme found in the gastric juice, is approximately 2, whereas the optimum pH of trypsin, an enzyme found in the pancreatic juice, is near 8.2. If the pH of a substrate is too far from the optimum pH required by the enzyme, that enzyme cannot function at all. However, since body fluids contain buffers, the pH usually does not vary too far from the optimum values.

Concentration

As with all chemical reactions, the speed is increased with an increase in concentration of reactants. With an increased concentration of substrate, the rate of the reaction will increase until the available enzyme becomes saturated with substrate. Also with an increase in the amount of enzyme, the rate of reaction will increase, assuming an unlimited supply of substrate.

Activators and Inhibitors

Inorganic substances that tend to increase the activity of an enzyme are called *activators*. For example, the magnesium ion, Mg^{2+}, is an inorganic activator for the enzyme phosphatase, and the zinc ion, Zn^{2+}, is an activator for the enzyme carbonic anhydrase (see page 367).

Substances that tend to decrease the activity of enzymes are called *inhibitors*. Inhibitors may act by combining directly with the enzyme and so effectively remove it from the substrate, or they may react with the activator so that it in turn cannot activate the enzyme. Heat, changes in pH, strong acids, alcohol, and alkaloidal reagents all can denature protein. These are examples of *nonspecific inhibitors*; they affect all enzymes in the same manner. *Specific inhibitors* affect one single enzyme or group of enzymes. In this category are most poisonous substances, such as cyanide, CN^-, which inhibits the activity of the enzyme cytochrome oxidase (see page 453).

Mode of Enzyme Activity

How do enzymes act? Why are they so specific toward certain substrates?

Each enzyme contains an "active site," that section of the molecule at which combination with the substrate takes place. The active site consists of amino acids from different parts of the protein chain (the enzyme). These amino acids are brought close together by the folding and bending of the protein chain (the secondary and tertiary structures), so that the active site occupies a relatively small area. The fact that enzymes (proteins) can be denatured by heat (changed in three-dimensional configuration) indicates the importance of structural arrangement (see Figure 24–3).

It is believed that enzyme activity occurs in two steps. First, the active site of the enzyme combines with the substrate to form an enzyme-substrate complex (see Figure 24–4). This enzyme-substrate complex then breaks up to form the products and the free enzyme that can react again. According

Figure 24–3. *A*. Representation of an active site in an enzyme. *B*. Denatured enzyme showing parts of active site no longer in close proximity.

A   B

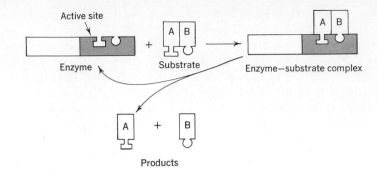

Figure 24–4. Interaction of enzyme and substrate.

to this theory (the lock-and-key method), the substrate must "fit" into the active site of the enzyme—hence the specificity of that enzyme. For substrates that have an optically active site, it is believed that there must be three points of attachment between the substrate and the active site of the enzyme in order that only one of the two optical isomers "fits." In the body, enzymes are specific for the L-amino acids and the D-carbohydrates.

If some other substance should fit into the active site of the enzyme, it could prevent that enzyme from reacting with the substrate. Such a substance is called a competitive inhibitor (see Figure 24–5).

Apoenzymes and Coenzymes

Many enzymes contain a protein part and a nonprotein part. Both parts must be present before the enzyme can function. The protein part is called the *apoenzyme* and the nonprotein part the *coenzyme*.

Figure 24–5. Enzyme-inhibitor complex.

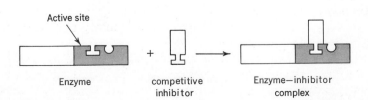

Coenzymes

Coenzymes are not proteins and so are not inactivated by heat. Examples of coenzymes are the vitamins or compounds derived from vitamins.

The reaction involving a coenzyme may be written as follows

$$coenzyme + apoenzyme \longrightarrow enzyme$$

Coenzyme A (CoA) is essential in the metabolism of carbohydrates, lipids, and proteins in the body. It also functions in certain acetylation reactions. Hydrolysis of coenzyme A yields pantothenic acid (a B vitamin), adenine (a purine), ribose (a sugar), phosphoric acid, and mercaptoethanolamine (a sulfur compound).

Acetyl coenzyme A (acetyl CoA) has an acetyl group attached to the sulfur atom. Acetyl CoA is used in the oxidation of food in the Krebs cycle (see Chapter 26).

coenzyme A

Nicotinamide is a very important constituent of two coenzymes—nicotinamide adenine dinucleotide (NAD^+) and nicotinamide adenine dinucleotide phosphate ($NADP^+$). These coenzymes are involved in most oxidation-reduction reactions in the mitochondria. They also take part in the citric acid cycle. The structure of NAD^+ is indicated below.

When NAD^+ is reduced to NADH, another hydrogen atom is attached to the carbon atom indicated by an asterisk. The charge on the nitrogen atom is eliminated, and the double bond between that nitrogen and the marked carbon is changed to a single bond.

In $NADP^+$, another phosphate group has been added in the position marked with a double asterisk in NAD^+ so that $NADP^+$ contains three phosphate groups.

nicotinamide adenine dinucleotide, NAD$^+$

Coenzyme Q is found in the mitochondria and has a structure similar to vitamin K and vitamin E (see pages 448 and 449). The structure of coenzyme Q is indicated below.

coenzyme Q

Coenzyme Q functions in electron transport and in oxidative phosphorylation.

Coenzymes frequently contain B vitamins as part of their structure. Many coenzymes involved in the metabolism of amino acids contain pyridoxine, vitamin B_6. Other B vitamins, such as riboflavin, pantothenic acid, nicotinamide, thiamine, and lipoic acid, are found in coenzymes involved in oxidation-reduction reactions. B vitamins such as folic acid and cyanocobalamin (vitamin B_{12}) are part of different coenzymes.

TABLE 24–1

Enzymes Named Under the Older System

Enzyme	Substrate
Rennin	Casein
Pepsin	Protein
Trypsin	Protein
Ptyalin	Carbohydrate

Nomenclature

Formerly enzymes were given names ending in -*in*, with no relation being indicated between the enzyme and the substance it affects—the substrate. Some of the enzymes named under this system are listed in Table 24–1.

The current system for naming enzymes utilizes the name of the substrate or the type of reaction involved, with the ending -*ase*. Table 24–2 lists some enzymes and substrates named under the preferred system.

Classification

The Commission on Enzymes of the International Union of Biochemistry has classified enzymes into six divisions. Each of these divisions can be further subdivided into several classes. Then following paragraphs indicate these divisions and some of the classes. The older names of the enzymes are still in common use and are given throughout this text merely because of simplicity. It is much easier to write *sucrase* than α-*glucopyrano-β-fructofurano-hydrolase*.

TABLE 24–2

Enzymes and Substrates

Enzyme	Substrate
Maltase	Maltose
Urease	Urea
Proteases	Protein
Carbohydrases	Carbohydrates
Lipases	Lipids
Hydrolases	Hydrolysis reactions
Deaminases	Deaminizing reactions
Dehydrogenases	Removing hydrogens

Oxidoreductases

Oxidoreductases are enzymes that catalyze oxidation-reduction reactions between two substrates. The enzymes that catalyze oxidation-reduction reactions in the body are important in that these reactions are responsible for the production of heat and energy. Recall that oxidation-reduction requires a transfer of electrons. Many of these enzymes are present in the mitochondria, which are the most prominent structural and functional units in animal cells, except for the nucleus (see Figure 24–6). They are sausage-shaped objects ranging in size from 0.2 to 5 micrometers (μm) (1 μm is one-millionth of a meter). The cytoplasm of a typical cell contains from 50 to 50,000 mitochondria.

Inside the mitochondrion is a double-layered membrane which is folded back and forth to form sacs called cristae. These cristae have a tremendous amount of surface area. On the surface of these membranes are enzyme-containing particles. Coenzymes are present in the fluid between the membranes. The mitochondria are highly organized and hold their enzymes in a definite spatial arrangement for the most efficient control of various cellular processes.

The mitochondria play an important part in oxidative phosphorylation in the citric acid cycle, which relates to carbohydrate metabolism. Mitochondria contain enzymes needed for the oxidation of fatty acids. They also are involved in the metabolism of amino acids.

In addition to these functions, the mitochondria manufacture adenosine triphosphate (ATP), the main energy-supplying substance in the cell.

Figure 24–6. A typical cell.

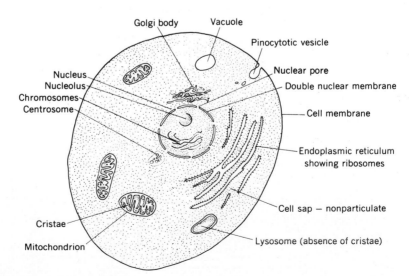

Dehydrogenases. Dehydrogenases catalyze the removal of hydrogen from a substrate.

Oxidases. Oxidases activate oxygen so that it will readily combine with a substrate.

Catalases. Catalases catalyze the decomposition of hydrogen peroxide to water and oxygen.

Peroxidases. Peroxidases also catalyze the decomposition of organic peroxides to hydrogen peroxide and water.

Transferases

Transferases are enzymes that catalyze the transfer of a functional group between two substrates.

Hydrolases—the Hydrolytic Enzymes

The hydrolytic enzymes catalyze the hydrolysis of carbohydrates, esters, and proteins. They are named for the substrate upon which they act. Some hydrolytic enzymes are present in the cytoplasm in organelles called lysosomes.

Carbohydrases. Enzymes that catalyze the hydrolysis of carbohydrates into simple sugars are called carbohydrases. The various carbohydrases are as follows

1. *Ptyalin*, or salivary amylase, for the hydrolysis of starch to dextrins and maltose.
2. *Sucrase*, for the hydrolysis of sucrose to glucose and fructose. Sucrase occurs in the intestinal juice.
3. *Maltase*, for the hydrolysis of maltose to glucose. Maltase occurs in the intestinal juice.
4. *Lactase*, for the hydrolysis of lactose to glucose and galactose. Lactase occurs in the intestinal juice.
5. *Amylopsin*, or pancreatic amylase, for the hydrolysis of starch to dextrin and maltose. Amylase occurs in the pancreatic juice.

Esterases. Esterases are enzymes that catalyze the hydrolysis of esters into acids and alcohols. The various types of esterases are as follows

1. *Gastric lipase*, for the hydrolysis of fats to fatty acids and glycerol.
2. *Steapsin*, or pancreatic lipase, for the hydrolysis of fats to fatty acids and glycerol.
3. *Phosphatases* for the hydrolysis of phosphoric acid esters to phosphoric acid.

Proteases. Proteases are enzymes that catalyze the hydrolysis of protein to derived protein and amino acids. These are of two types—proteinases and peptidases.

The proteinases, for the hydrolysis of proteins to peptides and amino acids, are as follows

1. *Pepsin*, found in the gastric juice, for the hydrolysis of protein to polypeptides.
2. *Trypsin*, found in the pancreatic juice, for the hydrolysis of protein to polypeptides.
3. *Chymotrypsin*, found in the pancreatic juice, for the hydrolysis of protein to polypeptides.

The peptidases, for the hydrolysis of polypeptides to amino acids, are

1. *Aminopeptidases*, from the intestinal juices.
2. *Carboxypeptidases*, from the pancreatic juice.

Nucleases. Enzymes that catalyze the hydrolysis of nucleic acids are called nucleases. Examples are ribonuclease and deoxyribonuclease.

Lyases

Lyases are enzymes that catalyze the removal of groups from substrates by means other than hydrolysis, usually with the formation of double bonds. An example is fumarase, which catalyzes the change of fumaric acid to L-malic acid in the citric acid cycle (see page 367).

Isomerases

Isomerases are enzymes that catalyze the interconversion of optical, geometric, or structural isomers. One example is retinene isomerase, which catalyzes the conversion of 11-*trans*-retinene to 11-*cis*-retinene (see page 445). Another example is alanine racemase, which catalyzes the conversion of L-alanine to D-alanine.

Ligases

Ligases are enzymes that catalyze the coupling of two compounds with the breaking of pyrophosphate bonds. One example is the enzyme that catalyzes the formation of malonyl CoA from acetyl CoA during lipogenesis (see page 383).

Chemotherapy

Chemotherapy is the use of chemicals to destroy infectious microorganisms and cancerous cells without damaging the host's cells. These chemicals function by inhibiting certain cellular enzyme reactions. Among the chemotherapeutic agents are the antibiotics and the antimetabolites.

Antibiotics

Antibiotics are compounds produced by one microorganism that are toxic to another microorganism. They function by inhibiting enzymes that are essential to bacterial growth. Among the most commonly used antibiotics are penicillin and tetracycline, whose structures are indicated below.

penicillin G

tetracycline

Various strains of penicillin contain different groups attached to the cysteine-valine combination. Recall that both of these substances are amino acids.

Antimetabolites

Antimetabolites are chemicals that have structures closely related to that of the substrates which enzymes act on, thus inhibiting enzyme activity. One example of an antimetabolite is sulfanilamide (see page 240), whose structure is similar to that of p-aminobenzoic acid. Bacteria require folic acid as a coenzyme for their growth, and they synthesize it from p-aminobenzoic acid. Because their structure is similar to that of p-aminobenzoic acid, the sulfa drugs prevent the formation of folic acid and so inhibit the growth of the bacteria. This accounts for the use of sulfa drugs to fight bacterial infections.

Some chemotherapeutic agents used in the treatment of cancer (antineoplastic agents) are antimetabolites (mercaptopurine used in the treatment of leukemias); some are antibiotics (adriamycin used in the treatment of Hodgkin's disease); others are either alkylating agents, hormones, or natural products.

Clinical Significance of Plasma Enzyme Concentrations

The measurement of plasma enzyme levels can be of great diagnostic value. For example, the levels of glutamic pyruvic transaminase increase with infectious hepatitis; the levels of trypsin increase during acute disease of the pancreas; ceruloplasmin levels decrease during Wilson's disease (see page 508); and glutamic oxaloacetic transaminase levels rise rapidly after myocardial infarction. Many other plasma enzymes are useful in the diagnosis of different diseases.

Summary

Catalysts alter the speed of a chemical reaction but do not themselves change. Catalysts are nonspecific and affect the speed of many different reactions.

Enzymes are biological catalysts. Each enzyme will catalyze only one type of reaction.

Enzymes are proteins and will undergo all the reactions of proteins. The enzymes in the body function best at about 40°C. Temperatures above or below body temperature will decrease the activity of enzymes.

Each enzyme has a certain pH at which it can best function.

An increase in the amount of enzyme will increase the rate of reaction. An increase in the amount of substrate will increase the rate of the reaction.

Inorganic compounds that increase the activity of an enzyme are called activators. Compounds that interfere with the activity of an enzyme are called inhibitors.

Enzymes contain an "active site" that binds to the substrate to form an enzyme-substrate complex. This complex yields the products and regenerates the enzyme.

Many enzymes contain two parts—a protein part and a nonprotein part. The protein part of an enzyme is called the apoenzyme.

Some enzymes require the presence of a substance called a coenzyme before they can act effectively. Coenzymes frequently contain the B vitamins or compounds derived from the B vitamins.

Under the older system of naming enzymes the substrate was not mentioned; the newer system indicates the substrate being acted upon. The names of enzymes under this system end in -ase.

Enzymes may be classified as oxidoreductases (enzymes that catalyze oxidation-reduction reactions between two substrates), transferases (which catalyze the transfer of a functional group between two substrates), hydrolases (which catalyze hydrolysis reactions), lyases (which catalyze the removal of groups from substrates by means other than hydrolysis), isomerases (which catalyze the interconversion of optical, geometric, or structural isomers), and ligases (which catalyze the coupling of two compounds with the breaking of pyrophosphate bonds).

Some hydrolytic enzymes are found in the lysosomes of the cytoplasm. The cytoplasm also contains mitochondria. These structural and functional units contain most of the oxidative enzymes and are deeply involved in the electron transport system of oxidation-reduction. The mitochrondria also produce ATP, the cells' chief source of energy.

Chemotherapy is the use of chemicals to destroy infectious microorganisms and cancerous cells without damaging the host's cells. Among chemotherapeutic agents are antibiotics and antimetabolites.

Abnormal plasma enzyme concentrations are of clinical significance in the diagnosis of certain diseases.

Questions and Problems

1. How do nonbiologic catalysts differ from enzymes?
2. Most enzymes are found to be what kind of compounds?

3. What is the effect of temperature upon an enzyme?
4. What is meant by the *optimum temperature* of an enzyme?
5. What is the effect of pH upon the activity of an enzyme?
6. What is the effect of concentration of enzyme upon the rate of the reaction?
7. What is the effect of concentration of substrate upon the rate of the reaction?
8. What is an apoenzyme? A coenzyme?
9. What is CoA used for? What products does it yield upon hydrolysis?
10. Compare the newer and older systems of naming enzymes.
11. What are the general classifications of enzymes?
12. What are carbohydrases? Name three. Indicate where they are found in the body and what substrate they act upon.
13. What are esterases? Name two. Where are they found? What substrate do they act upon?
14. Name two proteinases. What substrate do they act upon? What products are formed?
15. What is a nuclease?
16. Name three types of oxidation-reduction enzymes.
17. What is a transferase? An isomerase?
18. Describe the appearance of the mitochondria. What functions do they serve?
19. What is coenzyme Q? What is its function?
20. How do enzymes function? Where do they get their specificity?
21. How do enzymes distinguish between optical isomers?
22. What is a competitive inhibitor? How does it work?
23. Distinguish between specific and nonspecific inhibitors.
24. What effect does denaturing have upon the "active site" of an enzyme? Why?
25. What is chemotherapy? Give an example of two chemotherapeutic agents.
26. How does an antimetabolite function?
27. How may plasma enzyme concentration be used to diagnose disease?

References

Baum, S. J., and Scaife, C. W.: *Chemistry: A Life Science Approach.* Macmillan Publishing Co., Inc., New York, 1975, Chap. 26.

Bronk, J. R.: *Chemical Biology.* Macmillan Publishing Co., Inc., New York, 1973, Chaps. 7, 17.

Harper, H. A.: *Review of Physiological Chemistry*, 15th ed. Lange Medical Publications, Los Altos, Calif., 1975, Chap. 8.

Lehninger, A. H.: *Biochemistry*, 2nd ed. Worth Publishers, Inc., New York, 1975, Chaps. 8, 9.

Miller, J. J.: Mineral deficiencies in chronic disease. *Journal of the International Academy of Preventive Medicine*, 1:80–97 (Spring), 1974.

Montgomery, R.; Dryer, R. L.; Conway, T. W.; and Spector, A. A.: *Biochemistry.* C. V. Mosby Co., St. Louis, 1974, Chap. 3.

Racker, E.: The membrane of the mitochondrion. *Scientific American*, 218:32–39 (Feb.), 1968.

Rafelson, M. E.; Binkley, S. B.; and Hayashi, J. A.: *Basic Biochemistry*, 3rd ed. Macmillan Publishing Co., Inc., New York, 1971, Chap. 3.

Chapter 25

Digestion

Most foods (carbohydrates, fats, and proteins) are composed of large molecules that are usually not soluble in water. Before these foods can be absorbed through the alimentary canal, they must be broken down into smaller soluble molecules. *Digestion* is the process by which food molecules are broken down into simpler molecules that can be absorbed into the blood through the intestinal walls (see Figure 25–1).

Digestion involves the use of hydrolases—the hydrolytic enzymes. The hydrolases catalyze the hydrolysis of carbohydrates to monosaccharides, fats to fatty acids and glycerol, and proteins to amino acids.

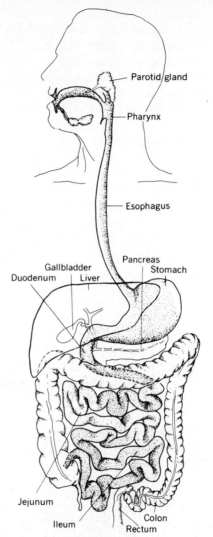

Figure 25–1. The digestive tract.

However, not all foods require digestion. Monosaccharides are already in their simplest form and do not require digestion. Inorganic salts and vitamins also do not require digestion.

Digestion of food takes place in the mouth, the stomach, and the small intestine, each area having its own particular enzyme or enzymes that catalyze the hydrolytic reactions.

Salivary Digestion

During chewing, the food is mixed with saliva. The saliva moistens the food, so that swallowing is easier. Saliva contains approximately 99.5 per cent water. The remaining 0.5 per cent contains mucin, a glycoprotein that acts as a lubricant, several inorganic salts that act as buffers, and salivary amylase, an enzyme that catalyzes the hydrolysis of starches. Saliva also acts as an excretory fluid for certain drugs, such as morphine and alcohol, and for certain inorganic ions, such as K^+, Ca^{2+}, HCO_3^-, and SCN^-. Saliva has a pH range of 5.75 to 7.0, with an optimum pH of 6.6.

Approximately 1500 ml of saliva are secreted daily by three pairs of glands—the parotid, the sublingual, and the submaxillary. The flow of saliva is controlled by the nervous system. It may be increased by chewing food or even by chewing a nonnutritive object such as a piece of wax. The flow of saliva may also be increased by smelling food, by seeing food, or even by thinking of food. The flow of saliva may be decreased when a person is anxious, afraid, or emotionally upset.

Gastric Digestion

When food is swallowed, it passes down the esophagus into the stomach where it is mixed with the gastric juice. The gastric juice is secreted by glands in the walls of the stomach. The flow of gastric juice is regulated by the nervous system. This flow may be stimulated by the sight, smell, or even the thought of food, just as is the flow of saliva. Likewise, the flow of gastric juice may be diminished during fright or anger. When food enters the stomach, it causes the production of the hormone gastrin. Gastrin diffuses into the bloodstream,

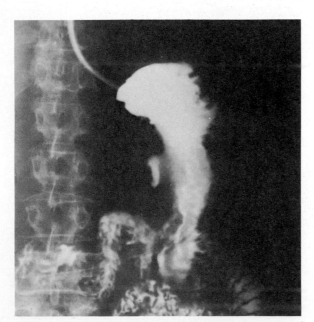

Figure 25–2. X-ray of gastric ulcer. (Courtesy of X-ray Department, Michael Reese Hospital and Medical Center, Chicago, Ill.)

which carries it back to the stomach where it then stimulates the flow of gastric juice.

Approximately 2 to 3 liters of gastric juice are secreted daily. Gastric juice is normally a clear, pale yellow liquid with a pH of 1 to 2. Gastric juice contains 97 to 99 per cent water and up to 0.5 per cent free hydrochloric acid. It is the presence of this hydrochloric acid that causes the gastric juice to have such a low pH. In certain pathologic conditions, the acidity of the stomach may be less than normal. Such a condition is known as *hypoacidity* and is commonly associated with stomach cancer and pernicious anemia. *Hyperacidity* is a condition in which the stomach has too high an acid concentration. It is indicative of gastric ulcers (Figure 25–2), hypertension, or gastritis (inflammation of the stomach walls).

The gastric juice contains the precursor enzyme pepsinogen and the enzyme (gastric) lipase.

In addition to these enzymes, a substance known as the *intrinsic factor* is secreted by the parietal cells in the walls of the stomach. Vitamin B_{12} must undergo a reaction with this intrinsic factor before it can be absorbed into the bloodstream. A lack of the intrinsic factor is associated with pernicious anemia.

Intestinal Digestion

After the food leaves the stomach, it is passed into the small intestine by small wavelike contractions called peristalsis. The food material leaving the stomach is called *chyme*. The small intestine is divided into three sections —the duodenum, the jejunum, and the ileum.

Three different digestive juices enter the duodenum. These digestive juices are alkaline and neutralize the acid chyme coming from the stomach. The three digestive juices entering the duodenum are

1. Pancreatic juice from the pancreas
2. Intestinal juice from the intestinal mucosa
3. Bile from the gallbladder

Pancreatic Juice

Pancreatic juice contains several enzymatic substances. Among these are trypsinogen, chymotrypsinogen, carboxypeptidase, pancreatic lipase, and pancreatic amylase.

Intestinal Juice

Intestinal juice contains several enzymes. Among these are sucrase, maltase, lactase, aminopeptidase, and dipeptidase. The intestinal juice also contains enzymes that catalyze the hydrolysis of phosphoglycerides, nucleo-proteins, and organic phosphates.

Bile

Bile is produced in the liver and stored in the gallbladder. The gall-bladder absorbs some of the water and other substances from the liver bile and so changes its composition slightly. When meat or fats enter the small intestine, they cause it to secrete a hormone, cholecystokinin-pancreozymin (CCK-PZ). This hormone enters the bloodstream and is carried to the gallbladder where it causes that organ to contract and empty into the duodenum through the bile duct.

Bile is a yellowish-brown to green viscous liquid with a pH of 7.8 to 8.6. Because it is alkaline, it serves to neutralize the acid chyme entering from the stomach. Primarily, bile contains bile salts, bile pigments, and cholesterol. Bile contains no digestive enzymes.

Bile Salts. Sodium glycocholate and sodium taurocholate are the two most important bile salts. They are both derived from cholic acid, a steroid similar to cholesterol in structure (see page 297)

cholic acid

Bile salts have the ability to lower surface tension and so aid in the emulsification of fats. They also increase the effectiveness of the pancreatic lipase (steapsin) in its digestive action on emulsified fats.

In addition, bile salts aid the absorption of fatty acids through the walls of the intestine. After absorption of these fatty acids, the bile salts are removed and carried back by portal circulation to the liver where they are again returned to the bile. Bile salts also help stimulate intestinal motility.

Bile Pigments. The average red blood cell lasts about 125 days and then is destroyed. The hemoglobin is broken down into globin and heme. The body removes the iron from the heme and reuses it. The heme, with the iron removed, becomes biliverdin. Biliverdin is reduced in the reticuloendothelial cells of the liver, spleen, and bone marrow to form bilirubin, the main bile pigment excreted into the bile by the liver. In the intestines, bilirubin is converted to urobilinogen and then to urobilin, a pigment that gives the feces

its characteristic yellow-brown color. Some urobilin and urobilinogen are
absorbed from the intestinal tract and appear in the urine, giving that fluid
its characteristic color. These reactions may be written as follows

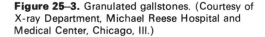

hemoglogin $\longrightarrow$ heme $\longrightarrow$ biliverdin $\longrightarrow$ bilirubin $\longrightarrow$ urobilinogen $\longrightarrow$ urobilin
$+$ $\qquad$ $+$
globin $\qquad$ iron

As the urobilinogen and urobilin (this mixture in urine is called urochrome)
are waste products, bile can be classified as an excretion as well as a secretion
(see page 396).

If the bile duct is blocked, the bile then goes into the bloodstream,
producing jaundice. This disorder is recognizable by the yellow pigmentation
of the skin. If the bile duct is blocked, no bile pigments can enter the intestine
and the feces will appear clay-colored or nearly colorless.

Cholesterol. The body's excess cholesterol is excreted by the liver and
carried to the small intestine in the bile. Sometimes the cholesterol precipitates in the gallbladder, producing gallstones. Figure 25–3 shows gallstones
present in a gallbladder.

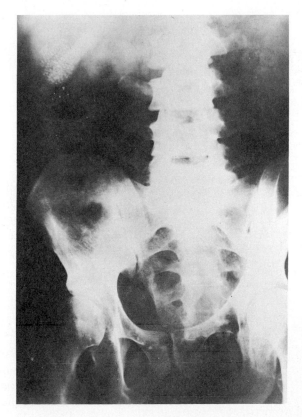

Figure 25–3. Granulated gallstones. (Courtesy of
X-ray Department, Michael Reese Hospital and
Medical Center, Chicago, Ill.)

Digestion of Carbohydrates

In the Mouth

Digestion begins in the mouth, as chewing action reduces the size of the food particles. Thus they will have more surface area in contact with the digestive enzymes.

Saliva contains salivary amylase (ptyalin), which catalyzes the hydrolysis of starch into maltose. However, this enzyme becomes inactive at a pH below 4 so that its activity ceases when it is mixed with the contents of the stomach, where the pH falls to about 1.5. Salivary amylase does not serve a very important function in digestion because the food does not remain in the mouth long enough for any appreciable hydrolysis to take place. Some hydrolysis of carbohydrates catalyzed by salivary amylase may take place in the stomach before the food is thoroughly mixed with the contents of the stomach, but this is of little importance because there are intestinal enzymes capable of hydrolyzing starch and maltose. The principal function of saliva is to lubricate and moisten the food so that it can be easily swallowed.

In the Stomach

The stomach contains no carbohydrases, so that no digestion of carbohydrates occurs there except for that catalyzed by salivary amylase. However, the activity of salivary amylase ceases as soon as it becomes mixed with the acid contents of the stomach.

In the Small Intestine

The major digestion of carbohydrates takes place in the small intestine through the action of enzymes in the pancreatic and intestinal juices.

The pancreatic juice contains the enzyme pancreatic amylase, which catalyzes the hydrolysis of starch and dextrins into maltose. The maltose thus produced is hydrolyzed to glucose through the activity of the enzyme maltase from the intestinal juice. The optimum pH of pancreatic amylase is 7.1. The intestinal juice also contains the enzymes sucrase and lactase, which catalyze the hydrolysis of sucrose and lactose, respectively.

$$\underset{\text{starch}}{(C_6H_{10}O_5)_n} + H_2O \xrightarrow{\underset{\text{amylase}}{\text{pancreatic}}} \underset{\text{maltose}}{C_{12}H_{22}O_{11}}$$

$$\underset{\text{maltose}}{C_{12}H_{22}O_{11}} + H_2O \xrightarrow{\text{maltase}} \underset{\text{glucose}}{2C_6H_{12}O_6}$$

If a monosaccharide such as glucose is eaten, digestion is not necessary because the monosaccharide is already in its simplest form and can easily undergo absorption into the bloodstream.

As there are no lipases present in the saliva, no digestion of fats takes place in the mouth.

In the Stomach

Although gastric lipase is present in the stomach, very little if any digestion of fats takes place because the pH of the stomach (1 to 2) is far below the optimum pH of that enzyme (7 to 8); also fats must be emulsified before they can be digested by lipase and there is no mechanism for emulsification of fats in the stomach. However, if emulsified fats are eaten, a very small amount of hydrolysis may take place in the stomach.

In the Small Intestine

In the small intestine, the pancreatic lipase catalyzes the hydrolysis of fats into fatty acids and glycerol. This action is aided by the bile, which emulsifies the fats so that they can be acted upon readily by pancreatic lipase.

$$\text{fat + water} \xrightarrow{\text{pancreatic lipase}} \text{fatty acid + glycerol}$$

Digestion of Proteins

As the saliva contains no enzymes for the hydrolysis of protein there is no digestion of protein in the mouth.

In the Stomach

The precursor enzyme pepsinogen is converted to pepsin when it is mixed with the hydrochloric acid of the stomach. Pepsin catalyzes the hydrolysis of protein to polypeptides.

$$\text{protein + water} \xrightarrow{\text{pepsin}} \text{polypeptides}$$

Rennin, an enzyme present in the gastric juice of infants, coagulates casein to form paracasein, which is precipitated by the calcium ions present in milk. Coagulated milk remains in the stomach longer than uncoagulated milk and so is more readily digested there. According to modern theories, rennin is not present in the gastric juice of an adult.

In the Small Intestine

In the small intestine, the precursor enzyme trypsinogen from the pancreatic juice is changed into trypsin by the intestinal enzyme enterokinase.

Trypsin in turn changes chymotrypsinogen, another pancreatic precursor enzyme, into chymotrypsin. Both trypsin and chymotrypsin catalyze the hydrolysis of protein, proteoses, and peptones to polypeptides. The optimum pH of trypsin and chymotrypsin is 8 to 9.

The intestinal enzymes aminopeptidase and dipeptidase catalyze the hydrolysis of polypeptides and dipeptides into amino acids. Carboxypeptidase, an enzyme of the pancreatic juice, also catalyzes the hydrolysis of polypeptides to amino acids. Carboxypeptidase contains the element zinc.

Absorption of Carbohydrates

As we have seen, the principal digestion of carbohydrates takes place in the small intestine, where the polysaccharides and disaccharides are hydrolyzed into monosaccharides—glucose, fructose, and galactose. The monosaccharides are transported through the walls of the small intestine directly into the bloodstream by means of the capillaries of the villi in the lining of the small intestine. The blood carries the monosaccharides to the liver and then into the general circulation to all parts of the body. The monosaccharides may be oxidized to furnish heat and energy. Some of the monosaccharides are converted to glycogen, a polysaccharide, and stored in the liver or the muscles, and the rest converted to fat and stored in the adipose tissue.

Absorption of Fats

The digestion of fats takes place primarily in the small intestine. The end products of digestion of fats—mono- and diglycerides, fatty acids, and glycerol—pass through the intestinal mucosa where they are reconverted into triglycerides and phosphoglycerides, which then enter the lacteals, the lymph vessels in the villi in the walls of the small intestine. From the lacteals these products pass into the thoracic duct (a main lymph vessel) and then into the bloodstream.

Bile salts are necessary for this absorption process. After the absorption of the fatty acids, the bile salts are returned to the liver to be excreted again into the bile. Fatty acids of less than 10 to 12 carbon atoms are transported through the intestinal walls directly into the bloodstream, as is any free glycerol present.

Absorption of Proteins

The end products of the hydrolysis of protein are amino acids. The L-amino acids are absorbed more rapidly than the D-isomers and pass

TABLE 25–1

Summary of Digestion

Type of Digestion	Location of Digestion	Digestive Juice and Enzymes	Substrate	Product
Salivary	Mouth	Saliva		
		salivary amylase (ptyalin)	Starch	Dextrins
Gastric	Stomach	Gastric juice		
		Pepsin	Protein	Polypeptides
		Lipase	Fats	Fatty acids + glycerol
		Rennin (infants only)	Casein	Paracasein
		Hydrochloric acid	Pepsinogen	Pepsin
Intestinal	Small intestine	Intestinal juice		
		Aminopeptidase	Polypeptides	Amino acids
		Dipeptidase	Peptides	Amino acids
		Maltase	Maltose	Glucose
		Sucrase	Sucrose	Glucose + fructose
		Lactase	Lactose	Glucose + galactose
		Pancreatic juice		
		Trypsin	Protein	Polypeptides
		Chymotrypsin	Protein	Polypeptides
		Pancreatic amylase	Starch + dextrins	Maltose
		Pancreatic lipase	Fats	Fatty acids + glycerol
		Carboxypeptidase	Polypeptides	Amino acids

through the capillaries of the villi directly into the bloodstream, which carries them to the tissues to be used to build or replace tissue. The amino acids may also be oxidized to furnish energy. Although the body can store carbohydrate and fat, it cannot store protein.

Formation of Feces

After the absorption of monosaccharides, glycerol, fatty acids, and amino acids, the remaining contents of the small intestine pass into the large intestine. The large intestine contains undigestible material (such as cellulose), undigested food particles, unused digestive juices, epithelial tissues from the walls of the digestive system, bile pigments, bile salts, and inorganic salts. The material passing into the large intestine is a semifluid and contains much water. Most of this water and some salts are reabsorbed through the walls of the large intestine, leaving behind a residue called feces. Little or no digestion takes place in the large intestine.

The conditions in the large intestine are ideal for the growth of bacteria and usually one quarter to one half of the feces consists of bacteria. The bacteria cause the fermentation of carbohydrates to produce hydrogen, carbon dioxide, and methane gases, as well as acetic, butyric, and lactic acid. The gases can cause distention and swelling of the intestinal tract, producing a feeling of discomfort. The acids may be irritating to the intestinal mucosa and cause diarrhea. Particularly in infants, the acids may cause excoriated buttocks. Infants who have this condition are usually given a high protein–low carbohydrate diet, since the fermentation bacteria act on carbohydrates and not on protein.

Some of the amino acids undergo decarboxylation because of the action of intestinal bacteria to produce toxic amines called ptomaines (see page 395). For example, decarboxylation produces cadaverine from the amino acid lysine; putrescine from ornithine; and histamine from histidine. These toxic substances are reabsorbed from the large intestine, carried to the liver where they are detoxified, and then excreted in the urine.

The amino acid tryptophan undergoes a series of reactions to form the compounds indole and skatole, which are primarily responsible for the odor of the feces.

tryptophan   indole   skatole (methylindole)

Summary

Digestion is the process by which foods are broken down (hydrolyzed) into simple molecules that can then be absorbed through the intestinal walls. Digestion takes place in the mouth, the stomach, and in the small intestine.

Digestion begins in the mouth. Food is mixed with saliva which moistens the food and makes swallowing easier. Saliva contains enzymes that begin the hydrolysis of carbohydrates (starch). Saliva has a pH of 5.75 to 7.0 and contains approximately 99.5 per cent water. The flow of saliva is controlled by the nervous system.

In the stomach the food is mixed with the gastric juices. The gastric juices contain hydrochloric acid and have a pH of 1 to 2. The gastric juice contains the precursor enzyme pepsinogen and the enzyme gastric lipase. Upon contact with hydrochloric acid, pepsinogen is converted into pepsin, which then catalyzes the hydrolysis of protein. Gastric lipase is not an important enzyme because it can act only on emulsified fats and there is practically no emulsified fat present in the stomach. Rennin is another enzyme of the gastric juice; however, it is present only in infants.

The food leaving the stomach is called chyme. When the chyme enters the small intestine, it is mixed with three digestive juices—those from the pancreas, those from the intestinal walls, and with bile from the gallbladder.

Bile is produced in the liver and stored in the gallbladder. Bile is alkaline and neutralizes the chyme entering the small intestine from the stomach. Bile salts lower surface tension and help emulsify fats; they aid in the absorption of fatty acids; they

help stimulate intestinal motility.

Bile also contains pigments which come from the breakdown of hemoglobin. These bile pigments give the feces and the urine their characteristic color. If cholesterol crystallizes in the gallbladder, gallstones are produced.

Digestion of carbohydrates begins in the mouth. No digestion of carbohydrate takes place in the stomach except that catalyzed by salivary amylase from the saliva. Even this activity ceases when the food from the mouth reaches the low pH of the stomach. The major digestion of carbohydrates takes place in the small intestine with the aid of the enzymes pancreatic amylase, maltase, lactase, and sucrase.

No digestion of fats takes place in the mouth. No digestion of fats takes place in the stomach unless the fats are already emulsified. In that case their hydrolysis is catalyzed by gastric lipase, although to a very limited extent. Fats are emulsified in the small intestine by the action of bile and then acted on by the enzyme pancreatic lipase.

Digestion of protein begins in the stomach with the aid of pepsin. The digestion of protein continues in the small intestine with the aid of the enzymes trypsin, chymotrypsin, carboxypeptidase, aminopeptidase, and dipeptidase.

The monosaccharides produced by the digestion of carbohydrates pass through the villi of the small intestine and enter the bloodstream.

Fatty acids and glycerol, products of the digestion of fats, are converted into glycerides in the intestinal mucosa by the action of bile and pass through the lacteals into the thoracic duct and then into the bloodstream. Fatty acids of less than 10 to 12 carbon atoms pass directly into the bloodstream through the villi.

Amino acids, from the digestion of protein, pass through the villi of the small intestine into the bloodstream.

The undigested food, the undigestible foods, unused digestive juices, epithelial tissues from the walls of the digestive system, bile salts, inorganic salts, and water pass from the small intestine into the large intestine. Most of the water and some salts are reabsorbed. The remaining material is excreted as feces.

Questions and Problems

1. Why is digestion necessary?
2. Do all foods require digestion? Why?
3. Where does digestion take place?
4. What does saliva contain? What does it do?
5. How much saliva is secreted daily?
6. What is the pH of saliva?
7. What controls the flow of saliva?
8. What is the pH of gastric juice? What causes this pH?
9. What controls the flow of the gastric juice?
10. What is hypoacidity? What may cause it?
11. What is hyperacidity? What may cause it?
12. Gastric juice contains what enzymes?
13. What is the difference between pepsinogen and pepsin?
14. What converts pepsinogen into pepsin?
15. Why is rennin no longer considered an important enzyme in the body?
16. Why is gastric lipase considered an unimportant enzyme?

17. What is the intrinsic factor? What disease is associated with a lack of this substance?
18. What is chyme?
19. What are the three sections of the small intestine?
20. Three different digestive juices enter the small intestine. What are they?
21. What enzymes are present in the pancreatic juice? What does each do?
22. What enzymes are present in the intestinal juice? What does each do?
23. Where is bile produced? Where is it stored?
24. What is the function of cholecystokinin-pancreozymin? Where is it produced?
25. What is the pH of bile? What does it contain?
26. What are the functions of bile salts?
27. How are bile pigments formed? Where are they excreted?
28. What is urochrome?
29. What happens when the bile duct is blocked?
30. What may cause gallstones?
31. Describe the digestion of carbohydrate. Where are the end products absorbed?
32. Describe the digestion of fats. Where are the end products absorbed?
33. Describe the digestion of protein. Where are the end products absorbed?
34. Describe the formation of feces.
35. What are the functions of the large intestine?
36. What may cause excoriated buttocks in infants? What may be done to overcome this effect?
37. What causes the characteristic odor of feces?
38. Where are some of the toxic putrefactive products detoxified? Where are they excreted?
39. What is enterokinase? Where does it function?
40. What is paracasein? How is it formed?

References

Allison, A.: Lysosomes and disease. *Scientific American*, **217**:62–72 (Nov.), 1967.

Harper, H. A.: *Review of Physiological Chemistry*, 15th ed. Lange Medical Publications, Los Altos, Calif., 1975, Chap. 12.

Rafelson, M. E.; Binkley, S. B.; and Hayashi, J. A.: *Basic Biochemistry*, 3rd ed. Macmillan Publishing Co., Inc., New York, 1971, Chap. 4.

Carbohydrate Metabolism

Concentration of Sugar in the Blood: Blood Sugar Level

The end products of carbohydrate digestion are the monosaccharides glucose, fructose, and galactose. Both fructose and galactose are converted to glucose in the liver so that the major monosaccharide remaining in the bloodstream is glucose.

The amount of glucose present in the blood will vary considerably, depending upon whether the measurements were taken $\frac{1}{2}$ hr after eating, 1 hr after eating, or during a period of fasting.

The normal quantity of glucose present in 100 ml of blood taken after a period of fasting is 70 to 100 mg. This value is called the *normal fasting blood sugar*.

Soon after eating a meal, the blood sugar level may rise to 120 mg per 100 ml of blood or even higher. However, the level soon drops so that after $1\frac{1}{2}$ to 2 hr the blood sugar level again returns to its normal fasting value.

During the time of fasting, even though the body is continuously using glucose for the production of heat and energy, the amount of glucose present in the blood remains fairly constant. How does the body regulate the amount of glucose present in the blood and what happens when these control mechanisms do not function properly?

When the glucose level in the blood rises, as happens after the digestion of a meal, the liver removes the excess glucose and converts it to glycogen, a polysaccharide. This process is called *glycogenesis*. This glycogen may be stored in the liver and in muscle tissue. However, only a certain amount of glycogen may be stored in the liver and muscle; the rest is changed to fat and stored as such.

Glucose is also removed from the blood by the normal oxidative reactions that take place continuously throughout the body.

Glucose does not normally appear in the urine except in amounts too small to be detected by Benedict's solution. However, if the blood sugar level rises above 170 to 180 mg per 100 ml of blood, the sugar "spills over" into the urine. The point at which the sugar spills over into the urine is called the *renal threshold*. The presence of glucose in the urine is called *glycosuria*.

Thus, the factors that remove excess glucose from the blood are (1) glycogenesis and storage as glycogen, (2) conversion to fat, (3) normal oxidation reactions in the body, and (4) excretion through the kidneys when the renal threshold is exceeded.

If the blood sugar level falls too low, the liver converts glycogen back to glucose. This process is called *glycogenolysis*.

Source of Energy

Where does a muscle get its energy to contract? Where does the body get the energy necessary to synthesize protein, to send nerve impulses, to perform countless other functions?

The energy necessary for the body functions comes from certain high-energy compounds, compounds that yield a large amount of energy on hydrolysis. The key compound of this type is adenosine triphosphate (ATP). Hydrolysis of ATP to adenosine diphosphate (ADP) and inorganic phosphate liberates about 7600 cal per mole. This hydrolysis breaks one of the high-energy phosphate bonds, designated by $\sim$. These two molecules—ATP and ADP—are shown opposite.

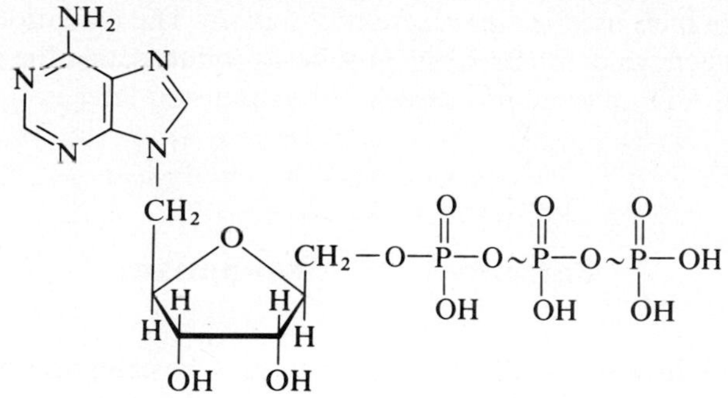

adenosine triphosphate, ATP

This formula for ATP may be abbreviated as

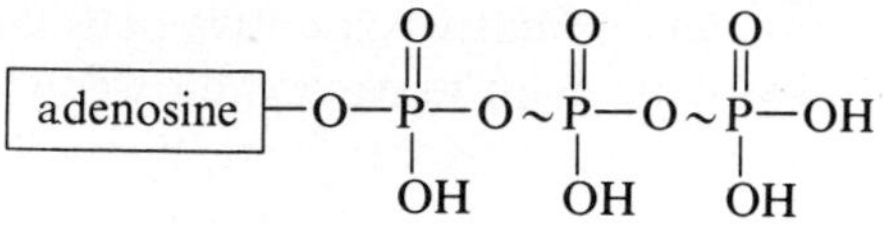

$$\boxed{\text{adenosine}} - \overset{\overset{\displaystyle O}{\|}}{\underset{\underset{\displaystyle OH}{|}}{P}} - O \sim \overset{\overset{\displaystyle O}{\|}}{\underset{\underset{\displaystyle OH}{|}}{P}} - O \sim \overset{\overset{\displaystyle O}{\|}}{\underset{\underset{\displaystyle OH}{|}}{P}} - OH \quad \text{or} \quad A - P \sim P \sim P$$

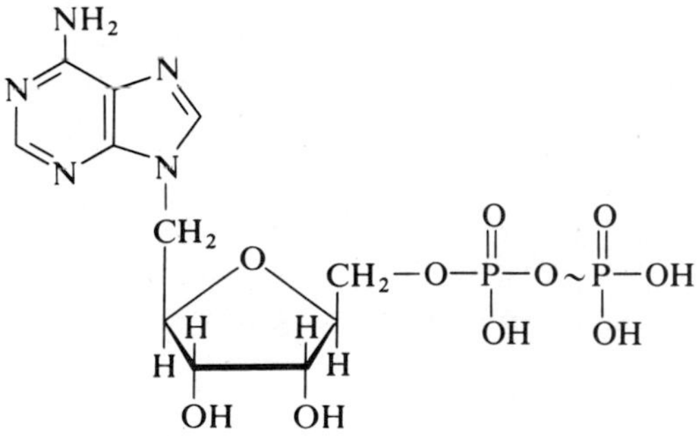

adenosine diphosphate (ADP)

This formula for ADP may be abbreviated as

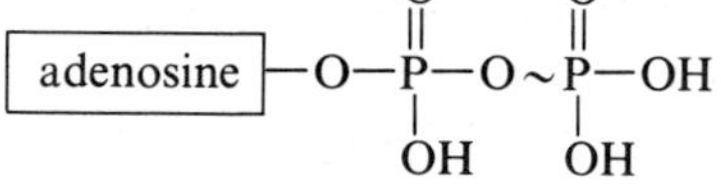

$$\boxed{\text{adenosine}} - O - \overset{\overset{\displaystyle O}{\|}}{\underset{\underset{\displaystyle OH}{|}}{P}} - O \sim \overset{\overset{\displaystyle O}{\|}}{\underset{\underset{\displaystyle OH}{|}}{P}} - OH \quad \text{or} \quad A - P \sim P$$

However, the supply of ATP in the body is limited. There must be some mechanism of regenerating this substance so that this high-energy compound is available for continued use. How does the body change ADP back to ATP? To accomplish this, a high-energy phosphate group must be added again to the ADP. Such a process is called phosphorylation and may be indicated by the equation

ADP + phosphate ion + fuel $\longrightarrow$ ATP + fuel residue

One of the fuels used in this reaction is glucose. The oxidation of glucose, the steps and enzymes required, the products produced, and the involvement of ADP and ATP in these processes will be discussed later in this chapter.

Carbohydrate Metabolism

The metabolism of carbohydrates in man may be categorized as follows:

1. *Glycogenesis*, the synthesis of glycogen from glucose.
2. *Glycogenolysis*, the breakdown of glycogen to glucose.
3. *Glycolysis*, the oxidation of glucose or glycogen to pyruvic or lactic acids.
4. *Hexose monophosphate shunt*, an alternative oxidative path for glucose.
5. *Citric acid cycle*, the final oxidative path to carbon dioxide and water.
6. *Gluconeogenesis*, the formation of glucose from noncarbohydrate sources.

Glycogenesis

Glycogenesis is the formation of glycogen from glucose. This process occurs primarily in the liver and in the muscles. The liver may contain up to 5 per cent glycogen after a high carbohydrate meal but may contain almost no glycogen after 18 hr of fasting.

The overall conversion of glucose to glycogen and vice versa may be written as

$$nC_6H_{12}O_6 \xrightleftharpoons[\text{glycogenolysis}]{\text{glycogenesis}} (C_6H_{10}O_5)_n + nH_2O$$

$$\text{glucose} \qquad\qquad\qquad \text{glycogen}$$

However, the reaction is by no means as simple as this equation indicates. There are several steps involved, each one being catalyzed by its own particular enzyme.

The first step in glycogenesis involves the conversion of glucose to glucose-6-phosphate (abbreviated as glucose-6-P), a phosphate ester of glucose. Adenosine triphosphate (ATP) from the liver cells serves as a source of the phosphate group. After the loss of the phosphate group, ADP is left. The enzyme *glucokinase* is necessary to catalyze this reaction. Insulin is involved in the phosphorylation of glucose by glucokinase.

$$\text{glucose} + \text{ATP} \xrightarrow[\text{insulin}]{\text{glucokinase}} \text{glucose-6-P} + \text{ADP}$$

Glucose-6-phosphate is then rearranged so that the phosphate group moves from the number 6 position to the number 1 position, producing glucose-1-phosphate. The enzyme required for this reaction is *phosphoglucomutase*:

$$\text{glucose-6-P} \xrightarrow{\text{phosphoglucomutase}} \text{glucose-1-P}$$

Glucose-1-phosphate then reacts with uridine triphosphate (UTP) to form uridine diphosphate glucose (UDPG). (Uridine triphosphate is similar to adenosine triphosphate except that uracil takes the place of adenine.) The enzyme is UDPG pyrophosphorylase.

$$\text{glucose-1-P} + \text{UTP} \xrightarrow{\text{UDPG pyrophosphorylase}} \text{UDPG} + \text{pyrophosphate}$$

Then, the glucose molecules in UDPG (activated glucose molecules) are joined together to form glycogen. The enzymes necessary here are a branching enzyme and glycogen synthetase. The latter enzyme is regulated by both insulin and cyclic-AMP.

$$\text{UDPG} \xrightarrow[\text{branching enzyme}]{\text{glycogen synthetase}} \text{glycogen} + \text{UDP}$$

The UDP formed in this reaction then reacts with ATP to regenerate UTP.

$$\text{UDP} + \text{ATP} \longrightarrow \text{UTP} + \text{ADP}$$

A diagram of the overall reaction is shown below.

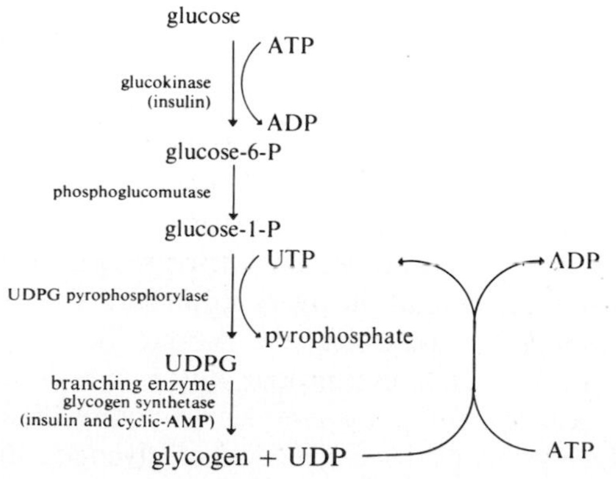

The same reactions, written structurally, are as follows

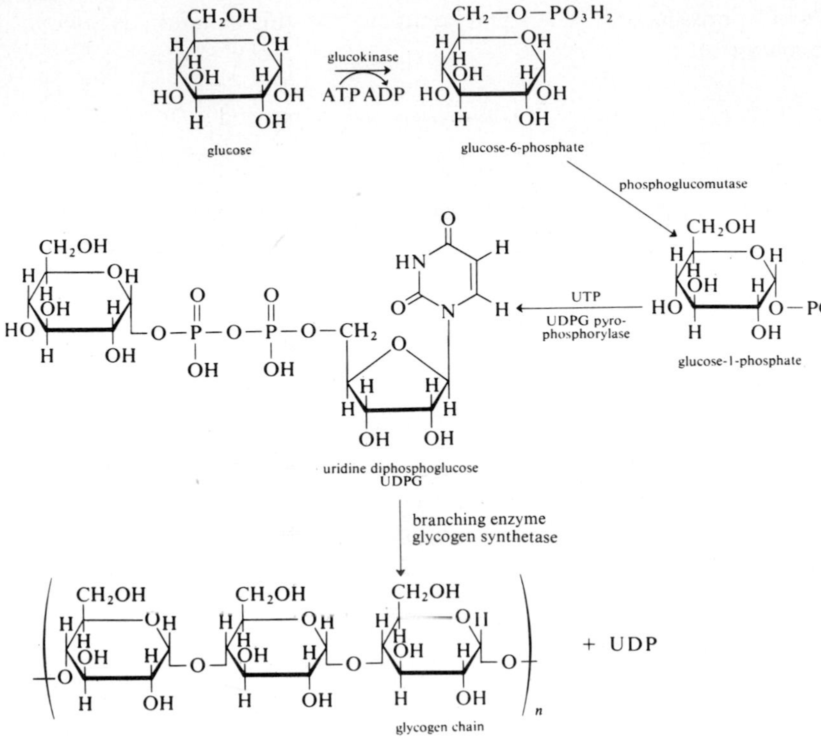

CH_2OH ... glucokinase ... ATP ADP ... $CH_2-O-PO_3H_2$

glucose $\longrightarrow$ glucose-6-phosphate

phosphoglucomutase

UTP $\longleftarrow$ UDPG pyro-phosphorylase

glucose-1-phosphate ... CH_2OH ... $O-PO_3H_2$

uridine diphosphoglucose
UDPG

branching enzyme
glycogen synthetase

$\left(\text{glycogen chain} \right)_n$ $+$ UDP

glycogen chain

Glycogenolysis

In glycogenolysis we might expect the reverse of all of the above reactions, but it should be noted that the first reaction, that involving glucokinase, is *not* a reversible reaction.

In glycogenolysis, glycogen is converted to glucose-1-phosphate by the enzyme phosphorylase *a* and then to glucose-6-phosphate by the enzyme phosphoglucomutase.

Glucose-6-phosphate is then converted to glucose by the enzyme glucose-6-phosphatase, an enzyme found in the liver but not in the muscle. Therefore, muscle glycogen cannot serve as a source of blood glucose.

Cyclic-AMP (see page 323) is involved in the conversion of glycogen into glucose-6-phosphate in both the liver and the muscles. When the body is under stress, it produces the hormone epinephrine (see page 479). This hormone is carried by the bloodstream to the liver cells, where it activates the enzyme adenyl cyclase, which in turn causes the production of cyclic-AMP from ATP. Cyclic-AMP then activates a protein kinase, which in turn activates a phosphorylase *b* kinase. This phosphorylase *b* kinase then activates phosphorylase *a*, which triggers the conversion of glycogen into glucose. The

same type of reaction is involved in the conversion of muscle glycogen into glucose-6-phosphate.

In the adipose tissue cyclic-AMP activates a protein kinase, which activates a second enzyme, triglyceride lipase, which in turn begins the breakdown of fat into fatty acids and glycerin.

Glycogenesis and glycogenolysis may be summarized as follows

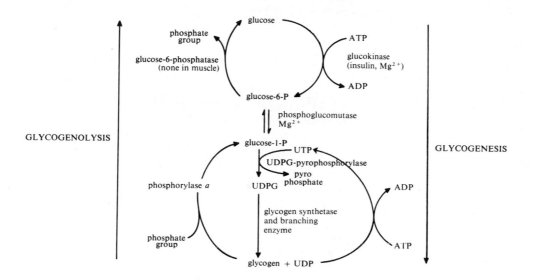

Glycolysis

The breakdown of glycogen resupplies the energy used up during muscle contraction. This breakdown involves a series of steps, each catalyzed by a particular enzyme.

The breakdown of glycogen to lactic acid, called the Embden-Meyerhof pathway, is the first step involved in muscle contraction. This process is also called *glycolysis* and is an anaerobic process—that is, each step takes place without oxygen. Glycolysis supplies the ATP needed for muscle contraction.

The overall reaction of glycolysis may be summarized as follows

glucose-6-P (from glycogen or from glucose) + 3 ADP $\longrightarrow$ 2 lactic acid + 3 ATP

The ATP formed is available for muscular work. As the ATP is used, it is changed to ADP and must then be regenerated. This regeneration can be accomplished through the above pathway or by the use of another anaerobic sequence

creatine phosphate + ADP $\longrightarrow$ creatine + ATP

The steps in glycolysis and the structural formulas of the intermediary products are given in the following chart.

Glycogen is changed to glucose-1-P by the catalytic action of the enzyme phosphorylase a. Glucose-1-P is then changed to glucose-6-P by the enzyme phosphoglucomutase. Glucose-6-P could also be formed directly from glucose with the action of ATP and the enzyme glucokinase.

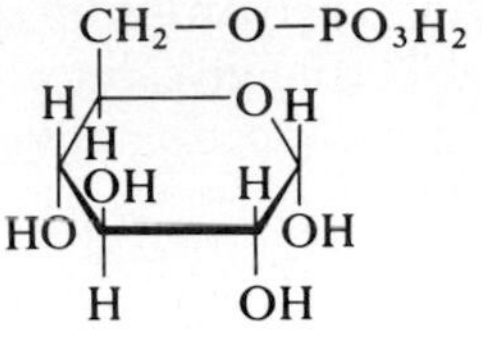

$$CH_2-O-PO_3H_2$$

glucose-6-phosphate

phosphohexose isomerase

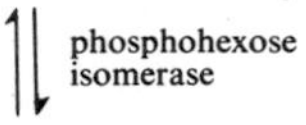

Glucose-6-P is then changed to fructose-6-P by the action of the enzyme phosphohexose isomerase.

$$CH_2-O-PO_3H_2 \qquad CH_2OH$$

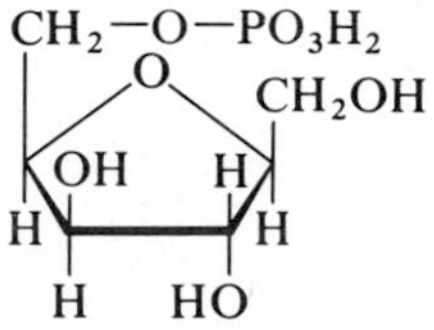

fructose-6-phosphate

Fructose-6-P is changed to fructose-1,6-diphosphate by the action of the enzyme phosphofructokinase. Note that during this reaction, ATP is converted to ADP. (The reverse reaction is catalyzed by the enzyme phosphatase.)

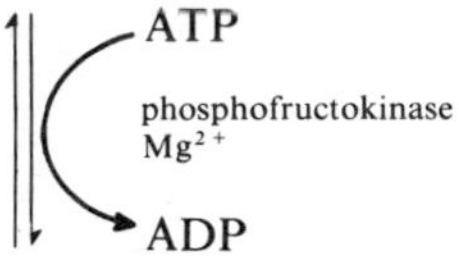

ATP

phosphofructokinase Mg^{2+}

ADP

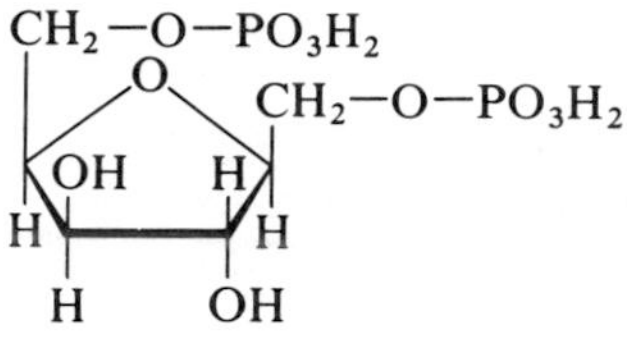

$$CH_2-O-PO_3H_2 \qquad CH_2-O-PO_3H_2$$

fructose-1,6-diphosphate

Fructose-1,6-diphosphate is changed to the three-carbon compounds, glyceraldehyde-3-P and dihydroxyacetone-P. The enzyme involved is aldolase. Dihydroxyacetone-P is converted to glyceraldehyde-3-P by the action of the enzyme phosphoglyceroisomerase.

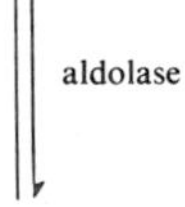

aldolase

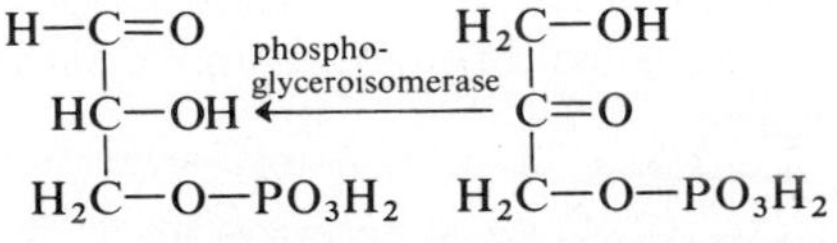

$$
\begin{array}{lll}
H-C=O & & H_2C-OH \\
HC-OH & \xleftarrow[\text{glyceroisomerase}]{\text{phospho-}} & C=O \\
H_2C-O-PO_3H_2 & & H_2C-O-PO_3H_2
\end{array}
$$

glyceraldehyde-3-phosphate   dihydroxyacetonephosphate

Glyceraldehyde-3-P is changed to 1,3-diphosphoglyceric acid by the action of the enzyme glyceraldehyde-3-P-dehydrogenase. During this reaction, nicotinamide adenine dinucleotide (NAD$^+$) is reduced to NADH +H$^+$.

$$\text{NAD}^+ \qquad\qquad \text{glyceraldehyde-3-P-dehydrogenase} \qquad\qquad \text{NADH} + \text{H}^+$$

$$
2\quad
\begin{array}{l}
\overset{\displaystyle O}{\overset{\|}{C}}\text{—O—PO}_3\text{H}_2 \\
\text{HC—OH} \\
\text{H}_2\text{C—O—PO}_3\text{H}_2
\end{array}
$$

1,3-diphosphoglyceric acid

Next, 1,3-diphosphoglyceric acid is changed to 3-phosphoglyceric acid by the action of the enzyme phosphoglycerokinase. In this reaction, two ADP (one for each three-carbon compounds) are changed to two ATP.

$$\text{ADP} \qquad\qquad \text{phosphoglycerokinase} \quad \text{Mg}^{2+} \qquad\qquad \text{ATP}$$

$$
2\quad
\begin{array}{l}
\overset{\displaystyle O}{\overset{\|}{C}}\text{—OH} \\
\text{HC—OH} \\
\text{H}_2\text{C—O—PO}_3\text{H}_2
\end{array}
$$

3-phosphoglyceric acid

phosphoglyceromutase

Then, 3-phosphoglyceric acid is changed to 2-phosphoglyceric acid through the enzyme phosphoglyceromutase.

$$
2\quad
\begin{array}{l}
\overset{\displaystyle O}{\overset{\|}{C}}\text{—OH} \\
\text{HC—O—PO}_3\text{H}_2 \\
\text{H}_2\text{C—OH}
\end{array}
$$

2-phosphoglyceric acid

enolase

Next, 2-phosphoglyceric acid is changed to phosphoenol pyruvic acid through the enzyme enolase.

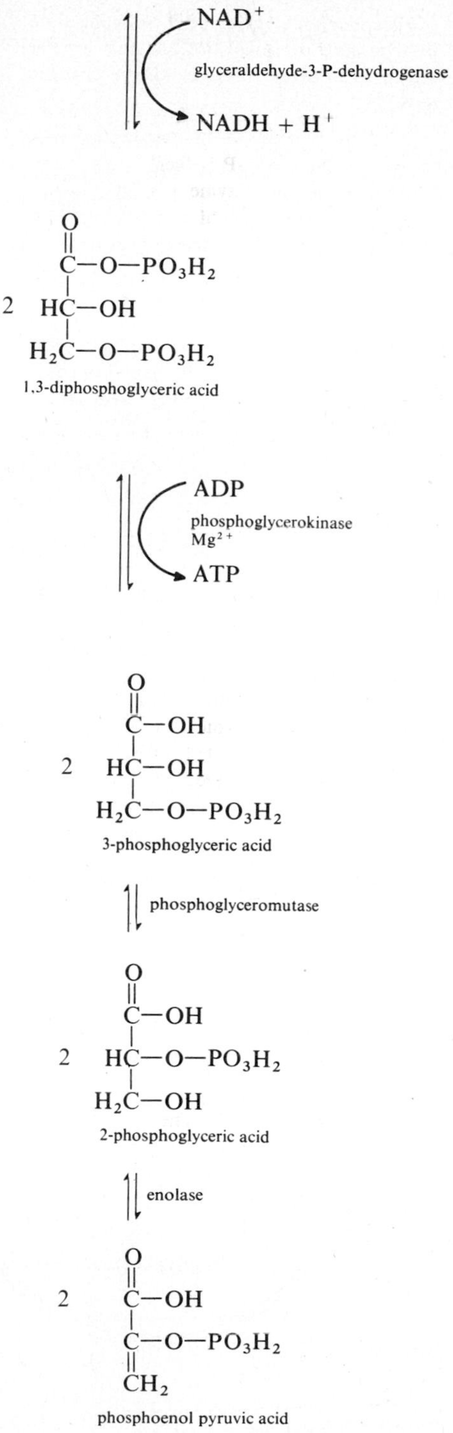

$$
2\quad
\begin{array}{l}
\overset{\displaystyle O}{\overset{\|}{C}}\text{—OH} \\
\text{C—O—PO}_3\text{H}_2 \\
\text{CH}_2
\end{array}
$$

phosphoenol pyruvic acid

Phosphoenol pyruvic acid is changed to pyruvic acid through the enzyme pyruvic kinase. During this reaction, ADP is changed to ATP.

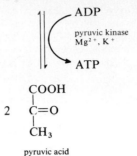

$$2 \quad \begin{array}{c} COOH \\ | \\ C=O \\ | \\ CH_3 \end{array}$$

pyruvic acid

Pyruvic acid is changed to lactic acid through the enzyme lactic dehydrogenase. At the same time, NADH and H^+ are changed to NAD^+.

$$2 \quad \begin{array}{c} COOH \\ | \\ HO-C-H \\ | \\ CH_3 \end{array}$$

lactic acid

The sequence of reactions involved in glycolysis may be summarized as shown below.

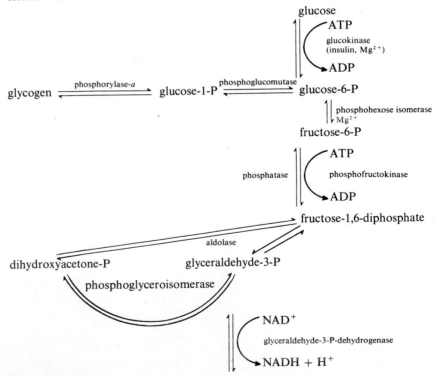

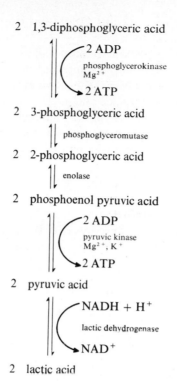

2   1,3-diphosphoglyceric acid

2 ADP

phosphoglycerokinase
Mg^{2+}

2 ATP

2   3-phosphoglyceric acid

phosphoglyceromutase

2   2-phosphoglyceric acid

enolase

2   phosphoenol pyruvic acid

2 ADP

pyruvic kinase
Mg^{2+}, K^+

2 ATP

2   pyruvic acid

$NADH + H^+$

lactic dehydrogenase

NAD^+

2   lactic acid

According to these reactions, muscle glycogen is changed into lactic acid. However, only about one fifth of the lactic acid thus formed is oxidized to carbon dioxide and water, resupplying the energy used up during muscle contraction. The other four fifths of the lactic acid is changed back to glycogen, reversing the above reactions. Part of the lactic acid is changed back to glycogen in the muscle. The rest of the lactic acid is carried to the liver by the bloodstream where it is converted to liver glycogen.

The oxidation of some of the lactic acid, the aerobic sequence, produces a large amount of energy. The oxidation of one molecule of lactic acid converts 18 molecules of ADP to ATP.

After a muscle contracts and relaxes, the net change is a partial loss of glycogen. This deficiency in glycogen can be replenished by the conversion of blood glucose to muscle glycogen (muscle glycogenesis).

This conversion of glycogen to lactic acid and partial reconversion to glycogen, the lactic acid cycle, may be shown as follows:

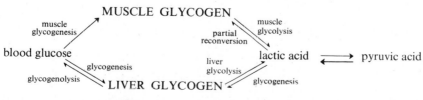

Other Oxidative Pathways

The oxidation of glucose to lactic acid can also proceed through a series of reactions called the *hexose monophosphate shunt*, or the *pentose shunt*. This sequence is important because it provides the five-carbon sugars needed for the synthesis of nucleic acids and nucleotides and because it makes available NADPH, the reduced form of nicotinamide adenine dinucleotide phosphate ($NADP^+$), a coenzyme necessary for the synthesis of fatty acids.

The overall reaction may be written as follows

$$3 \text{ glucose-6-P} + 6NADP^+ \xrightarrow{\text{enzymes}} 3CO_2 + 2 \text{ fructose-6-P} + \text{glyceraldehyde-3-P}$$
$$+ 6(NADPH + H^+)$$

The fructose-6-P may be converted back to glucose-6-P by the enzyme fructohexose isomerase, which is part of the Embden-Meyerhof pathway.

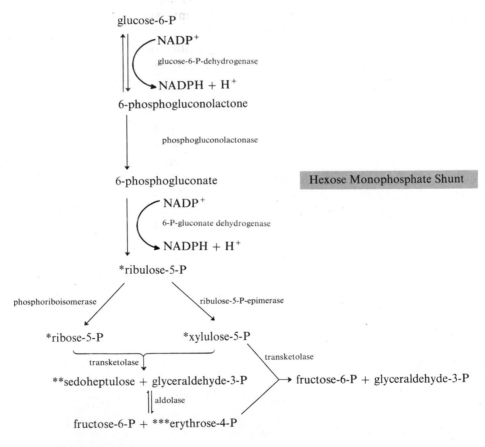

* Five-carbon sugars.
** Seven-carbon sugar.
***Four-carbon sugar/

The Aerobic Sequence (Krebs Cycle)

The aerobic sequence converts lactic and pyruvic acids (from anaerobic glycolysis) through a series of steps to carbon dioxide and water. This series of reactions is called the *Krebs cycle*, or *citric acid cycle*. The citric acid cycle uses oxygen transported to the cells by hemoglobin, hence the term *aerobic*. This cycle takes place in the mitochondria (see page 336).

During the complete oxidation of 1 mole of glucose, 38 high-energy phosphate bonds are produced. Four of these are formed during glycolysis. Two more are formed from the substrate level of the citric acid (Krebs) cycle, but two are used up during the phosphorylation of glucose in the glycolytic mechanism. The remaining 34 high-energy phosphate bonds are formed during the Krebs cycle. Thus, most of the energy for the body comes from the Krebs cycle.

Assuming 7600 cal per high-energy phosphate bond, the overall sum is 38×7600 or 288,800 cal. Theoretically, 686,000 cal should be produced from 1 mole of glucose; thus, the efficiency of conversion is approximately 42 per cent.

The first step in the aerobic process is the formation of active acetate from pyruvic acid. This active acetate is the acetyl derivative of coenzyme A, or *acetyl CoA*. Acetyl CoA is the converting substance in the metabolism of carbohydrates, fats, and protein. Acetyl CoA becomes the "fuel" for the Krebs cycle. As will be noted, acetyl CoA reacts with oxaloacetic acid and goes through the cycle. At the end of the cycle, oxaloacetic acid is regenerated and picks up another molecule of acetyl CoA to carry it through the sequence. During the cycle, acetyl CoA is oxidized to carbon dioxide and water as indicated. At the same time NADH and $FADH_2$ are produced. These enter into the electron transport chain that functions on the inner membranes of the mitochondria. In the electron transport chain (also called the electron transport system), electrons are transferred from NADH through a series of steps to oxygen, with the regeneration of NAD^+, FAD, and H_2O, and the energy yielded used for the formation of ATP. The overall reaction is also called oxidative phosphorylation and proceeds because large amounts of NADH and $FADH_2$ are produced from the citric acid cycle and also because there is a plentiful supply of oxygen in the tissues. The overall reaction is:

$$NADH + H^+ + 3P + 3ADP + \tfrac{1}{2}O_2 \longrightarrow NAD^+ + 3ATP + H_2O$$

Involved in this electron transport system are NAD, FMN, FAD, coenzyme Q, and several cytochromes, which are complexes containing heme (recall that heme is part of the hemoglobin molecule).

One of the cytochromes, cytochrome oxidase, is a complex that binds oxygen, reduces it with electrons received from other cytochromes in the electron transport system, and finally converts that oxygen to water. Note that the oxygen in this sequence (which is therefore an aerobic sequence)

Krebs Cycle

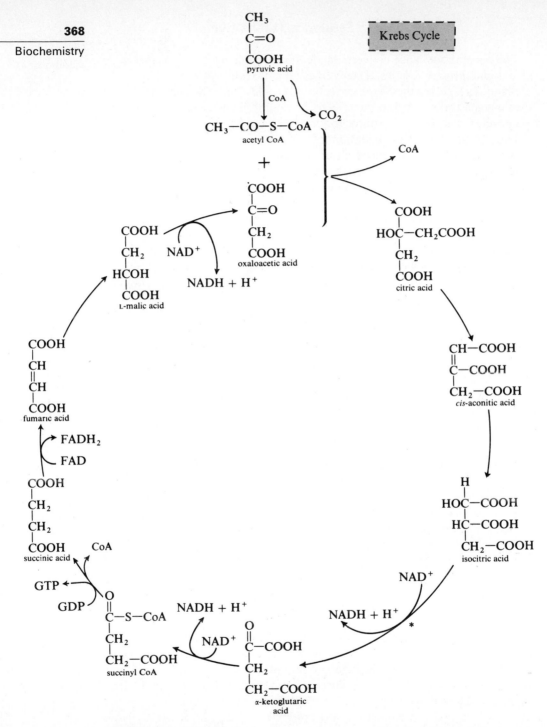

* An enzyme-bound complex of oxalosuccinic acid occurs here as an intermediary.

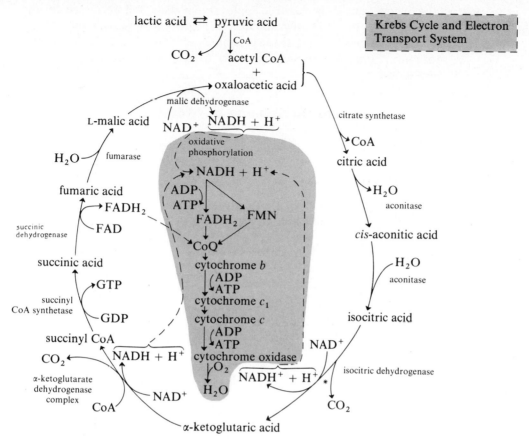

* See previous page.

reacts only at the last step. One unusual fact about cytochrome oxidase is that it contains two metals—iron and copper. This particular enzyme is mainly responsible for the introduction of oxygen into the metabolic processes of the citric acid cycle and so is considered to be absolutely vital to life. If the cytochrome oxidase function is blocked, as with cyanide poisoning, all cellular activity stops very quickly. It is believed that the cytochrome oxidase functions by transferring electrons from copper to iron to oxygen.

Note that another high-energy compound produced by the citric acid cycle is GTP, guanosine triphosphate.

Gluconeogenesis   *fasting*

Gluconeogenesis is the formation of glucose from noncarbohydrate substances such as amino acids and glycerol. This process takes place primarily in the liver, although a small amount also takes place in the kidneys.

369

Gluconeogenesis is increased in high-protein diets and decreased in high-carbohydrate diets. During starvation, gluconeogenesis supplies glucose from the amino acids of the tissue protein. In severe diabetes, gluconeogenesis not only from food protein but also from tissue protein may lead to emaciation.

Overall Scheme of Metabolism

The interrelationship of glycogenesis, glycogenolysis, glycolysis, gluconeogenesis, the hexose monophosphate shunt, and the Krebs (citric acid) cycle, along with the metabolism of fats and proteins, is indicated in the following diagram.

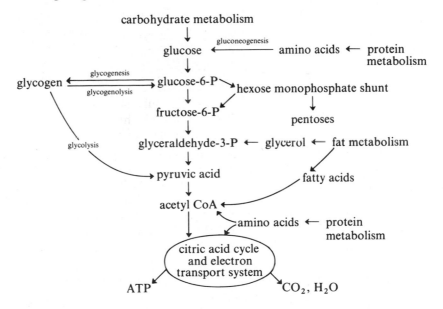

It should be noted that acetyl CoA is the key intermediate in fat and lipid synthesis. However, acetyl CoA does not serve as a source of glucose because the reaction of pyruvic acid to acetyl CoA is not readily reversible.

Hormones Involved in Regulating Blood Sugar

The liver plays a vital function in controlling the normal blood sugar level by removing sugar from and adding sugar to the blood. The activity of the liver in maintaining the normal blood sugar level is in turn controlled by several different hormones. Among these are insulin, epinephrine, and glucagon. The hormones of the anterior pituitary, the adrenal cortex, and the thyroid also have a definite effect upon carbohydrate metabolism.

Insulin

Insulin is a hormone produced by the beta cells of the islets of Langerhans in the pancreas. Insulin performs the following functions.

1. It accelerates the oxidation of glucose in the cells.
2. It increases the transformation of glucose to glycogen (glycogenesis) in the muscle and also in the liver. (Insulin controls the phosphorylation of glucose to glucose-6-P by means of the enzyme glucokinase; see page 358.)
3. It depresses the production of glucose (glycogenolysis) in the liver.
4. It promotes the formation of fat from glucose.

Thus, the principal function of insulin may be said to be the removal of glucose from the bloodstream and so a consequent lowering of the blood sugar level.

Epinephrine

Epinephrine is a hormone secreted by the medulla of the adrenal glands. It stimulates the formation of glucose from glycogen in the liver (glycogenolysis) (see page 360) and so has an action opposite to that of insulin. Insulin removes glucose from the bloodstream, whereas epinephrine increases the amount of glucose present in the blood.

During periods of strong emotional stress, such as anger or fright, epinephrine is secreted into the bloodstream where it promotes glycogenolysis in the liver. This increases the amount of glucose in the blood, making that glucose readily available as the body needs it to meet the emergency situation. The amount of glucose may then exceed the renal threshold (hyperglycemia) and sugar will appear in the urine. This is one example of how the presence of sugar in the urine may be due to a condition other than diabetes.

Glucagon

Glucagon is a hormone produced by the alpha cells of the pancreas. Its effects are opposite to those of insulin. Glucagon raises blood sugar levels by stimulating the activity of the enzyme phosphorylase in the liver, which changes liver glycogen to glucose. The activity of phosphorylase depends upon cyclic-AMP. Glucagon also increases gluconeogenesis from amino acids and from lactic acid. Glucagon has no effect on phosphorylation in the muscles.

Prompt injection of insulin will alleviate the symptoms accompanying high blood sugar. Persons suffering from diabetes mellitus can lead normal lives provided that they receive insulin as needed. Since insulin is a protein, it cannot be taken orally (it would be digested as are all proteins) and so must be administered by injection.

Glucose Tolerance Test

A positive Benedict's test on a urine specimen indicates that the patient *may* have diabetes mellitus. However, this test is by no means conclusive proof because the presence of sugar in the urine may be due to other conditions.

If a patient is suspected of being diabetic, he is given a glucose tolerance test. In this test, after a 12-hr fast he is given approximately 1 g of glucose for each kilogram of body weight. Then his blood sugar level is checked by withdrawing blood samples at regular intervals for several hours. These samples are chemically analyzed and the concentration of sugar in the blood is plotted in relation to the time intervals. In a normal person, the blood sugar level rises from about 80 mg per 100 ml of blood to 130 mg per 100 ml of blood in about 1 hr. Then the blood sugar level gradually returns to normal after a period of about $2\frac{1}{2}$ hr. In a diabetic patient, because no insulin is being secreted, the blood sugar level rises to an even higher level than in the normal patient and remains there for a much longer period of time. Then it slowly begins to return toward its normal value.

These results may be seen in the graphs of the glucose tolerance test, as shown in Figure 26–1.

Also, in the diabetic patient, large amounts of glucose appear in the urine because of the spillover when the renal threshold is exceeded.

Figure 26–1. Criteria used for interpretation of glucose tolerance tests. (Modified from Pansky, B.: *Dynamic Anatomy and Physiology*. Macmillan Publishing Co., Inc., New York, 1975.)

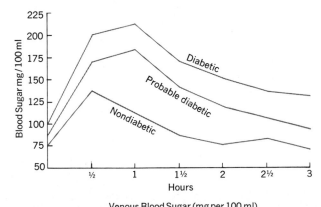

	Venous Blood Sugar (mg per 100 ml)					
	Fasting	½ hour	1 hour	1½ hours	2 hours	2½ hours
Normal	<100	<160	<160	–	<100-110	<100
Probable	<110-120	130-159	160-180	>135	110-120	<100-120
Diabetes	>110	>150-160	>160	>140	>120	>110

The glucose tolerance test is a valuable diagnostic tool since it indicates the ability of the body to utilize carbohydrate. A decreased utilization may indicate diabetes, while an increased utilization may indicate Addison's disease, hypopituitarism, or hyperinsulinism.

Diabetes Mellitus

If the amount of insulin is decreased or eliminated (either because of decreased activity of the islets of Langerhans or by the degeneration of these cells), the blood sugar level will rise. Increased blood sugar level (hyperglycemia) leads to glycosuria (glucose in the urine) because the renal threshold is exceeded.

Also, the lack of insulin in diabetes leads to an increased oxidation of fatty acids as a source of ATP. Increased oxidation of fatty acids leads to an accumulation of acetoacetic acid, beta-hydroxybutyric acid, and acetone. These substances, commonly known as ketone bodies, form faster than they can be oxidized and removed and so accumulate in the blood (and urine). A higher-than-normal concentration of these substances in the blood is known as ketosis.

The presence of ketone bodies affects the pH of the blood since two of the three compounds are acids. If the ketone bodies accumulate and lower the pH of the blood, a condition known as acidosis exists. A decreased pH reduces the ability of hemoglobin to carry oxygen so that acidosis can be very serious. Prolonged acidosis first causes nausea, then depression of the central nervous system, severe dehydration, deep coma (known as diabetic coma), and finally death.

Summary

The end products of carbohydrate digestion are the monosaccharides. The major monosaccharide in the bloodstream is glucose. Some of the blood glucose is converted to glycogen in the liver and in the muscle. This process is called glycogenesis. Other glucose is constantly being oxidized to furnish energy for the body. If the blood sugar level rises too much, the excess spills over into the urine. The presence of glucose in the urine is known as glycosuria.

The oxidation of glucose produces energy, which is stored in high-energy compounds, especially adenosine triphosphate (ATP). Hydrolysis of ATP to adenosine diphosphate (ADP) and inorganic phosphate liberates about 7600 cal.

Glycogenesis—the formation of glycogen from glucose—takes place in the liver and in muscle tissue. There are many intermediate steps involved in this conversion, each being catalyzed by a specific enzyme.

Glycogenolysis—the conversion of glycogen to glucose—takes place primarily in the liver.

The breakdown of muscle glycogen to lactic acid, a process that requires no oxygen (anaerobic), is termed glycolysis. Glycolysis supplies most of the ATP needed for muscle contraction. Glycolysis proceeds through a series of steps, each catalyzed by a specific enzyme. In this process three molecules of ADP are converted to ATP, which is then available for muscular work.

About one fifth of the lactic acid formed in glycolysis is oxidized to carbon dioxide and water. The other four fifths is converted back to liver glycogen. The cycle of glucose-glycogen-lactic acid-glycogen is known as the lactic acid cycle.

The oxidation of glucose to lactic acid may also proceed through a series of reactions called the hexose monophosphate shunt or the pentose shunt. This sequence is important because it provides five-carbon sugars needed for the synthesis of nucleic acids and nucleotides and also because it makes available the reduced form of $NADP^+$, a coenzyme necessary for the synthesis of fatty acids.

The aerobic sequence for the oxidation of lactic and pyruvic acids is called the citric acid or Krebs cycle. Most of the energy from the oxidation of glucose comes from this cycle. This cycle takes place in the mitochondria.

The first step in the Krebs cycle is the formation of acetyl CoA from pyruvic acid. This acetyl CoA, also called active acetate, is the fuel for the citric acid cycle. Acetyl CoA reacts with oxaloacetic acid and then goes through a series of steps, each catalyzed by a particular enzyme. At the completion of the cycle, oxaloacetic acid is regenerated and then picks up another molecule of acetyl CoA to carry through the same cycle again. The NADH and $FADH_2$ produced in the citric acid cycle enter the electron transport chain in the mitochondria. The overall reaction, called oxidative phosphorylation, involves oxygen and produces ATP. Several different coenzymes and cytochromes are involved.

Glucose may also be formed from noncarbohydrate substances such as amino acids, fatty acids, and glycerol. Such a process is termed gluconeogenesis.

Thus carbohydrate metabolism interrelates the processes of glycogenesis, glycogenolysis, glycolysis, gluconeogenesis, and the Krebs cycle.

The liver controls the blood sugar level. This activity is governed by several hormones. Among these are insulin, epinephrine, and glucagon.

Insulin, a hormone secreted by the pancreas, accelerates oxidation of glucose in the cells, increases glycogenesis, decreases glycogenolysis, and promotes the formation of fat from glucose. Thus insulin removes glucose from the bloodstream.

Epinephrine, a hormone of the adrenal medulla, changes liver glycogen to glucose and muscle glycogen to lactic acid. Epinephrine is secreted into the bloodstream during periods of emotional stress.

Glucagon, a hormone secreted by the pancreas, has an effect opposite to that of insulin. It raises blood sugar levels.

The glucose tolerance test is given to a patient suspected of being a diabetic. He is fed glucose and his blood sugar level checked for several hours. In a normal person the blood sugar level rises and then gradually returns to normal after about $2\frac{1}{2}$ hr. In a diabetic the blood sugar level rises to a much higher level and remains there for a longer period of time.

Questions and Problems

1. Which monosaccharide is the principal one remaining in the bloodstream after passing through the liver?
2. What is meant by the term *normal fasting blood sugar*?

3. What is glycogenesis? Where does it occur?
4. What happens to excess carbohydrate that cannot be immediately utilized or converted to glycogen?
5. What is meant by the term *renal threshold*?
6. What is glycosuria?
7. How may glucose be removed from the bloodstream?
8. What types of compounds does the body use to store energy?
9. Indicate the hydrolysis reaction of ATP. How much energy is produced in this reaction?
10. Name two ways in which ADP may be converted to ATP.
11. Describe the process of glycogenesis, indicating the intermediary products and enzymes necessary.
12. What is glycogenolysis?
13. Are glycogenesis and glycogenolysis reversible in both muscle and liver? Why?
14. What is glycolysis? Is it an aerobic or an anaerobic sequence?
15. Does glycolysis supply most of the body's energy?
16. Indicate the steps in glycolysis and the enzymes required at each step.
17. Is all the lactic acid formed in glycolysis oxidized to carbon dioxide and water?
18. What happens to most of the lactic acid formed during glycolysis?
19. After a muscle contracts and then relaxes, how does it replenish its glycogen?
20. Diagram and label the lactic acid cycle.
21. What is the hexose monophosphate shunt? Why is it important?
22. Indicate the steps and enzymes in the hexose monophosphate shunt.
23. Explain briefly what happens in the Krebs cycle (citric acid cycle).
24. What is the function of acetyl CoA in this cycle?
25. How much energy is produced by the oxidation of glucose? What is the efficiency of conversion?
26. How does the energy produced during the Krebs cycle compare with that produced during glycolysis?
27. Diagram and label the Krebs cycle.
28. What is gluconeogenesis? What might increase it?
29. Show the interrelationship of glycolysis, gluconeogenesis, the Krebs cycle, glycogenesis, and glycogenolysis.
30. Why is acetyl CoA not considered to be a source of glucose?
31. What are the functions of insulin?
32. When is epinephrine secreted? What does it do?
33. Where is glucagon formed and what is its function?
34. Describe the glucose tolerance test. What is its purpose?
35. Where does the electron transport system function? Describe the steps involved.
36. What is cytochrome oxidase? What does it do? What is unusual about it? Why is it considered essential to life?

References

Baum, S. J., and Scaife, C. W.: *Chemistry: A Life Science Approach*. Macmillan Publishing Co., Inc., New York, 1975, Chap. 28.

Bronk, J. R.: *Chemical Biology*. Macmillan Publishing Co., Inc., New York, 1973, Chap. 10.

Cheraskin, E.; Ringsdorf, W. M.; and Hicks, B. S.: The sweet sickness syndrome: the refined carbohydrate consumption. *Journal of the International Academy of Preventive Medicine*, **1**:107–120 (Fall), 1974.

Fernstrom, J. D., and Wurtman, R.: Nutrition and the brain. *Scientific American*, **230**:84–91 (Feb.), 1974

Harper, H. A.: *Review of Physiological Chemistry*, 15th ed. Lange Medical Publications, Los Altos, Calif., 1975, Chap. 13.

Lehninger, A. H.: *Biochemistry*, 2nd ed. Worth Publishers, Inc., New York, 1975, Chaps. 16, 17, 18.

Lieber, C. S.: The metabolism of alcohol. *Scientific American*, **234**:25–33 (March), 1976.

Margaria, R.: The sources of muscular energy. *Scientific American*, **226**:84–91 (March), 1972.

Montgomery, R.; Dryer, R. L.; Conway, T. W.; and Spector, A. A.: *Biochemistry*, C. V. Mosby Co., St. Louis, 1974, Chaps. 6, 7.

Chapter 27

Metabolism of Fats

Absorption of Fat

The digestion of fats takes place primarily in the small intestines with the hydrolysis to fatty acids and glycerol. Prior to their digestion, the fats have been emulsified by the bile salts. The products of fat digestion pass through the lacteals of the villi into the lymphatics where they appear as resynthesized fats. From the lymphatics, the fats flow through the thoracic duct into the bloodstream and then to the liver.

After a meal the fat content of the blood rises and remains at a high level for several hours, then gradually decreases to the fasting level.

In the liver some of the fats are changed to phospholipids, so that the blood leaving the liver contains both fats and phospholipids. These phospholipids, such as sphingomyelin and lecithin (see page 293), are necessary for the formation of nerve and brain tissue. Lecithins (phosphatidyl cholines) are also involved in the transportation of fat to the tissues. Cephalin, another phospholipid, is involved in the normal clotting of the blood. From the liver, some fat goes to the cells where it is oxidized to furnish heat and energy. The fat in excess of what the cells need is stored as adipose tissue.

Lipolysis, the hydrolysis of triacylglycerols (triglycerides) to fatty acids and glycerol, is under the control of cyclic-AMP and various hormones. Glucagon and epinephrine stimulate the production of cyclic-AMP and so increase lipolysis. However, the prostaglandins (see page 285) depress the levels of cyclic-AMP and so decrease the rate of lipolysis.

Oxidation of Fat

The oxidation of fat (triacylglycerol) actually involves the oxidation of the two hydrolysis products—glycerol and fatty acids.

Oxidation of Glycerol

The glycerol part of a fat is oxidized to dihydroxyacetone phosphate, as is indicated below. Recall that dihydroxyacetone phosphate is part of the glycolysis sequence (see page 361). This compound may be converted into glycogen in the liver or muscle tissue or into pyruvic acid, which enters the citric acid (Krebs) cycle. Thus, the glycerol part of a fat is metabolized through the carbohydrate sequence.

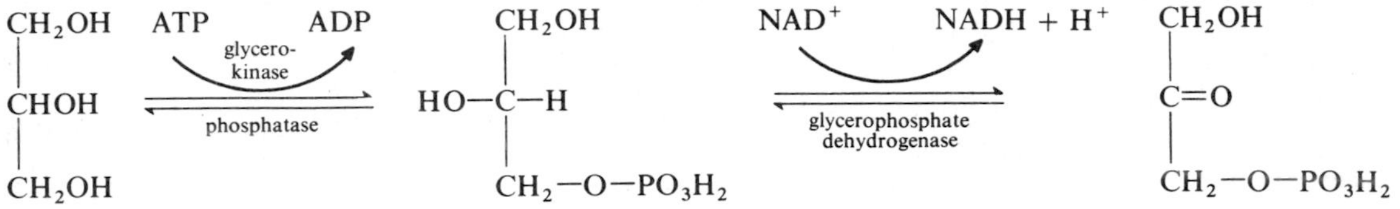

$$
\begin{array}{c}
CH_2OH \\
| \\
CHOH \\
| \\
CH_2OH \\
\text{glycerol}
\end{array}
\quad
\underset{\text{phosphatase}}{\overset{ATP \quad \overset{\text{glycero-}}{\text{kinase}} \quad ADP}{\rightleftharpoons}}
\quad
\begin{array}{c}
CH_2OH \\
| \\
HO-C-H \\
| \\
CH_2-O-PO_3H_2 \\
\text{α-glycerophosphate}
\end{array}
\quad
\underset{\substack{\text{glycerophosphate} \\ \text{dehydrogenase}}}{\overset{NAD^+ \quad NADH + H^+}{\rightleftharpoons}}
\quad
\begin{array}{c}
CH_2OH \\
| \\
C=O \\
| \\
CH_2-O-PO_3H_2 \\
\text{dihydroxyacetone} \\
\text{phosphate}
\end{array}
$$

Oxidation of Fatty Acids

There are several theories about the oxidation of fatty acids. The original one, proposed by Knoop in 1905 and still preferred today, is called the beta-oxidation theory. This theory involves the oxidation of the second carbon atom from the acid end of the fatty acid molecule—the beta carbon atom. In this process beta oxidation removes two carbon atoms at a time from the fatty acid chain. That is, an 18-carbon fatty acid is oxidized to a 16-carbon fatty acid, then to a 14-carbon fatty acid, and so on, until the oxidation process is complete. A simplified version of such an oxidation is illustrated below. Note that in this oxidative process acetyl CoA, a derivative of acetic acid, is formed, which then enters the Krebs cycle.

The first step in this sequence is the reaction of a fatty acid molecule with coenzyme A (here written as CoA—S—H) in the presence of magnesium ions and the enzyme thiokinase to form an active fatty acid. During this step, ATP is converted to AMP and pyrophosphate (PP):

$$- - - - -CH_2-CH_2-CH_2-\overset{\overset{O}{\|}}{C}-OH + CoA-S-H + ATP \xrightarrow[Mg^{2+}]{thiokinase}$$

fatty acid

$$- - - - -CH_2-CH_2-CH_2-\overset{\overset{O}{\|}}{C}-S-CoA + AMP + PP$$

active fatty acid

The active fatty acid in the presence of the enzyme acyl CoA-dehydrogenase is changed to a *trans-α,β*-unsaturated fatty acid. During this reaction, the coenzyme flavin adenine dinucleotide (FAD) is reduced to FADH$_2$.

$$- - - -CH_2-CH_2-CH_2-\overset{\overset{O}{\|}}{C}-S-CoA \xrightarrow[\substack{FAD \quad\quad FADH_2}]{\substack{acyl\ CoA-\\dehydrogenase}} - - - - -CH_2-\overset{\overset{H}{|}}{C}=\overset{\overset{H}{|}}{C}-\overset{\overset{O}{\|}}{C}-S-CoA$$

trans-α,β-unsaturated fatty acid

The alpha,beta-unsaturated fatty acid is changed to a beta-hydroxy fatty acid. Note that this is a hydrolysis reaction involving the enzyme enoyl-CoA-hydratase.

$$- - - - -CH_2-\overset{\overset{H}{|}}{C}=\overset{\overset{H}{|}}{C}-\overset{\overset{O}{\|}}{C}-S-CoA \xrightarrow{\substack{enoyl-CoA-\\hydratase}} - - - - -CH_2-\overset{\overset{OH}{|}}{C}H-CH_2-\overset{\overset{O}{\|}}{C}-S-CoA$$

β-hydroxy fatty acid

Next, the beta-hydroxy fatty acid is oxidized to a beta-keto fatty acid in the presence of the enzyme beta-hydroxyacyl CoA-dehydrogenase.

$$- - - - -CH_2-\overset{\overset{OH}{|}}{C}H-CH_2-\overset{\overset{O}{\|}}{C}-S-CoA \xrightarrow[\substack{NAD^+ \quad\quad NADH + H^+}]{\substack{β-hydroxy-acyl\\CoA-dehydrogenase}} - - - - -CH_2-\overset{\overset{O}{\|}}{C}-CH_2-\overset{\overset{O}{\|}}{C}-S-CoA$$

β-keto fatty acid

Finally, the beta-keto fatty acid is changed to acetyl CoA plus a molecule of active fatty acid containing two less carbon atoms than it had originally. This reaction proceeds in the presence of the enzyme beta-ketothiolase.

$$- - - - -CH_2-\overset{\overset{O}{\|}}{C}-CH_2-\overset{\overset{O}{\|}}{C}-S-CoA \xrightarrow[\substack{β-ketothiolase}]{CoA-S-H} CH_3\overset{\overset{O}{\|}}{C}-S-CoA + - - - - -CH_2-\overset{\overset{O}{\|}}{C}-S-CoA$$

acetyl CoA

active fatty acid
(with two less C's)

The acetyl CoA thus produced enters the citric acid cycle and the new molecule of active fatty acid goes through the same sequence again, each time losing two carbon atoms until the entire fatty acid molecule has been oxidized. This sequence presupposes the presence of fatty acids containing an even number of carbon atoms, a condition usually encountered in nature. Note that the end product of the beta-oxidation is acetic acid, which is formed as a CoA derivative.

The beta-oxidation cycle for the oxidation of fatty acids with an even number of carbon atoms may be indicated diagrammatically as follows.

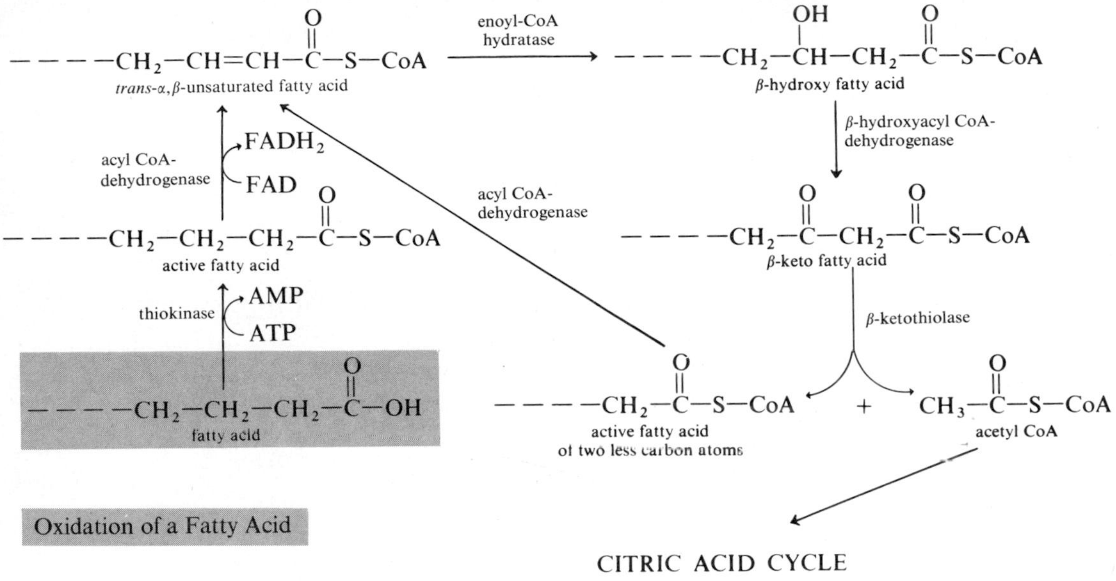

Oxidation of a Fatty Acid

CITRIC ACID CYCLE

If fatty acids containing an odd number of carbon atoms are oxidized, they follow the same steps except that the final products are acetyl CoA and propionyl CoA. The propionyl CoA is changed in a series of steps to succinyl CoA, which then enters the citric acid cycle, as does the acetyl CoA. These reactions require the presence of cobamide, a derivative of the coenzyme vitamin B_{12}, and also biotin.

The unsaturated fatty acids are metabolized slowly. They must first be reduced by some of the dehydrogenases found in the cells. Then they can follow the fatty acid cycle for oxidation.

Energy Produced by Oxidation of Fatty Acids

The oxidation of 1 g fat produces more than twice as much energy as the oxidation of 1 g carbohydrate. Let us see why.

The oxidation of acetyl CoA through the citric acid cycle yields 12 high-energy phosphate bonds (ATP) per molecule of acetyl CoA. If we consider the oxidation of palmitic acid, a 16-carbon fatty acid, 8 two-carbon units will be formed during the beta-oxidation cycle. These eight-carbon units will yield 8×12 or 96 ATP's. However, two ATP's are used up in the initial activation of the fatty acid. In addition, it has been calculated that palmitic acid will produce 35 ATP's as it goes through the fatty acid cycle. That is, the net number of ATP molecules produced will be $96 - 2 + 35$, or 129.

Considering each ATP molecule as requiring 7600 cal for formation, 129×7600 cal or 980 Cal (kcal) are needed. The theoretic yield from one molecule of palmitic acid is 2340 kcal, so that the efficiency of conversion is 980/2340, or 42 per cent, with the remainder of the energy being produced as heat. (Other fatty acids and glycerol are also oxidized, so that the net result is that fats produce much more energy than do carbohydrates.)

Ketone (Acetone) Bodies

In a diabetic patient, carbohydrate metabolism is restricted and acetyl CoA cannot be properly metabolized. When this occurs, the acetyl CoA is changed to acetoacetyl CoA which is converted into acetoacetic acid in the liver by the enzyme deacylase. Acetoacetic acid may be changed into acetone and beta-hydroxybutyric acid, as shown in the following reactions.

These three substances—acetoacetic acid, beta-hydroxybutyric acid, and acetone—are commonly called acetone bodies, or ketone bodies. They are carried by the blood to the muscles and tissues where they are converted back to acetoacetyl CoA and then oxidized normally. However, during diabetes, the production of these ketone bodies by the liver exceeds the ability of the muscles and tissues to oxidize them so that they accumulate in the blood.

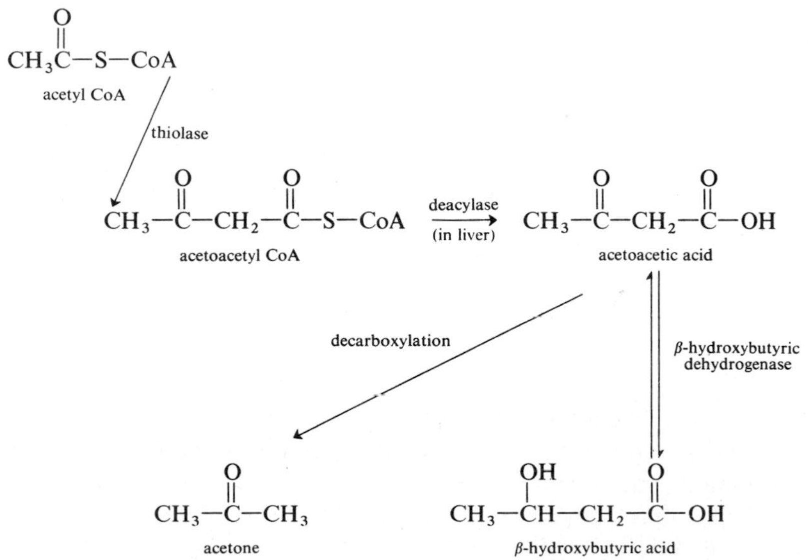

Acetone (Ketone) Bodies

The excess accumulation of ketone bodies in the blood is called ketonemia. The excess accumulation of ketone bodies in the urine is called ketonuria. The overall accumulation of ketone bodies in the blood and the urine is called ketosis. During ketosis acetone may be detected on the patient's breath because it is a volatile compound and is easily excreted through the lungs.

Ketosis occurs when the metabolism of carbohydrates is not normal because the presence of glucose (to maintain the citric acid cycle) is necessary for the oxidation of fats. Ketosis may occur with diabetis mellitus, in starvation or severe liver damage, or on a diet high in fats and low in carbohydrates.

During diabetes mellitus the body is unable to oxidize carbohydrates and instead oxidizes fats, leading to an accumulation of ketone bodies in the blood and the urine. These ketone bodies are acidic and tend to decrease the pH of the blood. The lowering of the pH of the blood is termed acidosis and may lead to a fatal coma. During acidosis an increased amount of water intake is needed to eliminate the products of metabolism. Unless the water intake of a diabetic is increased, dehydration will occur. Dehydration of diabetics may also be caused by polyuria due to an increased amount of glucose in the urine.

Likewise, during prolonged starvation or on a high-fat, low-carbohydrate diet, the body tends to burn fat instead of carbohydrate, leading to ketosis and acidosis.

In severe liver damage, the liver cannot store glycogen in the required amounts so that carbohydrates are not available for the normal oxidation of fats, leading to ketosis.

Storage of Fat

The fat in excess of that required for the normal oxidative processes of the body is stored as adipose tissue under the skin and around the internal organs. This stored fat serves several important purposes.

1. Reserve supply of food
2. Support for the internal organs
3. Shock absorber for the internal organs
4. Insulation of interior of the body against sudden external changes in temperature

The fat stored in the body is in equilibrium with that in the bloodstream. That is, the fats stored in the adipose tissue do not merely remain there as inert compounds until they are needed. They are continuously being used and replaced so that there is always a dynamic transfer of fats between the bloodstream and the storage tissues.

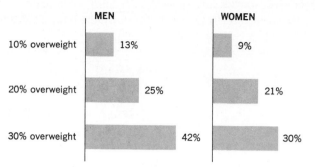

Figure 27–1.
Mortality
table based on obesity.
(Courtesy of
Metropolitan Life
Insurance Co., New
York, N.Y.)

*Compared with mortality of Standard risks
(Mortality ratio of Standard risks = 100%)

Obesity is a condition in which excess fat is deposited as adipose tissue. An obese person eats more food than his body can burn up so that the excess is converted to fat and stored as adipose tissue. For every 9 Cal of food eaten in excess of the body's requirements, 1 g of fat is deposited.

Most people have a tendency to become obese as they grow older. This is because they require less food for the maintenance of their body, especially because they exercise less than younger people.

In general obesity leads to a shortened life expectancy, as indicated in Figure 27–1.

The answer to obesity lies in proper dieting under the supervision of a doctor, because the metabolism of the body is a highly intricate mechanism that can very easily be disturbed.

Lipogenesis

Lipogenesis—the conversion of glucose to fats—takes place in the liver and in the adipose tissue, with the latter place predominating. Insulin is necessary for lipogenesis both in the liver and in the adipose tissue. Lipogenesis is reduced during fasting or during a high-fat diet, whereas it is increased during a high-carbohydrate diet.

The synthesis of fatty acids occurs both inside and outside the mitochondria. The process inside the mitochondria involves the lengthening of fatty acid chains of moderate length. While this process involves the reversal of the steps in the beta-oxidation process, it is responsible for only a small fraction of the fatty acids synthesized in the tissues.

The synthesis of a majority of the fatty acids (lipogenesis) occurs in the cell outside the mitochondria. Malonyl CoA, the main reactant, is primed with acetyl CoA and goes through the cycle indicated on the following page.

Each time around the cycle, two more carbon atoms (from malonyl CoA) are added, until a fatty acid-ACP* complex is produced. The free fatty acid is then released from the complex by the enzyme deacylase.

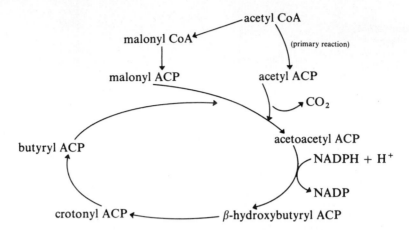

Synthesis of Phospholipids

The phospholipids are very important because they form (along with protein) the framework of most of the cell membrane system (see page 296).

The first step in the synthesis of phospholipids is the phosphorylation of glycerol with ATP to form L-α-glycerophosphate.

$$
\begin{array}{ccc}
\text{CH}_2\text{OH} & & \text{CH}_2\text{OH} \\
| & & | \\
\text{CHOH} + \text{ATP} \longrightarrow & & \text{HO}-\overset{|}{\text{C}}-\text{H} + \text{ADP} \\
| & & | \\
\text{CH}_2\text{OH} & & \text{CH}_2\text{OP} \\
\text{glycerol} & & \text{L-α-glycerophosphate}
\end{array}
$$

Next, L-α-glycerophosphate reacts with active fatty acid to form 1,2-glycerophosphatidic acid.

$$
\begin{array}{ccc}
\text{CH}_2\text{OH} & & \overset{\displaystyle O}{\overset{\|}{\text{CH}_2\text{O}-\text{C}-\text{R}}} \\
| & \overset{\displaystyle O}{\overset{\|}{}} & | \\
\text{HO}-\text{C}-\text{H} \;+\; 2\text{R}-\text{C}-\text{S}-\text{CoA} \longrightarrow & \text{R}-\overset{\|}{\text{C}}-\text{O}-\text{C}-\text{H} & +\; 2\text{CoA}-\text{S}-\text{H} \\
| & & | \\
\text{CH}_2-\text{O}-\text{P} & & \text{CH}_2-\text{O}-\text{P} \\
\text{L-α-glycerophosphate} & \text{active fatty acid} & \text{1,2-glycerophosphatidic acid}
\end{array}
$$

* ACP = acyl carrier protein.

The 1,2-glycerophosphatidic acid then loses its phosphate to form a

1,2-diglyceride (1,2-diacylglycerol).

$$
\begin{array}{ccc}
& \overset{\displaystyle O}{\overset{\|}{CH_2O-C-R}} & \\
\overset{\displaystyle O}{\overset{\|}{R-C-O}}-\overset{|}{C}-H & & \\
& \overset{|}{CH_2-O-P} &
\end{array}
\longrightarrow
\begin{array}{ccc}
& \overset{\displaystyle O}{\overset{\|}{CH_2O-C-R}} & \\
\overset{\displaystyle O}{\overset{\|}{R-C-O}}-\overset{|}{C}-H & & \\
& \overset{|}{CH_2OH} &
\end{array}
$$

1,2-glycerophosphatidic acid   1,2-diglyceride (1,2-diacylglycerol)

The 1,2-diglyceride may combine with another molecule of active fatty acid to form a triglyceride or fat (a triacylglycerol).

$$
\begin{array}{c}
\overset{\displaystyle O}{\overset{\|}{CH_2O-C-R}} \\
\overset{\displaystyle O}{\overset{\|}{R-C-O}}-\overset{|}{C}-H \\
\overset{|}{CH_2OH}
\end{array}
+ \overset{\displaystyle O}{\overset{\|}{R-C-S-CoA}}
\longrightarrow
\begin{array}{c}
\overset{\displaystyle O}{\overset{\|}{CH_2O-C-R}} \\
\overset{\displaystyle O}{\overset{\|}{R-C-O}}-\overset{|}{C}-H \\
\overset{\displaystyle O}{\overset{\|}{CH_2O-C-R}}
\end{array}
+ CoA-S-H
$$

1,2-diglyceride   active fatty acid   a triglyceride or fat   coenzyme A

The 1,2-diglyceride may also combine with cytidine diphosphate-choline (CDP-choline) to form lecithin (phosphatidyl choline), or it may combine with CDP-ethanolamine to form cephalin (phosphatidyl ethanolamine).

Cholesterol

Cholesterol is found in all cells of the body but particularly in brain and nerve tissue. Cholesterol occurs in animal fat but not in plant fat. About 0.3 g cholesterol is ingested daily from such foods as egg yolk, meat fats, liver, and liver oils. In addition it has been estimated that the body manufactures about 1 g of cholesterol daily.

Cholesterol normally is eliminated in the bile. However, sometimes it settles out in the gallbladder as gallstones. If cholesterol deposits in the walls of the larger arteries, the condition is known as atherosclerosis, a type of hardening of the arteries. When this occurs, there is a decrease in the usable diameter of the blood vessels. The elasticity of the arterial walls decreases. There is an interference with the rate of blood flow because there is greater friction due to the irregular lining of the blood vessels. This irregular lining may also cause clots as the blood flows over that type of surface.

Cholesterol is important as a precursor of several important steroids such as vitamin D, the sex hormones, and the adrenocortical hormones (the hormones of the cortex of the adrenal glands).

Synthesis

Cholesterol is synthesized primarily in the liver, but the adrenal cortex, skin, testes, aorta, and intestines are also able to synthesize it. This synthesis takes place in the microsomal and cytosomal fraction of the cell. Note that acetyl CoA is the starting material and is also the source of all the carbon atoms in cholesterol. The steps in the synthesis are outlined below; the specific enzymes for each step are omitted.

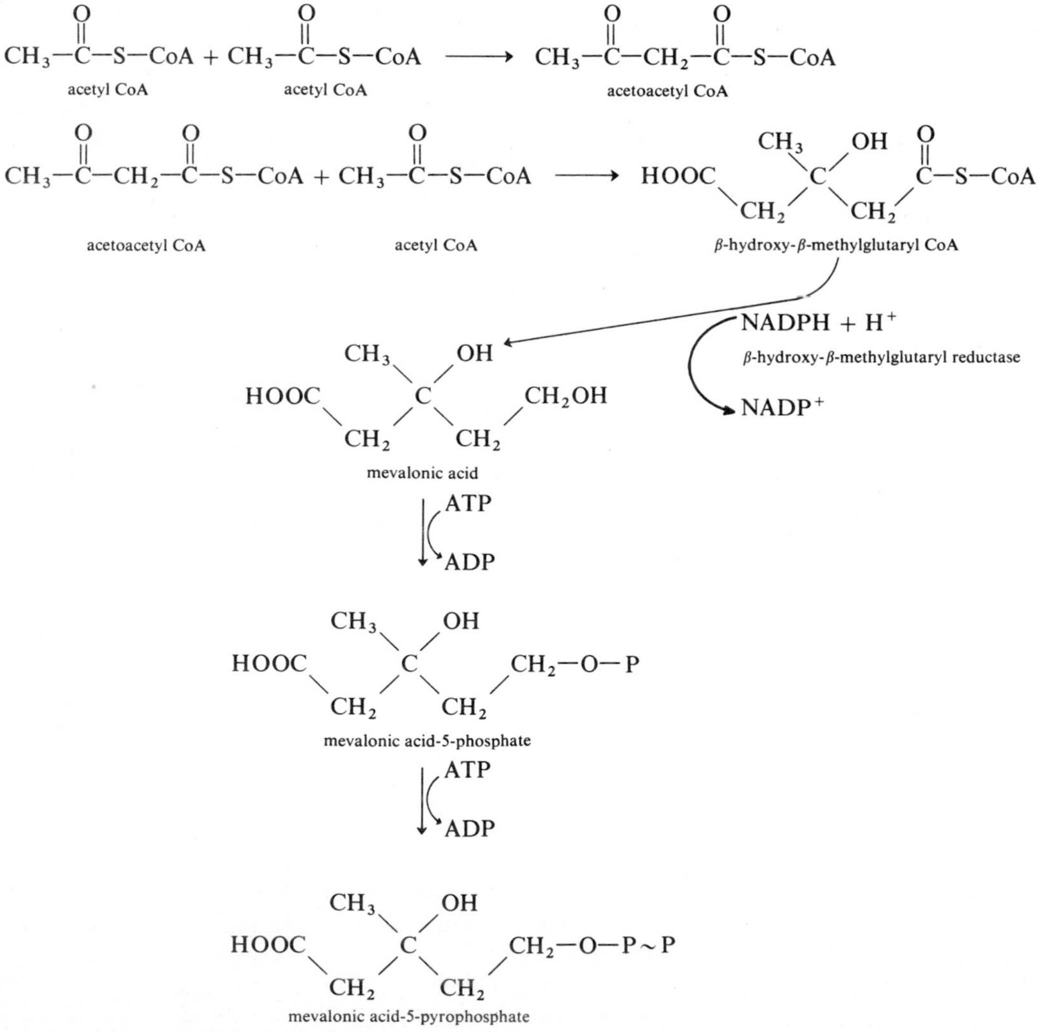

$$CH_3-\overset{O}{\underset{\|}{C}}-S-CoA + CH_3-\overset{O}{\underset{\|}{C}}-S-CoA \longrightarrow CH_3-\overset{O}{\underset{\|}{C}}-CH_2-\overset{O}{\underset{\|}{C}}-S-CoA$$

acetyl CoA   acetyl CoA   acetoacetyl CoA

$$CH_3-\overset{O}{\underset{\|}{C}}-CH_2-\overset{O}{\underset{\|}{C}}-S-CoA + CH_3-\overset{O}{\underset{\|}{C}}-S-CoA \longrightarrow$$

acetoacetyl CoA   acetyl CoA

β-hydroxy-β-methylglutaryl CoA

mevalonic acid

ATP → ADP

mevalonic acid-5-phosphate

ATP → ADP

mevalonic acid-5-pyrophosphate

β-hydroxy-β-methylglutaryl reductase

NADPH + H⁺ → NADP⁺

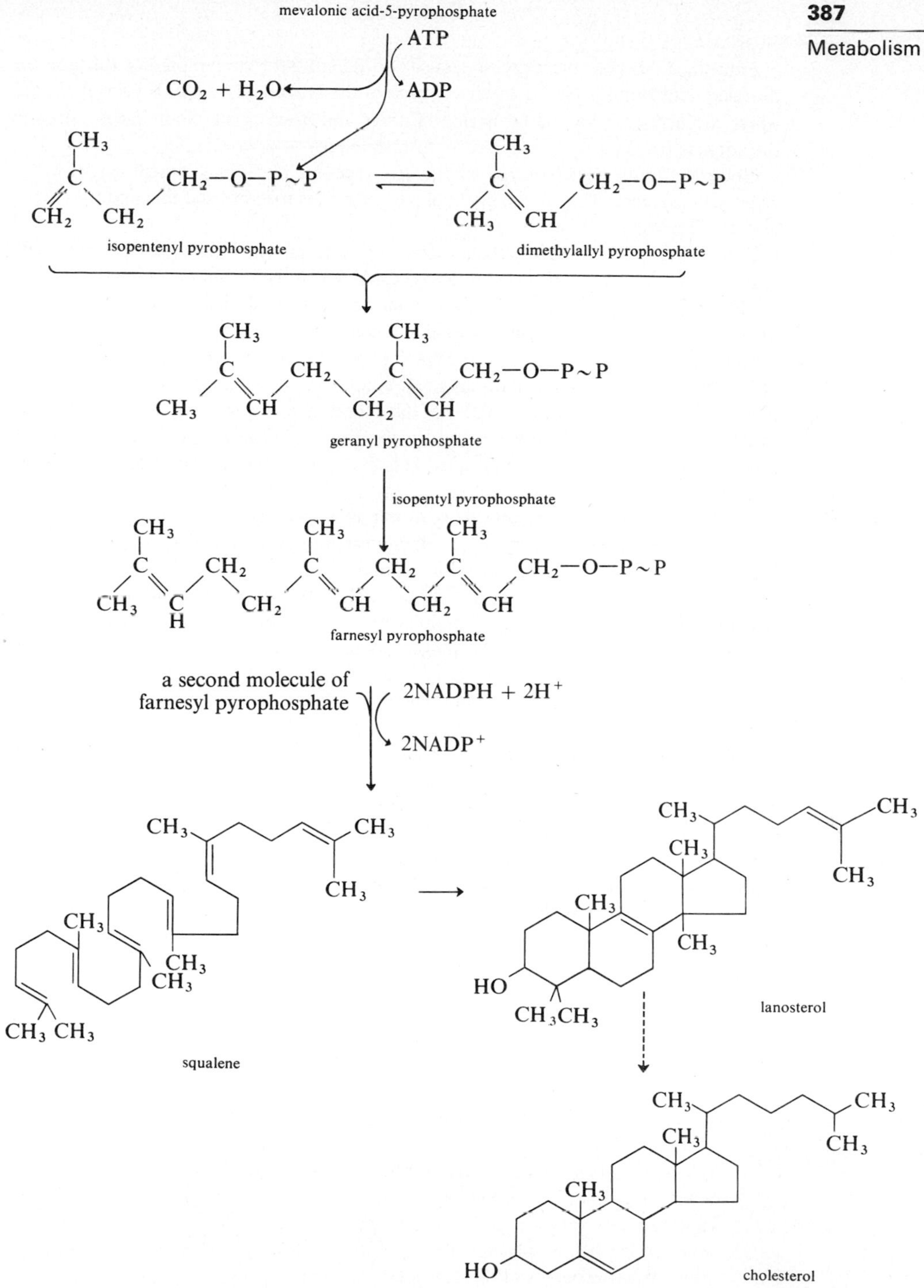
mevalonic acid-5-pyrophosphate
ATP
ADP
CO2 + H2O
CH3
C
CH2—O—P~P
CH2
CH2
isopentenyl pyrophosphate
CH3
C
CH2—O—P~P
CH3
CH
dimethylallyl pyrophosphate
CH3
C
CH2
CH3
CH
CH2
CH
CH2—O—P~P
geranyl pyrophosphate
isopentyl pyrophosphate
CH3
C
CH2
CH2
CH
CH2—O—P~P
H
farnesyl pyrophosphate
a second molecule of
farnesyl pyrophosphate
2NADPH + 2H+
2NADP+
CH3
squalene
CH3
HO
CH3CH3
lanosterol
cholesterol

Summary

Emulsified fats pass through the lacteals of the villi into the lymphatics through the thoracic duct to the liver. In the liver some of the fats are changed to phospholipids, which are necessary for the formation of nerve and brain tissue. Some fat is stored in the adipose tissue; some is oxidized to furnish energy.

Glycerol from a fat is oxidized to dihydroxyacetone phosphate, which is part of the glycolysis sequence. That is, the glycerol part of a fat is metabolized through the carbohydrate sequence.

Fatty acids pass through a beta-oxidation cycle in which two carbon atoms are removed at a time and converted to acetyl CoA, which then enters the citric acid cycle.

The oxidation of one molecule of a 16-carbon fatty acid yields a total of 129 ATP molecules, with an efficiency of about 42 per cent.

Ketone (acetone) bodies are normally produced in the beta-oxidation process. However, they are produced only in small amounts and do not normally accumulate. If excess ketone bodies accumulate in the blood, a condition known as ketonemia exists. Accumulation of ketone bodies in the urine is termed ketonuria. The overall accumulation of ketone bodies is called ketosis; it occurs in the abnormal metabolism of carbohydrates.

Fat in excess of the body's needs is stored as adipose tissue. Stored fat serves as a reserve supply of food, as a support for the internal organs, as a shock absorber for the internal organs, and as an insulator for the body.

If excess fat is deposited in the adipose tissue, the resulting condition is termed obesity. Obesity may be due to a glandular disorder or simply to overeating.

The conversion of glucose to fats, lipogenesis, takes place in the liver and in the adipose tissue.

The synthesis of fatty acids occurs both inside and outside the mitochondria, with the latter being the predominant site.

Phospholipids may be synthesized from glycerol. If the 1,2-diglyceride formed from glycerol reacts with cytidine diphosphate choline, then the phospholipid lecithin is formed. If instead the 1,2-diglyceride combines with cytidine diphosphate ethanolamine, then the phospholipid cephalin is formed.

Cholesterol is found in all cells of the body but particularly in nerve tissue. Cholesterol is normally eliminated in the bile but if it settles out in the gallbladder, gallstones are formed. If cholesterol deposits on the walls of the arteries, atherosclerosis occurs.

Cholesterol is an important precursor for vitamin D, for the sex hormones, and for the adrenocortical hormones.

Cholesterol is synthesized primarily in the liver beginning with acetyl CoA and proceeding through a series of steps.

Questions and Problems

1. Where are fats absorbed? How do they get to the liver?
2. What happens to fats in the liver?
3. Describe the oxidation of the glycerol part of a fat.
4. What is the beta-oxidation theory of fats? What are the end products of this cycle? What becomes of these products?

5. What types of products are formed during the oxidation of even and odd number carbon fatty acids?
6. What happens to unsaturated fatty acids before they can be oxidized?
7. Compare the energy produced by the oxidation of 1 g carbohydrate and 1 g fat.
8. In the oxidation of a fat, in which part of the sequence is most of the energy produced?
9. The oxidation of one molecule of a 16-carbon fatty acid produces how many ATP molecules?
10. What are the acetone bodies? How are they interrelated structurally?
11. Where are acetone bodies formed?
12. What is ketonemia? Ketonuria? Ketosis? Under what conditions might ketosis occur?
13. Where is excess fat stored? What functions does this fat have?
14. What conditions lead to obesity?
15. What is lipogenesis? Where does it occur?
16. How does lipogenesis in the mitochondria compare with the oxidation of a fat?
17. Describe lipogenesis in the cytoplasm.
18. Why are phospholipids important?
19. Describe the synthesis of lecithin.
20. How does the synthesis of lecithin compare with that of cephalin?
21. Where is cholesterol found in the body? Where is it produced?
22. Why is cholesterol important in the body?
23. If cholesterol is deposited in the walls of the large arteries, what is the condition termed?
24. If cholesterol is deposited in the gallbladder, what is formed?

References

Baum, S. J., and Scaife, C. W.: *Chemistry: A Life Science Approach.* Macmillan Publishing Co., Inc., New York, 1975, Chap. 29.

Bronk, J. R.: *Chemical Biology.* Macmillan Publishing Co., Inc., New York, 1973, Chap. 10.

Green, D. E.: Synthesis of fat. *Scientific American,* **202**:46-51 (Feb.), 1960.

Harper, H. A.: *Review of Physiological Chemistry*, 15th ed. Lange Medical Publications, Los Altos, Calif., 1975, Chap. 14.

Lehninger, A. H.: *Biochemistry*, 2nd ed. Worth Publishers, Inc., New York, 1975, Chaps. 20, 24.

Montgomery, R.; Dryer, R. L.; Conway, T. W.; and Spector, A. A.: *Biochemistry.* C. V. Mosby Co., St. Louis, 1974, Chaps. 8, 10.

Chapter 28

Metabolism of Proteins

Functions of Protein in the Body

During digestion, proteins are hydrolyzed into amino acids, which are then absorbed into the bloodstream through the villi of the small intestine. These amino acids are then transported to the tissues of the body. After a meal, the amino acid concentration of the blood rises but soon falls to its normal fasting value.

The amino acids from the digested food serve many different purposes. Among these are the following:

1. Conversion to tissue protein to build new tissue
2. Conversion to tissue protein to replace old tissue
3. Use in the formation of hemoglobin
4. Use in the formation of some hormones

5. Use in the formation of enzymes
6. Source of the different amino acids that the body needs but does not have
7. Source of energy when catabolized into a nitrogen part and a non-nitrogen part with subsequent oxidation

Nitrogen Balance

The body can store carbohydrates (as glycogen in the liver and the muscles), and can store fats (as adipose tissue and in tissues around the internal organs). However, the body cannot store protein. The amino acids that result from the digestion of protein are used either for the synthesis of new tissues, for the replacement of old tissues, and for the formation of the various required body substances such as hormones and enzymes, or they are converted to fat or are oxidized to furnish energy.

Because the body cannot store protein and because protein contains nitrogen, the amount of nitrogen intake in the food per day should usually equal the amount of nitrogen excreted per day (for a normal adult whose weight remains constant). This takes into consideration the normal replacement of worn-out tissue where the reaction is merely one of exchange between one amino acid and another.

A person whose body excretes as much nitrogen per day as he takes in with his food is said to be in nitrogen balance. Children exhibit a positive nitrogen balance because they take in more nitrogen in their food than they excrete. This is because the child has a greater need for amino acids to build his growing body's tissues. Any body condition marked by the growth of new tissues will exhibit a positive nitrogen balance. An example of this is seen in persons recovering from a wasting illness where the body needs to rebuild tissues and so does not excrete as much nitrogen (amino acids) as it takes in.

Conversely, if the body excretes more nitrogen than it takes in in the food, a negative nitrogen balance exists. Conditions that can produce a negative nitrogen balance are starvation, malnutrition, prolonged fever, and various wasting illnesses. However, a person can survive for a reasonable period of time in negative nitrogen balance, as when dieting, since body protein will be used for essential purposes.

Synthesis of Protein

As was discussed in Chapter 23, proteins are synthesized from amino acids through the various intermediaries such as peptides and polypeptides. The body takes the amino acids produced by the digestion of protein and recombines them into the protein that it needs in the various parts of the body. (See Chapter 33 for the mechanism of protein synthesis.)

Some amino acids that the body needs may be synthesized from other amino acids. However, there are certain amino acids that the body needs but cannot synthesize. These amino acids must be supplied in the food if the body is to function normally. These essential amino acids are listed on page 306.

Some proteins contain all of the essential amino acids; one that does not contain all the essential amino acids is called an *incomplete protein*. Two common incomplete proteins are gelatin, which is lacking in tryptophan, and zein (from corn), which is lacking in both tryptophan and lysine.

Body's Requirement of Protein

A certain minimum daily amount of protein is required for the normal replacement of body tissues. This amount, however, may be greatly increased by increased metabolism such as occurs in high fevers. However, for the normal adult keeping a constant weight, the recommended daily intake of protein is approximately 0.8 g per kilogram of body weight. This amounts to approximately 46 g protein per day for the adult female and 56 g per day for the adult male.

Catabolism of Amino Acids

The amino acids that the body does not need for tissue building or that are not of the correct type for this purpose are broken down to ammonia, carbon dioxide, and water, at the same time producing heat and energy.

Deamination

Deamination (also called oxidative deamination) is a catabolism reaction whereby the alpha-amino group of an amino acid is removed, forming an alpha-keto acid and ammonia. Deamination occurs primarily in the liver and the kidneys under the catalysis of the enzyme amino acid oxidase.

$$\underset{\text{α-amino acid}}{CH_3-\underset{\underset{NH_2}{|}}{CH}-COOH} \xrightarrow{\text{amino acid oxidase}} \underset{\text{α-keto acid}}{CH_3-\underset{\underset{O}{\|}}{C}-COOH} + \underset{\text{ammonia}}{NH_3}$$

The alpha-keto acid produced by this process can undergo several different types of reactions.

1. It can be catabolized to carbon dioxide, water, and energy in the citric acid cycle.
2. It can be converted to carbohydrates (glycogen) or to fat.
3. It may be reconverted to a different amino acid by a process called transamination.

Transamination is a reaction whereby an amino group from an amino acid is transferred to a keto acid. This reaction is catalyzed by the enzymes called transaminases. By this process the body can manufacture the amino acids that it needs and does not have. An essential part of the active site of transaminases is pyridoxal phosphate, the coenzyme form of vitamin B_6.

An example of transamination is the reaction of glutamic acid (an alpha-amino acid) and pyruvic acid (an alpha-keto acid) to form alpha-ketoglutaric acid (another alpha-keto acid) and alanine (another alpha-amino acid).

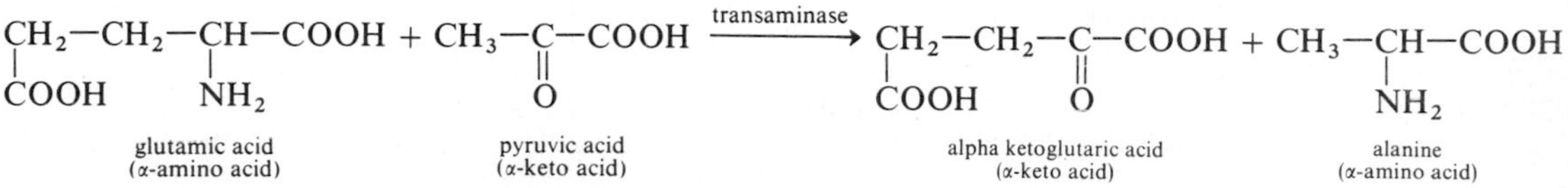

$$\underset{\substack{\text{glutamic acid} \\ (\alpha\text{-amino acid})}}{CH_2-CH_2-\underset{\underset{NH_2}{|}}{\overset{\overset{\displaystyle|}{COOH}}{CH}}-COOH} + \underset{\substack{\text{pyruvic acid} \\ (\alpha\text{-keto acid})}}{CH_3-\underset{\underset{O}{\|}}{C}-COOH} \xrightarrow{\text{transaminase}} \underset{\substack{\text{alpha ketoglutaric acid} \\ (\alpha\text{-keto acid})}}{CH_2-CH_2-\underset{\underset{O}{\|}}{\overset{\overset{\displaystyle|}{COOH}}{C}}-COOH} + \underset{\substack{\text{alanine} \\ (\alpha\text{-amino acid})}}{CH_3-\underset{\underset{NH_2}{|}}{CH}-COOH}$$

Formation of Urea

The ammonia formed from the deamination of amino acids combines with carbon dioxide to form urea and water. This process takes place in the liver. The overall reaction may be written as follows

$$\underset{\text{ammonia}}{2NH_3} + \underset{\text{carbon dioxide}}{CO_2} \xrightarrow{\text{enzymes}} \underset{\text{urea}}{NH_2CONH_2} + \underset{\text{water}}{H_2O}$$

However, this process is certainly not as simple as that shown above. It consists of a series of steps, each catalyzed by an appropriate enzyme.

Ammonia is a toxic by-product of the deamination of amino acids and as such must be removed from the body, predominantly in the form of the compound urea. In the conversion of ammonia to urea, three different amino acids are involved. These are arginine, citrulline, and ornithine. The pathway for the conversion of ammonia to urea is called the Krebs ornithine cycle.

The first step in this cycle is the reaction of ammonia and carbon dioxide to form carbamoyl phosphate. In this reaction, ATP is converted to ADP. This reaction is catalyzed by magnesium ions and the presence of N-acetyl-glutamic acid and the enzyme carbamoyl phosphate synthetase.

$$NH_3 + CO_2 + H_2O + 2\,ATP \xrightarrow[\substack{N\text{-acetylglutamic} \\ \text{acid} \\ \text{carbamoyl phosphate} \\ \text{synthetase}}]{Mg^{2+}} \underset{\substack{\text{carbamoyl} \\ \text{phosphate}}}{H_2N-\overset{\overset{\displaystyle O}{\|}}{C}-O-P} + 2\,ADP + P$$

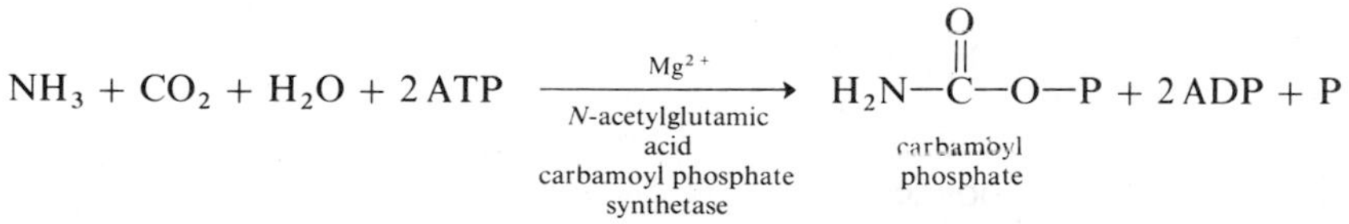

In the second step, carbamoyl phosphate combines with ornithine to form citrulline. This reaction is catalyzed by the liver enzyme ornithine trans-carbamoyl transferase.

$$CH_2-NH_2 \qquad \qquad \qquad \qquad \qquad \qquad \qquad \qquad \qquad \qquad O$$
$$| \qquad \qquad \qquad \qquad \qquad \qquad \qquad \qquad \qquad \qquad \qquad \qquad ||$$
$$(CH_2)_2 \qquad \qquad O \qquad \qquad \qquad \text{ornithine} \qquad \qquad \qquad CH_2-NH-C-NH_2$$
$$| \qquad \qquad \qquad || \qquad \qquad \textit{trans}\text{-carbamoyl} \qquad \qquad |$$
$$H-C-NH_2 + H_2N-C-O-P \qquad \xrightarrow{\quad \text{transferase} \quad} \qquad (CH_2)_2$$
$$| \qquad \qquad \qquad \qquad \qquad \qquad \qquad \qquad \qquad \qquad \qquad \qquad |$$
$$COOH \qquad \qquad \qquad \qquad \qquad \qquad \qquad \qquad \qquad \qquad H-C-NH_2$$
$$\qquad \qquad \qquad \qquad \qquad \qquad \qquad \qquad \qquad \qquad \qquad \qquad |$$
$$\qquad \qquad \qquad \qquad \qquad \qquad \qquad \qquad \qquad \qquad \qquad \qquad COOH$$

ornithine   carbamoyl   citrulline
      phosphate

Next, citrulline reacts with aspartic acid to form arginine succinic acid. This reaction takes place in the presence of ATP, magnesium ions, and an enzyme arginosuccinate synthetase:

$$\begin{array}{c} O \\ || \\ CH_2-NH-C-NH_2 \end{array} + \begin{array}{c} COOH \\ | \\ H_2N-C-H \end{array} \xrightleftharpoons[\substack{Mg^{2+}, \\ \text{arginosuccinate} \\ \text{synthetase}}]{ATP \quad AMP} \begin{array}{cc} NH & COOH \\ || & | \\ CH_2-NH-C-N-C-H \\ & H \end{array}$$

$$(CH_2)_2 \qquad \qquad \qquad CH_2 \qquad \qquad \qquad \qquad \qquad \qquad (CH_2)_2 \qquad CH_2$$
$$| \qquad \qquad \qquad \qquad | \qquad \qquad \qquad \qquad \qquad \qquad \qquad | \qquad \qquad |$$
$$H-C-NH_2 \qquad \qquad COOH \qquad \qquad \qquad \qquad \qquad H-C-NH_2 \qquad COOH$$
$$| \qquad \qquad \qquad \qquad \qquad \qquad \qquad \qquad \qquad \qquad \qquad \qquad |$$
$$COOH \qquad \qquad \qquad \qquad \qquad \qquad \qquad \qquad \qquad \qquad \qquad COOH$$

citrulline   aspartic acid   arginine succinic acid

Arginine succinic acid is cleaved (split) hydrolytically into arginine and fumaric acid. The fumaric acid enters the citric acid cycle.

$$\begin{array}{cc} NH & COOH \\ || & | \\ CH_2-NH-C-N-C-H \\ & H \end{array} \xrightarrow{\substack{\text{arginosuccinate} \\ \text{lyase}}} \begin{array}{c} NH \\ || \\ CH_2-NH-C-NH_2 \end{array} + \begin{array}{c} HC-COOH \\ || \\ HOOC-CH \end{array}$$

$$(CH_2)_2 \qquad \qquad CH_2 \qquad \qquad \qquad \qquad \qquad (CH_2)_2$$
$$| \qquad \qquad \qquad | \qquad \qquad \qquad \qquad \qquad \qquad \qquad |$$
$$H-C-NH_2 \qquad COOH \qquad \qquad \qquad \qquad H-C-NH_2$$
$$| \qquad \qquad \qquad \qquad \qquad \qquad \qquad \qquad \qquad \qquad |$$
$$COOH \qquad \qquad \qquad \qquad \qquad \qquad \qquad \qquad COOH$$

arginine succinic acid   arginine   fumaric acid

Finally, arginine is split hydrolytically by the liver enzyme arginase into ornithine and urea. The ornithine then can go through the cycle again and the urea is excreted.

$$\begin{array}{c} NH \\ || \\ CH_2-NH-C-NH_2 \end{array} \qquad \qquad \qquad \begin{array}{c} CH_2-NH_2 \end{array}$$
$$| \qquad \qquad \qquad \qquad \qquad \qquad \qquad \qquad \qquad \qquad (CH_2)_2 \qquad \qquad O$$
$$(CH_2)_2 \qquad \qquad \xrightarrow{\quad \text{arginase} \quad} \qquad | \qquad \qquad \qquad \qquad ||$$
$$| \qquad \qquad \qquad \qquad \qquad \qquad \qquad \qquad \qquad H-C-NH_2 + H_2N-C-NH_2$$
$$H-C-NH_2 \qquad \qquad \qquad \qquad \qquad \qquad \qquad |$$
$$| \qquad \qquad \qquad \qquad \qquad \qquad \qquad \qquad \qquad COOH$$
$$COOH$$

arginine   ornithine   urea

To summarize, the Krebs-ornithine cycle is shown diagrammatically on the following page (note that some fumaric acid may be converted back to aspartic acid and some enters the citric acid cycle).

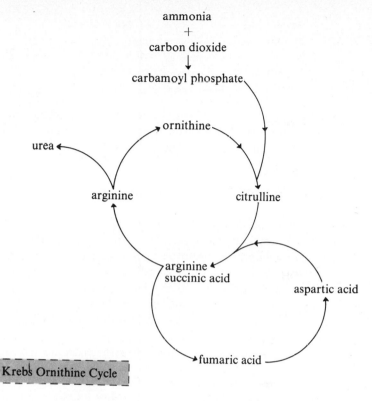

Krebs Ornithine Cycle

The blood picks up the urea from the liver and carries it to the kidneys where it is excreted in the urine. Urea is the principal end product of protein metabolism and contains a large percentage of the total nitrogen excreted by the body.

Decarboxylation

The decarboxylation (removal of a —COOH group) of an amino acid yields a primary amine. The carboxyl group that is removed is converted to carbon dioxide. The enzyme involved in a decarboxylation reaction requires pyridoxal phosphate as a coenzyme. The decarboxylation reaction may be summarized as follows

$$
\underset{\substack{\alpha\text{-amino acid}}}{R-\underset{\substack{| \\ NH_2}}{\overset{\substack{H \\ |}}{C}}-COOH} \quad \xrightarrow[\text{pyridoxal phosphate}]{\text{amino acid decarboxylase}} \quad \underset{\substack{\text{primary amine}}}{R-CH_2-NH_2 + CO_2}
$$

Several naturally occurring amines are formed by the decarboxylation of amino acids. The following list includes some of these.

Amino Acid	decarboxylation	Primary Amine
histidine	$\longrightarrow$	histamine
lysine	$\longrightarrow$	cadaverine
ornithine	$\longrightarrow$	putrescine
tyrosine	$\longrightarrow$	tyramine

Decarboxylation reactions are brought about by intestinal bacterial attack on amino acids, to produce toxic amines called ptomaines (see page 352). This process is common in the spoilage of food protein.

Metabolism of Hemoglobin

A red blood cell has a life-span of about 120 days. After that period of time, the hemoglobin is catabolized. The globin (protein) part is metabolized as is any other protein. The heme is metabolized and excreted as waste products, but the iron is reused. The normal diet supplies about 12 to 15 mg of iron per day, but of this amount only about 1 mg per day may be absorbed. When hemoglobin is metabolized, 20 to 25 mg of iron are released per day. This amount must be reused or else the body would suffer a serious loss of iron.

When an erythrocyte (a red blood cell) ruptures, the hemoglobin ring is broken and the products formed are globin, ferrous ions, and biliverdin, a blue-green pigment. This process takes place in the reticuloendothelial cells of the liver, spleen, and bone marrow. Biliverdin is rapidly reduced to bilirubin, an orange-yellow pigment, by the enzyme bilirubin reductase, also in the reticuloendothelial cells. From there the bilirubin is transported to the liver as a bilirubin-albumin complex with the aid of serum albumin. In the liver, bilirubin is converted to bilirubin diglucuronide, which is then excreted into the bile. The bile flows into the small intestine. In the small intestine the bilirubin diglucuronide is changed to stercobilinogen and then to stercobilin for excretion into the stool and also into urobilinogen and then to urobilin for excretion into the urine.

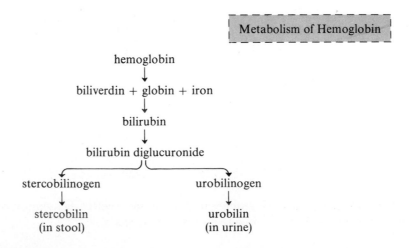

Metabolism of Hemoglobin

hemoglobin
↓
biliverdin + globin + iron
↓
bilirubin
↓
bilirubin diglucuronide
↓   ↓
stercobilinogen   urobilinogen
↓   ↓
stercobilin   urobilin
(in stool)   (in urine)

Jaundice is the condition in which abnormal amounts of bilirubin accumulate in the blood. Patients with jaundice exhibit a characteristic yellow-colored skin, a color due to the presence of bilirubin.

If hemolysis takes place at an abnormally high rate, so that bilirubin accumulates in the blood, the condition is termed *hemolytic jaundice*. If the bile duct is obstructed so that bile cannot enter the intestinal tract, bilirubin again accumulates in the blood. This condition is termed *obstructive jaundice* and is characterized by white or clay-colored stools because decomposition products of bilirubin are not present. If the liver is damaged in such diseases as infectious hepatitis or cirrhosis, bilirubin cannot be removed and a jaundiced condition results.

Metabolism of Nucleoproteins

As was discussed in Chapter 23, nucleoproteins are composed of proteins conjugated with nucleic acids. DNA and RNA, two nucleoproteins, are essential constituents of chromosomes, viruses, and cell nuclei, as well as being involved in protein synthesis.

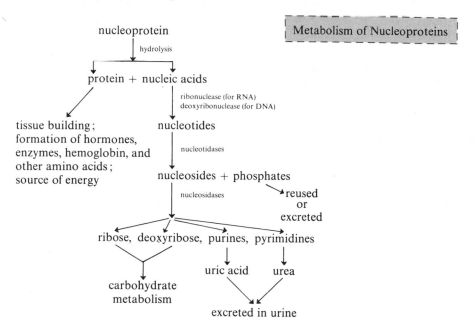

Metabolism of Nucleoproteins

Metabolism of Proteins, Carbohydrates, and Lipids

The metabolism of protein produces compounds, such as pyruvic acid, that can enter the citric acid cycle, other products that can enter the

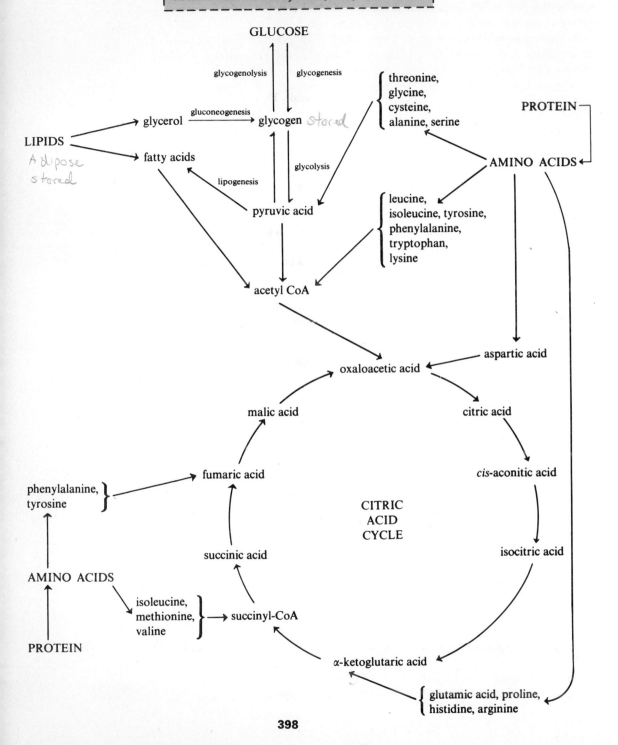

Metabolism of Carbohydrates, Fats, and Proteins

GLUCOSE

glycogenolysis · glycogenesis

threonine, glycine, cysteine, alanine, serine

PROTEIN

gluconeogenesis

glycerol → glycogen *Stored*

LIPIDS

Adipose stored

fatty acids

AMINO ACIDS

glycolysis

lipogenesis

pyruvic acid

leucine, isoleucine, tyrosine, phenylalanine, tryptophan, lysine

acetyl CoA

aspartic acid

oxaloacetic acid

malic acid

citric acid

fumaric acid

cis-aconitic acid

CITRIC ACID CYCLE

phenylalanine, tyrosine

isocitric acid

succinic acid

AMINO ACIDS

isoleucine, methionine, valine

succinyl-CoA

PROTEIN

α-ketoglutaric acid

glutamic acid, proline, histidine, arginine

398

glycogen-forming cycle, and still others that may enter the lipogenesis cycle. The interrelationships of the metabolism of carbohydrates, lipids, and proteins are indicated in the diagram on page 398.

Summary

During digestion, proteins are hydrolyzed into amino acids. These amino acids may be used for the synthesis of new tissue, for the replacement of old tissue, and for the formation of enzymes and hormones, or they may be oxidized to furnish energy. The body cannot store protein (amino acids).

A person who excretes as much nitrogen daily as he takes in is said to be in nitrogen balance. Children have a positive nitrogen balance because they need extra protein for growth and so excrete less. Malnutrition and prolonged fever may lead to a negative nitrogen balance.

Amino acids that the body cannot synthesize and that must be supplied in the food are called essential amino acids. A protein that does not contain all the essential amino acids is called an incomplete protein.

For the normal adult, 46 to 56 g of protein are required per day.

Amino acids that the body does not need are catabolized into carbon dioxide and water. This catabolism may be oxidative deamination (deamination) whereby the alpha-amino group is removed, forming an alpha-keto acid and ammonia. The alpha-keto acid may then be catabolized to carbon dioxide and water and energy through the citric acid cycle; it may be converted to glycogen or to fat; or it may undergo transamination, whereby a different amino acid is formed.

The ammonia formed from the deamination of an amino acid unites with carbon dioxide and water to form urea. This process, which takes place in the liver, proceeds through a cycle called the Krebs ornithine cycle.

Amino acids may also be decarboxylated to form primary amines.

When hemoglobin is metabolized, the protein part is metabolized as is any other protein. The iron is used over again and the remaining heme part goes through a series of steps eventually ending up as urobilinogen and urobilin in the urine and stercobilinogen and stercobilin in the feces.

When excessive amounts of bilirubin, one of the intermediate products of the metabolism of hemoglobin, accumulate in the blood, the condition is known as jaundice.

The end product of the metabolism of purines is uric acid, which is excreted in the urine.

The metabolism of amino acids is interrelated to that of carbohydrates and fats.

Questions and Problems

1. What are the functions of protein in the body?
2. Can the body store protein? Carbohydrate? Fat?
3. What is meant by the term *nitrogen balance*?
4. What might cause a positive nitrogen balance?
5. What might cause a negative nitrogen balance?
6. What is an essential amino acid? List the essential amino acids.

7. What is an incomplete protein?
8. What is the normal daily requirement of protein?
9. What is deamination? Where does it occur in the body?
10. What products are produced by a deamination reaction?
11. What is transamination? What is its function in the body?
12. What various processes may an alpha-keto acid undergo?
13. Where is urea formed?
14. Diagram and explain the ornithine cycle for the production of urea.
15. What is decarboxylation? What type of product is produced by the decarboxylation of an amino acid?
16. What are ptomaines? How are they produced?
17. What is biliverdin? Where is it formed?
18. What are the end products of the metabolism of hemoglobin? Which are found in the urine? In the feces?
19. What is jaundice? What may cause it?
20. Indicate the relationship between the metabolism of carbohydrates, lipids, and proteins.
21. Describe the hydrolysis of nucleoprotein. What happens to the hydrolysis products?

References

Baum, S. J., and Scaife, C. W.: *Chemistry: A Life Science Approach.* Macmillan Publishing Co. Inc., New York, 1975, Chap. 30.

Bronk, J. R.: *Chemical Biology.* Macmillan Publishing Co., Inc., New York, 1973, Chap. 10.

Harper, H. A.: *Review of Physiological Chemistry*, 15th ed. Lange Medical Publications, Los Altos, Calif., 1975, Chaps. 15, 16.

Janick, J.; Noller, C. H.; and Rhykerd, C. L.: The cycles of plant and animal metabolism. *Scientific American*, **235**:75–86 (Sept.), 1976.

Lehninger, A. H.: *Biochemistry*, 2nd ed. Worth Publishers, Inc., New York, 1975, Chap. 21.

Kappas, A., and Albares, A. P.: How the liver metabolizes foreign substances. *Scientific American*, **232**:22–31 (June), 1975.

Montgomery, R.; Dryer, R. L.; Conway, T. W.; and Spector, A. A.: *Biochemistry.* C. V. Mosby Co., St. Louis, 1974, Chaps. 9, 11.

Stetten, D.: Gout and metabolism. *Scientific American*, **198**:73–81 (June), 1958.

Body Fluids: Urine

Excretion of Waste Material

The waste products of the body are excreted through the lungs, the skin, the intestines, and the kidneys. The liver also excretes waste products—the bile pigments and cholesterol.

The lungs eliminate water and carbon dioxide through the expired air. The skin eliminates water in the form of perspiration. Included in the perspiration are small amounts of inorganic and organic salts. The feces, excreted from the large intestine, contain undigested and undigestible material plus the excretory products from the liver—the bile pigments and cholesterol—some water, and some organic and inorganic salts. The primary excretory organs of the body, however, are the kidneys, which excrete water and water-soluble compounds including nitrogen compounds from the catabolism of amino acids.

401

The kidneys are important not only for their excretory function but also because they play an important role in the control and regulation of water, electrolyte, and acid-base balance in the body.

Formation of Urine

Blood flows to the kidneys through the renal arteries. From the renal arteries the blood passes into the arterioles and then into the capillaries of the kidneys. These capillaries coil up to form a *glomerulus*, a rounded ball of capillaries. Around the glomerulus is a structure called *Bowman's capsule* (see Figure 29–1). Each Bowman's capsule is connected by a tubule to a larger tube, which in turn carries the urine to the bladder where it is stored until it is excreted.

As blood flows into the kidney, the various soluble components diffuse into the glomeruli (there are over a million glomeruli in each kidney). The protein material in the blood cannot pass through the membranes (recall that proteins are colloids and colloids do not pass through membranes). The driving force for this diffusion of fluid through the walls of the glomerulus is the blood pressure. The liquid in the glomerulus thus has approximately the same composition as does the blood plasma except for the protein material.

As the fluid in the glomerulus passes down the tubule, a large proportion of the water is reabsorbed into the bloodstream. Also reabsorbed are the glucose, amino acids, and most inorganic ions. The remaining liquid, containing urea and other waste products, flows to the collecting tubules and then to the bladder.

Thus, the kidneys act as a very efficient filter removing the waste materials but not the needed nutrients from the blood. Approximately 1 liter of blood is filtered through the kidneys every minute. Of this amount, most of the water is reabsorbed so that the amount of urine excreted per day is less than

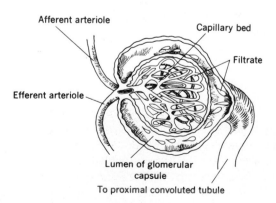

Figure 29–1. Glomerulus and Bowman's capsule in the kidney.

1 per cent of the total amount of liquid filtered. If the kidneys are not functioning normally, an artificial kidney machine may be used (see page 149).

General Properties of Urine

Volume

A normal adult excretes 600 to 2500 ml of urine per day. The amount depends on the liquid intake and also on the weather conditions. In hot weather, more water is lost through perspiration so that the amount of urine formed is less. Conversely, a greater amount of urine is formed during cold weather or when the humidity is high so that little evaporation of perspiration takes place. Drugs, such as caffeine (in coffee or tea) and also alcoholic beverages have a diuretic effect—that is, they increase the flow of urine.

A decreased flow of urine is called *oliguria*. Such a condition may occur during a high fever when most of the water lost by the body is in the form of perspiration. Certain kidney diseases may also cause oliguria.

Anuria means a total lack of urine excretion. Anuria indicates extensive kidney damage such as may be caused by a blood transfusion of the wrong type. In this condition the blood cells disintegrate, releasing hemoglobin which clogs the glomeruli and does not allow any excretion of urine. Bichloride of mercury also affects the kidneys and may cause oliguria or anuria.

Polyuria is a condition in which the amount of urine excreted is much greater than normal. It may be due to excessive intake of water or to certain pathologic conditions. Polyuria may be caused by such diuretics as alcohol or caffeine. Urea, a normal constituent of urine, is also a diuretic. A person on a high-protein diet will excrete more urea, which in turn causes the formation of more urine.

During diabetes insipidus the hormone vasopressin, which controls the reabsorption of water in the kidneys, is lacking or deficient, so that the amount of urine is greatly increased, sometimes as high as 30 liters per day.

Patients with diabetes mellitus show a definite polyuria. This excessive water loss may lead to dehydration.

Specific Gravity

The specific gravity of the urine depends upon the concentration of the solutes. The greater the concentration of the solutes, the greater the specific gravity. A normal range of the specific gravity of the urine is 1.003 to 1.030. In cases of diabetes mellitus, the specific gravity will be higher because of a high concentration of sugar in the urine. In cases of diabetes insipidus, the specific gravity of the urine will be very low (close to 1.000), owing to the large amounts of water being excreted.

pH

Urine is normally slightly acidic with a pH range of 4.7 to 8.0 and an average value of about 6.0. However, the pH of the urine varies with the diet.

Protein foods, such as meats, increase the acidity of the urine (lower the pH) because of the formation of phosphates and sulfates. The acidity of the urine is also increased during acidosis and with fever.

Conversely, the urine may tend to become alkaline on a diet high in vegetables and fruits or because of alkalosis, a condition that may be produced by excess vomiting.

Color

Normal urine is pale yellow or amber. The color, however, varies with the amount of urine produced and also with the concentration of the solutes in the urine. The larger the volume of urine excreted, the lighter the color. The greater the concentration of solutes, the darker the color. The color of the urine is due to urobilin and urobilinogen (see page 396). Various other components of the urine may cause it to have different colors. The presence of blood in the urine gives it a reddish color. Homogentisic acid (an intermediary in the metabolism of phenylalanine and tyrosine) colors the urine brown. The drug methylene blue colors the urine green.

Freshly voided urine is clear and usually contains no sediment. However on standing it may become cloudy and develop sediment because of the precipitation of calcium phosphate.

Odor

Fresh urine has a distinctive odor, but this odor may be modified by the presence of other substances. In patients having ketosis the odor of acetone may be detected. Diet may also modify the odor of urine. For example, when asparagus is eaten, the urine may have a sulfurlike odor.

Normal Constituents

Approximately 50 to 60 g of solid material are excreted daily in the urine of the average person. This solid material may be subdivided into inorganic and organic constituents (see Table 29–1). The inorganic constituents of the urine make up approximately 45 per cent of the total solids; the organic constituents comprise the other 55 per cent.

Organic Constituents

Urea. The principal end product of the metabolism of protein, urea, comprises about one half of the total solids in the urine.

$$H_2N-\overset{\overset{\displaystyle O}{\displaystyle \|}}{C}-NH_2$$

urea

TABLE 29–1

Constituents of Urine
(Amount Excreted Per Day)

Constituent	Amount, g	Approximate Percentages
Organic		
Urea	25–30	40–50
Uric acid	0.7	1
Creatinine	1.4	2.5
Creatine	0.06–0.15	0.1–0.25
Others	0.1–1	0.1–1
Inorganic		
Chloride ion	9–16	15–25
Sodium ion	4	6
Phosphates	2	3
Sulfates	2.5	4
Ammonium ions	0.7	1
Other ions and inorganic constituents	2.5	4

Uric Acid. Uric acid, a product of the metabolism of purines (see page 397) from nucleoprotein, is only slightly soluble in water and is excreted primarily as urate salts. The structure of uric acid is shown below.

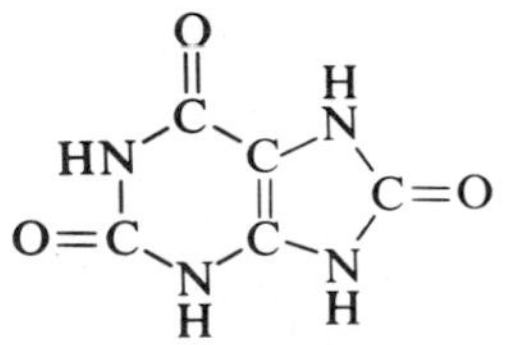

uric acid

When urine is allowed to stand, uric acid may crystallize and settle out, since it is only slightly soluble in water or acid solution.

The average daily excretion of uric acid is about 0.7 g but an increase in nucleoproteins in the diet will cause an increased excretion of uric acid.

The output of uric acid will be increased in leukemia, in severe liver disease, and in various stages of gout. Deposits of urates and uric acid in the joints and tissues are also characteristic of gout, so that this disease appears to be a form of arthritis (see Figure 29–2).

Under certain conditions uric acid or urates crystallize in the kidneys and are called kidney stones, or calculi (Figure 29–3).

Creatinine. Creatinine is a product of the breakdown of creatine. The amount of creatinine excreted per day is fairly constant regardless of the protein intake. The number of milligrams of creatinine excreted in the urine within a 24-hr period per kilogram of body weight is the *creatinine coefficient*

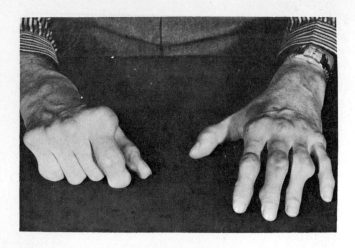

Figure 29–2. The ravaging effects of gout are manifest in the deformed hands of a patient afflicted with this crippling disease. (Courtesy of the National Institute of Arthritis and Metabolic Diseases, National Institutes of Health, Bethesda, Md.)

of that individual. The creatinine coefficient of the normal male is 20 to 26; that of females is 14 to 22.

$$
\begin{array}{c}
\underset{\overset{\|}{C}-NH_2}{\overset{NH}{\|}} \\
\underset{\overset{|}{CH_3}}{N-CH_2-COOH}
\end{array}
\longrightarrow
\begin{array}{c}
\underset{\overset{\|}{C}\text{————}NH}{\overset{NH}{\|}} \\
\underset{\overset{|}{CH_3}}{N-CH_2-C=O} + H_2O
\end{array}
$$

creatine   creatinine

The average adult excretes 1.4 g of creatinine daily and about 0.06 to 0.15 g of creatine in the same period of time. Formation of creatinine appears necessary for the excretion of most of the creatine.

Creatine. Creatine is produced in the body from three amino acids—arginine, methionine, and glycine. Creatine is normally present in muscle, brain, and blood, both as free creatine and as creatine phosphate. Recall the reaction

$$\text{creatine-P} + \text{ADP} \longrightarrow \text{creatine} + \text{ATP}$$

Creatinuria is a condition in which abnormal amounts of creatine occur in the urine. It may occur during starvation, diabetes mellitus, prolonged fevers, wasting diseases, and hyperthyroidism. Creatinuria may also occur in pregnancy.

Other Organic Constituents. Also present in small amounts are amino acids, alantoin (from partial oxidation of uric acid), hippuric acid, urobilin, and other pigments. The presence of cyclic-AMP (see page 323) helps diagnose parathyroid function. Patients with hyperparathyroidism excrete significantly more cyclic-AMP and those with hypoparathyroidism excrete

406

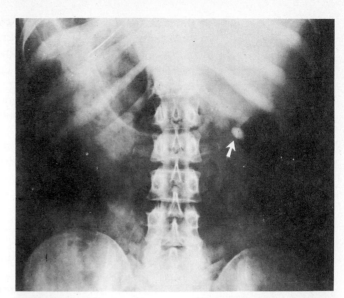

A

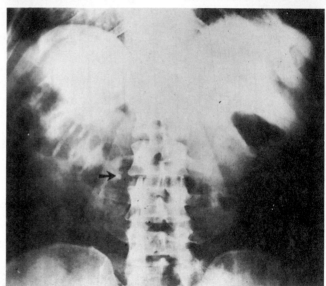

B

Figure 29–3. *A*. Stone in right kidney. *B*. Stone in left ureter. (Courtesy of X-ray Department, Michael Reese Hospital and Medical Center, Chicago, Ill.)

significantly less cyclic-AMP than persons with normal parathyroid glands (see page 476).

Inorganic Constituents

The inorganic constituents of urine are the various positive and negative ions that make up the inorganic compounds being excreted. Among these are the following.

407

Chloride Ions. Between 9 and 16 g of chloride ion are excreted daily, mostly as sodium chloride. The amount of chloride ion varies with the intake, which is primarily sodium chloride. The excretion of sodium chloride is decreased in fevers and in some stages of nephritis.

Sodium Ions. The amount of sodium ion excreted varies with the intake and the body's requirement. However it is usually about 4 g per day.

Phosphates. The amount of phosphates present in the urine also depends upon the diet; the amount is higher when the diet contains foods high in phosphorus (nucleoproteins and phospholipids). An increase in excreted phosphates is found in certain bone diseases and in hyperparathyroidism. A decrease in phosphates is found in hypoparathyroidism, in renal diseases, and during pregnancy.

Sulfates. The sulfates in the urine are derived from the metabolism of sulfur-containing proteins, so that the amount of sulfur is influenced by the diet. Sulfates are found both as organic and inorganic salts.

Ammonium Ions. The hydrolysis of urea produces ammonium ions in the urine. Ammonium compounds are present as chlorides, sulfates, and phosphates (see page 395).

Other Ions. In addition to the sodium and ammonium ions, other positive ions present in the urine are calcium, potassium, and magnesium.

Other Organic Constituents. In addition to these constituents, the urine also normally contains very small amounts of vitamins, hormones, and enzymes. Urinalysis for these substances is of diagnostic value.

Abnormal Constituents

Protein

Because proteins are colloids and because colloids cannot pass through membranes, urine should not normally contain protein. *Proteinuria* denotes the presence of protein in the urine. Sometimes it is termed *albuminuria* because albumin is the smallest plasma protein and is the protein most frequently found in the urine. In cases of kidney disease, such as nephritis and nephrosis, and in severe heart disease, protein appears in the urine. The presence of protein due to such disorders is frequently called renal proteinuria or renal albuminuria to distinguish it from false albuminuria, which is a temporary harmless condition. False albuminuria, often called orthostatic albuminuria, is found in certain patients who stand for a long period of time. It is due to the constriction of the kidneys' blood vessels and disappears when the patient lies down. Small amounts of protein may also be found in the urine after severe muscular exercise, but they soon disappear.

The tests for the presence of protein in the urine are based on the fact that protein coagulates when heated. When a sample of urine is heated, any protein (albumin) present will precipitate out as a white cloud. However, phosphates may also precipitate when the urine is heated. To prove that the cloudy substance is albumin, the urine, after heating, is acidified with dilute acetic acid. The acid will dissolve the phosphates but not the protein so that a cloudy precipitate in the urine after heating and acidification is a verification of the presence of protein.

Glucose

The presence of glucose in the urine is called *glycosuria*. Normally there is always a very small amount of glucose present in the urine, but this amount is too small to give a positive test with Benedict's solution.

Glucose may be found in the urine after severe muscular exercise, but this condition clears up when the body returns to normal. Glucose may also be found in the urine after a meal high in carbohydrates.

Glycosuria may be due to such diseases as diabetes mellitus or renal diabetes or to liver damage.

Other Sugars

Lactose and galactose may occur in the urine during pregnancy and lactation. Both of these sugars give a positive Benedict test.

Pentoses may occur in the urine after eating foods such as plums, grapes, and cherries which contain large amounts of these carbohydrates.

Acetone Bodies

Acetone (ketone) bodies are present in the urine during diabetes mellitus and in starvation. The excretion of the acetone bodies, which are acidic compounds, requires alkaline compounds. This results in a depletion of the alkaline reserve of the blood and leads to acidosis. The kidneys produce more ammonia to neutralize these acetone bodies, thus saving sodium and potassium ions for the control of the pH in the blood buffers.

The test for the presence of acetone bodies in the urine is performed by adding sodium nitroprusside to a sample of urine and then making the mixture alkaline with ammonium hydroxide. The presence of acetone bodies is indicated by a pink-red color. Normal urine gives no color with this test.

Blood

The presence of blood in the urine is called *hematuria*. It may result from lesions or stones in the kidneys or urinary tract. The presence of free hemoglobin in the urine, *hemoglobinuria*, results from hemolysis of the red blood cells caused by an injection of hypotonic solution, severe burns, or blackwater fever.

Large amounts of blood in the urine may be detected by the reddish color. Small amounts do not color the urine enough to show any color change but may be detected by adding benzidine and hydrogen peroxide to the urine. The presence of blood is indicated by the appearance of a blue color.

Bile

Normally, bile is excreted by the liver into the small intestine and eventually ends up in the feces. The presence of bile in the urine indicates obstruction to the flow of bile to the intestines. Bile in the urine is indicated by a greenish-brown color. Bile in the urine is also indicated by the presence of a yellow foam when the urine is shaken.

Phenolsulfonphthalein Test

Phenolsulfonphthalein (PSP) is a red dye used to test how well the kidneys are functioning. The dye is administered by intravenous or intramuscular injection, usually the former. Urine specimens are collected at frequent intervals—after 15, 30, 60, and 120 min. If the 15-min specimen contains 25 per cent or more of PSP, then the kidneys are functioning normally; 40 to 60 per cent of the dye should be excreted within 1 hr; 20 to 25 per cent more in the second hour.

Other kidney-function tests are based upon the amount of urea eliminated in the urine compared to the amount present in the blood (urea clearance test) or upon the change in the specific gravity of the urine after the patient's fluid intake is restricted (concentration test). Mannitol is also used as a diagnostic agent to measure kidney function.

Summary

The principal excretory organs of the body are the kidneys, which also control and regulate the water balance, electrolyte balance, and pH of body fluids.

The waste materials in the blood are picked up by the kidneys and are excreted in the urine.

Approximately 600 to 2500 ml of urine are excreted daily, the amount depending on fluid intake, weather conditions, humidity, and certain diuretic substances.

A decreased flow of urine is called oliguria; anuria is a total lack of urine; polyuria is excess urine formation.

The specific gravity of the urine varies between 1.003 and 1.030 and the pH ranges from 4.7 to 8.0, with an average value of 6.0. Urine is normally pale yellow or amber, but certain components may cause another color.

Approximately 50 to 60 g of solid material, both organic and inorganic, are excreted daily. The principal organic constituent of urine is urea, the end product of protein metabolism. Another important constituent is uric acid, a product of the metabolism of nucleoproteins. Gout is characterized by an increase of uric acid in the urine and blood and the deposition of uric acid or urate salts in the joints and tissues. Uric acid and urates may also crystallize in the kidneys as kidney stones.

The average adult excretes 0.06 to 0.15 g of creatine and 1.4 g of creatinine daily. Creatine is produced from three amino acids—arginine, methionine, and glycine. Creatinine is produced from creatine by a dehydration reaction. Creatinuria is the condition in which abnormal amounts of creatine occur in the urine.

Inorganic constituents of urine are the following ions: chloride, sodium, phosphate, sulfate, ammonium, and some small amounts of calcium, potassium, and magnesium.

Abnormal constituents in the urine are protein (proteinuria, or albuminuria) due to kidney disease; glucose (glycosuria) due to diabetes mellitus or liver damage; acetone bodies due to diabetes mellitus or starvation; blood (hematuria) due to lesions in the kidneys or urinary tract; and bile due to an obstruction to the flow of bile to the intestines.

Various tests may be performed to see if the kidneys are functioning normally. Among these is the phenolsulfonphthalein test (PSP) in which a red dye is administered intravenously and the amount of color measured in the urine at specified intervals of time.

Questions and Problems

1. How does the body excrete waste products?
2. What are the principal excretory organs of the body? What additional functions do they have?
3. What is a glomerulus? Bowman's capsule?
4. Describe the formation of urine. Where is the urine stored?
5. What forces the fluids through the membranes in the kidneys?
6. What happens to the nutrients and the water that also filter through the membranes in the kidneys?
7. What volume of urine does an adult excrete daily? What might affect this amount?
8. What is a diuretic?
9. What is oliguria? What might cause this condition?
10. What is anuria? What might cause this condition?
11. What might cause polyuria?
12. What does the hormone vasopressin do? If this hormone is lacking, what will be the effect on the body?
13. What is the pH range of urine? What might affect the pH of the urine?
14. What might affect the color of the urine?
15. What might cause urine to become cloudy upon standing?
16. How much solid material is excreted daily in the urine?
17. What is the principal organic constituent of the urine? Where does it come from? What is its structural formula?
18. Where does the uric acid in the urine come from?
19. Draw the structure of uric acid.
20. How much uric acid is excreted daily? What might affect this amount?

21. What is gout? What other conditions may cause an increased amount of uric acid in the urine?
22. What might cause kidney stones?
23. What is the creatinine coefficient?
24. How does the structure of creatine compare with that of creatinine?
25. Where does the body obtain its creatine?
26. Where is creatine normally present in the body?
27. What is creatinuria and what causes it?
28. Name several inorganic ions normally found in the urine.
29. How is the amount of phosphates in the urine affected by various diseases?
30. What is proteinuria?
31. What might cause albuminuria?
32. What is false albuminuria?
33. Describe the test for the presence of protein in the urine. Is it necessary to acidify the urine during this test? Why?
34. What is glycosuria? What causes this condition?
35. Could other sugars besides glucose be present in the urine? Under what conditions?
36. What might cause the presence of acetone (ketone) bodies in the urine?
37. Describe the test for the presence of acetone bodies in the urine.
38. What is hematuria? What might cause this condition?
39. The presence of bile in the urine might indicate what condition?
40. Describe the PSP test for kidney function.

References

Harper, H. A.: *Review of Physiological Chemistry*, 15th ed. Lange Medical Publications, Los Altos, Calif., 1975, Chap. 18.

Mazur, A., and Harrow, B.: *Textbook of Biochemistry*, 10th ed. W. B. Saunders Co., Philadelphia, 1971, Chap. 16.

Newton, M., and Leblond, C. P.: The Golgi apparatus. *Scientific American*, **220**: 100–107 (Feb.), 1969.

Body Fluids: The Blood

Functions

Blood has been called a circulating tissue. It carries oxygen, minerals, and
food to the cells and carries carbon dioxide and other waste products away

from the cells. It also carries hormones, enzymes, and blood cells. Blood regulates body temperature by carrying heat from the interior to the surface capillaries. The blood buffers maintain the pH of the body at its optimum value. Blood contains a clotting system that protects the body against hemorrhage and a defense mechanism against infection (the antibodies).

Composition

Blood consists of two parts—the suspended particles and the suspending liquid, the plasma. The suspended particles in the blood are the red blood cells, white blood cells, and platelets.

Red Blood Cells (Erythrocytes)

Normally there are 4.5 to 5.0 million red blood cells per cubic millimeter of blood. Since there are about 6 quarts of blood in the human body, the total number of red blood cells is approximately 30 trillion (30,000,000,000,000). An excess of red blood cells is called *polycythemia*; a shortage is called *anemia*. Each day 200 billion new red blood cells are formed in the bone marrow.

Erythropoietin, a hormone that stimulates red blood cell formation, is a glycoprotein with a molecular weight of approximately 35,000. Erythropoietin is formed by the action of a substance produced by the kidneys (renal erythropoietic factor) on a globulin in the blood plasma. The production of this hormone is increased by hypoxia in the kidneys, by cobalt salts, and by androgens. If the kidneys do not function properly, the patient may become anemic (see page 418).

White Blood Cells (Leukocytes)

The number of white blood cells normally present in the blood ranges from 5 to 10 thousand per cubic millimeter. White blood cells are larger than red blood cells. White blood cells have a nucleus; red blood cells do not. There are several types of white blood cells; among these are the basophils, lymphocytes, monocytes, eosinophils, and neutrophils.

The white blood cells attack and destroy harmful microorganisms and thus serve as one of the body's defenses against infection. A white blood count above normal usually indicates an infection. For diagnostic purposes, a *differential count* is sometimes ordered. This count gives the percentages of each of the various types of leukocytes present.

Platelets (Thrombocytes)

The number of thrombocytes or platelets ranges from 250 to 400 thousand per cubic millimeter of blood. Platelets are smaller than red blood cells and do not have a nucleus. Platelets contain cephalin (phosphatidyl ethanol-amine), a phospholipid that is involved in the clotting of the blood.

Blood Plasma

Approximately 92 per cent of the plasma is water. The solids dissolved or colloidally dispersed in the blood make up the other 8 per cent. The most important of all plasma solids are the blood proteins. These include serum albumin, the globulins, and fibrinogen.

The plasma proteins (primarily albumin) maintain the osmotic pressure of the blood, thus regulating the water and acid-base balance in the body. The globulins (there are several) include the antibodies, which function against infection and diseases. Globulins also function in the transportation of lipids, steroids, and hormones in the plasma. Two particular globulins, transferrin and ceruloplasmin, transport iron and copper, respectively, in the plasma. Fibrinogen and prothrombin, another of the globulins, function in the clotting of the blood.

Blood plasma also contains a small amount of lipids and carbohydrates (glucose), inorganic salts, waste products (such as urea, uric acid, carbon dioxide, creatinine, and ammonia), enzymes, vitamins, hormones, and antibodies.

The inorganic ions present in the blood plasma serve to regulate the acid-base balance of the body.

On standing, freshly drawn blood soon forms a clot. When the clot settles, a yellowish liquid remains. This liquid is called *blood serum*. Blood serum is blood without the blood cells and without the fibrinogen necessary for the clotting of the blood. Blood plasma may be separated from the solid parts of the blood by centrifuging, during which the cells settle at the bottom of the test tube and the plasma remains above them. To keep the blood from clotting, an anticoagulant must be added before centrifuging.

General Properties of Blood

Oxygenated blood has a characteristic bright red color; deoxygenated blood has a dark purplish color.

The specific gravity of whole blood ranges from 1.054 to 1.060, while that of blood plasma ranges from 1.024 to 1.028.

Blood is normally slightly alkaline with a pH range of 7.35 to 7.45. If the pH of the blood falls slightly below 7.35, the condition is termed *acidosis*. If the pH of the blood rises slightly above 7.45, the condition is termed *alkalosis*. If the pH of the blood changes more than a few tenths from the normal values, the results are usually fatal.

The viscosity of the blood is approximately 4.5 times that of water and varies according to the number of cells, the quantity of protein, the temperature, and the amount of water present in the body.

TABLE 30–1

Normal Composition of Blood

Determination	Normal Range* (per 100 ml)	Clinical Significance	
		Increased in	Decreased in
Calcium	9–11 mg (4.5–5.5 mEq/L)†	Hyperparathyroidism, Addison's disease, malignant bone tumor, hypervitaminosis D	Hypoparathyroidism, rickets, malnutrition, diarrhea, chronic kidney disease, celiac disease
Cholesterol, total	150–280 mg	Diabetes mellitus, obstructive jaundice, hypothyroidism, pregnancy	Pernicious anemia, hemolytic jaundice, hyperthyroidism, tuberculosis
Uric acid	3–7.5 mg	Gout, leukemia, pneumonia, liver and kidney disease	
Urea nitrogen	8–20 mg	Mercury poisoning, acute glomerulonephritis, kidney disease	Pregnancy, low-protein diet, severe hepatic failure
Nonprotein nitrogen	15–35 mg	Kidney disease, pregnancy, intestinal obstruction, congestive heart failure	Low-protein diet
Creatine	3–7 mg	Nephritis, renal destruction, biliary obstruction, pregnancy	
Creatinine	0.7–1.5 mg	Nephritis, chronic renal disease	
Glucose	60–100 mg	Diabetes mellitus, hyperthyroidism, infections, pregnancy, emotional stress, after meals	Starvation, hyperinsulinism, Addison's disease, hypothyroidism, extensive hepatic damage
Chlorides	100–106 mEq/L†	Nephritis, anemia, urinary obstruction	Diabetes, diarrhea, pneumonia, vomiting, burns
Phosphorus, inorganic	3–4.5 mg	Hypoparathyroidism, Addison's disease, chronic nephritis	Hyperparathyroidism, diabetes mellitus
Sodium	136–145 mEq/L†	Kidney disease, heart disease, pyloric obstruction	Vomiting, diarrhea, Addison's disease, myxedema, pneumonia, diabetes mellitus
Potassium	2.5–5 mEq/L†	Addison's disease, oliguria, anuria, tissue breakdown	Vomiting, diabetic acidosis, diarrhea
Carbon dioxide	Adults, 24–29 mEq/L† Infants, 20–26 mEq/L	Tetany, vomiting, intestinal obstruction, respiratory disease	Acidosis, diarrhea, anesthesia, nephritis
Hemoglobin	Male, 14–18 g Female, 12–16 g	Polycythemia	Anemia

* mg/100 ml is also called mg per cent.

† mEq/L = milliequivalents per liter.

Blood Analysis

For most laboratory tests 5 ml of blood are collected from a vein in the arm before the patient is given breakfast. If blood plasma is to be used for the test, an anticlotting agent such as potassium oxalate is added to the blood sample. If blood serum is to be tested, the blood is allowed to clot and the serum poured off.

The usual blood chemistry tests, the normal ranges of the results, and the clinical significance of these tests are indicated in Table 30–1. Other tests include blood gas analysis (see Figure 30–1) and lactate/pyruvate determination.

Blood Volume

Approximately 8 to 9 per cent of the total body weight is blood. The volume of the blood in the body amounts to 3 to 7 L in the adult. The volume increases in fever and pregnancy and decreases during diarrhea and hemorrhaging.

Figure 30–1. Blood gas analysis. (Courtesy of Ayerst Laboratories, New York, N.Y.)

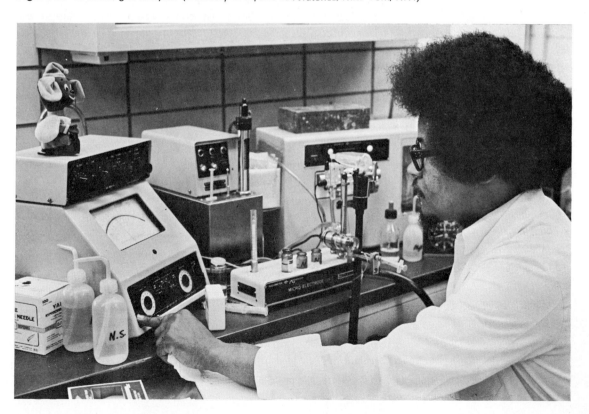

Since blood volume can be rapidly replaced, a small loss of blood because of bleeding or because of donating blood has no serious effect upon the body.

Blood volume may be determined by injecting a suitable dye into the bloodstream, waiting about 10 min, and then withdrawing a sample to determine the concentration of the dye. A radioactive tracer may also be injected into the bloodstream and the blood volume determined by the amount of dilution of the radioactivity in a sample taken a short time later.

Hemoglobin

Hemoglobin is a conjugated protein made up of a protein part, globin, and an iron-containing part, heme. Heme contains four pyrrole groups joined together with an iron ion in the center. The structure of pyrrole is

$$\text{HC} \underset{\underset{\text{HC}}{\|}}{\overset{}{\rule{2em}{0.4pt}}} \text{CH}$$

pyrrole

The structure of heme is indicated in Figure 30–2. Note that various hydrocarbon side chains are attached to the pyrrole rings in this compound. Four heme molecules combine with one globin molecule to form one molecule of hemoglobin.

The structures of cytochrome c, part of the oxidative phosphorylation sequence of the citric acid cycle (see page 367), and that of chlorophyll, a plant pigment, are shown in Figure 30–2 for comparison with that of heme. Note that while both heme and cytochrome c have an Fe in the center of four pyrrole rings, chlorophyll has an Mg.

Anemia

If the hemoglobin content of the blood falls below normal, the condition is called anemia. Anemia may result from a decreased rate of production of red blood cells, from an increased destruction of red blood cells, or from an increased loss of red blood cells.

A decreased rate of production of red blood cells may be due to various diseases that destroy or suppress the activity of the blood-forming tissues. Among these diseases are leukemia, multiple myeloma, and Hodgkin's disease. Radiation and certain drugs such as benzene and gold salts also decrease the activity of the blood-forming tissues. Another frequent cause of decreased red blood cell production is a diet lacking in iron and protein, particularly in infancy and childhood and during pregnancy. Anemia may also be related to a genetic defect that affects the production of hemoglobin (see Chapter 33).

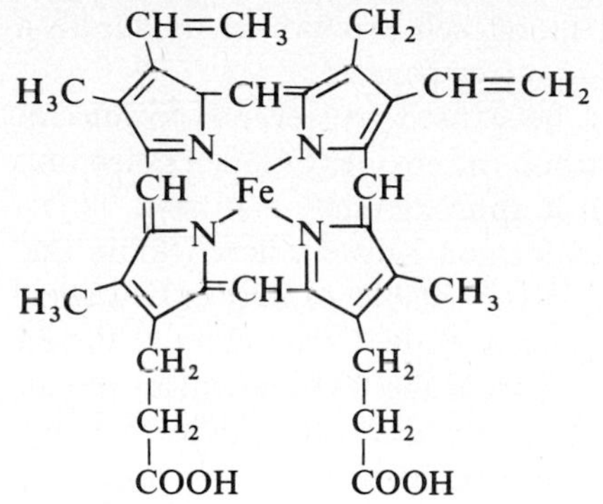

heme

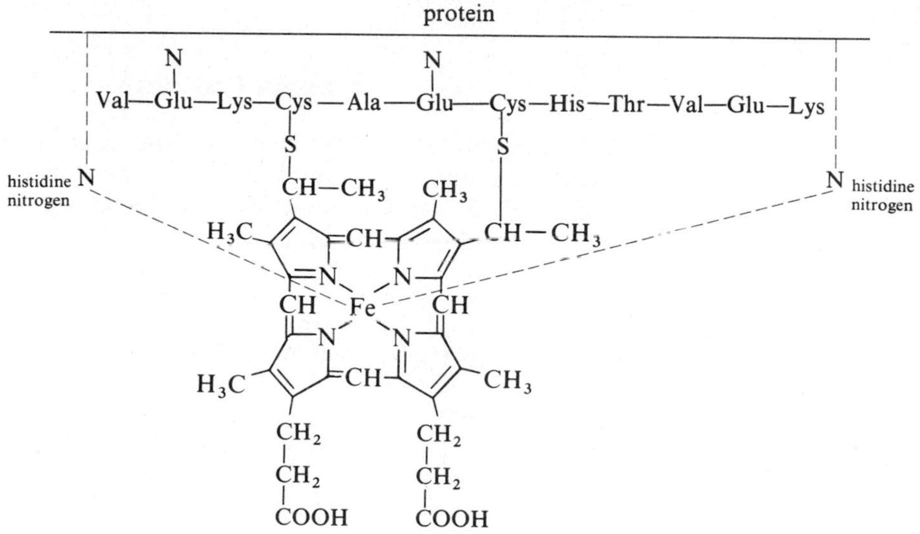

cytochrome *c*

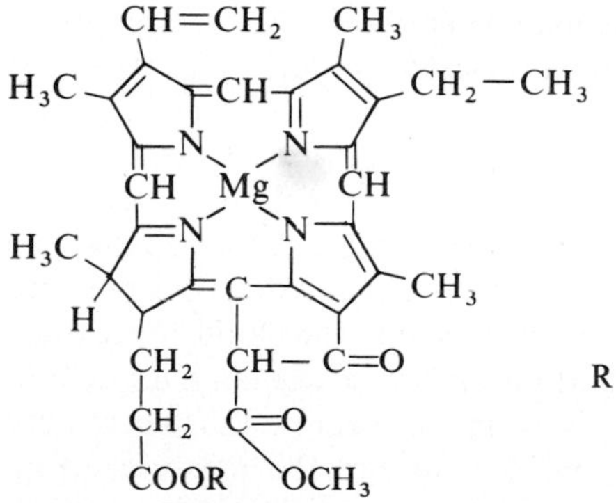

chlorophyll *a*

$$R = CH_2CH{=}\underset{CH_3}{\underset{|}{C}}{-}(CH_2)_3{-}\underset{CH_3}{\underset{|}{CH}}{-}(CH_2)_3{-}\underset{CH_3}{\underset{|}{CH}}{-}(CH_2)_3{-}\underset{CH_3}{\underset{|}{CH}}{-}CH_3$$

Figure 30–2. Structural formulas of heme, cytochrome *c*, and chlorophyll *a*.

Pernicuous anemia, a failure of red blood cell production, is due to a lack of vitamin B_{12} or of the intrinsic factor (see page 345).

Destruction of red blood cells may be caused by several poisons or infections that may cause hemolysis. Carbon monoxide (CO) is a poisonous gas because hemoglobin combines with it approximately 210 times as fast as it does with oxygen. The compound formed between hemoglobin and carbon monoxide, carboxyhemoglobin, is very stable so that only a small amount of hemoglobin is left to carry oxygen. If the CO content of the air is 0.02 per cent, nausea and headache occur; if the CO content of the air rises to 0.1 per cent, unconsciousness will occur within 1 hr and death within 4 hr. Other poisonous gases such as hydrogen sulfide (H_2S) and hydrocyanic acid (HCN) have similar effects on hemoglobin.

An increased loss of hemoglobin may be due to hemorrhaging.

Plasma Proteins

The plasma proteins constitute about 7 per cent of the plasma and are usually divided into three groups—albumin, the globulins, and fibrinogen. Approximately 55 per cent of the plasma proteins is albumin, 38.5 per cent globulins, and 6.5 per cent fibrinogen.

Albumin

Albumin in the blood functions in the regulation of the osmotic pressure of that liquid. The control of osmotic pressure in turn affects the water balance in the body.

Albumin, as well as other plasma proteins, cannot pass through the walls of the blood vessels (because they are colloids and colloids cannot pass through membranes). Since albumin is the principal plasma protein and since it is the smallest plasma protein both in size and weight, it accounts for most of the colloid osmotic pressure of the blood.

The effect of albumin (and other plasma proteins) on water balance (both filtration and reabsorption) has been hypothesized by Starling, as indicated below and also in Figure 30–3.

When blood enters the arterial end of a capillary, it exerts a hydrostatic (blood) pressure of 35 mm, forcing fluid outward from the blood vessel. At the same time, the colloid osmotic pressure of the plasma, 25 mm, pulls fluid back into the blood vessel. The interstitial fluid exerts a hydrostatic pressure of 2 mm, which forces fluid out of the tissues back into the blood. Since there should be no protein in the tissue fluids at the end of the capillary, the colloid osmotic pressure of the interstitial fluid is 0 mm. The net result of the pressure acting outward from the blood (35 and 0 mm) and the pressure acting inward toward the blood (25 and 2 mm) causes a net pressure of 8 mm, forcing fluid out of the blood at the arterial end of the capillary. Thus, there is a net outward filtration from the capillary.

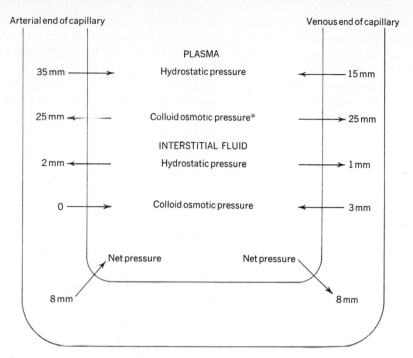

* Osmotic pressures are indicated for proteins (colloids) only because the concentrations of electrolytes and organic compounds in the plasma and the tissues are essentially the same and, therefore, the osmotic pressure they exert is the same.

Figure 30–3. Effect of albumin (and other plasma proteins) on water balance.

At the venous end of the capillary, the following conditions exist. The hydrostatic pressure of the blood is 15 mm, forcing fluid outward. (Note that the hydrostatic pressure of the blood at the venous end is less than at the arterial end of the capillary.) The colloid osmotic pressure of the plasma remains at 25 mm, pulling fluid inward. The interstitial fluid hydrostatic pressure of 1 mm forces fluid out of the tissues back into the blood, while the interstitial fluid colloid osmotic pressure of 3 mm causes fluid to flow back to the tissues.[1] The net result of two pressures acting outward at the venous end of the capillary (15 and 3 mm) and two pressures acting inward (25 and 1 mm) gives a net pressure of 8 mm inward, causing reabsorption of materials. This reabsorption is aided by the lymphatics.

If the plasma proteins (primarily albumin) are present in decreased amounts (as in nephritis or during a low-protein diet), the osmotic pressure of the plasma decreases. This decreased osmotic pressure of the blood causes a greater net pressure outward at the arterial end of the capillary and a lower net inward venous pressure at the venous end of the capillary. When this occurs,

[1] This colloid osmotic pressure is caused by small amounts of plasma protein which pass through the capillary membranes and tend to accumulate at the venous end of the capillaries.

water (fluid) accumulates in the tissues. Such a condition is known as *edema*. Edema may also occur because of heart disease, whereby there is an increase in venous hydrostatic pressure. In many terminal illnesses, edema results. This becomes a serious problem so that tapping and draining may be necessary. Concentrated albumin infusions (25 g in 100 ml diluent) are helpful in the treatment of shock, to increase blood volume, and to remove fluid from the tissues.

The amount of albumin present in the blood is lowered in liver disease because albumin is formed in that organ.

Globulins

The globulins present in the plasma may be separated into different groups by a process known as electrophoresis, whereby the charged protein particles migrate at varying rates to electrodes of opposite charge, with albumin migrating the fastest. The distribution of the plasma proteins is shown in Figure 30–4. As can be seen in the illustration, the globulins are subdivided into alpha (α), beta (β), and gamma (γ). Alpha, beta, and gamma globulins form complexes (loose combinations) with such substances as carbohydrates (mucoprotein and glycoprotein), lipids (lipoprotein), and metal ions (transferrin for iron and ceruloplasmin for copper). The amount of transferrin is decreased in such diseases as pernicious anemia and liver disease. The amount of ceruloplasmin is decreased in Wilson's disease (see page 508). These complexes can be transported to all parts of the body.

The gamma globulins (immunoglobulins) include the antibodies with which the body fights infectious diseases. Gamma globulin has been found to contain as many as 20 different antibodies for immunity against such diseases as measles, infectious hepatitis, poliomyelitis, mumps, and influenza. The most important use of serum electrophoresis is as an aid in the diagnosis of diseases in which abnormal proteins appear in the blood (multiple myeloma and macroglobulinemia), or when a protein component is either present in decreased amounts or lacking altogether (agammaglobulinemia; see next paragraph).

Some people lack the ability to make gamma globulin. These people are

Figure 30–4. Distribution of plasma proteins during electrophoresis.

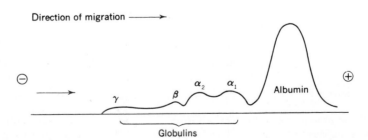

quite susceptible to infections because they have no antibodies to counteract such diseases. The lack of gamma globulin is called *agammaglobulinemia* and can be counteracted by the administration of gamma globulin.

Fibrinogen

Fibrinogen is the plasma protein involved in the clotting of the blood. Fibrinogen is manufactured in the liver, so that any disease which destroys liver tissue causes a decrease in the amount of fibrinogen.

Blood Clotting

When the skin is ruptured, blood flows out and soon forms a clot. When blood is taken from a vein and placed in a test tube, it soon forms a clot. Why does blood clot when it is removed from its normal place in the circulatory system? Why doesn't it clot in the blood vessels themselves?

When blood clots, a series of reactions occur in which the soluble plasma protein fibrinogen is converted into insoluble fibrin. Fibrin precipitates in the form of long threads that cling together to form a spongy mass which entraps and holds the blood cells, forming a clot.

When a blood vessel is cut, the blood comes in contact with the tissues. The contact of the blood with the tissues liberates thromboplastin from the platelets, from the plasma, and from the tissues themselves. It should be noted that thromboplastin is not a single substance but instead refers to substances that catalyze the conversion of prothrombin to thrombin. There are many "factors" with thromboplastic activity. In the presence of thromboplastin and calcium ions, prothrombin (a globulin present in the plasma) is changed into thrombin.

The thrombin in turn acts on the fibrinogen to convert it into fibrin, which forms the clot.

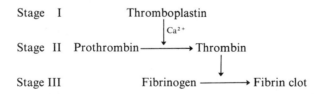

The above reactions are merely a simplified version of a very complex sequence of reactions. There are many factors involved, particularly among those having thromboplastic activity. The International Committee for the Standardization of the Nomenclature of Blood Clotting Factors has designated a system of numbering for such factors. For more detailed information, consult an advanced biochemistry textbook.

Calcium ions are necessary for the clotting of blood. The coagulation of freshly drawn blood samples may be prevented by adding a substance (such as potassium oxalate) that removes the calcium ions from solution. However, since oxalates are poisonous, this method can be used only if the blood is to be analyzed in a laboratory.

To prevent clotting in blood used for transfusions, sodium citrate is added. This substance removes the calcium ions from the blood by forming calcium citrate, a compound that is almost completely nonionized.

A deficiency of vitamin K reduces the production of prothrombin, without which the blood cannot clot.

Anticoagulant drugs such as bishydroxycoumarin (Dicumarol) and heparin reduce the conversion of prothrombin to thrombin and so keep the blood from clotting rapidly. Anticoagulant drugs may be used after surgery to prevent clots from forming in the cut blood vessels. A clot formed in a blood vessel is called a *thrombus*. A thrombus in a blood vessel does no harm if it remains where it was formed because it is slowly reabsorbed. However, if the clot breaks loose and travels through the blood vessels it may lodge in and obstruct a blood vessel leading to the heart or brain, causing paralysis or death.

Respiration

The tissues require oxygen for their normal metabolic processes; they must also eliminate carbon dioxide. Oxygen is carried from the lungs to the tissues by the hemoglobin of the blood. In the tissues the oxygen is given up by the hemoglobin and the waste carbon dioxide from the tissues is picked up and carried to the lungs.

The inspired air has a higher concentration of oxygen than does the blood in the alveoli of the lungs. Gases always flow from an area of high concentration to one of lower concentration so that the oxygen flows from the lungs (high concentration) into the blood (lower concentration). In the blood the oxygen combines with the hemoglobin. Very little oxygen is actually dissolved (uncombined) in the blood. When the oxygen-rich blood reaches the tissues, it gives up its oxygen to the cells because those cells are using up oxygen and have a lower concentration of that gas than the blood. At the same time, the cells have a higher concentration of carbon dioxide than the blood; therefore, that gas flows from the cells into the bloodstream. The blood carries the carbon dioxide to the lungs. There it is in contact with air, which has a lower carbon dioxide concentration, and it passes from the blood into the lungs where it is exhaled (see page 426).

The transportation of gases in respiration includes the nine steps described in the next section.

Transportation of Oxygen and Carbon Dioxide

1. As oxygen passes from the alveoli of the lungs into the bloodstream, some of it dissolves in the blood plasma. However, most of it reacts

with hemoglobin (here represented by the formula HHb) to form oxyhemoglobin, HbO_2^-.

$$HHbO_2 + O_2 \longrightarrow HbO_2^- + H^+$$

2. In the tissues, oxyhemoglobin reacts with hydrogen ions to yield oxygen and hemoglobin (HHb).

$$HbO_2^- + H^+ \longrightarrow HHb + O_2$$

The release of oxygen to the tissues is enhanced by a decrease in the pH and by the presence of 2,3-diphosphoglyceric acid. Most of the hemoglobin travels back to the lungs to pick up more oxygen. But some of the hemoglobin reacts with carbon dioxide to form carbamino-hemoglobin (here represented as $HHbCO_2$).

$$HHb + CO_2 \longrightarrow HHbCO_2$$

3. At the same time, carbon dioxide flows from the tissues into the blood. A small amount dissolves directly in the plasma, but most of the CO_2 reacts with water in the red blood cells to form carbonic acid. This reaction takes place rapidly under the influence of the enzyme carbonic anhydrase.

$$CO_2 + H_2O \xrightarrow{\text{carbonic anhydrase}} H_2CO_3$$

4. The carbonic acid thus formed ionizes to yield hydrogen and bicarbonate ions

$$H_2CO_3 \longrightarrow H^+ + HCO_3^-$$

The hydrogen ions thus produced react with oxyhemoglobin, as indicated in step 2.

5. The bicarbonate ions from step 4 cannot remain in the red blood cells because those cells can hold only a small amount of that ion. So, the excess bicarbonate ions diffuse outward into the blood plasma. Red blood cells cannot stand a loss of negative ions, so that, to counteract this, chloride ions from the blood plasma flow into the red blood cells. This process is called the *chloride shift*.

red blood cell | blood plasma

$HCO_3^- \longrightarrow$

$\longleftarrow Cl^-$

6. In the lungs, the bicarbonate ions react with the hydrogen ions produced according to step 1 to form carbonic acid.

$$HCO_3^- + H^+ \longrightarrow H_2CO_3$$

7. The carbonic acid thus formed rapidly decomposes into carbon dioxide and water under the influence of the enzyme carbonic anhydrase.

$$H_2CO_3 \xrightarrow{\text{carbonic anhydrase}} H_2O + CO_2$$

8. As the bicarbonate ions are used up, in steps 6 and 7, more bicarbonate ions from the plasma flow into the red blood cells. At the same time, chloride ions migrate outward from the red blood cells. This is a *reverse chloride shift*.

red blood cell | blood plasma

$$Cl^- \rightleftarrows HCO_3^-$$

9. At the same time, the carbaminohemoglobin, formed as indicated in step 2, decomposes in the lungs to yield hemoglobin and carbon dioxide.

$$HHbCO_2 \longrightarrow HHb + CO_2$$

The overall reactions are summarized below.

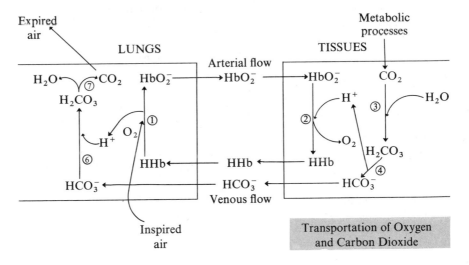

Transportation of Oxygen and Carbon Dioxide

Acid-Base Balance

The normal pH range of the blood is 7.35 to 7.45. When the pH falls below this range, the condition is termed *acidosis*. *Alkalosis* occurs when the pH rises above its normal value. Acidosis is more common than alkalosis because many of the metabolic products produced during digestion are acidic in nature. The ability of the blood buffers to neutralize acid is called the *alkaline reserve* of the blood. In acidosis, the alkaline reserve decreases; during alkalosis, it increases.

How does the blood maintain the pH when acid or basic substances are continuously being added to it?

The blood retains its fairly constant pH because of the presence of buffers. These buffers are present both in the blood plasma and in the red blood cells. Those in the plasma are primarily sodium buffers; those in the blood cells are mainly potassium buffers. Recall that buffers are substances (usually a weak acid and a salt of a weak acid) that resist changes in pH (see page 172). The blood buffers consist of the following

1. Bicarbonate buffers
2. Phosphate buffers
3. Protein buffers (including hemoglobin and oxyhemoglobin)

Bicarbonate Buffers

The bicarbonate buffer system in the red blood cells consists of carbonic acid, H_2CO_3, and potassium bicarbonate, $KHCO_3$. If we assume that a strong acid (such as HCl) is added to a sample of blood, it will react with the $KHCO_3$ part of the buffer and undergo the following reaction

$$HCl + KHCO_3 \longrightarrow H_2CO_3 + KCl$$

The carbonic acid produced (H_2CO_3) is part of the original buffer. Note that the strong acid, HCl, has been replaced by a very weak one, H_2CO_3. The other product, KCl, is a neutral salt and will not affect the pH of the system.

If we assume that a strong base such as KOH is added to a sample of blood, the following reaction will occur with the bicarbonate buffer system.

$$KOH + H_2CO_3 \longrightarrow KHCO_3 + H_2O$$

The $KHCO_3$ produced is part of the original buffer system and the water produced is neutral, so that the pH again is unaffected.

If these two reactions are compared

$$HCl + KHCO_3 \longrightarrow H_2CO_3 + KCl$$

$$KOH + H_2CO_3 \longrightarrow KHCO_3 + H_2O$$

we will see that, in each case, more of the buffer is produced plus a neutral compound.

The bicarbonate buffers and the blood protein buffers play a major part in the control of the pH; the phosphate buffers have a less important role.

Phosphate Buffers

The phosphate buffers consist of mixtures of K_2HPO_4 and KH_2PO_4 (also Na_2HPO_4 and NaH_2PO_4), which function similarly to the bicarbonate buffers in neutralizing excess acid and base.

$$HCl + K_2HPO_4 \longrightarrow KH_2PO_4 + \quad KCl$$

$$KOH + KH_2PO_4 \longrightarrow K_2HPO_4 + \quad H_2O$$

more buffer   neutral compound

Hemoglobin Buffers

The hemoglobin buffers account for more than one half of the total buffering action in the blood. There are hemoglobin buffers and oxyhemoglobin buffers.

Hemoglobin Buffer	*Oxyhemoglobin Buffer*
HHb	$HHbO_2$
KHb	$KHbO_2$

These buffers, as well as other proteins that act as buffers in the bloodstream, pick up excess acid or base to help keep the pH of the blood within its normal range.

Function of the Kidneys in Controlling Acid-Base Balance

The kidneys help maintain the acid-base balance of the blood by excreting or absorbing phosphates and also by forming ammonia. The acid substances in the blood combine with the ammonium ions and are excreted as salts through the kidneys, saving sodium and potassium ions for the buffer systems.

Function of the Lungs in Controlling Acid-Base Balance

If too much acid enters the bloodstream too rapidly, the pH will tend to fall. At the same time, more carbonic acid is produced by the bicarbonate (HCO_3^-) ions. The carbonic acid in turn decomposes to carbon dioxide and water.

$$HCO_3^- + H^+ \rightleftharpoons H_2CO_3 \rightleftharpoons H_2O + CO_2$$

The excess carbon dioxide thus produced stimulates the respiratory center of the brain, making the person breathe faster thus removing more carbon dioxide from the blood. The increased rate of respiration continues until the bicarbonate ion–carbonic acid ratio returns to normal. When this occurs, the amount of carbon dioxide is too small to further stimulate the respiratory center of the brain, and breathing returns to its normal rate.

Water Balance

Normally, in man, water intake is balanced by water output. If the intake of water exceeds the output, *edema* results (see page 421). If the output of water exceeds the input, *dehydration* may occur (see Figure 30–5).

The body replenishes its water supply in three ways:

1. By the ingestion of liquids (which are primarily, if not wholly, water).
2. By the ingestion of foods such as meats, vegetables, and fruits, all of which contain a very high percentage of water.
3. By metabolic processes taking place normally in the body. When carbohydrates, fats, and proteins are metabolized, water is produced. Approximately 14 ml of water are formed for every 100 Cal of energy released by the oxidation of foods. The oxidation of 100 g of carbohydrate produces 55 g of H_2O; the oxidation of 100 g fat produces 107 g of H_2O; the oxidation of 100 g protein produces 41 g of H_2O.

The total normal input of water in the body is approximately 2500 ml per day.

The body loses water in several ways:

1. Through the kidneys, as urine.
2. Through the skin, as perspiration.
3. Through the lungs, as exhaled moisture.
4. Through the feces.

The total of these methods of water loss should approximate that of the water intake, 2500 ml per day. However, the amount of water lost by these individual methods may vary considerably. For example, the amount of moisture lost through the skin (by the evaporation of perspiration) and through the lungs increases

1. During vigorous muscular exercise.
2. With an increased respiratory rate.
3. In a hot, dry environment.
4. During a fever.
5. When the skin receives severe burns.

Figure 30–5. Water balance in the body.

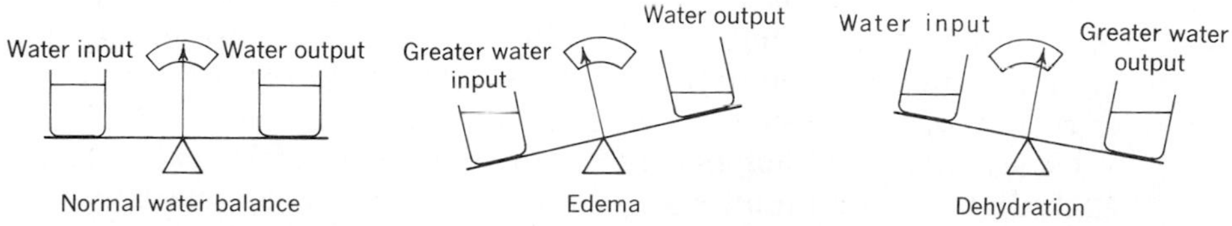

Water input / Water output — Normal water balance

Greater water input / Water output — Edema

Water input / Greater water output — Dehydration

Conversely, the amount of water lost through the kidneys increases following the ingestion of large amounts of water within a short period of time. It also increases because of the presence of large amount of waste products in the bloodstream. However, when the body loses excess water through vomiting and/or diarrhea, the output of water through the kidneys is immediately lessened.

Distribution of Water in the Body

The water of the body is considered to be distributed in two major areas—intracellular (within the cells), 55 per cent, and extracellular (outside the cells), 45 per cent. The extracellular water, in turn, may be further divided into four areas:

1. Intravascular (plasma): fluid within the heart and blood vessels (7.5 per cent).
2. Interstitial and lymph: fluids outside the cells (20 per cent).
3. Dense connective tissue, cartilage, and bone (15 per cent).
4. Transcellular fluids: extracellular fluid collections, including the salivary glands, thyroid gland, gonads, mucous membranes of the respiratory and gastrointestinal tracts, kidneys, liver, pancreas, cerebrospinal fluid, and the fluid in the spaces within the eyes (2.5 per cent).

Approximately 55 per cent of the weight of an adult male and 50 per cent of the weight of an adult female is water.

The principal difference between the blood plasma (the intravascular fluid) and the interstitial fluid is in the protein content. Proteins cannot pass through membranes and so they remain in the blood vessels. Thus, the protein content of the interstitial fluid is very low compared to that of the intravascular fluid, although the soluble electrolytes in each are approximately the same.

There are distinct differences in salt concentration in the intracellular and extracellular fluids.

The extracellular water in the dense connective tissue, cartilage, and bone and the extracellular water in the transcellular fluids do not readily interchange fluids and electrolytes with the rest of the body water. The rest of the body water—intracellular, intravascular, interstitial, and lymph—freely moves from one area to another within the body.

Electrolyte Balance

Table 30–2 indicates the electrolyte concentrations in the intravascular, interstitial, and intracellular fluids. Note that the ions are divided into two groups—cations (positively charged ions) and anions (negatively charged ions).

Concentrations of anions and cations are expressed in units of milliequivalents per liter (mEq/L); see page 137. The unit milliequivalent measures

TABLE 30–2

431

Body Fluids:
The Blood

Electrolyte Concentrations of Body Fluids (mEq/L)

	Intravascular	Interstitial	Intracellular
Cations			
Na^+	142	145	10
K^+	4	4	158
Mg^{2+}	3	2	35
Ca^{2+}	5	3	2
	154	154	205
Anions			
Cl^-	103	115	2
HCO_3^-	27	30	8
HPO_4^{2-}	2	2	140
SO_4^{2-}	1	1	—
Protein$^-$	16	1	55
Organic acids$^-$	5	5	—
	154	154	205

the chemical and physiologic activity of an ionized substance, the electrolyte. Since milliequivalents are based upon ions, the term milliequivalents represents the number of charged particles or the number of both positive and negative charges present in a solution of an electrolyte. Recall that the number of positive charges must always equal the number of negative charges.

Note that the electrolyte concentrations of the interstitial fluid are similar to those of the intravascular fluid, except that there is more chloride ion and less protein concentration. Note also that in both the intravascular and the interstitial fluid, sodium is the principal cation and chloride the principal anion.

The intracellular fluid differs in concentrations from the intravascular and interstitial fluids in that potassium is the principal cation and phosphate the principal anion. Electrolyte concentrations in intracellular fluids are only approximate because they vary slightly from tissue to tissue.

Concentrations of electrolytes are frequently measured and expressed in the units mg/100 ml. This may be changed to mEq/L by means of the following formula:

$$\frac{\text{mg/100 ml} \times 10 \times \text{ionic charge}}{\text{atomic weight}} = \text{mEq/L}$$

For example, suppose a laboratory reports the serum potassium concentration as 15.6 mg/100 ml. The concentration of potassium in mEq/L may be calculated by using the above formula and the atomic weight (39) and the ionic charge (1), or

$$\frac{15.6 \times 10 \times 1}{39} = 4 \, \text{mEq/L}$$

Osmotic activity depends upon the number of particles present in a solution regardless of whether they carry a charge or not. Thus, sodium ions, potassium ions, and chloride ions cause osmotic pressure, but so do glucose and urea, both of which are nonelectrolytes (that is, they carry no charge).

Osmotic activity is expressed by the unit *milliosmol*, which is a measure of the amount of work that dissolved particles can do in drawing a fluid through a semipermeable membrane. Osmotic activity is measured by means of an instrument called an osmometer (Figure 30–6).

Clinical Importance of Cations and Anions

Sodium Ions.   Sodium ions are the primary cations of the extracellular fluids. The principal functions of sodium ions are:

1. To maintain the osmotic pressure of the extracellular fluid.
2. To control water retention in tissue spaces.
3. To help maintain blood pressure.
4. To maintain the body's acid-base balance by means of the bicarbonate buffer system.

Figure 30–6. Osmometer. (Courtesy of Ayerst Laboratories, New York, N.Y.)

5. To regulate the irritability of the nerve and muscle tissue and of the heart.

The average daily adult intake of sodium, as NaCl, is 5 to 15 g. About 95 per cent of the sodium lost by the body passes through the kidneys. The body's sodium ion concentration is influenced by aldosterone, a hormone of the adrenal cortex. This hormone promotes the reabsorption of sodium ions in the kidney tubules. The antidiuretic hormone, ADH, promotes water absorption in the kidneys and so has a definite effect on extracellular sodium ion concentration.

Hyponatremia, a lower than normal serum sodium ion concentration, may be due to such causes as vomiting, diarrhea, excessive sweating, starvation, extensive skin burns, loss of sodium ions because of kidney damage, or because of diuretics. The clinical symptoms of hyponatremia are cold, clammy extremities, lowered blood pressure, weak and rapid pulse, oliguria, muscular weakness, and cyanosis (a dark purplish discoloration of the skin and mucous membranes due to decreased oxygenation of the blood). In addition, due to an increased plasticity of the tissues, hyponatremia frequently shows as fingerprinting over the sternum. In hyponatremia the specific gravity of the urine is less than 1.010.

Hypernatremia, a higher-than-normal serum sodium ion concentration, may be due to such causes as deficient water intake, excessive water output, poor kidney excretion, rapid administration of sodium salts, hyperactivity of the adrenal cortex (as in Cushing's disease), and some cases of cerebral disease. The clinical symptoms of hypernatremia are dry, itchy mucous membranes, intense thirst, oliguria or anuria, rough dry tongue, and elevation of temperature. The specific gravity of the urine rises above 1.030. In an extreme case of hypernatremia the symptoms include tachycardia (rapidly beating heart), edema, and cerebral disturbances.

Potassium Ions.　　Potassium ions are the principal cations of the intracellular fluid. Since the kidneys do not conserve potassium ions as well as they preserve sodium ions and since the body cannot store potassium ions, a depletion of this substance occurs readily in patients whose diets are low in potassium or who are excreting more potassium than they take in.

The principal functions of potassium ions in the body are

1. To maintain the osmotic pressure of the cells.
2. To maintain the electrical potential of the cells.
3. To maintain the size of the cells.
4. To maintain proper contraction of the heart.
5. To maintain proper transmission of nerve impulses.

Potassium ions move into the cells during anabolic activity and move out of the cells during catabolic activity. The concentration of potassium ion is usually measured in terms of serum potassium because this is a much easier quantity to measure than cellular potassium concentration.

Hypokalemia, a lower than normal serum potassium ion concentration, can occur under the following conditions.

1. Too low an intake of potassium ions
   (a) During starvation or malnutrition
   (b) In a diet deficient in potassium
   (c) During intravenous infusions of fluids low or lacking in potassium ions
2. Too great an output of potassium ions
   (a) Because of the use of diuretics
   (b) Use of corticosteroids (these hormones promote retention of sodium ions at the expense of potassium ions)
   (c) Because of prolonged vomiting
   (d) Because of gastric suction and intestinal drainage
   (e) Because of diarrhea
   (f) With polyuria

In addition, hypokalemia may be caused by a sudden shift of potassium ions from the extracellular fluid to the intracellular fluid. This could occur in such a case as the treatment of diabetic acidosis with insulin and glucose.

In general, hypokalemia occurs most frequently in conjunction with some other pathologic condition.

The general symptoms of hypokalemia are

1. General feeling of being ill.
2. Lack of energy.
3. Muscular weakness.
4. Numbness of fingers and toes.
5. Apathy.
6. Dizziness on rising.
7. Cramps, particularly in the calf muscles.

As hypokalemia develops to a greater extent, symptoms relating to the heart become evident. Among these are weak pulse, falling blood pressure, faint heart sounds, and changes in the ECG—first a flattening of the T wave, later inverted T waves with a sagging ST segment and AV block, and finally cardiac arrest.

Hypokalemia may be treated or prevented by giving the patient potassium intravenously (in the form of a potassium salt) or orally by the use of high-potassium foods such as veal, chicken, beef, pork, bananas, orange and pineapple juices, broccoli, and potatoes. Note that although there are many other foods high in potassium, they are usually also high in sodium. A patient on a high-potassium diet usually also has a low-sodium requirement, so that these other types of foods are not recommended.

Hyperkalemia, an increased serum potassium ion level, occurs

1. If the intake of potassium ions is too great because of too rapid an infusion of potassium ions or the administration of excess potassium ions intravenously.
2. If the output of potassium ions is too low because of renal failure or because of acute dehydration.
3. If there is a sudden shift of potassium ions from the intracellular fluid to the extracellular fluid because of severe burns or crush injuries (both of these could release potassium ions from the cells into the bloodstream).

The symptoms of hyperkalemia are a general feeling of ill-being, muscular weakness, listlessness, mental confusion, slower heart beat, poor heart sounds, bradycardia, and eventually cardiac arrest. Characteristic changes in the ECG are elevated T waves, widening of the QRS complex, gradual lengthening of the P–R interval, and final disappearance of the P wave.

Removal of excess potassium ions may be accomplished either by dialysis or by the administration of glucose and insulin.

Calcium Ions. Most of the body's calcium is found in the bones and the teeth in the form of calcium carbonate and calcium phosphate. If the blood calcium ion concentration falls, it can readily be replenished from the bone. Conversely, if the blood calcium ion concentration rises, the amount replenished from the bones decreases (see pages 473 and 476).

The daily intake (adult) for calcium varies from 200 to 1500 mg and comes primarily from milk and milk products. Ionized calcium is present in body fluids and is important in blood coagulation (see page 423), in the regulation of membrane permeability, and in the normal functioning of nerve, heart, and muscle tissue.

Because of the great amount of calcium present in the bones, calcium is not required during intravenous therapy. In addition, calcium-containing solutions are not suitable for infusions because, if mixed with citrated blood, they may cause a clot in the drip tube.

Hypocalcemia, a low serum calcium concentration, may be due to a hypoactive parathyroid gland (see page 476), the surgical removal of the parathyroid glands, or a large infusion of citrated blood. The symptoms of hypocalcemia include tingling of the fingertips, abdominal and muscle cramps, and tetany.

Hypercalcemia, an increased serum calcium concentration, may be caused by an overactive parathyroid or by a tumor of that gland. It may also be caused by the administration of excess vitamin D. The symptoms of hypercalcemia include hypotonicity of muscles, kidney stones, deep bone pain, and bone cavitation.

The serum calcium ion concentration decreases during hypoparathyroidism and rises during hyperparathyroidism (see page 476).

Magnesium Ions.   Magnesium ion, like potassium ion, is found primarily in the intracellular fluid. Magnesium ions are essential for the proper functioning of the neuromuscular system.

Magnesium, as an activator, catalyzes more enzymes than any other metal ion in the body; it is necessary for over 100 metabolic reactions. The normal intake of magnesium is 15 to 30 mEq per day. Another unusual property of magnesium is that it is the only positively charged ion that has a higher concentration in the cerebrospinal fluid than in the blood serum.

A deficiency of serum magnesium ions, *hypomagnesemia*, is unusual due to dietary intake because magnesium is a necessary element for chlorophyll, which is found in all green plant foods. A lower than normal magnesium ion concentration may be due to such causes as

1. Chronic alcoholism
2. Diabetic acidosis
3. Prolonged intravenous infusion without magnesium ions
4. Hypoparathyroidism
5. Prolonged nasogastric suction
6. Acute pancreatitis
7. Severe malabsorption

The clinical symptoms of a deficiency of magnesium ions are

1. Muscular tremors
2. Convulsions
3. Delirium
4. Delusions
5. Disorientation
6. Hyperirritability
7. Elevated blood pressure

An excess of serum magnesium ions may be caused by severe dehydration or renal insufficiency. Excess magnesium ions act as a sedative. Extreme excesses may cause coma, respiratory paralysis, or cardiac arrest.

Chloride Ions.   The chloride ion is the primary anion of the extracellular fluid. The body's intake of chloride ion is closely related to that of the sodium ion (see page 432).

One of the principal functions of the chloride ion is as a component of gastric hydrochloric acid. The chloride ion also serves an important function in the transportation of oxygen and carbon dioxide in the blood (see page 425).

Hypochloremia, a lower than normal serum chloride ion concentration, occurs after prolonged vomiting, profuse sweating, and diarrhea. This condition causes an alkalosis because of an increased concentration of bicarbonate ions. Hypochloremia may also occur when there is a marked loss of potassium ions.

Phosphate Ions. The phosphate ion is the primary anion of the intra-cellular fluid. Diets that are adequate in calcium usually contain more than enough phosphorus for the body's needs.

Most of the body's phosphate is present in the bones as calcium phosphate, a substance that gives the bodes their rigidity, but phosphate is found in every cell of the body. Phosphate ions are important in the acid-base balance of the body. They constitute one of the body's buffer systems. Phosphates are also of great importance in the production of ATP, the body's principal energy compound. Serum phosphate levels are low in hyperparathyroidism and high in hypoparathyroidism and celiac disease.

Iron Ions. Iron ions are involved almost exclusively in cellular respiration. Iron is part of hemoglobin, myoglobin, and cytochromes as well as several oxidative enzymes. The formation of hemoglobin requires the presence of traces of copper. The best dietary sources of iron are the "organ meats," such as liver, heart, and kidneys. Other sources are egg yolk, fish, beans, and spinach.

Most of the iron present in the food we eat is in the form of ferric ions (Fe^{3+}). In the digestive system, ferric ions are reduced to ferrous ions (Fe^{2+}), which are then absorbed into the bloodstream from the stomach and duodenum. In the blood plasma, ferrous ions are oxidized to ferric ions, which then become part of a specific protein—transferrin, a β-globulin. The conversion of ferrous ions to ferric ions in the blood plasma is catalyzed by the enzyme ceruloplasmin, a copper-containing compound (see following paragraphs).

The liver, spleen, and bone marrow are able to extract the iron from transferrin and to store that iron in the form of two proteins—ferritin and hemosiderin. The bone marrow is also able to extract the iron from trans-ferrin and to use that iron for the production of hemoglobin.

A deficiency of iron (*iron-deficiency anemia*) may result from a low intake of iron because of a diet high in cereal and low in meat, because of poor absorp-tion of iron due to gastrointestinal disturbances or diarrhea, or because of excessive loss of blood. This type of anemia may be treated with a daily dose of ferrous sulfate in the diet, if absorption is normal.

Copper Ions. In addition to being necessary for the synthesis of hemoglobin, copper ions are necessary for certain enzymes, such as cytochrome oxidase (part of the oxidative phosphorylation sequence; see page 369) and uricase (which catalyzes the oxidation of uric acid to allantoin).

Copper is found in the brain in the form of cerebrocuprein, in the blood cells as erythrocuprein, and in the blood plasma as ceruloplasmin, an α-globulin. In *Wilson's disease*, there is a decreased concentration of cerulo-plasmin in the blood. This disease is characterized by the presence of large amounts of copper in the brain along with an excessive urinary output of copper.

Copper aspirinate, a complex of copper and aspirin, has been found to be 20 times as effective as aspirin itself in the treatment of arthritis in animals. It is hoped that tests on humans will show similar results.

The average daily diet contains about 2.5 to 5 mg of copper, an amount that is considered to be adequate for the normal adult. The richest sources of copper are liver, nuts, kidney, raisins, and dried legumes.

Other Ions. Zirconium chlorhydrate has been used in antiperspirant sprays, but its use has been discontinued. It was found that this compound could reach the user's lungs through inhalation. In the lungs, zirconium chlorhydrate could induce granulomas, which in turn cause growth of tumors. Aluminum chlorhydrate is still in use in antiperspirant sprays and appears to have no adverse effect upon the body.

Lithium salts are being used to treat manic-depressive psychoses and as antidepressants for some psychiatric patients. Previously, lithium chloride had been employed as a sodium substitute, with dangerous and frequently fatal results. Patients on a therapeutic dosage of lithium salts may complain of fatigue, muscular weakness, nausea, and diarrhea. Slurred speech and hand tremors are noticeable. In larger doses, the central nervous system is affected and the patient may become unconscious or even go into a coma. Abnormalities in the electroencephalogram are also common.

Summary

Blood carries oxygen, minerals, and food to the cells. Blood carries carbon dioxide and other waste products from the cells. Blood also carries hormones, enzymes, antibodies, and blood cells. Blood regulates body temperature and maintains the pH of the body fluids.

Blood consists of two parts: the suspended particles, red blood cells, white blood cells, and platelets; and the suspending liquid, the plasma. When freshly drawn blood is allowed to clot and settle, the yellow liquid remaining is called blood serum.

Oxygenated blood has a bright red color, whereas deoxygenated blood has a dark purplish color. Blood has a pH range of 7.35 to 7.45. If the pH falls below 7.35, the condition is called acidosis. If the pH rises above 7.45, the condition is called alkalosis. The volume of blood in the body is 5 to 6 liters.

The chemical analysis of blood samples is of great clinical significance. Increased or decreased amounts of some substances may indicate a certain disease.

Blood contains hemoglobin, a conjugated protein containing iron. Hemoglobin is composed of heme and globin. Cytochrome c, part of the oxidative phosphorylation sequence in the citric acid cycle, has a structure similar to that of hemoglobin. Chlorophyll a has a structure similar to heme but with a magnesium ion at the center instead of the iron.

If the hemoglobin content of the blood falls below normal, the condition is called anemia. Anemia may result from a decreased rate of production of red blood cells, an increased destruction of red blood cells, or an increased loss of red blood cells.

Plasma proteins are divided into three groups: albumin, globulins, and fibrinogen. Albumin regulates the osmotic pressure of the blood and controls the water balance of the body. Alpha and beta globulins form loose combinations with carbohydrates, metal ions, and lipids so that these substances can be transported to all parts of the body. Gamma globulins contain the antibodies with which the body fights infectious diseases. Fibrinogen is the plasma protein involved in the clotting of the blood.

When a blood vessel is cut, thromboplastin is liberated. Thromboplastin, in the presence of calcium ions, changes prothrombin in the plasma to thrombin. Thrombin in turn acts on fibrinogen to convert it to fibrin, which forms the clot.

In respiration, the hemoglobin picks up oxygen to form oxyhemoglobin in the lungs. The oxyhemoglobin is carried to the tissues where it gives up its oxygen. At the same time the blood picks up carbon dioxide and other waste products from the cells. Bicarbonate ions are involved in the hemoglobin-oxyhemoglobin cycle and also in the carbon dioxide removal cycle. Excess bicarbonate ion in the blood cells is shifted to the plasma in a process called the chloride shift whereby chloride ions take the place of the bicarbonate ions. In the lungs, as carbon dioxide is removed from the blood, a reverse chloride shift takes place.

The blood maintains a constant pH by the use of buffers, which react with acid (or base) substances entering the blood and form new substances that are either neutral or more buffer salts.

Buffer systems in the blood are the bicarbonate buffers, the phosphate buffers, the hemoglobin buffers, and protein.

The kidneys help control the acid-base balance of the body by excreting or absorbing phosphate and also by forming ammonia. The lungs control the amount of carbon dioxide by a mechanism that regulates the rate of breathing. If too much carbon dioxide is present in the blood, the pH falls and the respiratory center of the brain is stimulated to make the person breathe faster and so remove the excess carbon dioxide. When the excess carbon dioxide is removed, the stimulation ceases and the rate of breathing returns to normal.

Water intake must be balanced by water output. If water intake is greater than water output, edema occurs. If water output is greater than water intake, dehydration may occur.

Water in the body is considered to be distributed in two major areas—intracellular and extracellular. Extracellular water is further subdivided into interstitial water; intravascular and lymph water; water in dense connective tissues, cartilage, and bone; and water in transcellular fluids.

Each of the body's water compartments has its own concentration of electrolytes, all concentrations being expressed as milliequivalents per liter. An increase or decrease in the concentration of any one of the ions will have some effect on the body.

Serum osmotic pressure is affected primarily by the concentration of sodium ions, bicarbonate ions, and chloride ions. Osmotic activity is measured in the unit milliosmols.

Questions and Problems

1. What are the functions of the blood?
2. What is the normal concentration of red blood cells in the body?
3. What is the normal concentration of white blood cells in the body?
4. What is polycythemia? Anemia?

5. What is the function of the leukocytes?
6. Name several types of white blood cells.
7. What is a differential count? What is the function of the thrombocytes?
8. What are the functions of the inorganic ions present in the plasma?
9. What is the difference between blood plasma and blood serum?
10. What is the normal pH range of the blood? The specific gravity range? The viscosity?
11. Under what conditions might the blood volume increase? Decrease?
12. How may blood volume be measured?
13. What type of compound is hemoglobin?
14. What is heme? Diagram its structure.
15. How does the structure of heme compare with that of chlorophyll *a*? With cytochrome *c*?
16. What conditions might cause a decreased production of red blood cells?
17. What causes pernicious anemia?
18. What conditions might cause the destruction of red blood cells?
19. What might cause an increased loss of hemoglobin?
20. Name the three groups of plasma proteins. What is the function of each?
21. Describe the effect of albumin on osmotic pressure in terms of Starling's hypothesis.
22. What is edema? What might cause such a condition?
23. What are the functions of the different types of globulins?
24. What is agammaglobulinemia?
25. Describe the clotting mechanism of the blood.
26. How may blood clotting be prevented in samples taken for laboratory analysis?
27. Why should oxalates never be used to prevent blood from coagulating when that blood is to be used in a transfusion?
28. What is the effect of a deficiency of vitamin K on blood clotting?
29. What is a thrombus? What might result from the presence of a thrombus in the brain?
30. Describe the process whereby the blood carries oxygen to the tissues.
31. How does the blood carry carbon dioxide from the tissues?
32. What is the function of carbonic anhydrase?
33. What is the chloride shift? The reverse chloride shift?
34. What is the alkaline reserve of the blood?
35. What types of buffers are present in the blood?
36. Where are potassium buffers located? Sodium buffers?
37. Describe the reaction (in equation form) of an acid and a base with a bicarbonate buffer.
38. How do the lungs help control the acid-base balance of the body?
39. How do the kidneys help in controlling the acid-base balance of the body?
40. How does the body replenish its water supply?
41. How does the body normally lose water?
42. Into what areas may the body's water be considered as being subdivided?
43. What are the principal functions of sodium ions in the extracellular fluid?
44. What may cause hyponatremia? What are its symptoms?
45. What may cause hypernatremia?
46. What are the principal functions of potassium ions in the body?
47. What may cause hypokalemia? What are its symptoms?

48. What may cause hyperkalemia? What are its symptoms?
49. Under what conditions may the serum calcium level change?
50. What are the symptoms of a deficiency of magnesium ions? What might cause this deficiency?
51. What are the principal functions of the chloride ion in the body?
52. What are the principal functions of the phosphate ion in the body? Iron ions? Copper ions?
53. A lab reports the magnesium ion concentration as 3.6 mg/100 ml. Express this as mEq/L.
54. If the colloid osmotic pressure of the plasma drops to 20 mm, all other pressure remaining the same, what will be the effect on fluid flow? What will this condition be called?

References

Baum, S. J., and Scaife, C. W.: *Chemistry: A Life Science Approach.* Macmillan Publishing Co., Inc., New York, 1975, Chap. 31.

Goodman, L. S., and Gilman, A. (eds.): *The Pharmacological Basis of Therapeutics,* 5th ed. Macmillan Publishing Co., Inc., 1975, Sec. VII.

Harper, H. A.: *Review of Physiological Chemistry*, 15th ed. Lange Medical Publications, Los Altos, Calif., 1975, Chap. 9.

Hughes, D. T. D.: Gaseous exchange in the lungs. *Nursing Mirror,* **139**:68–69 (Oct. 3), 1974.

Ingram, M., and Preston, K. J.: Automatic analysis of blood cells. *Scientific American,* **223**:72–82 (Nov.), 1970.

Lamb, C.: Potassium imbalance: walking the mEq tightrope: part 2. *Patient Care,* **9**:128 (Feb. 1), 1975.

Montgomery, R.; Dryer, R. L.; Conway, T. W.; and Spector, A. A.: *Biochemistry.* C. V. Mosby Co., St. Louis, 1974, Chap. 4.

Nossal, G. J.: How cells make antibodies. *Scientific American,* **211**:106–115 (Dec.), 1964.

Perutz, M. F.: The hemoglobin molecule. *Scientific American,* **211**:64–76 (Nov.), 1964.

Zucker, M. B.: Blood platelets. *Scientific American,* **204**:58–64 (Feb.), 1961.

Zuckerkandl, E.: Evolution of hemoglobin. *Scientific American,* **212**:110–118 (May), 1965.

Chapter 31

Vitamins

Animals fed on a diet consisting only of purified carbohydrates, fats, proteins, minerals, and water will lose weight and develop certain deficiency diseases. Something else must be administered to sustain normal life. The additional substances required are called vitamins.

The name *vitamin* was originally *vitamine* because the first one that was found was an amine, hence the name *vital amine*, or *vitamine*. Subsequent studies of other such substances showed that they were not all amines so the "e" was dropped.

Vitamins are similar to hormones in many ways. Vitamins and hormones are carried by the bloodstream to the various parts of the body where they are needed. Vitamins and hormones are required by the body only in extremely small amounts. Neither vitamins nor hormones furnish energy by themselves, although vitamins function with certain enzymes to control energy changes in the body. One important difference between vitamins and hormones is that most vitamins must be supplied in the diet whereas hormones are synthesized by the body.

Vitamins are divided into two major groups—those that are soluble in fat solvents (the fat-soluble vitamins), and those soluble in water (the water-soluble vitamins). The fat-soluble vitamins include vitamins A, D, E, and K and are usually found associated with lipids in natural foods. The water- soluble vitamins include vitamins B and C. Although the vitamins have a letter designation and sometimes a subscript in addition, such as vitamin B_1, the chemical names are being more and more widely used. For example, the chemical name for vitamin B_1 is thiamine.

More and more people eat a balanced diet, take vitamins, or are under medical care. This has helped to produce a healthier population. One piece of evidence that points in this direction is the constant overthrowing of older athletic records.

Fat-Soluble Vitamins

Vitamin A

Source. Vitamin A is found in fish liver oils, butter, milk, and to a small extent in kidneys, fat, and muscle meats. The precursor of vitamin A (the substance from which vitamin A may be made) is called provitamin A and is found in yellow fruits and vegetables such as peaches, apricots, sweet potatoes, carrots, tomatoes, and leafy green vegetables.

Structure. Vitamin A is a high-molecular-weight alcohol. Two different forms of this vitamin are known: retinol (vitamin A_1) and 3-dehydroretinol (vitamin A_2). Vitamin A_2 differs from A_1 in that it has one additional double bond in the ring. It is interesting to note that vitamin A_1 is found in salt-water fish and A_2 in fresh-water fish. Both of the A vitamins have an all-*trans* structure (see page 260), but the potency of vitamin A_2 is only 30 per cent of that of vitamin A_1.

retinol, vitamin A_1

Provitamin A is a compound that can be converted into vitamin A. One such provitamin A is beta-carotene. The conversion of beta-carotene into retinal (vitamin A_1 aldehyde) and then to retinol (vitamin A_1) is indicated in the reaction shown on the following page.

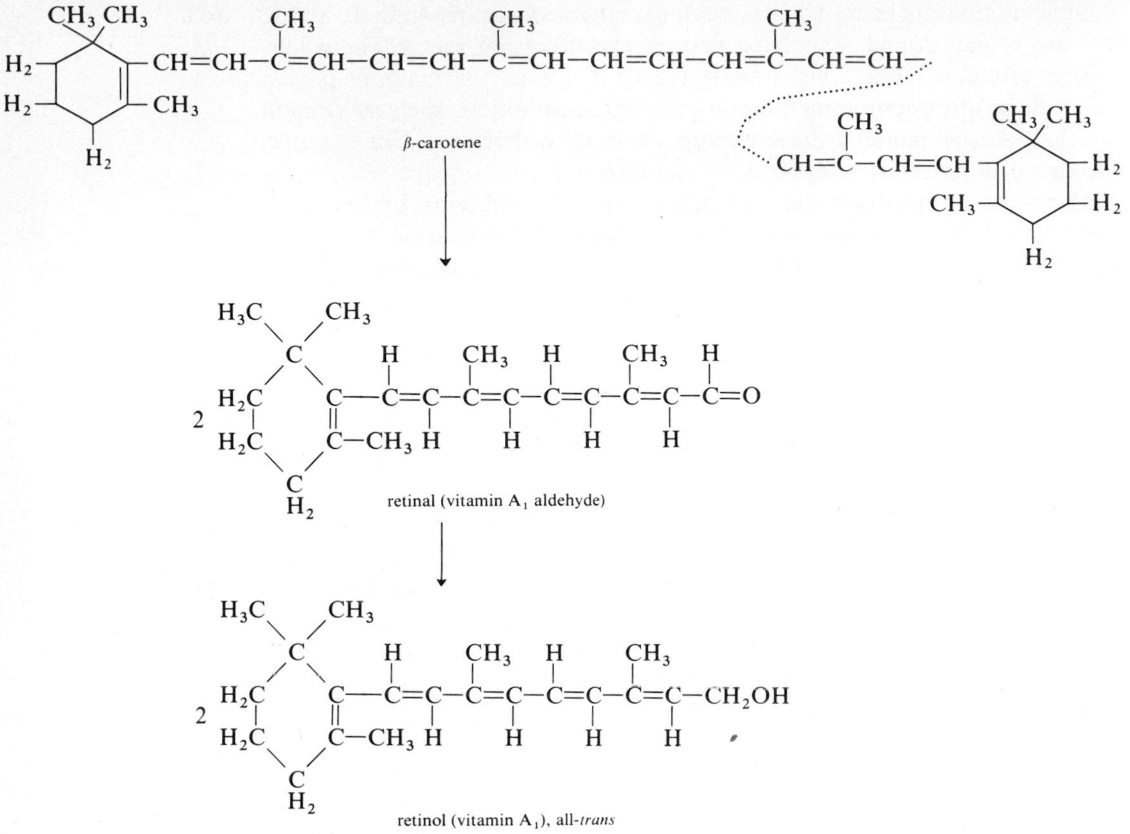

β-carotene

retinal (vitamin A₁ aldehyde)

retinol (vitamin A₁), all-*trans*

It should be noted that one molecule of beta-carotene produces two molecules of vitamin A_1. Alpha- and gamma-carotene yield only one molecule of vitamin A_1 because those compounds are not symmetrical as is beta-carotene.

The provitamins A are transformed into vitamin A in the intestinal walls of animals, such as rats and pigs, but in the liver in man.

Properties. Vitamin A is soluble in fats but not in water. It is stable to heat, acid, and alkalis but is destroyed by oxidation (recall that alcohols are usually quite susceptible to oxidation). Ordinary cooking does not destroy vitamin A. The vitamin A present in butter is destroyed when the butter turns rancid (becomes oxidized).

Daily Requirement. The recommended daily dosages of vitamin A for the normal adult male and female are 1000 and 800 retinol equivalents (5000 and 4000 IU), respectively. One retinol equivalent is 1 microgram of retinol or 6 micrograms of beta-carotene. One International Unit (IU) of vitamin A is equivalent to 0.3 microgram of retinol or 0.6 microgram of beta-carotene. The daily requirements of vitamin A are increased to 1000 retinol equivalents (5000 IU) during pregnancy and to 1200 retinol equivalents (6000 IU) during lactation. A child requires 400 to 700 retinol equivalents (2000 to 3300 IU) daily.

Vitamin A is necessary for normal growth and development, reproduction, and lactation. It is necessary for the synthesis of the membranes around the lysosomes and the mitochondria and acts to regulate membrane permeability. Vitamin A plays an important role in the functioning of the retina and in the maintenance of the integrity of epithelial tissues.

Effect of Deficiency. Vitamin A was discovered by the observation that certain animals did not grow on a diet low in some animal fats. However, this effect on growth is characterized by a lack of other vitamins as well. A lack of vitamin A causes a shrinking and hardening of the epithelial tissues of the membranes in the eyes, digestive tract, respiratory tract, and genitourinary tract. Such a hardening is called *keratinization.*

When keratinization occurs in the lining of the respiratory tract, the patient is more likely to suffer from colds, pneumonia, and other respiratory infections because of the drying of the membranes.

When keratinization occurs in the eyes, the tear ducts become keratinized and are no longer able to secrete tears to wash the eyes. When this occurs, bacteria are able to attack the corneal tissue of the eyes producing an infection called *xerophthalmia.* In this disease, the cornea becomes cloudy and does not allow light to pass through, so sight is lost permanently.

An early symptom of the lack of vitamin A is *nyctalopia,* or night blindness. A person with night blindness cannot see very well in dim light because of a lack of visual purple (rhodopsin) in the retina of the eyes. This pigment is acted on and changed by light and then regenerated in the presence of vitamin A. If there is a lack or deficiency of vitamin A, the visual purple is regenerated very slowly; thus, there is an impairment in vision at night.

The role of vitamin A in the visual process and the part it plays in the regeneration of rhodopsin (visual purple) are illustrated in the following cycle:

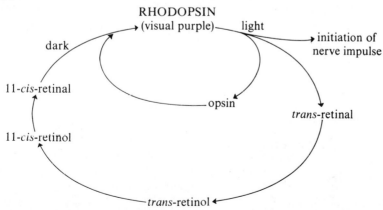

A lack of vitamin A may also produce keratinization of the epithelial cells of the genital system and might cause sterility. A lack of vitamin A causes deformities in the teeth of young animals.

All of these symptoms may be cured by the administration of vitamin A.

Storage and Absorption. Vitamin A is stored in the liver until it is needed. Mother's milk contains 10 to 100 times as much vitamin A as does ordinary milk.

Hypervitaminosis A, an intake of vitamin A far in excess of normal daily requirements, has such early symptoms as irritability, loss of appetite, fatigue, and itching. These symptoms usually disappear within a week after the withdrawal of the vitamin.

Acute cases of hypervitaminosis A have been observed following ingestion of polar bear liver, which contains up to 35,000 IU of vitamin A per gram. Symptoms are drowsiness, sluggishness, vomiting, severe headache, and generalized peeling of the skin after 24 hours. Membranes have increased permeability and decreased stability, leading to mitochondrial swelling, lysosomal rupture, and eventually death.

Since vitamin A is fat soluble, it cannot be absorbed from the intestinal tract without the presence of bile. Any interference with the flow of bile to the intestines will also cause a lack of this fat-soluble vitamin.

Vitamin D

Source. The richest sources of vitamin D are oils from such fish as cod or halibut and the flesh of such oily fish as sardines, salmon, and mackerel. Milk is not a very good source of vitamin D, although its vitamin D content may be increased by irradiation with ultraviolet light. The vitamin D content of the body may be increased by exposure of the skin to ultraviolet rays from the sun, but care must be taken to avoid overexposure and consequent sunburn. For this reason, vitamin D is sometimes called "the sunshine vitamin."

Structure. The D vitamins are a group of sterols with considerable differences in their potency. The two most important vitamins in the D group are vitamin D_2 (calciferol, or activated ergosterol) and vitamin D_3 (activated 7-dehydrocholesterol).

The structure of vitamin D_2, calciferol, is

calciferol (vitamin D_2)

Vitamin D_3 is similar to vitamin D_2, the difference being in the structure of the side chain. Vitamin D_3 is made by the irradiation of 7-dehydrocholesterol so it is frequently called "activated 7-dehydrocholesterol."

7-dehydrocholesterol

vitamin D_3

Ergosterol, the precursor of vitamin D_2 occurs in plants, whereas 7-dehydrocholesterol, the precursor of vitamin D_3, occurs in the skin of humans and animals. 7-Dehydrocholesterol in the skin is isomerized to vitamin D_3 under the influence of ultraviolet light. Vitamin D_3 must then be hydroxylated in the liver and then in the kidney before it can function. Vitamin D_1, originally called vitamin D, was found to be a mixture of vitamins D_2 and D_3.

Properties. The D vitamins are soluble in fats and insoluble in water. They are stable to heat, are resistant to oxidation, and are unaffected by cooking. Vitamin D_2 has a greater potency in man than does vitamin D_3, whereas the reverse is true in chickens.

Daily Requirement. The daily requirement for children is 400 IU. Daily requirements for adults are not stated because, in most instances, exposure to sunlight is sufficient to supply the body's need. However, 400 IU are recommended daily for women during pregnancy and lactation. One international unit of vitamin D is defined as the biologic activity of 0.025 microgram of calciferol.

Physiologic Action. The principal action of vitamin D is to increase the absorption of calcium and phosphorus from the small intestines. It also functions in the deposition of calcium phosphate in the bones and teeth, and is necessary for normal growth and development. Vitamin D is required for the proper activity of the parathyroid hormone (see page 476) and so is used therapeutically in the treatment of hypoparathyroidism.

Effect of Deficiency. Rickets, a disease primarily of infancy and childhood, was previously thought to be due to a deficiency of vitamin D. This disease is now believed to be due to a lack of sunshine on the skin. Why? Because sunshine is necessary for the synthesis of calciferol in the skin.

Rickets is characterized by an inability to deposit calcium phosphate in the bones. The bones become soft and pliable; they bend and become deformed. The joints become enlarged and the ribs become beaded. The knobby or beaded appearance of the ribs is called *rachitic rosary*. When a child has rickets, he does not grow. He develops such symptoms as nervousness, irritability, a bulging abdomen, loss of weight, loss of appetite, anemia, and a delayed development of the teeth. Injection of small amounts of calciferol, or one of its derivatives, or adequate exposure to sunlight can prevent or cure rickets.

A diet low in phosphorus and vitamin D may produce *osteomalacia*, or adult rickets. Since adult rickets is a rare condition, this suggests that adults need less vitamin D than children do. When an adult develops rickets, there is no bulging of the joints since the growth of the bones in an adult is already complete. There is, however, some softening of the bones with accompanying deformities. This disease occurs most often in women after repeated pregnancies during which there has been a deficiency of vitamin D.

A lack of calcium and vitamin D in the diet may cause *osteoporosis* in the adult. This disease, like osteomalacia, is characterized by decalcification and softening of the bones but to a much greater extent.

Effect of Excess. Hypervitaminosis D may show such early symptoms as weakness, lassitude, fatigue, nausea, vomiting, and diarrhea, all of which are associated with hypercalcemia. Later symptoms include calcification of soft tissues, including the kidneys and the lungs. Treatment of hypervitaminosis D consists of the immediate withdrawal of the vitamin, increased fluid intake, a diet low in calcium, and the administration of glucocorticoids.

Vitamin E

Source. Vitamin E is found in milk, eggs, fish, muscle meats, cereals, leafy vegetables such as lettuce, spinach, and parsley, and in plant oils such as cottonseed oil, corn oil, palm oil, and peanut oil. Wheat germ oil is particularly rich in vitamin E.

Structure. There are several vitamins E. The most important is called alpha-tocopherol. Others are beta-, gamma-, and delta-tocopherol.

As indicated by the ending of the name, *-ol*, vitamin E is an alcohol. The structure of alpha-tocopherol is shown on the following page.

$$HO-\text{(ring with } CH_3, H_3C, CH_3 \text{ substituents, } O, CH_2-CH_2-CH_2-CH-(CH_2)_3-CH-(CH_2)_3-CH-CH_3)$$

α-tocopherol

Properties. Alpha-tocopherol, the most important and the most active of the E vitamins, is a colorless to pale yellow oil. It is soluble in fats and fat solvents but insoluble in water. Vitamin E is stable to heat but is destroyed by ultraviolet light and by oxidizing agents.

The activity of the tocopherols appears to be due to their antioxidant properties. They are very effective in preventing the oxidation of vitamin A and unsaturated fatty acids. There is also evidence that vitamin E functions as a cofactor in oxidative phosphorylation reactions. Vitamin E is believed to protect the lung tissues from damage by oxidants present in polluted air.

Premature infants suffering from hemolytic anemia show very low levels of alpha-tocopherol in their blood. When given vitamin E supplements, hemolytic anemia is greatly improved. It is now believed that vitamin E is essential in infant metabolism. The United States Food and Drug Administration requires that commercial milk substitutes sold as infant foods contain adequate amounts of vitamin E. For normal, full-term babies fed on human or cow's milk, there is no need for a vitamin E supplement.

Daily Requirement. The international unit of vitamin E is defined as the activity of 1 mg of *d-l*-alpha-tocopherol acetate. The recommended daily requirement of vitamin E is 15 IU for the adult male and 12 IU for the adult female.

Effect of Deficiency. Vitamin E is known to prevent sterility in animals. A deficiency of vitamin E in male rats produces sterility because the animal is unable to produce spermatozoa. (Mated female rats lacking in vitamin E will produce embryos that develop normally for a few weeks then die and are reabsorbed.)

Some animals on a vitamin E-deficient diet develop muscular dystrophy, resulting in paralysis. Administration of vitamin E helps these animals overcome such effects. Vitamin E has not been found effective in treating muscular dystrophy in humans, and there is no evidence that vitamin E has any therapeutic use at all.

Vitamin K

Source. Vitamin K is found in the green leafy tissues of such plants as spinach, cabbage, and alfalfa. Vitamin K is also found in putrefied fish meal, in liver, eggs, and cheese. Fruits and cereals contain very little vitamin K.

Structure. There are three K vitamins—K_1, K_2, and K_3. Vitamin K_3 is the most active on a weight basis. The structures of vitamin K_1 and K_3 are

vitamin K₁

vitamin K₃

Vitamins K_1 and K_2 differ from K_3 in that they possess a second side chain on the right ring. Vitamin K_3 is a synthetic vitamin; vitamin K_1 is produced in plants and vitamin K_2 by intestinal bacteria.

Compare the structure of the K vitamins with that of coenzyme Q (page 334), which functions in oxidative phosphorylation.

Properties. The K vitamins are soluble in fats and insoluble in water. They are stable to heat but are destroyed in acid and alkaline solutions. They are also unstable to light and oxidizing agents.

Daily Requirement. There is no generally accepted figure for the daily human requirement of vitamin K, but the amount needed appears to be extremely small. The average diet supplies sufficient vitamin K; in addition, intestinal bacteria are able to synthesize this vitamin for their host.

Effect of Deficiency. Vitamin K is known as the antihemorrhaging vitamin. It is necessary for the production of prothrombin in the liver. When there is a deficiency of vitamin K, there is a lack of prothrombin and thus a prolonged clotting time for the blood. Vitamin K is also necessary as a cofactor for oxidative phosphorylation reactions.

Vitamin K is absorbed from the small intestine with the help of bile. In conditions in which bile does not enter the small intestine, such as in obstructive jaundice, vitamin K is not absorbed. This condition leads to a tendency to bleed for a long period of time after an injury or when undergoing surgery. This effect may be overcome by administering both bile and vitamin K to the patient.

Some infants are born with a deficiency of prothrombin and are subject to bleeding. Without proper care such infants may die because of brain hemorrhage. This condition may be alleviated by administering vitamin K to the mother before delivery or to the infant shortly after birth.

Vitamin K is also used therapeutically as an antidote for anticoagulant drugs such as Dicumarol.

Water-Soluble Vitamins

The water-soluble vitamins include vitamins B and C. Vitamin B represents a whole series of vitamins, many of which act as cofactors in various oxidative reactions. Each of the B vitamins has a different physiologic activity. The vitamin B family, also called the B complex, contains the following vitamins

1. Vitamin B_1—thiamine
2. Vitamin B_2—riboflavin
3. Niacin
4. Pyridoxine
5. Pantothenic acid
6. Lipoic acid
7. Biotin
8. Folic acid
9. Inositol
10. Para-aminobenzoic acid
11. Cyanocobalamin—vitamin B_{12}

Vitamin B_1—Thiamine

Source. Thiamine occurs in yeast, milk, eggs, meat, nuts, and whole grains. Vegetables and fruits contain very little vitamin B_1. Synthetic vitamin B_1 is now being added to enrich the vitamin content of flour and bread.

Structure. Thiamine has been crystallized as a hydrochloride with the following structure

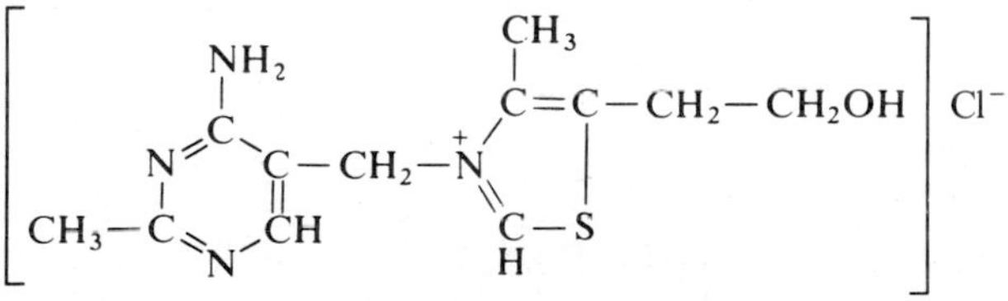

thiamine hydrochloride

Properties. Thiamine is soluble in water and also in alcohol up to 70 per cent. It is insoluble in fats and fat solvents. Thiamine is stable in acid solution but is destroyed in alkaline and neutral solutions. Thiamine is quite stable to heat; it can be sterilized for 30 min at 120°C without appreciable loss in activity. Thiamine hydrochloride, a salt produced by treating thiamine with hydrochloric acid, is more soluble than thiamine itself and so is generally used whenever this vitamin is required.

Daily Requirement. The amount of thiamine required daily is a difficult figure to determine. It depends upon several factors. The body's requirement of thiamine increases during a fever, increased muscular activity, hyperthyroidism, pregnancy, and lactation. The thiamine requirement of the body also increases during a diet high in carbohydrates, whereas it decreases with a diet high in fat and protein.

The recommended thiamine intake according to the Food and Nutrition Board of the National Research Council is 0.5 mg per 1000 Cal. For a 3000-Cal diet, the daily requirement of thiamine would be 1.5 mg.

Effect of Deficiency. A deficiency of vitamin B_1 (thiamine) causes a lack of appetite, failure of growth, and loss in weight.

As the lack of thiamine continues, a disease called beriberi develops in man (in animals this disease is called polyneuritis). Beriberi occurs mainly in the Orient where fish and polished rice (both lacking in vitamin B_1) are the chief diet. In beriberi there is a degeneration of certain nerves leading to the muscles. When pressure is applied along these nerves, severe pain is felt. The muscles served by these nerves become stiff and atrophy from disuse. Cardiovascular symptoms also occur. There are palpitation, tachycardia, an enlarged heart, and an abnormal electrocardiogram. Finally, death may occur because of heart failure. Beriberi can be prevented or treated by a diet containing thiamine.

Thiamine is necessary for the normal metabolism of carbohydrates. The vitamin is changed in the liver to thiamine pyrophosphate, which acts as a coenzyme (cocarboxylase) for the decarboxylation of pyruvic and alpha-keto acids and also acts in transketolase reactions. In the Krebs cycle, cocarboxylase is necessary for the conversion of pyruvic acid to acetyl CoA and also for the conversion of alpha-ketoglutaric acid to succinyl CoA (see page 368). During a deficiency of thiamine, pyruvic acid accumulates in the blood and carbohydrates are not properly metabolized.

Thiamine also functions in the utilization of pentoses in the hexose-monophosphate shunt and in some amino acid syntheses.

Vitamin B_2—Riboflavin

Source. Riboflavin occurs in many of the same sources as thiamine. It is found in yeast, milk, liver, kidneys, heart meats, and leafy vegetables. Cereals contain very little riboflavin unless it is added artificially.

Structure. Riboflavin or vitamin B_2 has been found to consist of a five-carbon sugar alcohol (ribitol) and a pigment (flavin). Its structure is shown below.

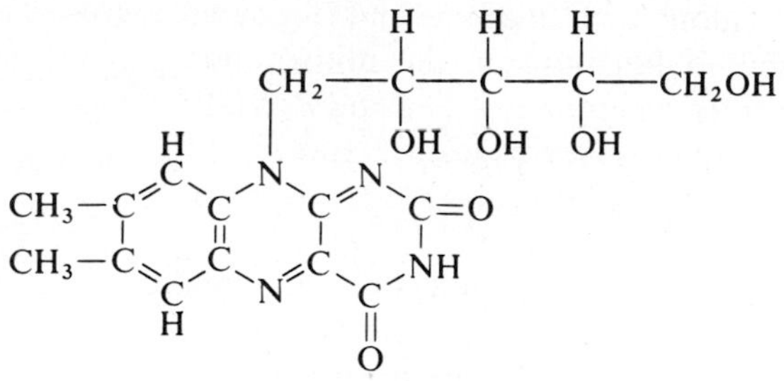

riboflavin

Properties. Riboflavin is an orange-red crystalline solid, slightly soluble in water and alcohol but insoluble in fats and fat solvents. In water solution riboflavin forms a greenish-yellow fluorescent liquid. Riboflavin is destroyed by light and alkaline solutions but is fairly stable to heat and so is not destroyed by cooking.

Riboflavin acts as a coenzyme in two different forms—flavin adenine dinucleotide (FAD) (see page 333) and flavin mononucleotide (FMN). These riboflavin coenzymes act as acceptors for the transfer of protons between NAD^+ and $NADP^+$ and the cytochromes which transport electrons in the mitochondria (see page 367).

Effect of Deficiency. A deficiency of riboflavin in man produces lesions in the corners of the mouth (cheilitis), inflammation of the tongue (glossitis), and lesions on the lips, and around the eyes and nose. There is also an inflammation of the skin (dermatitis) and a clouding of the cornea of the eye.

In rats a deficiency of riboflavin produces dermatitis, clouding of the corneas, and loss of hair.

Niacin

Source. Niacin (formerly known as nicotinic acid or vitamin B_5) is widely distributed in plants and animals. It is found in liver, kidney, and heart meats as well as in yeast, peanuts, and wheat germ. Milk, eggs, and fruit contain some niacin but are generally classified as poor sources of that vitamin.

Structure and Properties. Niacin and niacinamide, which it readily forms in the body, have the following structures.

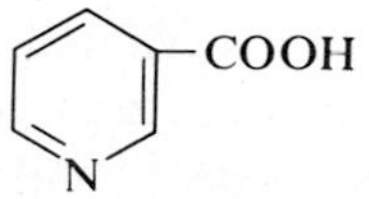

niacin 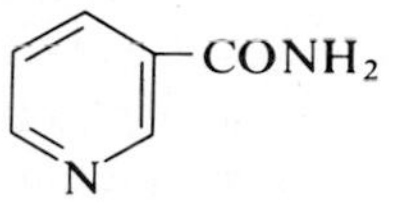   niacinamide

Niacin is slightly soluble in water but quite soluble in alkali. It is insoluble in fats. Niacin is stable to alkalis and acid, to heat and light, and is not destroyed by cooking.

Niacinamide, along with thiamine and riboflavin, serves as a coenzyme in tissue oxidation. It functions in the mitochondria in the form of NAD (nicotinamide adenine dinucleotide) and $NADP^+$ (nicotinamide adenine dinucleotide phosphate) (see pages 333 and 334).

Daily Requirement. The recommended daily intake of niacin is 6.6 mg per 1000 Cal, but not less than 13 mg if the caloric intake is less than 2000 Cal. There is a slight increase in requirements for adolescents and during pregnancy and lactation. However, these requirements can be greatly affected by the protein of the diet because the amino acid tryptophan can supply much of the body's needed niacin (60 mg of tryptophan equals 1 mg of niacin). Some niacin may also be synthesized by intestinal bacterial action and thus become available for use in the body.

Effect of Deficiency. Niacin was originally called nicotinic acid or the antipellagra factor. The word *pellagra* comes from the Italian words *pelle agra* meaning rough skin. A deficiency of niacin in humans produces serious consequences, although a deficiency in this vitamin is usually accompanied by a deficiency in other substances also. In pellagra there is a dermatitis (skin rash or lesions) and an inflammation of the mouth and tongue (glossitis). These symptoms are accompanied by diarrhea and then dementia.

A lack of niacin in dogs produces tongue lesions called black tongue. A diet containing niacin is effective in curing pellagra in man and also niacin deficiency diseases in animals.

Pellagra was quite common in the South where the diet consisted chiefly of corn and fat pork. Corn has a low tryptophan content and can give very little niacin. Fat pork also has very little niacin. Thus there was a deficiency of this vitamin, leading to pellagra. With an improvement in the diet, especially with the addition of foods containing niacin (or tryptophan), pellagra is not as common in the United States as it was previously.

Pyridoxine

Structure. Pyridoxine was originally called vitamin B_6, or the rat antidermatitis factor. Subsequent work showed that vitamin B_6 was a mixture of pyridoxine, pyridoxal, and pyridoxamine. The generally accepted term for these compounds is pyridoxine since these compounds are readily interconvertible. The structures are

CHO   CH₂OH   CH₂NH₂

$$\text{pyridoxal} \qquad \text{pyridoxine} \qquad \text{pyridoxamine}$$

Pyridoxine is found in yeast, liver, egg yolk, and the germ of various grains and seeds. It is also found to a limited extent in milk and leafy vegetables.

Daily Requirement. The recommended requirement of pyridoxine for the adult is 2 mg per day.

Effect of Deficiency. A deficiency of pyridoxine in rats produces dermatitis in the paws, nose, and ears. A deficiency in dogs and pigs produces anemia. If the deficiency of pyridoxine is continued for a long period of time, these animals suffer from epileptiform fits.

A deficiency of pyridoxine in infants produces convulsions. A deficiency of this vitamin in adults produces such symptoms as dermatitis, sore tongue, irritability, and apathy.

A diet containing pyridoxine (or vitamin B_6) will alleviate the above symptoms.

Pyridoxal phosphate and pyridoxamine phosphate serve as coenzymes for the decarboxylation of amino acids, taking part in the reactions occurring primarily in the gray matter of the central nervous system. It is believed that a deficiency of these coenzymes interferes with decarboxylation reactions in the central nervous system and so leads to epileptiform seizures. Pyridoxal phosphate and pyridoxamine phosphate also serve as coenzymes in amino acid metabolism.

Pantothenic Acid

Sources. Pantothenic acid has a widespread distribution in nature. Its name comes from the Greek word meaning "from everywhere." Good sources of pantothenic acid are egg yolk, yeast, kidney, and lean meats. Other fairly good sources are lean beef, skimmed milk, broccoli, sweet potatoes, and molasses.

Structure and Properties. Pantothenic acid is a viscous yellow oil, soluble in water but insoluble in fat solvents such as chloroform. It is stable in acid and alkaline solution. Pantothenic acid is one of the constituents of CoA which is involved in the metabolism of carbohydrates, fats, and proteins and in the synthesis of cholesterol (see pages 333 and 385).

The structures of pantothenic acid and of CoA are shown below.

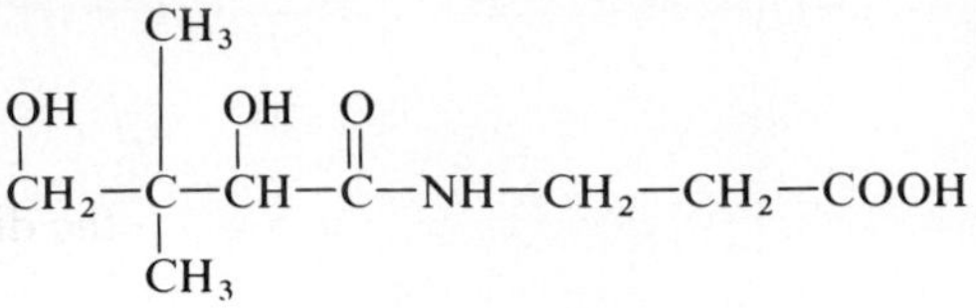

$$CH_2(OH)-C(CH_3)(CH_3)-CH(OH)-C(=O)-NH-CH_2-CH_2-COOH$$

pantothenic acid

pyrophosphate

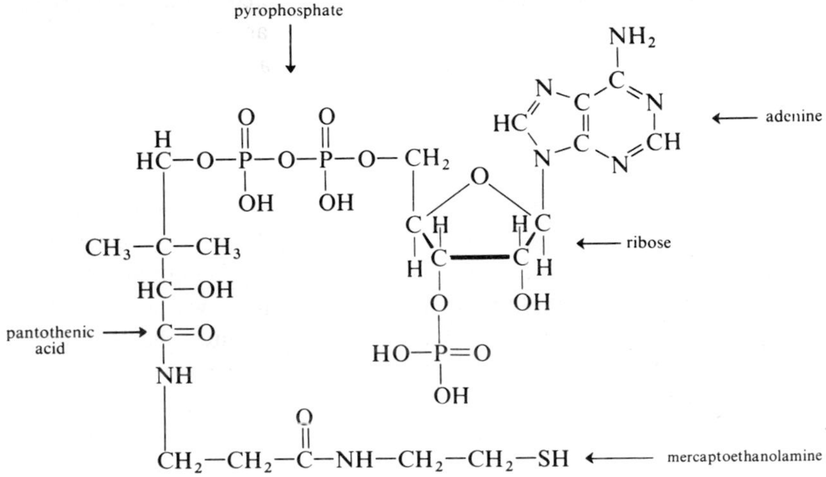

coenzyme A

The daily human requirement of pantothenic acid is not known, but estimates of 5 to 10 mg are usually considered adequate and are easily met from an ordinary diet.

Effect of Deficiency. Pantothenic acid was originally known as "chick antidermatitis factor" because it was a substance that prevented dermatitis in chicks.

Rats and dogs who were given a diet deficient in pantothenic acid showed a loss of pigmentation from their hair. The black hair of such animals turned gray but returned to its original black color upon the addition of pantothenic acid to the diet. There is no evidence that this vitamin is of significant value in restoring hair color in man. There is little evidence of pantothenic acid deficiency in humans.

A deficiency of pantothenic acid in animals causes degeneration in the adrenal cortex and a failure in reproduction.

Lipoic Acid

Lipoic acid was first detected in the studies of the growth of lactic acid bacteria. It is fat soluble and so was called lipoic acid. Its structure is

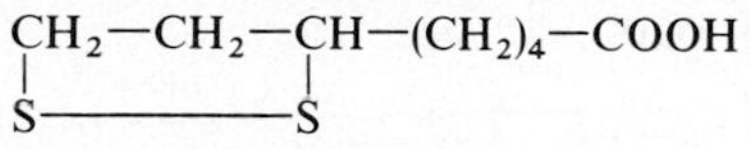

$$CH_2-CH_2-CH-(CH_2)_4-COOH$$
$$S\text{————}S$$

lipoic acid

As far as is known, lipoic acid is not required in the diet of higher animals and no deficiency effects have been noted.

Actually, lipoic acid is not a true vitamin; however, because its coenzyme function in carbohydrate metabolism is closely related to that of thiamine, it is classified along with the B-vitamin group. Lipoic acid functions, along with thiamine, in the initial decarboxylation of α-keto acids to form acetyl CoA for the citric acid cycle.

Biotin

Biotin, another member of the vitamin B complex, is widely distributed in nature. Rich sources of this vitamin are liver, egg yolk, kidneys, yeast, and milk. Biotin was formerly known as the anti-egg-white injury factor. This name was given to it because rats fed raw egg white failed to grow and also developed dermatitis. Raw egg white contains a protein, avidin, which combines with biotin and renders it unavailable to the animal.

The structure of biotin is

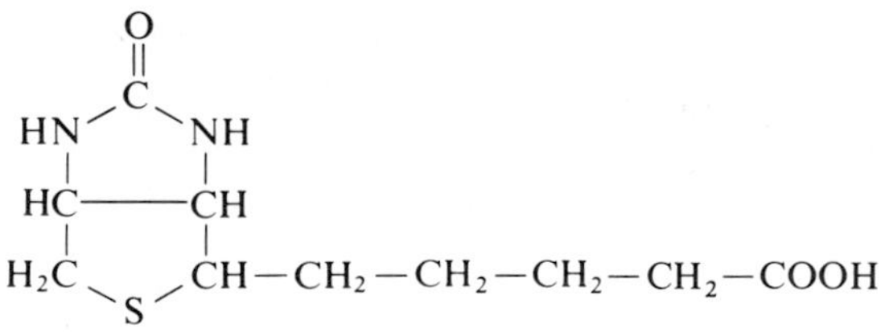

biotin

An artificially produced deficiency of biotin in man causes scaly dermatitis, nausea, muscle pains, and depression. These symptoms are rapidly relieved by the administration of a diet containing biotin.

Biotin is supplied by the action of intestinal bacteria, in man as well as in animals, so that a deficiency of this vitamin is unlikely in most cases, except on a severely restricted diet. Biotin functions as a coenzyme for carboxylation reactions in the formation of fatty acids (see page 380).

Folic Acid

Folid acid (folacin) occurs in green leaves, yeast, liver, kidneys, and cauliflower. The structure of folic acid (which contains a pteridine nucleus, p-aminobenzoic acid, and glutamic acid) is shown below.

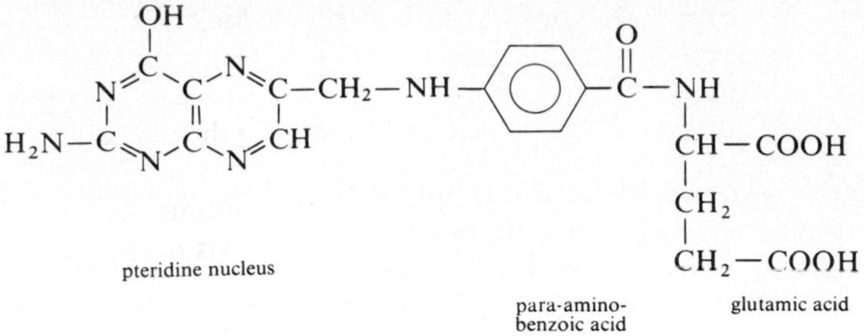

pteridine nucleus   para-amino-benzoic acid   glutamic acid

Folic Acid (Folacin)

Actually, the above compound, which is also called pteroylglutamic acid (PGA) is only one of several related compounds in the folic acid group. Others are pteroic acid, pteroyltriglutamic acid, and pteroylheptaglutamic acid.

Pernicious anemia, macrocytic anemia (presence of giant red corpuscles), and sprue are greatly benefited by the addition of folic acid to the diet.

Folic acid in its reduced form, tetrahydrofolic acid, acts as a coenzyme for the transfer of methyl groups in the formation of such compounds as choline and methionine.

Para-aminobenzoic Acid

Para-aminobenzoic acid (PABA) is a growth factor for certain microorganisms. It forms part of the folic acid molecule and is believed to be necessary for the formation of that vitamin. However, man is incapable of using para-aminobenzoic acid to produce folic acid. Para-aminobenzoic acid is formed upon the hydrolysis of folic acid. The structure of para-aminobenzoic acid is shown below.

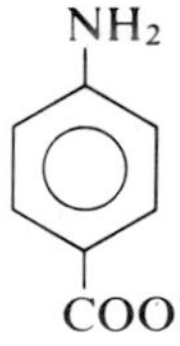

para-aminobenzoic acid

Compare the structure of sulfanilamide with that of para-aminobenzoic acid given above.

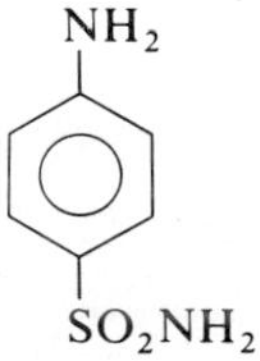

sulfanilamide

Sulfonamides exert their antibacterial action by acting as antimetabolites to para-aminobenzoic acid. One theory is that the sulfa drugs are similar in structure to para-aminobenzoic acid and so attach themselves to an enzyme required in the metabolism of PABA, thus blocking the use of that substance (see page 241).

Para-aminobenzoic acid is not regarded as a vitamin for man, but it is included here because it is necessary for the synthesis of folic acid by some microorganisms.

Inositol

Inositol is found in liver, milk, vegetables, yeast, whole grains, and fruits. The molecular formula for inositol is $C_6H_{12}O_6$, so it is an isomer of glucose. Its structure is shown below.

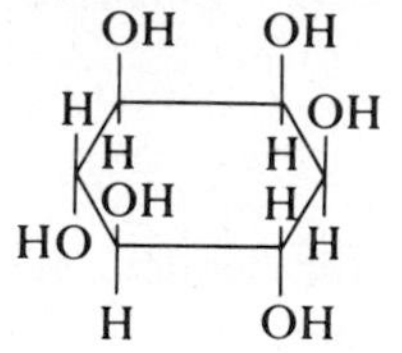

inositol (hexahydroxycyclohexane)

It was found that mice, on a synthetic diet containing all of the known vitamins, still failed to grow. Also there was an effect on their hair and impaired lactation. Pantothenic acid, which affects hair in mice, did not help these animals. However, these symptoms were overcome by the addition of a compound obtained from cereal grain. This compound was inositol. Inositol is required for the growth of yeasts, mice, rats, guinea pigs, chickens, and turkeys.

Inositol, along with folic acid and para-aminobenzoic acid, is not a true vitamin. However, it is included with the B group even though no specific role in human nutrition has been established. Recall that inositol is a component of the phospholipid phosphatidylinositol (see page 294).

Cyanocobalamin—Vitamin B_{12}

Source. Plants do not contain this vitamin. Microorganisms are able to synthesize it. The best sources of cyanocobalamin are liver and kidneys. Animal and fish muscle contain some of this vitamin.

Structure and Properties. Cyanocobalamin is an odorless, tasteless, reddish crystalline compound, soluble in water and alcohol and insoluble in fat solvents such as ether and acetone. One unusual property of this vitamin is that it contains the element cobalt (4.35 per cent). The structure of cyanocobalamin is shown below.

cyanocobalamin

Daily Requirement. The recommended daily amount of vitamin B_{12}, cyanocobalamin, is 3 micrograms.

Effect of Deficiency. Vitamin B_{12} (cyanocobalamin) is also called the anti-pernicious anemia factor. It is absorbed from the small intestine in the presence of hydrochloric acid and the intrinsic factor (see page 345). In the absence of the intrinsic factor, vitamin B_{12} cannot be absorbed, and this leads to pernicious anemia. Injection of a small amount of cyano-cobalamin will produce remarkable improvement in that disease. Pernicious anemia results from a lack of the intrinsic factor in the gastric juice rather than to a lack of vitamin B_{12}.

Since vitamin B_{12} occurs in most animal foods, a dietary deficiency is rare.

Derivatives. Coenzyme B_{12} is one of several coenzymes derived from vitamin B_{12}. Collectively they are called cobamides. Cobamides are required, as coenzymes, for hydrogen transfer and isomerization in the conversion of methyl malonate to succinate, thus involving both carbohydrate and fat metabolism.

Coenzyme B_{12} differs from vitamin B_{12} in that a 5-deoxyadenosine group replaces the CN group attached to the central cobalt.

Choline

Choline is an essential substance as far as the body is concerned although it is generally not classified as a vitamin because it can be synthesized in the required amounts in the body and it is required in much larger amounts than vitamins.

Choline is a viscous, colorless liquid, soluble in water and alcohol but insoluble in ether. Its structure is

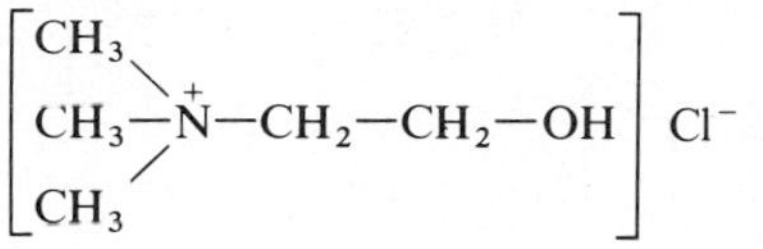

$$\left[\begin{array}{l} CH_3 \\ CH_3-\overset{+}{N}-CH_2-CH_2-OH \\ CH_3 \end{array} \right] Cl^-$$

choline

A deficiency in choline leads to such symptoms as fatty liver. Young rats on a diet deficient in choline also had hemorrhagic degeneration of the kidneys. Older rats who survived these symptoms developed cirrhosis. Chicks and young turkeys develop perosis or slipped tendon disease.

Choline is a constituent of lecithin (phosphatidyl choline) and sphingomyelin and is important in brain and nervous tissue. A deficiency of choline will not develop in a person on a high-protein diet because proteins supply the amino acids from which the body can synthesize this compound.

Choline is a constituent of acetylcholine, which is present in nerve cells and aids in the transmission of nerve impulses by the following means. When a nerve cell is stimulated, it releases acetylcholine. This acetylcholine in turn stimulates the adjacent nerve cell to also release acetylcholine. This process continues as a chain reaction until the impulse reaches the brain. Once a nerve cell has passed its impulse on to the next cell, the enzyme acetylcholinesterase hydrolyzes acetylcholine into acetic acid and choline. The nerve cells then use these two compounds to regenerate the acetylcholine for the next impulse.

Ascorbic Acid—Vitamin C

Source. Fresh fruits and vegetables such as oranges, lemons, grapefruit, berries, melons, tomatoes, and raw cabbage are excellent sources of vitamin C. Dry cereals, legumes, milk, meats, and eggs contain very little of this vitamin.

Structure and Properties. Vitamin C, ascorbic acid, is a white crystalline substance soluble in water and alcohol but insoluble in most fat solvents. Ascorbic acid is a strong reducing agent and is easily oxidized in air, especially in the presence of such metallic ions as Fe^{3+} or Cu^{2+}. Ascorbic acid is rapidly destroyed by heating. For this reason, cooking of foods in copper pots should be avoided because this will destroy the vitamin C content of the foods being cooked.

It can easily be oxidized to dehydroascorbic acid. Both ascorbic acid and dehydroascorbic acid are biologically active, and both have been synthesized in the laboratory. Their structures are shown below.

$$
\begin{array}{c}
\text{C}=\text{O} \\
\text{HO}-\text{C} \\
\text{HO}-\text{C} \\
\text{H}-\text{C} \\
\text{HO}-\text{C}-\text{H} \\
\text{CH}_2\text{OH}
\end{array}
\quad\rightleftharpoons\quad
\begin{array}{c}
\text{C}=\text{O} \\
\text{O}=\text{C} \\
\text{O}=\text{C} \\
\text{H}-\text{C} \\
\text{HO}-\text{C}-\text{H} \\
\text{CH}_2\text{OH}
\end{array}
$$

ascorbic acid   dehydroascorbic acid

Daily Requirement. The recommended daily intake of ascorbic acid for an adult is 45 mg; the amount is increased to 60 mg during pregnancy and 80 mg during lactation.

Effect of Deficiency. Plants and all animals except pigs, man, and other primates are able to synthesize ascorbic acid and are resistant to diseases caused by a lack of this vitamin. A deficiency of vitamin C produces a disease known as scurvy. The symptoms in man are swollen, bleeding gums, pain in the joints, decalcification of the bones, loss of weight, and anemia.

A deficiency of vitamin C prevents the body from forming and maintaining the intercellular substance that cements the tissues together. A lack of this intercellular substance in the capillaries leads to rupturing and subsequent hemorrhaging in these vessels; to the formation of weak bones and atrophy of the bone marrow, accompanied by anemia; and also accounts for loosening of the teeth and spongy gums. All of these symptoms are relieved by the addition of ascorbic acid to the diet.

Vitamin C is also believed to function in oxidation-reduction reactions and in cellular respiration.

Linus Pauling, Nobel prize winner in chemistry, has suggested that the common cold can be prevented or treated by the administration of large doses of ascorbic acid. To date there have been many efforts to prove or disprove this hypothesis, but since the cause of a cold itself is not known, the controversy still continues (see references at the end of this chapter).

Summary

The body requires vitamins in addition to carbohydrates, fats, and proteins but in much smaller amounts. Vitamins must be supplied in the diet, whereas hormones are synthesized by the body.

Vitamins are divided into two types—the fat soluble and the water soluble.

Vitamin A is found in fish liver oils, in butter, and in milk. Vitamin A is an alcohol and occurs in two forms—vitamin A_1 and vitamin A_2, with A_1 being more potent. The recommended daily adult dose of vitamin A is 1000 retinol equivalents (5000 IU) for the adult male and 800 retinol equivalents (4000 IU) for the adult female. Both A vitamins have an all-*trans* structure.

A lack of vitamin A produces keratinization in the membranes of the eyes, digestive tract, respiratory tract, and genitourinary tract. Nyctalopia is also due to a deficiency of vitamin A. Vitamin A is stored in the liver.

Vitamin D is found in fish liver oils such as cod or halibut. The vitamin D content of milk is increased by irradiation. The D vitamins are sterols with a structure similar to that of cholesterol. Vitamin D functions to increase the absorption of calcium and phosphorus from the small intestine. Vitamin D also functions in the deposition of calcium phosphate in the teeth and bones. A lack of vitamin D produces rickets.

Vitamin E is found in many foods; wheat germ is particularly rich in this substance. The E vitamins have antioxidant properties and act as a cofactor in oxidative phosphorylation reactions. Vitamin E prevents sterility in animals but its necessity in humans has not yet been determined.

Vitamin K is known as the antihemorrhaging vitamin. It is necessary for the production of the prothrombin in the liver.

Vitamin B_1 (thiamine) occurs naturally in yeast, milk, and whole grains and also may be made synthetically. A deficiency of this vitamin causes a lack of appetite, failure of growth, and loss in weight. A prolonged deficiency of vitamin B_1 leads to the disease known as beriberi (or polyneuritis in animals). Thiamine is also necessary for the normal metabolism of carbohydrates. This vitamin acts as a coenzyme for the decarboxylation of pyruvic acid. The coenzyme, known as cocarboxylase, is necessary in the Krebs cycle for the conversion of alpha-ketoglutaric acid to succinyl CoA.

Vitamin B_2 (riboflavin) is found in the same sources as thiamine. Riboflavin acts as a coenzyme in two different forms flavin adenine dinucleotide (FAD) and flavin mononucleotide (FMN). These coenzymes act as acceptors for the transfer of protons between NAD^+ and $NADP^+$ and the cytochromes.

Niacin (or nicotinic acid) is widely distributed in nature. A deficiency of this vitamin produces a condition known as pellagra. Niacin is an important constituent of two coenzymes—nicotinamide adenine dinucleotide (NAD^+) and nicotinamide adenine dinucleotide phosphate ($NADP^+$). These coenzymes are involved in most oxidation-reduction reactions in the mitochondria.

Pyridoxine is involved in the decarboxylation of amino acids and is also required for certain transaminase reactions.

Pantothenic acid is one of the constituents of CoA, which is involved in the metabolism of carbohydrates, fats, and proteins as well as in the synthesis of cholesterol.

Lipoic acid is closely related to thiamine in the initial oxidation of alpha-keto acids.

Biotin functions in the activation of carbon dioxide for carboxylation reactions in the formation of fatty acids.

Folic acid is concerned with the transfer of methyl groups in the formation of such compounds as choline and methionine.

Para-aminobenzoic acid is formed upon the hydrolysis of folic acid. This substance is a growth factor for certain microorganisms.

Inositol is required for the growth of yeasts, mice, and rats, and for phosphatidyl inositol.

Vitamin B_{12}, cyanocobalamin, is called the anti-pernicious anemia factor. It is also involved in the synthesis of certain amino acids and of choline, and is necessary for the formation of coenzyme B_{12}.

Choline is a constituent of lecithin and so is important in brain and nerve tissue. Choline is also a constituent of acetylcholine, which aids in the transmission of nerve impulses.

Vitamin C, ascorbic acid, may be synthesized by plants and most animals but not by man. A lack of this vitamin produces a disease known as scurvy. A lack of ascorbic acid also causes a lack of the intercellular substance that cements the tissues together.

Questions and Problems

1. Compare vitamins with hormones.
2. Which vitamins are water soluble? Fat soluble?
3. What are the sources of vitamin A?
4. What is the structural difference between vitamin A_1 and A_2? Which is more potent? What arrangement do the A vitamins have around their double bonds?
5. What is beta-carotene? How is it used in the body?
6. What are the effects of a deficiency of vitamin A?
7. Where is vitamin A stored in the body? How is it absorbed?
8. Why does the body need vitamin A? What causes hypervitaminosis A? What are the symptoms?
9. Describe the role of vitamin A in the visual process.
10. What are the sources of vitamin D?
11. Why is vitamin D called the sunshine vitamin?
12. What type of compound is vitamin D? Are all the D vitamins equally potent?
13. What are the daily adult requirements of vitamin A? D?
14. What are the functions of vitamin D in the body?
15. What are the effects of a lack of vitamin D? Of an excess?
16. What are the sources of vitamin E? What are its properties? Functions?
17. What are the sources of vitamin K? What are its properties? Functions?
18. What are the sources of vitamin B_1? What are its properties? Functions?
19. What are the effects of a lack of thiamine?
20. What are the sources of riboflavin?
21. What are the sources of niacin? Pyridoxine?
22. What are the effects of a deficiency of niacin? Pyridoxine? How does the body use these substances?
23. What are the sources of pantothenic acid? Its properties? Effects of a deficiency?
24. What are the functions of lipoic acid, biotin, folic acid, p-aminobenzoic acid, inositol, and cyanocobalamin?
25. What are the sources of vitamin C? Its properties? Functions?
26. What are the effects of a deficiency of ascorbic acid?
27. How do sulfonamides exert their antibacterial effect?
28. What are cobamides? What do they do?
29. How are nerve impulses transmitted?

References

Butterworth, C. E., Jr.: Multivitamins, megavitamins, and orthomolecular psychiatry. *Hospital Formulary*, **10**:8–9 (Feb.), 1975.

Cheraskin, E.; Ringsdorf, W. M.; and Hicks, B. S.; Daily vitamin C consumption and reported cardiovascular findings. *Journal of the International Academy of Preventive Medicine*, **1**:31–44 (Spring), 1974.

Dowling, J. E.: Night blindness. *Scientific American*, **215**:78–84 (Oct.), 1966.

Goodman, L. S., and Gilman, Λ. (eds.): *The Pharmacological Basis of Therapeutics*, 5th ed. Macmillan Publishing Co., New York, 1975, Sec. XVIII.

Harper, H. A.: *Review of Physiological Chemistry*, 15th ed. Lange Medical Publications, Los Altos, Calif., 1975, Chap. 7.

Klenner, F. R.: The significance of high daily intake of ascorbic acid in preventive medicine. *Journal of the International Academy of Preventive Medicine*, **1**:45–69 (Spring), 1974.

Leslie, C. R.: The action of vitamin C on blood vessels. *Health*, **11**:13–15 (Winter), 1974/1975.

Loomis, W. F.: Rickets. *Scientific American*, **223**:76–91 (Dec.), 1970.

———: Zinc: a trace element essential in vitamin A metabolism. *Science*, **181**:954–955 (Sept. 7), 1973.

Scrimshaw, N. S., and Young, V. R.: The requirements of human nutrition. *Scientific American*, **235**:51–64 (Sept.), 1976.

Chapter 32

Hormones

Hormones exert a very important influence on the regulation of body processes. Hormones are produced in the endocrine glands (see Figure 32–1). These glands are also called the ductless glands because the hormones are secreted directly into the bloodstream instead of passing through ducts to the site of their need. Hormones may be proteins, polypeptides, amino acids,

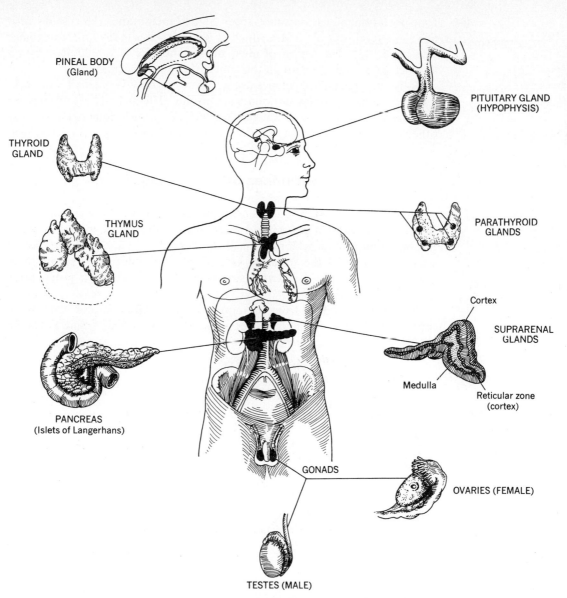

PINEAL BODY
(Gland)

PITUITARY GLAND
(HYPOPHYSIS)

THYROID
GLAND

PARATHYROID
GLANDS

THYMUS
GLAND

Cortex

SUPRARENAL
GLANDS

Medulla

Reticular zone
(cortex)

PANCREAS
(Islets of Langerhans)

GONADS

OVARIES (FEMALE)

TESTES (MALE)

Figure 32–1. Locations of endocrine glands in the body. (Reproduced from Pansky, B.: *Dynamic Anatomy and Physiology*. Macmillan Publishing Co., Inc., New York, 1975.)

or steroids. There is an internal balance and interaction among the various endocrine glands.

Hormones act in several different ways. Some stimulate RNA production in target cell nuclei and thus increase the production of enzymes. Some hormones stimulate enzyme synthesis in the ribosomes through the translation of information carried by messenger RNA. Others are involved in the

467

transportation of various substances across membranes. Hormones also affect the levels of cylic-AMP, which in turn activates many kinase enzymes (see page 323). The action of most protein hormones is inhibited by an absence or a large decrease in the concentration of calcium ions.

Both vitamins and hormones are necessary in only very small amounts; however, unlike vitamins, hormones are produced in the body. Hormones are produced in a particular organ and are carried by the bloodstream to some other body part, where they produce the desired specific physiologic effect.

Hormones of the Stomach

The pyloric mucosa produces a hormone called gastrin. *Gastrin* is absorbed into the bloodstream and carried back to the stomach where it stimulates the secretion of hydrochloric acid in the gastric juice and also produces a greater motility of the stomach. Gastrin also stimulates the secretion of pepsin and the intrinsic factor (see page 345). Gastrin is a polypeptide containing 17 amino acids.

Histamine, produced from the amino acid histidine, is also effective in increasing the flow of gastric juices. The structure of histamine is shown below.

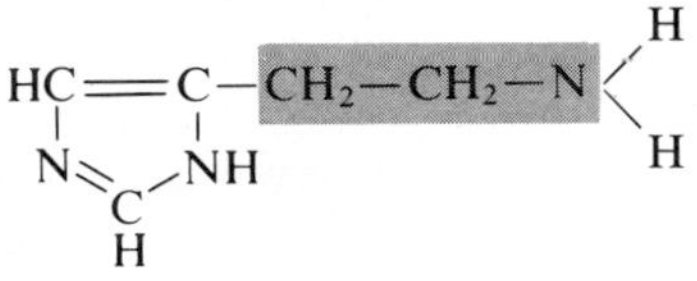

histamine

Histamine also causes the dilation of capillaries and small blood vessels. If this happens in the brain a severe histamine headache may occur. Histamine may also cause respiratory problems in persons suffering from pulmonary diseases such as bronchial asthma.

Antihistamines are drugs that reduce the effect of allergic reactions by opposing or limiting the effects of histamine. Compare the structures of two antihistamines with that of histamine (shown above).

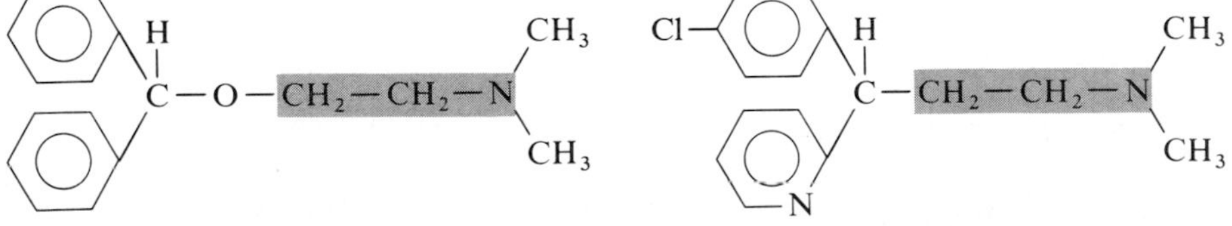

diphenhydramine

Antihistamines

chlorpheniramine

The antihistamines appear to function by occupying the sites normally taken by histamine and so blocking that substance from reacting in its normal fashion.

Secretin is formed in the mucosa of the small intestine when the acidic chyme enters the duodenum. Secretin stimulates the pancreas to release the pancreatic juices into the small intestine. Secretin is a polypeptide containing 27 amino acids.

Cholecystokinin-pancreozymin (CCK-PZ) is secreted when fat enters the duodenum. CCK-PZ stimulates the contraction and emptying of the gall bladder into the small intestine. It also stimulates pancreatic enzyme secretion. CCK-PZ is a polypeptide containing 33 amino acids.

Hormones of the Pancreas

The pancreas secretes digestive enzymes (see Chapter 25). It also secretes two hormones that affect carbohydrate metabolism. These hormones are insulin and glucagon.

Insulin

Insulin is a protein secreted by the beta cells of the islets of Langerhans in the pancreas. Insulin has been isolated and crystallized. Crystalline insulin contains a small amount of zinc, which it obtains from the zinc-rich tissues in the pancreas. Small amounts of chromium are also needed for the synthesis of insulin. The amount of chromium in the body decreases with age. The amino acid sequence in insulin is indicated on page 308.

Insulin increases the rate of oxidation of glucose, facilitates the conversion of glucose to glycogen in the liver and the muscles, and increases the synthesis of fatty acids, protein, and RNA. Insulin functions to decrease blood sugar to its normal fasting level after it has been increased by digestion of carbohydrates by facilitating the transportation of glucose through the membranes into the cells.

Normally, the blood sugar level rises to 120 to 160 mg per 100 ml of blood after a meal containing carbohydrates. This level is returned to its normal fasting value by the action of insulin. If the islets of Langerhans are underactive or degenerated, little or no insulin is produced so that the blood sugar level remains high. This condition is termed hyperglycemia and is associated with diabetes mellitus. In diabetes mellitus, there is an increased blood sugar level; glucose appears in the urine, and there is formation of acetone bodies accompanied by acidosis. Injection of insulin will produce a rapid recovery from these symptoms. It will not, however, cure diabetes because these same symptoms will reappear after a short time. Thus insulin must be taken by a diabetic for the rest of his life.

Since insulin is a protein, it cannot be taken orally because it would be digested. Therefore, insulin is given by subcutaneous injection (see Figure

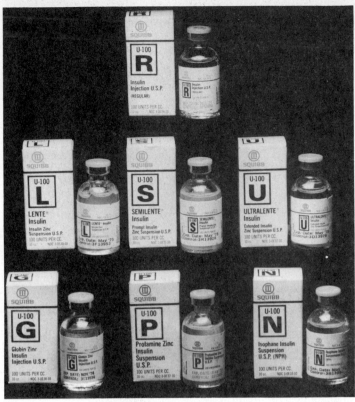

Reusable

A

Disp. 100

B

Figure 32–2. *A*. Insulin syringes. (Courtesy of Becton, Dickinson and Co., Rutherford, N.J.) *B*. Insulin preparations. (Courtesy of Squibb Professional Services Dept., Princeton, N.J.)

32–2, *A*). Injections of ordinary insulin must be given two or three times a day to a patient with diabetes mellitus. However, when insulin is combined with protamine (a protein), the product, called protamine zinc insulin, is absorbed much more slowly and is effective for more than 24 hr. Thus only one injection is needed daily.

Another hormone, somatostatin (see page 485), is being tried either by itself or in combination with insulin for the control of diabetes. Somatostatin inhibits the release of glucagon, one of the hormones that is responsible for maintaining high blood sugar levels.

If too large a dose of insulin is administered to a person, the blood sugar level falls far below its normal fasting value. This condition is called hypoglycemia and is characterized by such symptoms as dizziness, nervousness, blurring of vision, and then unconsciousness. Such a state is called insulin shock and may be relieved by the administration of sugar, either orally or by injection. Hypoglycemia also occurs when there is a tumor on the islets of Langerhans in the pancreas. Insulin is degraded primarily in the liver and kidneys by the enzyme glutathione insulin transhydroxylase.

Insulin is available in sterile solutions whose concentration is such that 1 ml contains 100 units (U-100) (see Figure 32–2, *B*). However, U-40 and U-80 insulin are still in use but are gradually being phased out. One unit of insulin is the amount required to reduce the blood sugar level of a normal 2-kg rabbit after a 24-hr fast from 120 to 45 mg per 100 ml.

Although insulin cannot be taken orally, there are certain hypoglycemic substances that will lower blood sugar level. These substances are effective in treating "adult diabetes" but not "juvenile diabetes." They function by stimulating the beta cells of the islets of Langerhans to produce insulin. Examples of such hypoglycemic agents are tolbutamide (Orinase) and chlorpropamide (Diabinese). The structural formula of tolbutamide is shown below.

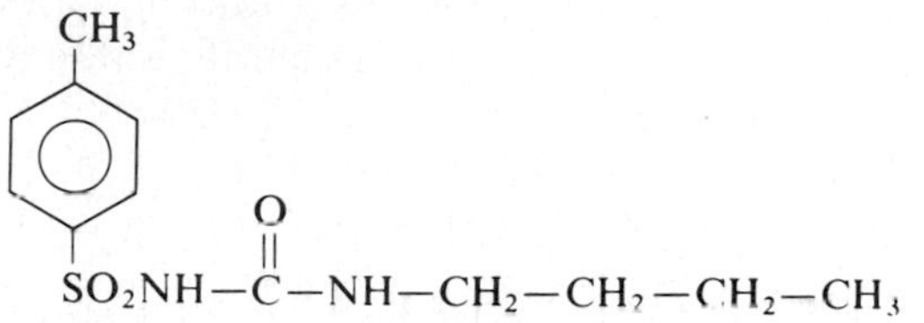

$$CH_3$$

$$SO_2NH-\overset{\overset{\displaystyle O}{\|}}{C}-NH-CH_2-CH_2-CH_2-CH_3$$

tolbutamide

Recent evidence indicates that these so-called hypoglycemic agents actually may be ineffective or even toxic. Adult patients with a mild type of diabetes frequently can control it just as well by means of a proper diet.

Glucagon

When crude insulin was first used to lower blood sugar levels, it was noted that a temporary hyperglycemia occurred first and then, soon afterward, came the hypoglycemic effects expected of insulin. The unknown substance present in the crude insulin that gave the hyperglycemic effect was called the hyperglycemic-glycogenolytic factor, or HGF. This factor is now known as glucagon. Glucagon has been isolated and crystallized. It is a polypeptide containing 29 amino acids in a straight chain.

Glucagon has a different arrangement of amino acids than docs insulin. It contains no disulfide bridges as does insulin. Glucagon contains methionine and tryptophan, which insulin does not. However, insulin contains cystine, proline, and isoleucine, whereas glucagon does not.

Glucagon is formed in the alpha cells of the islets of Langerhans in the pancreas.

Glucagon causes an increase in the sugar content of the blood by stimulating phosphorylase activity in the liver. The reactions involved are as follows

$$\text{glycogen + phosphate} \xrightarrow{\text{phosphorylase}} \text{glucose-1-phosphate}$$

$$\text{glucose-1-phosphate} \xrightarrow{\text{phosphoglucomutase}} \text{glucose-6-phosphate}$$

$$\text{glucose-6-phosphate} \xrightarrow{\text{phosphatase}} \text{glucose + phosphate}$$

Glucagon increases the formation of cyclic-AMP, which in turn activates phosphorylase, and so increases the glucose content of the blood, causing hyperglycemia.

Glucagon has been used to treat hypoglycemic effects due to an overdose of insulin or after the inducement of insulin shock in the treatment of psychiatric patients.

Hormones of the Thyroid Gland

The thyroid gland is an H-shaped gland consisting of one lobe on each side of the trachea with a piece of tissue connecting the two lobes. In the adult the thyroid gland weighs approximately 25 to 30 g.

The hormones of the thyroid gland regulate the metabolism of the body. They also affect the growth and development of the body.

The thyroid is filled with many small follicles that contain colloid. This colloid contains the thyroid gland's stored hormones. Actually, the thyroid is the only endocrine gland in the body that is capable of storing appreciable amounts of hormone.

The thyroid gland contains the element iodine—one of the elements necessary for the proper functioning of the body. Iodine is found in the body in two different forms—as iodide ions and in thyroid hormones. The body uses the inorganic iodide ions to manufacture the iodine-containing thyroid hormones. The hormonal iodine transported by the blood is called the *protein-bound iodine*, or PBI. The normal PBI is 4 to 8 μg per 100 ml of blood plasma and is of diagnostic value in determining whether the thyroid is functioning normally. A high PBI can indicate hyperthyroidism but may also be found in pregnancy and in burns. A low PBI usually indicates hypothyroidism. The PBI is preferred to total blood iodine (inorganic iodine plus protein-bound iodine) because in hypothyroidism the PBI value will be low even though the total blood iodine may be normal. If iodine has been taken orally or if iodine compounds have been used for diagnosis or treatment of a condition, then the PBI value is of no diagnostic significance. In this case, the hormonal iodide is extracted with butanol and then measured. This method is called the butanol-extractable iodine, or BEI.

These two tests were the first developed for testing thyroid function. They have since been largely replaced by radioimmunoassay methods, which are much more reliable.

The colloid of the thyroid contains the protein thyroglobin, a glycoprotein. This protein liberates triiodothyronine and thyroxine, the principal thyroid hormones. Some mono- and diiodotyrosine are also formed, but these compounds are quickly deiodinated in the bloodstream and the freed iodine is used to form more thyroglobin. The structures of these hormones are shown below.

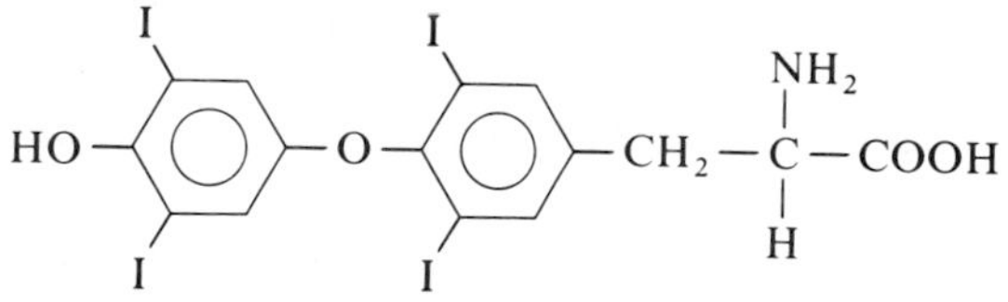

thyroxine (T$_4$)

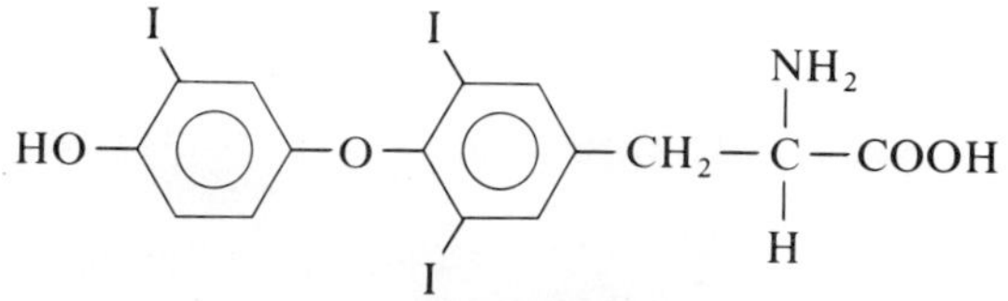

triiodothyronine (T$_3$)

The C cells in the thyroid gland produce the hormone calcitonin (thyrocalcitonin), which, along with the parathyroid hormone, regulates the calcium ions in the blood. It has been shown that the parathyroid hormone sustains the blood supply of calcium ions while calcitonin prevents the blood calcium ion concentration from rising above the required level.

Pure calcitonin was first isolated in 1968 and its structure determined shortly thereafter. This hormone is a polypeptide containing a single chain of 32 amino acids. Calcitonin has now been synthesized in the laboratory, and its mode of action in the body has been investigated. It produces its effect by inhibiting the release of calcium ions from the bone to the blood.

Hypothyroidism

Hypothyroidism refers to a condition in which the thyroid gland does not manufacture sufficient thyroxine for the body's needs. It is usually due to a lack of iodine in the diet, particularly in parts of the country where the water and foods contain little iodine. Hypothyroidism may also be due to a disease of the thyroid gland or to its congenital absence.

The symptoms of hypothyroidism are sluggishness, gain of weight, slower heartbeat, reduced metabolic rate, and loss of appetite. Hypothyroidism is easily remedied by the use of iodized salts as part of the normal diet.

Cretinism

If the thyroid gland is absent or fails to develop in an infant, the effects produced are called cretinism and the individual is called a cretin. Cretins have a greatly retarded growth, both physically and mentally. They are usually abnormal dwarfs with coarse hair, thick dry skin, obese with protruding abdomens. They are also underdeveloped mentally and sexually.

A cretin may develop normally if given thyroid hormones before he reaches adulthood (see Figure 32–3).

If the thyroid gland should atrophy after an individual reaches adulthood, the same symptoms as cretinism appear, except that the individual remains adult in size. One very noticeable symptom is in the development of thick, coarse, dry skin. Such a condition is known as *myxedema*. Persons with

Figure 32–3. Cretin B is much shorter than her twin A. After therapy with thyroid hormones the difference, although present, is greatly reduced. (Courtesy of Warner-Chilcott Laboratories, Morris Plains, N.J.)

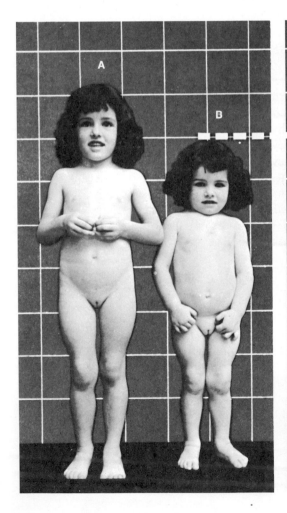
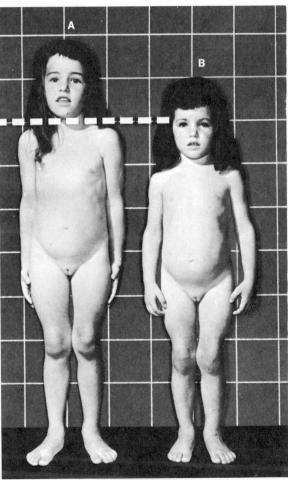

myxedema are also sluggish, have a lower pulse and metabolic rate and lower body temperature. They are also very sensitive to cold.

Myxedema may usually be cured by the administration of thyroxine.

Simple Goiter

Simple goiter, also called colloid goiter or endemic goiter, is a condition in which the thyroid gland enlarges, usually because of a lack of iodine in the diet. The decreased production of thyroid hormones causes an increased production of TSH, the thyroid-stimulating hormone, which in turn over-stimulates the thyroid gland. Simple goiter is accompanied by a definite increase in the amount of colloid material in the thyroid gland and also an increase in the size of the neck itself.

Simple goiter occurs in areas where there is a deficiency of iodine in the food and drinking water. The condition may be successfully prevented or cured by the addition of iodine compounds (usually iodized salt) to the diet.

Hyperthyroidism

Hyperthyroidism occurs when the thyroid gland produces excess thyroxine. The symptoms are an increased metabolic rate, bulging of the eyes (exophthalmos), nervousness, loss of weight, and a rapid, irregular heartbeat. Such a condition is also called Graves' disease, Basedow's disease, or exophthalmic goiter. Hyperthyroidism may also be due to a tumor in the thyroid gland (toxic adenoma or Plummer's disease).

Hyperthyroidism may be controlled or cured by surgical removal of part of the thyroid gland, by the oral administration of radioactive iodine, or by the use of antithyroid drugs. Hypertrophy of endocrine glands may lead to toxic adenomas with a malignant potential if not treated promptly.

Radioactive iodine, used in the treatment of hyperthyroidism, is usually administered in the form of sodium iodide, NaI. The body converts the inorganic radioactive iodide into thyroglobin in the thyroid gland, thus subjecting that gland to radiation that will cut down its activity.

Antithyroid Drugs

Certain drugs such as thiouracil, thiourea, propylthiouracil, and methimazole are given orally to counteract the effect of an overactive thyroid gland. These drugs perform their function by preventing the synthesis of thyroxine. The structure of two of these drugs is given below.

thiouracil   methimazole

Hormones of the Parathyroid Glands

There are four small parathyroid glands attached to the thyroid gland (see Figure 32–1). In man these glands are reddish brown and together weigh 0.05 to 0.3 g. In early experimental thyroidectomies (removal of the thyroid glands) in animals, the parathyroid glands were also inadvertently removed, which caused the death of the animals.

The parathyroid glands produce a hormone, parathormone, which influences the metabolism of calcium and phosphorus in the body. This hormone is a protein with a molecular weight of approximately 9500 and consists of a single polypeptide chain of 84 amino acids. The parathyroid gland cannot store this hormone, so it is synthesized and secreted continuously. Recall that calcitonin (page 473) is also involved in the regulation of calcium.

Surgical removal of the parathyroid glands causes hypoparathyroidism, characterized by such symptoms as muscular weakness, irritability, and tetany owing to a decrease in the calcium content of the blood plasma. Death occurs because of convulsions caused by the lack of calcium. At the same time as the calcium content of the plasma is decreasing, the calcium content of the urine is also decreasing and the phosphorus content of the plasma is increasing.

The symptoms of hypoparathyroidism may be relieved by treatment with vitamin D and/or calcium salts.

Hyperparathyroidism is an increase in the production of hormones by the parathyroid glands. It is usually due to a tumor of that gland (parathyroid adenoma) and produces such symptoms as decalcification of the bones followed by deformation and fractures of these bones, nausea, and polyuria. Deposits of calcium occur in soft tissues and renal stones frequently occur.

In hyperparathyroidism, the calcium content of the blood plasma is high and the phosphorus content low. The extra calcium in the blood is obtained by withdrawal of that element from the bones, thus causing those bones to become decalcified.

Hyperparathyroidism is usually treated by the surgical removal of the tumor of the parathyroid glands.

Hormones of the Adrenal Glands

The adrenal (suprarenal) glands are located close to the upper pole of the kidneys and weigh about 3 to 6 g each. The adrenal glands are divided into two distinct portions—the cortex, which is the outer portion, and the medulla, which is the inner portion (see Figure 32–4). Each of these portions is distinct both structurally and physiologically and each produces its own hormones.

Hormones of the Adrenal Cortex

The hormones of the adrenal cortex are steroid in nature and fall into three categories:

1. *Glucocorticoids,* which primarily affect the metabolism of carbohydrates, fats, and protein. Examples are corticosterone, cortisone (11-dehydroxycorticosterone), and cortisol (hydrocortisone).
2. *Mineralocorticoids,* which primarily affect the transportation of electrolytes and the distribution of water in the body. The most potent of this group is aldosterone.
3. *Androgens* or *estrogens,* which primarily affect secondary sex characteristics. The principal androgen is dehydroepiandrosterone.

The steroids of the adrenal cortex function in the cell nucleus for the synthesis of RNA and protein. The steroid nucleus contains four fused carbon rings, numbered as indicated in the formula below.

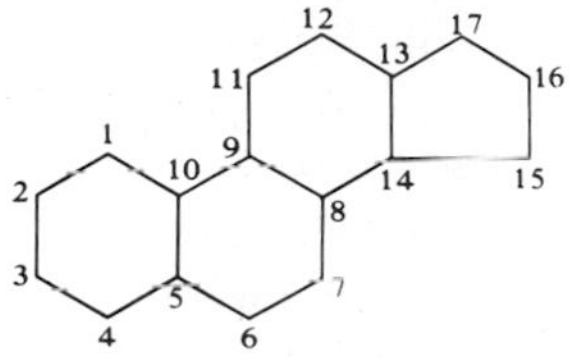

steroid nucleus

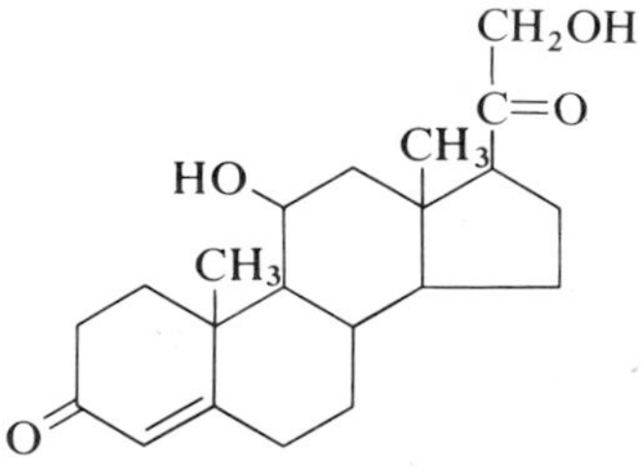

corticosterone

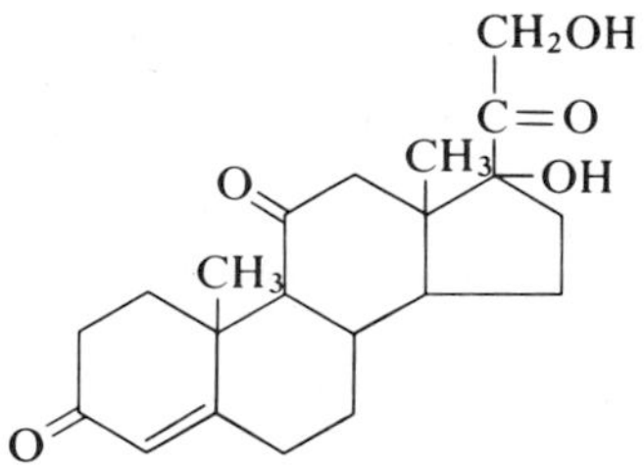

cortisone

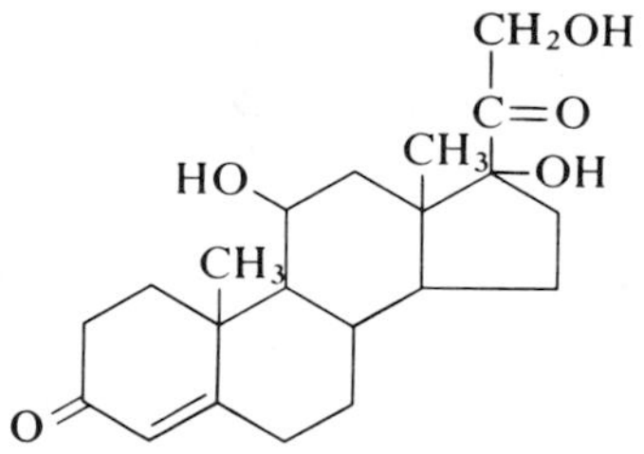

cortisol

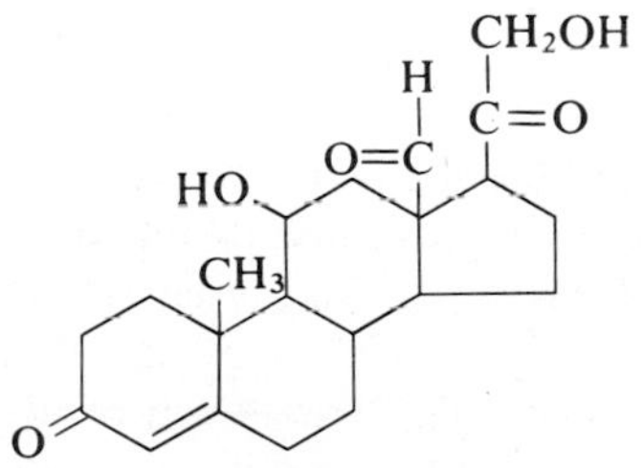

aldosterone

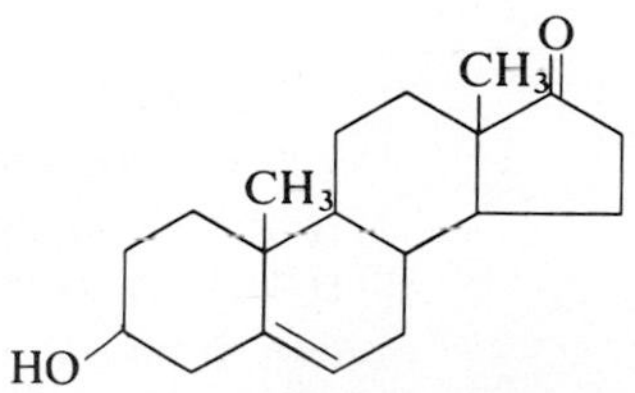

dehydroepiandrosterone

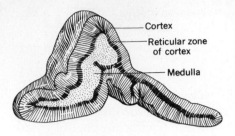

Cortex

Reticular zone
of cortex

Medulla

Figure 32–4. Adrenal glands showing cortex and medulla.

Effect of a Deficiency of the Hormones of the Adrenal Cortex. If the adrenal glands are removed from an animal, it will soon die because of a lack of the hormones produced by the adrenal cortex. In humans, a hypofunctioning of the adrenal glands, because of a tuberculosis of those glands, or in association with pernicious anemia, diabetes, or hypothyroidism, results in Addison's disease. It is characterized by an excessive loss of sodium chloride in the urine, low blood pressure, low body temperature, hypoglycemia, elevation of serum potassium, muscular weakness, a progressive brownish pigmentation of the skin, nausea, and loss of appetite.

The above symptoms, except for the pigmentation, are due to a lack of the salt- and water-regulating hormones—the mineralocorticoids, primarily aldosterone. These hormones stimulate the kidney tubules to reabsorb sodium ions. If there is a lack of these hormones, sodium ions are not reabsorbed and are eliminated in the urine. Along with the sodium ions, chloride ions are also eliminated so that the urine is abnormally high in sodium chloride. Accompanying the sodium chloride loss by the body is a loss of water through osmosis. This loss in water in turn decreases both blood volume and blood pressure. The blood becomes more concentrated and excretion of urea, uric acid, and creatinine is decreased, thereby increasing the concentration of these substances in the blood. Dehydration occurs rapidly and death occurs from circulatory collapse.

Another function of the hormones of the adrenal cortex is to stimulate the process of gluconeogenesis, the formation of glucose from amino acids. This process takes place in the liver. In the absence of the adrenocortical hormones, the *glucocorticoids*, the blood sugar falls and the glycogen stored in the liver and the muscles decreases considerably.

All of these symptoms may be relieved, except the skin discoloration, by the use of cortisonelike compounds—either naturally occurring extracts or synthetically prepared ones.

Effects of Hyperactivity of the Adrenal Cortex. Hyperactivity of the adrenal cortex (hyperadrenocorticism) is caused by a tumor on the adrenal cortex, by an overdosage of cortisone or ACTH, or by an increased production of ACTH (see page 482).

In children, hyperactivity of the adrenal cortex is manifested by early sexual development. Hyperactivity of the adrenal cortex in the adult female, which occurs more often than in the adult male, causes a decrease in feminine characteristics. The voice deepens, the breasts decrease in size, the uterus atrophies, and hair appears on the face. In the adult male, hyperactivity of the adrenal cortex is manifested by an increase in male characteristics— an increase in amount of body hair, increased size of sex organs, deeper voice.

Other effects of hyperadrenocorticism are hyperglycemia and glycosuria, retention of sodium ions and water, increased blood volume, edema, depletion of potassium ions (hypokalemia), and excessive gluconeogenesis.

Hormones of the Adrenal Medulla

Even though the medulla of the adrenal glands secretes two hormones— epinephrine (adrenaline) and norepinephrine (noradrenaline)—it is not essential to human life. That is, it may be removed without causing death. Norepinephrine is a precursor of epinephrine. The structure of these two hormones is shown below.

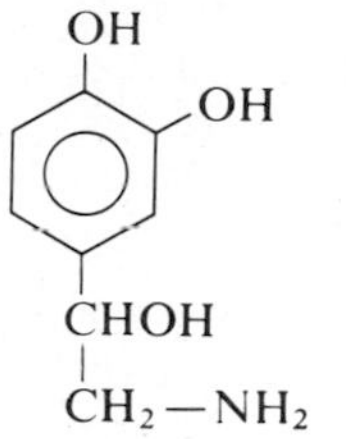

norepinephrine 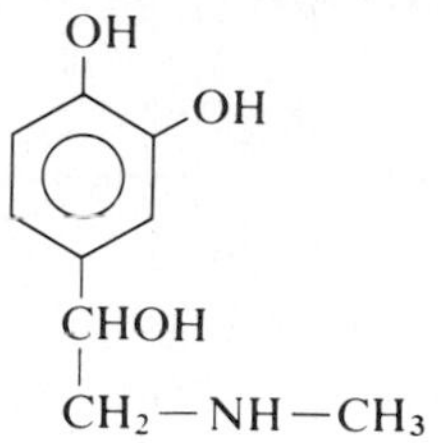   epinephrine

The L-form of these hormones is more physiologically active than the D-form. Epinephrine produces several marked effects on the body.

1. It relaxes the smooth muscles of the stomach, intestines, bronchioles, and bladder. This relaxing effect on the muscles of the bronchioles makes epinephrine especially useful in the treatment of asthma and hay fever.
2. It elevates blood pressure by stimulating the action of the heart and also by constricting the arterioles of the skin and dilating the arterioles of the skeletal muscles. Epinephrine is sometimes used during minor surgery when it is administered along with a local anesthetic. The constriction of the arterioles by epinephrine prevents the anesthetic from spreading too rapidly from the site of the injection. Epinephrine can be injected directly into the heart muscle when that organ stops beating or when it does not start to beat in a newborn baby.
3. It causes glycogenolysis in the liver, producing a rise in blood sugar level. This effect is due to the activation of adenylate cyclase, which increases the production of cyclic-AMP. Cyclic-AMP, in turn, activates a protein kinase, which results in increased phosphorylation, thus catalyzing glycogenolysis in the liver. Epinephrine production is

increased during anxiety, fear, or some other stress. This extra epine-phrine in turn causes a rise in blood sugar, frequently exceeding the renal threshold. In this case, glucose appears in the urine. Such a condition is called emotional glycosuria and disappears as soon as the stress is relieved.

Norepinephrine raises blood pressure by constricting the arterioles. It does not affect the heart itself and does not relax the muscles of the bronchioles as does epinephrine.

The medulla of the adrenal glands rarely becomes diseased. To date, no deficiency effects of its hormones are known. However, certain tumors of the medulla of the adrenal glands stimulate these glands to produce excess hormones. The symptoms caused are intermittent hypertension leading to permanent hypertension and eventually to death from such complications as coronary insufficiency, ventricular fibrillation, and pulmonary edema.

Hormones of the Pituitary Gland

The pituitary gland (hypophysis) is located at the base of the brain. It consists of two parts—the anterior and intermediate lobes (the adenohypo-physis) and the posterior lobe (the neurohypophysis). Each of these parts secretes or releases its own hormones. The pituitary gland has often been called the "master gland" of the body because it seemed to exert a direct influence upon most of the other endocrine glands. However, it is now known that almost all of the secretory activity of the pituitary gland is controlled by a small area of the brain known as the hypothalamus (see Figure 32–5).

Hormones of the Anterior Lobe

Six hormones have been isolated from the anterior lobe of the pituitary gland.

1. Growth hormone (GH)
2. Thyrotropic hormone (TSH)
3. Adrenocorticotropic hormone (ACTH)
4. Lactogenic hormone (LTH)
5. Luteinizing hormone (LH), also known as the interstitial cell-stimulating hormone (ICSH)
6. Follicle-stimulating hormone (FSH)

Growth Hormone. Growth hormone, GH, also called somatotropin, is a protein with a molecular weight of about 21,500 and containing 191 amino acids. It stimulates the growth of the long bones at the epiphyses, stimulates the growth of soft tissue, and increases the retention of calcium ions. The

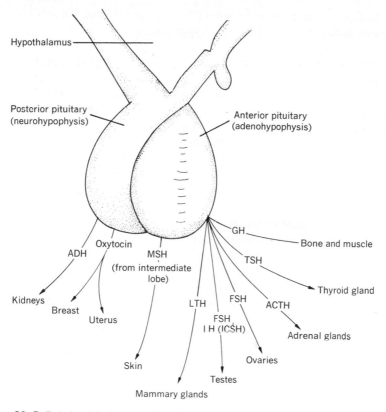

Figure 32–5. Relationship between hypothalamus and the pituitary gland.

growth hormone also increases protein synthesis, leading to a positive nitrogen balance. In the muscle, growth hormone increases amino acid transportation across membranes, also leading to increased protein, DNA, and RNA synthesis. The growth hormone causes mobilization of fatty acids from fat deposits, providing cellular fuel. In the muscles, growth hormone antagonizes the effects of insulin; it inhibits glucose metabolism by muscle tissue.

Underactivity of the anterior lobe of the pituitary gland in children leads to a deficiency of the growth hormone, causing dwarfism. These children develop normally but do not grow in size. Unlike cretins they are not mentally retarded, but they may be sexually underdeveloped.

An overactivity of the anterior lobe of the pituitary gland, possibly due to a tumor, results in an overproduction of the growth hormone. When this occurs in a child, growth is stimulated and gigantism results. If the overactivity of this gland occurs during adulthood, the individual does not grow in size but the bones in the hands, feet, and face grow, producing a condition known as acromegaly.

Secretion of the growth hormone is regulated by the growth hormone-releasing factor (GHRF) from the hypothalamus

Thyrotropic Hormone. The thyrotropic hormone is also called the thyroid-stimulating hormone, hence the abbreviation TSH. This hormone is a glycoprotein with a molecular weight of approximately 30,000. The release of TSH from the pituitary gland is regulated by TRF, the thyrotropic-releasing factor, from the hypothalamus.

A deficiency of TSH causes the thyroid gland to atrophy. As a result, thyroxine production ceases with a consequent drop in metabolic rate.

If this hormone is injected into an animal, the symptoms of hyperthyroidism appear—increased metabolic rate, increased heart rate, and exophthalmos.

Adrenocorticotropic Hormone. The adrenocorticotropic hormone, ACTH, is a polypeptide with a molecular weight of 4500. ACTH contains 39 amino acids, but it has been shown that only the first 23 are required for activity. The remaining 16 vary according to the animal source. The release of ACTH by the pituitary gland appears to be regulated by the corticotropin-releasing factor (CRF) of the hypothalamus in response to various biologic stresses. ACTH has been prepared synthetically and is used medically as indicated below.

ACTH stimulates the synthesis and release of the hormones by the adrenal cortex. ACTH, as do many other hormones, controls its target tissue through cyclic-AMP. Administration of ACTH to a normal person causes retention of sodium ions, chloride ions, and water, elevation of blood sugar, and increased excretion of potassium ions, nitrogen, phosphorus, and uric acid.

Lactogenic Hormone (Prolactin) or Luteotropin (LTH). This hormone was first identified by its property of stimulating the formation of "crop milk" in the crop glands of pigeons. The arrangement of the amino acids in prolactin has been determined. It is a protein with a molecular weight of about 23,000.

Prolactin initiates lactation. In mammals, a hormone produced by the placenta stimulates the growth of the mammary glands and, at the same time, inhibits the secretion of prolactin. At parturition the inhibiting effect of the placenta is not present so that prolactin is secreted and thus initiates lactation.

Gonadotropic Hormones. The gonadotropic hormones LH and FSH (and also prolactin) are secreted by the anterior lobe of the pituitary gland. Removal of the anterior lobe in a male causes atrophy of the testes, prostate gland, and the seminal vesicles. Removal in a female causes atrophy of the ovaries, uterus, and fallopian tubes.

LH and FSH as well as TSH consist of an α- and a β-chain of amino acids. The α-chain is the same in all three, so that biologic specificity must reside in the β-chain. Both LH and FSH stimulate cyclic-AMP synthesis in appropriate target organs.

Luteinizing Hormone (LH). LH is also known as the interstitial cell-stimulating hormone (ICSH). This hormone is a glycoprotein with a molecular weight of about 40,000. It is the first pituitary gonadotropin whose sequence of amino acids has been precisely determined.

LH stimulates the development of the testes in males and also causes an increased production of testosterone. In male animals who have had the pituitary gland surgically removed (hypophysectomy), administration of LH increases the weight of the seminal vesicles and also increases the weight of the ventral lobe of the prostate gland.

In females, LH plays an important role in causing ovulation. It not only causes the production of the corpus luteum but sustains it and stimulates the production of progesterone. In hypophysectomized females, an injection of LH stimulates repair of interstitial ovarian tissues and also increases ovarian weight.

Follicle-Stimulating Hormone (FSH). FSH is a glycoprotein with a molecular weight of 25,000. It stimulates and initiates the development of the follicles of the ovary and prepares those follicles for the action of LH. It also stimulates the secretion of estrogen. In males it causes the growth of the testes and stimulates the production of spermatozoa.

Hormone of the Intermediate Lobe

The pars intermedia or intermediate lobe of the pituitary gland secretes a hormone called intermedin or the melanocyte-stimulating hormone, MSH. It is a polypeptide with alpha and beta parts.

MSH increases the deposition of melanin in the human skin, thus producing darker pigmentation. When the adrenal cortex is underactive, as in Addison's disease, more MSH is produced. This leads to an increased synthesis of melanin with the accompanying brown pigmentation of the skin. Epinephrine and, even more strongly, norepinephrine inhibit the action of the melanocyte-stimulating hormone.

Hormones of the Posterior Lobe

The posterior lobe of the pituitary gland contains two hormones—vasopressin and oxytocin—which are produced in the hypothalamus and stored in this lobe of the pituitary gland.

Vasopressin. Vasopressin, also called the antidiuretic hormone, ADH, stimulates the kidneys to reabsorb water. When the water content of the body is high, very little of this hormone is secreted so that more water is eliminated. Conversely, when the water content of the body is low, more of this hormone is secreted, causing the kidney tubules to reabsorb more water. Thus this hormone serves to regulate the water balance in the body (see page 433).

Diabetes insipidus (see page 403) is caused by the absence of the anti-diuretic hormone, which causes excess daily elimination of water (up to 30 liters). Diabetes insipidus may be controlled by the administration of ADH.

Vasopressin, ADH, stimulates the peripheral blood vessels to constrict and causes an increase in blood pressure. Because of this effect, it has been used to overcome low blood pressure caused by shock following surgery.

Vasopressin is a polypeptide whose arrangement of amino acids is indicated below.

$$
\begin{array}{c}
\text{cys—tyr—phe} \\
| \qquad\qquad | \\
\text{cys—asn—gln} \\
| \\
\text{pro—arg—gly}
\end{array}
$$

vasopressin

Oxytocin. Oxytocin is also a polypeptide whose structure is similar to that of vasopressin except that isoleucine occurs in place of phenylalanine, and leucine in place of arginine.

Oxytocin contracts the muscles of the uterus and also stimulates the ejection of milk from the mammary glands. Oxytocin is used in obstetrics when uterine contraction is desired.

Hormones of the Hypothalamus

The hypothalamus secretes certain neurohormones, some of which act as stimulators and others of which act as inhibitors for the secretion of hormones by the anterior pituitary. These hormones are also called releasing factors. They have a relatively short life-span in the bloodstream. Their half-life is about 2 to 3 min compared to 10 to 15 min for the growth hormone. Therefore, levels of releasing factors are 100 to 1000 times lower in the bloodstream than for the growth hormone.

The first of the hypothalamic hormones to be isolated (and later synthesized) was the thyrotropic releasing hormone (TRH), also known as the thyrotropic releasing factor (TRF), which controls the release of the thyroid-stimulating hormone (TSH). TRF is a tripeptide containing three amino acids—glutamic acid, histidine, and proline. The structure of TRF is indicated below.

TRF, pyroglutamylhistidylproline amide

TRF is highly specific and causes an increase of TSH within 1 min. It may be administered orally. TRF is used to distinguish between lesions in the pituitary and the hypothalamus. TRH also affects the central nervous system. Synthetic analogs of TRH have been prepared that enhance the effect on the central nervous system while not affecting the thyrotropin-releasing influence.

The second hypothalamic releasing factor that was isolated was the luteinizing hormone releasing factor, LH-RF, which controls the release of the leuteinizing hormone from the anterior pituitary. LRF is a decapeptide (10 amino acid chain), whose amino acid sequence is:

$$\text{glu-his-trp-ser-tyr-gly-leu-arg-pro-gly}$$

An injection of LH-RF causes an increase in circulating LH in 1 to 2 min. LH-RF also increases the amount of FSH and so may be identical with the follicle-stimulating hormone releasing factor, FSH-RF.

Other releasing factors are corticotropin (ACTH) releasing factor, CRF; growth hormone releasing factor, GH-RF; prolactin releasing factor, PRF; and melanocyte-stimulating hormone releasing factor, MRF. The structures of these factors have not been determined, although they appear to be simple peptides.

In addition to the releasing factors, the hypothalamus also contains release-inhibiting factors—hormones that inhibit the release of the "releasing factors."

Among the release-inhibiting factors are the growth hormone release-inhibiting factor, GH-RIF (GIF). GIF, also called somatostatin, is a tetra-decapeptide (14 amino acid chain). This factor inhibits the release of the growth hormone. GIF also inhibits the release of insulin (see page 469), glucagon, gastrin, TSH, and FSH.

Release-inhibiting factors whose structures have not been determined are prolactin release-inhibiting factor, PRIF or PIF, and melanocyte-stimulating hormone release-inhibiting factor, MRIF or MIF. Other release-inhibiting factors have been postulated but have not been isolated or identified.

The hypothalamus also produces two hormones that are stored in and secreted by the posterior lobe of the pituitary gland. They are the antidiuretic hormone (ADH) and oxytocin (see page 483).

In addition, it has been found that the hypothalamus contains polypep-tides called endorphins which may be linked to schizophrenia and other forms of mental disease. α-Endorphin acted as a tranquilizer on laboratory animals; β-endorphin produced a long-lasting catatonic state; and α-endorphin led to extremely aggressive behavior.

The Female Sex Hormones

The ovary secretes two different types of hormones. The follicles of the ovary secrete the *follicular* or *estrogenic hormones*. The corpus luteum that forms in the ovary from the ruptured follicle secretes the *progestational hormones.*

Estrogenic or Follicular Hormones

The maturing follicles of the ovaries produce the estrogenic hormones, which are also called estrogens. These hormones are estradiol, estrone, and estriol. Of these three hormones, estradiol is the parent compound and also the most active. The other two hormones are derived from estradiol. Estriol is the main estrogen found in the urine of pregnant women and also in the placenta. Estrone is in metabolic equilibrium with estradiol. Note the similarities of structures of these three hormones.

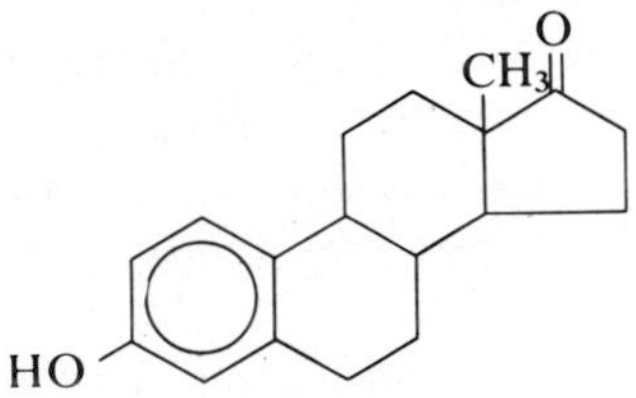

estrone

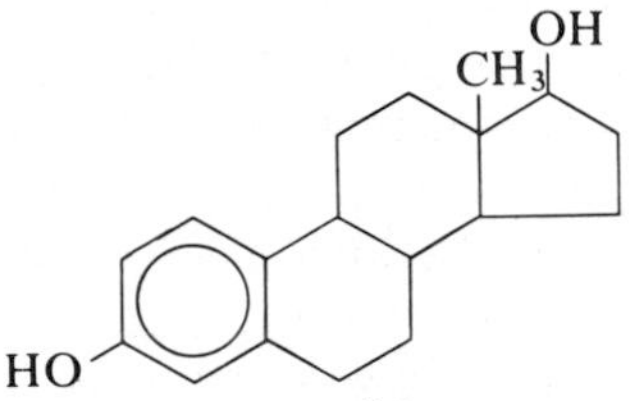

estradiol

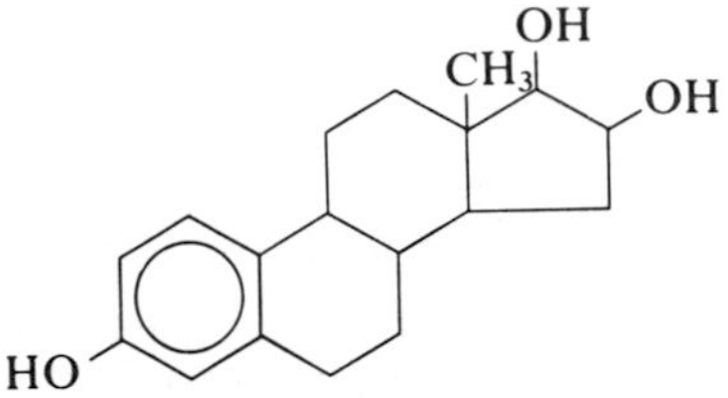

estriol

One International Unit (1 IU) of estrogen activity is equal to 0.1 mg of estrone.

Estradiol, estrone, and estriol are concerned with the maturation of the eggs (ova) and the maintenance of the secondary sex characteristics. In lower animals, the estrogens produce estrus, the urge for mating. The estrogens also suppress production of FSH, the follicle-stimulating hormone, by the pituitary gland. FSH initially starts the development of the follicle. Once the follicle begins to develop, further production of FSH is not needed and is inhibited by the estrogenic hormones. However, the estrogenic hormones stimulate the production of LH, the luteinizing hormone.

Synthetic estrogens such as diethylstilbestrol and ethinyl estradiol have been developed. These synthetics may be given orally, but the naturally occurring estrogens are destroyed in the digestive tract. Estrogens may be used in the treatment of underdeveloped female characteristics.

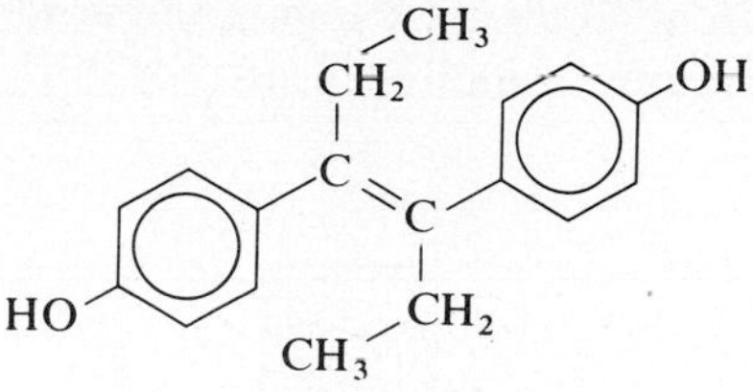

diethylstilbestrol

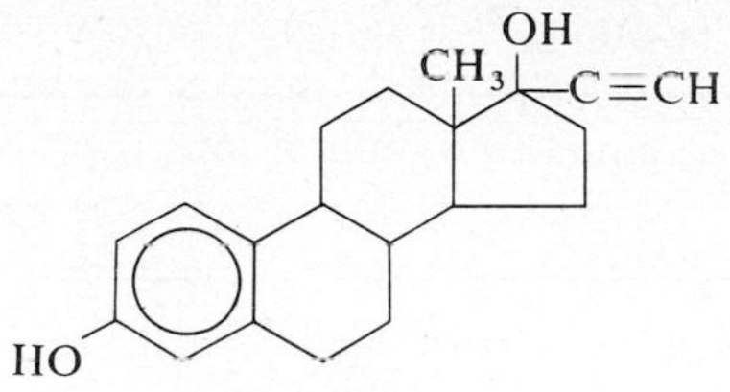

ethinyl estradiol

Progestational Hormones

The corpus luteum is produced in the follicle after the matured ovum is discharged into the uterus. The corpus luteum produces a hormone called progesterone. This hormone causes development of the endometrium of the uterus, preparing the uterus to receive and maintain the ovum, and it stimulates the mammary glands. Progesterone inhibits estrus, ovulation, and the production of LH, the hormone that initially stimulated ovulation, which led to the formation and maintenance of the corpus luteum. If the ovum is not fertilized, the corpus luteum breaks down and menstruation follows. If the ovum is fertilized, progesterone from the corpus luteum aids in the development of the placenta.

Progesterone is excreted as pregnanediol and found in the urine.

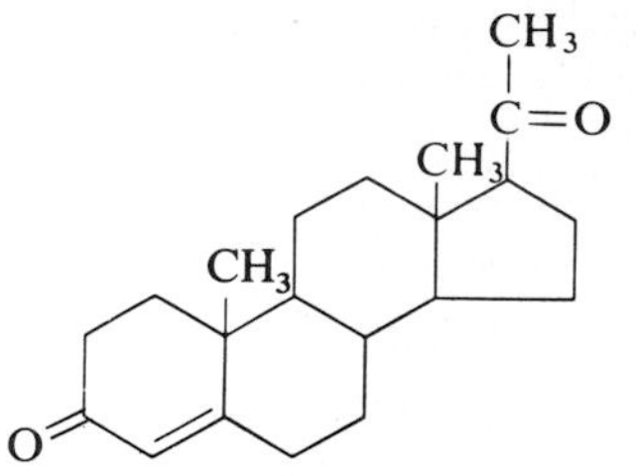

progesterone

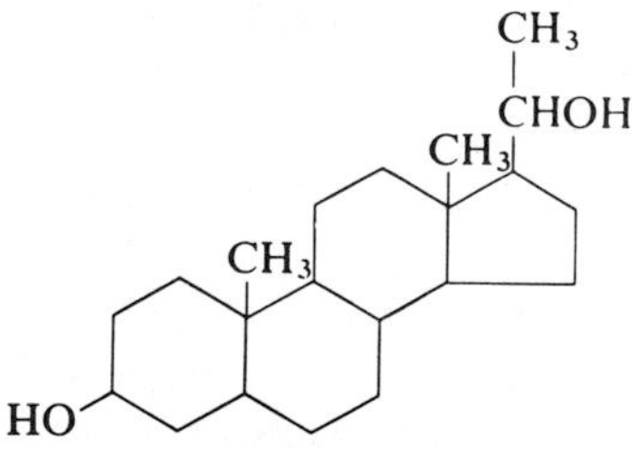

pregnanediol

In addition to progesterone, the corpus luteum also produces another hormone, relaxin, which is a polypeptide with a molecular weight of 9000. Relaxin also occurs in the placenta. It causes ligaments of the symphysis pubis to distend. It also increases dilation of the cervix in pregnant women at parturition, and also helps, along with estrogen and progesterone, to maintain gestation.

The Male Sex Hormones

Male hormones are produced primarily in the testes, although small amounts are also produced in the adrenal glands. The male hormones are called androgens. The principal male hormone is testosterone. This hormone is changed to another hormone, androsterone. Both are found in the urine. Androsterone is used as the international unit of androgen activity; I IU is equal to 0.1 mg of androsterone.

testosterone

androsterone

The male hormones are responsible for the development of the male sex organs and the secondary male characteristics. Note that the structures of the male sex hormones are similar to those of the female sex hormones. Both are derived from a common substance, cholesterol.

The Menstrual Cycle

The human menstrual cycle (or estrus cycle) covers approximately 28 days (see Figure 32–6). From the end of menstruation to the 14th day of the cycle, the ovum and follicle mature in the ovary. About the 14th day of the cycle, the follicle ruptures and the ovum escapes into the uterus. The ruptured follicle fills with a blood clot and forms the corpus luteum. If pregnancy does not occur, the corpus luteum is replaced by scar tissue and the cycle is repeated.

The cycle begins with the action of the follicle-stimulating hormone (FSH) of the anterior pituitary gland. FSH causes a graafian follicle in the ovary to develop into a mature follicle. This mature follicle ruptures about the 14th day of the cycle and expels an ovum into the uterus. While the follicle is maturing, it produces estrogenic hormones that cause regeneration in the endometrium of the uterus. This stage of the menstrual cycle is called the follicular stage.

The hormone made by the follicular cells, estradiol, is changed by the body into estriol and estrone and eliminated in the urine. These hormones are known as the estrogenic hormones because they cause a periodic estrus in the lower animals. The estrogenic hormones are responsible for the development of secondary sex characteristics—growth of the breasts, changes in uterus and vagina at puberty, and distribution of body hair.

When the follicle ruptures, the endometrium of the uterus is very thick and the secretion of the follicular hormones has reached its maximum. When ovulation occurs, the production of estrogenic hormones decreases rapidly.

The second stage of the menstrual cycle is known as the progestational or luteal stage. After the follicle ruptures, it fills with blood, which is replaced in a few days by large yellow cells forming the corpus luteum. The corpus

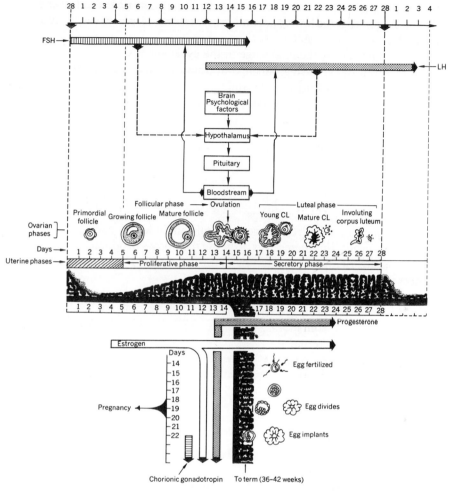

Figure 32–6. Summary of plasma hormone appearance, uterine changes, ovarian events during the menstrual cycle, and possible pregnancy. (Reproduced from Pansky, B.: *Dynamic Anatomy and Physiology*. Macmillan Publishing Co., Inc., New York, 1975.)

luteum secretes the hormone progesterone in large amounts. This hormone prepares the endometrium of the uterus for the implantation and nutrition of the fertilized ovum. It also suppresses estrus and ovulation and stimulates the mammary glands.

Progesterone is changed by the body to pregnanediol and is excreted in the urine. If pregnancy occurs, the corpus luteum continues to secrete progesterone. The placenta, which forms in the early stages of pregnancy, also secretes progesterone and other hormones that inhibit the normal menstrual cycle.

If conception does not occur, the corpus luteum degenerates and the secretion of estrogenic hormones and progesterone decreases suddenly about the 28th day of the cycle.

This is followed by the final stage of the cycle—menstruation, which lasts three to seven days, during which time there is menstrual bleeding and a rapid sloughing off of the uterine wall.

After menstruation, FSH stimulates a new graafian follicle and the cycle begins again.

It has been found that combinations of progestins and estrogens, when administered orally on days 5 to 25 of the menstrual cycle, prevent ovulation. This combination of drugs (called the "Pill") produces contraception because of its ability to suppress ovulation by means of a reduction in the output of FSH and LH by the anterior pituitary gland. Normal menstruation occurs upon the withdrawal of these drugs.

Small amounts of estrogen are normally included in such preparations because the progestins themselves decrease the body's production of estrogen. The structure of one combination of progestational and estrogenic substances used in the "Pill" is indicated below.

progestational substance

estrogenic substance

norethindrone

mestranol

Other Hormones

The thymus gland is often called "the master gland of immunity." This gland is believed to produce a hormone, called thymosin, that could control immune system malfunctions. Use of this hormone is still in the experimental stage on human patients, although it has been well tested on animals and on human cells in tissue cultures.

Light controls the synthesis of the hormone melatonin by the pineal gland. Melatonin blocks the action of the melanocyte-stimulating hormone MSH (see page 483) and also the adrenocorticotropic hormone ACTH (see page 482).

Summary

Hormones are produced in the body's endocrine glands and secreted directly into the bloodstream, which carries them to the body parts on which they produce their effects.

The pyloric mucosa produces gastrin, which stimulates the secretion of hydrochloric acid and also produces a greater motility of the stomach.

The mucosa of the small intestine produces secretin, which stimulates the pancreas to release the pancreatic juices, and cholecystokinin-pancreozymin, which stimulates the contraction and emptying of the gallbladder.

The pancreas secretes insulin, which increases the rate of oxidation of glucose and also facilitates the conversion of glucose to glycogen in the liver and the muscles.

The pancreas also secretes glucagon, which causes an increase in blood sugar content by stimulating phosphorylase activity in the liver.

The hormones of the thyroid gland regulate the metabolism of the body and also affect the growth and development of the body. The thyroid hormones contain the element iodine. The hormonal iodine carried by the blood is termed protein-bound iodine, or PBI. Hypothyroidism occurs when the thyroid gland does not manufacture sufficient thyroxine. The symptoms of hypothyroidism are sluggishness, gain in weight, slower heartbeat, reduced metabolic rate, and loss of appetite.

If the thyroid gland is absent or fails to develop in an infant, the effect produced is called cretinism. If the thyroid gland in an adult should atrophy, the effect produced is called myxedema. If the thyroid gland enlarges, the condition is called simple goiter.

If the thyroid gland produces excess thyroxine, hyperthyroidism results; the symptoms are increased metabolic rate, bulging of the eyes, nervousness, loss of weight, and a rapid, irregular heartbeat. Certain drugs may be given to counteract the effects of an overactive thyroid gland.

Another hormone of the thyroid gland, calcitonin, keeps the calcium concentration of the blood from becoming too high by inhibiting the release of calcium ions from the bone.

The parathyroid glands produce a hormone, parathormone, which influences the metabolism of calcium and phosphorus in the body.

The adrenal glands are divided into two portions—the cortex and the medulla.

The hormones of the adrenal cortex are divided into three categories—the glucocorticoids, which primarily affect the metabolism of carbohydrate, fat, and protein; the mineralocorticoids, which primarily affect the transportation of electrolytes and the distribution of water in the body; and the androgens or estrogens, which primarily affect secondary sex characteristics. A hypofunctioning of the adrenal glands produces Addison's disease, which is characterized by an excessive loss of NaCl in the urine, low blood pressure, elevation of serum potassium, muscular weakness, brownish pigmentation of the skin, nausea, and loss of appetite.

The medulla of the adrenal glands secretes epinephrine and norepinephrine. Norepinephrine is a precursor of epinephrine. Epinephrine relaxes the smooth muscles, elevates blood pressure, and causes glycogenolysis in the liver.

The pituitary gland consists of two parts—the anterior and posterior lobes. The flow of hormones from the pituitary gland is under the control of the hypothalamus.

The hormones of the anterior lobe of the pituitary gland are the growth hormone (GH), the thyrotropic hormone (TSH), the adrenocorticotropic hormone (ACTH), lactogenic hormone (LTH), luteinizing hormone (LH), and follicle-stimulating hormone FSH).

The posterior lobe of the pituitary secretes the hormones vasopressin and oxytocin. Vasopressin stimulates the kidneys to reabsorb water. Oxytocin contracts the muscles of the uterus and stimulates the ejection of milk from the mammary glands.

The pars intermedia of the pituitary gland secretes the hormone intermedin, which increases the deposition of melanin in the skin.

The hypothalamus secretes neurohormones, some of which stimulate and others of which inhibit the secretion of hormones by the anterior pituitary.

The ovaries secrete two different types of hormones. The follicles secrete the follicular or estrogenic hormones. The corpus luteum that forms in the ovary from the ruptured follicle secretes the protestational hormones.

The male sex hormones are produced in the testes. The principal male hormone is testosterone. Both male and female sex hormones are derived from cholesterol.

The menstrual cycle lasts approximately 28 days. The first stage, the follicular stage, occurs while the follicle is maturing. The second stage is called the progestational or luteal stage. During this stage the endometrium of the uterus is prepared for the implantation and nutrition of the fertilized ovum. The final stage is menstruation, after which the cycle begins again.

Questions and Problems

1. Compare hormones with vitamins.
2. In what different ways can hormones act?
3. Where is gastrin produced? What is its function?
4. What is the function of histamine?
5. Compare the structure of histamine with those of antihistamines.
6. Where is secretin formed? What is the function of secretin? CCK-PZ?
7. What hormones are produced by the pancreas?
8. Insulin contains what metallic element?
9. Why can insulin not be taken orally?
10. What is protamine zinc insulin?
11. What are the functions of insulin in carbohydrate metabolism?
12. What is hyperglycemia?
13. Will insulin cure diabetes mellitus? Why?
14. What might cause hypoglycemia? What will relieve such a symptom?
15. What is the function of glucagon?
16. Where is the thyroid gland located?
17. What unusual element is found in the thyroid gland?
18. What is PBI? What is it used for?
19. What is a BEI? What has replaced PBI and BEI tests?
20. List the symptoms of diabetes mellitus.
21. List the symptoms of hyperthyroidism.
22. What is a cretin? How may this defect be overcome?
23. What causes myxedema?
24. What are the symptoms of a goiter?
25. What are the symptoms of hypothyroidism?
26. How do antithyroid drugs function?
27. Where are the parathyroid glands located?
28. What function do the parathyroid hormones perform?
29. What are the symptoms of hypoparathyroidism? Of hyperparathyroidism?
30. What three types of hormones are produced by the adrenal cortex? What is their function? Give an example of each.

31. Where are the adrenal glands located?
32. What causes Addison's disease?
33. Indicate the symptoms of a lack of the adrenocortical hormones.
34. What are the effects of hyperadrenocorticism?
35. What are the hormones of the medulla of the adrenal glands? What is the function of each?
36. List the hormones of the anterior lobe of the pituitary gland. What is the function of each?
37. Compare dwarfism with cretinism.
38. What are the symptoms of an overactive anterior lobe of the pituitary gland?
39. What are the hormones of the posterior lobe of the pituitary gland? What is the function of each?
40. What is MSH? Where is it formed? What is its function?
41. What types of hormones are produced by the hypothalamus? What do they do?
42. What type of structure do the hypothalamic hormones have?
43. Where are the estrogenic hormones produced? The progestational hormones?
44. Where are the male sex hormones produced?
45. Discuss the menstrual cycle in terms of hormones and their effects on various tissues.
46. Why is the pituitary gland no longer considered to be "the master gland" of the body?
47. Where are vasopressin and oxytocin produced? Stored?
48. What does the "Pill" contain? How does it function?
49. What is the "master gland of immunity"? What hormone does it produce?
50. What controls the synthesis of melatonin? What does this hormone do?

References

Brazeau, P.; Vale, W.; Burgus, R.; Ling, N.; Butcher, M.; Rivier, J.; and Guillemin, R.: Hypothalamic polypeptide that inhibits the secretion of immunoreactive pituitary growth hormone. *Science*, **179**:77 (Jan. 5), 1973.

Goodman, L. S., and Gilman, A. (eds.): *The Pharmacological Basis of Therapeutics*, 5th ed. Macmillan Publishing Co., Inc., New York, 1975, Sec. XVII.

Guillemin, R., and Burgus, R.: The hormones of the hypothalamus. *Scientific American*, **227**:24–33 (Nov.), 1972.

Guillemin, R.; Burgus, R.; and Vale, W.: The hypothalamic hypophysiotropic thyrotropin releasing factor (TRF). *Vitamins and Hormones*, **29**:1, 1971.

Harper, H. A.: *Review of Physiological Chemistry*, 15th ed. Lange Medical Publications, Los Altos, Calif., 1975, Chap. 20.

McEwen, B. S.: Interactions between hormones and nerve tissue. *Scientific American*, **235**: 48–58 (July), 1976.

Montgomery, R.; Dryer, R. L.; Conway, T. W.; and Spector, A. A.: *Biochemistry*. C. V. Mosby Co., St. Louis, 1974, Chap. 13.

O'Malley, B. W., and Schrader, W. T.: The receptors of steroid hormones. *Scientific American*, **234**:32–43 (Feb.), 1976.

Peterson, R. E., and Guillemin, R.: The hormones of the hypothalamus. *American Journal of Medicine*, **57**:591–600 (Oct.), 1974.

Pike, J. E.: The prostaglandins. *Scientific American*, **225**:84–92 (Nov.), 1971.

Chapter 33

Heredity

A human being grows from one fertilized egg cell. As this organism divides and grows, it develops gradually into recognizable form. How do the cells "know" where the head should be, where the arms should be, that there should be two arms and not one or three? Why not claws instead of fingers, scales instead of skin?

Originally it was believed that the chromosomes somehow transmitted information from one generation to the next. The next step forward in the study of heredity was the theory that the chromosomes contained genes—each gene giving a particular characteristic to that individual; that is, there were genes for tallness or shortness, for blue eyes or brown eyes, for light hair or dark hair.

Recent studies have shown that genes are composed of deoxyribonucleic acid, DNA. In 1953 J. D. Watson (see Figure 33–1) and F. H. C. Crick of

Figure 33–1. Dr. J. D. Watson, Harvard University, who was awarded the Nobel Prize for Medicine and Physiology in 1962, along with Dr. F. H. C. Crick and Dr. M. H. F. Wilkins for their discovery of the molecular structure of deoxyribonucleic acid (DNA).

Cambridge University proposed a double helix structure for the DNA molecule. Before we discuss this structure and explain what a double helix is, let us review a little about nucleic acids.

Nucleic Acids

Nucleic acids are built up of many small units called nucleotides. These nucleotides in turn consist of a five-carbon sugar, a phosphate group, and an organic base (a heterocyclic nitrogen compound).

Nucleic acids contain either deoxyribose or ribose as the five-carbon sugar component of the nucleotide. The nucleic acids based on deoxyribose are called deoxyribonucleic acid (DNA). Those based on ribose are called ribonucleic acid (RNA).

The prefix *deoxy-* means "without oxygen." Note in the following structural diagrams that deoxyribose contains one less oxygen atom than does ribose.

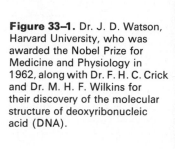

The DNA molecule contains four different heterocyclic nitrogen compounds. They are adenine, thymine, cytosine, and guanine.

| adenine | thymine | cytosine | guanine |

Adenine and thymine are always present in the same $1:1$ ratio in the DNA molecule. This suggests that these two molecules were somehow joined together. Likewise, guanine and cytosine occurred in a $1:1$ ratio and are believed to be paired.

The Watson-Crick Model of DNA

The Watson-Crick model of DNA proposed a double-coiled chain consisting of two strands intertwined around one another. They also proposed that an adenine of one chain was bonded to a thymine of the opposite chain and a guanine of one chain was bonded to a cytosine of the other chain.

The Watson-Crick model of DNA may also be shown as indicated in Figure 33–2, representing it as a spiral ladder with the two sides being held together by rungs consisting of heterocyclic nitrogen compounds (adenine, thymine, guanine, and cytosine).

If the chain is untwisted and straightened out, it may be represented as follows

The solid lines indicate ordinary chemical bonding and the dotted lines indicate hydrogen bonding. Recall that hydrogen bonds are much weaker than ordinary chemical bonds. Note that there are two hydrogen bonds between adenine and thymine and three hydrogen bonds between guanine and cytosine.

The DNA Code

All DNA molecules have the same sequence of deoxyribose and phosphates in the ladder part of the chain. The difference lies in the order of the adenine,

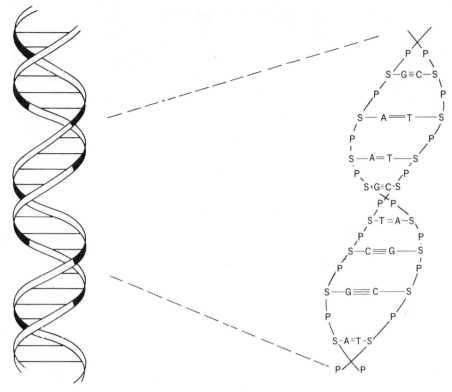

Figure 33–2. Structure of DNA. *P*, phosphate; *S*, sugar; *G*, guanine; *C*, cytosine; *A*, adenine; *T*, thymine.

thymine, cytosine, and guanine parts of the chain. This difference in sequence of the heterocyclic compounds constitutes the genetic code.

This code consists of only four letters—A, T, G, and C, representing adenine, thymine, guanine, and cytosine, respectively. In the DNA molecule these letters of the code are grouped in threes (which we shall discuss later). Thus the code might be ATG, TAC, AAA, CCT, and so on.

Simple bacterial viruses contain about 5500 nucleotides in their DNA molecules; 5500 nucleotides, grouped in threes can produce 5500/3 or approximately 1800 coded pieces of information. These 1800 coded pieces of information are sufficient to describe that particular virus. If each piece of information corresponded to one letter of our alphabet, it would take 1800 letters to describe, genetically, that virus; 1800 letters corresponds roughly to the number of letters on one page of this book.

One DNA molecule in a human contains approximately 5,000,000,000 nucleotides, or enough to form 1,700,000,000 coded pieces of information (of three nucleotides to a group). If each of these coded pieces corresponded to a letter of our alphabet, it would take approximately 2500 volumes, each the size of this book, to describe a man genetically.

Replication of DNA

When a cell divides, it produces two new cells with identical characteristics. This means that the DNA originally present must duplicate (or replicate) itself. How does the DNA molecule direct the synthesis of another identical DNA molecule?

Watson and Crick theorized that the DNA molecule unwinds, and each half acts as a template for nucleotide units to collect on and form a new chain. Recall that if adenine (A) is present on one chain, it can attract and hold only thymine (T), and cytosine (C) can attract and hold only guanine (G). Thus each half of the chain is highly specific in what it attracts. It forms complementary chains which coil up again and form two new DNA molecules.

This may be represented diagrammatically, as shown in Figure 33–3. This representation is necessarily quite general and approximate. The actual mechanism of DNA unwinding and untwisting for replication is quite complex, involving several enzymes such as DNA polymerase and DNA ligase, and is beyond the scope of this book.

Transfer of Information

The preceding paragraph indicates how the DNA molecule duplicates itself, but how does the DNA molecule transfer its coded information to the cell's ribosomes where the actual production of the protein called for by the code takes place?

Figure 33–3. Unwinding of DNA.

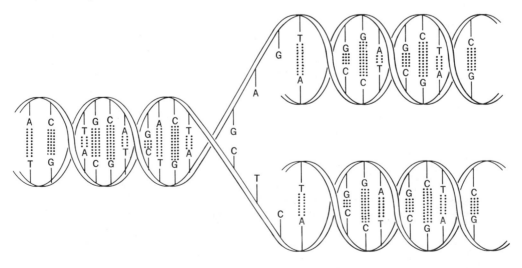

The information contained in the DNA molecule is carried by another molecule called messenger RNA. During replication, DNA "unzips" and synthesizes two new chains to produce two identical molecules. However, when the DNA molecule synthesizes messenger RNA, the DNA again "unzips"; however, only one part of one strand of DNA (and always the same strand) acts as a template for the formation of messenger RNA (see Figure 33–4). The enzyme required is called RNA polymerase.

When messenger RNA is produced, there is one major difference in the attraction of nucleotides between it and DNA. In RNA, the partner of adenine (A) is uracil (U) and not thymine (T). Another difference is that RNA contains the pentose ribose, while DNA contains the pentose deoxyribose.

There are at least three types of RNA. They are (1) messenger RNA, (2) ribosomal RNA, and (3) transfer RNA. Messenger RNA (mRNA) is concerned with the transmission of genetic information from DNA to the site of protein synthesis, the ribosomes; it is found in the nucleus and the cytoplasm of the cell. Ribosomal RNA (rRNA), which is the major fraction of the total RNA, combines with protein to form the ribosomes; therefore, ribosomes are an example of nucleoprotein. It is in the ribosomes that the

Figure 33–4. Formation of messenger RNA from DNA.

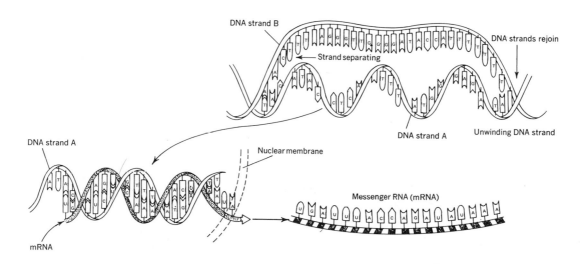

transcription of the DNA message takes place. Transfer RNA (tRNA) holds a specific amino acid for incorporation into a protein molecule, as indicated in the following paragraphs. A fourth type of RNA, heterogeneous RNA (hRNA), may be a precursor for messenger RNA. Heterogeneous RNA is found in the nucleus of the cell.

Messenger RNA acts as a template for the synthesis of protein, but only a small amount of the total information in DNA is used at one time to produce a certain type of protein. What determines whether messenger RNA is formed from a certain segment of the DNA? What "turns on" the DNA and what "turns it of"? It has been postulated that histones block the transcription of DNA to RNA and that nonhistone proteins remove the inhibiting effect so that DNA can be transcribed into RNA.

Let us see how messenger RNA and transfer RNA function in the ribosomes in the synthesis of a protein. The messenger RNA moves from the nucleus to the cytoplasm and then to the ribosomes, where it acts as a template for the formation of protein. The ribosomes contain transfer RNA, which is a spiral form of RNA containing relatively few nucleotides. At one end of the transfer RNA are three nucleotides that are specific for a certain code on the messenger RNA. Amino acids in the cytoplasm are activated by ATP and are coupled to the other end of transfer RNA. This may be designated as shown below.

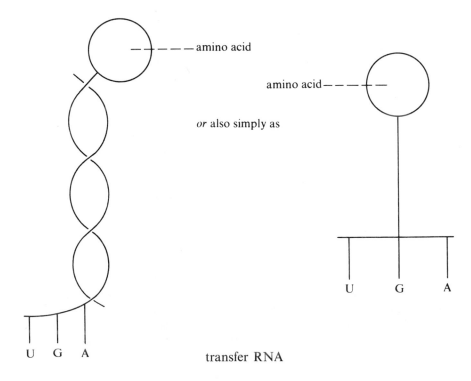

transfer RNA

If messenger RNA contains the coded units ACU, the transfer RNA that would attach itself must have the code UGA. [Recall that in RNA the adenine (A) is bonded to uracil (U) and cytosine (C) to guanine (G).] If the coded message in messenger RNA is UGC, the corresponding code in transfer RNA must be ACG:

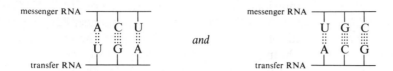

As the messenger RNA travels across the ribosome, the first coded group (of three letters) picks up and holds a corresponding transfer RNA. As the second coded group of the messenger RNA passes, it too picks up a corresponding transfer RNA. Attached to the opposite end of each transfer RNA is an amino acid. The amino acid attached to the end of the second transfer RNA becomes bonded to the amino acid at the end of the first transfer RNA. The amino acid at the end of the third transfer RNA in turn becomes bonded to the second amino acid.

Thus, as each transfer RNA attaches itself to the messenger RNA, the transfer RNA gives up its amino acid to form a chain of amino acids, or a protein. After giving up its amino acid, the transfer RNA leaves the messenger RNA and goes in search of another amino acid that it can pick up and use to repeat the sequence.

Thus, as the messenger RNA passes through the ribosome, it directs the gathering, in specified sequence, of the transfer RNAs, which in turn bond together their amino acids to form a protein.

This effect may be illustrated diagrammatically as shown in Figure 33–5.

In these diagrams, the first coded group in the messenger RNA—group CAG—attracts transfer RNA coded GUC. Attached to transfer RNA coded GUC is an amino acid that we have simply labelled #1. As the messenger RNA passes further along into the ribosome, the next coded group, code UUA, attracts transfer RNA coded AAU. The amino acid attached to the end of this transfer RNA is labelled #2. Amino acid #2 bonds itself to amino acid #1 as illustrated. At the same time, the transfer RNA coded GUC that held amino acid #1 goes off in search of another amino acid labelled #1 so it can be used again whenever the code calls for it. The third coded group in the messenger RNA, code UGA, attracts transfer RNA, coded ACU, with its attached amino acid #3. Then amino acid #3 bonds to amino acid #2, which is already bonded to amino acid #1. So the chain of amino acids begins and continues as illustrated in the fourth figure. This process continues until the end of the coded groups of the messenger RNA. At this time, the protein is complete and moves out of the ribosome into the cytoplasm.

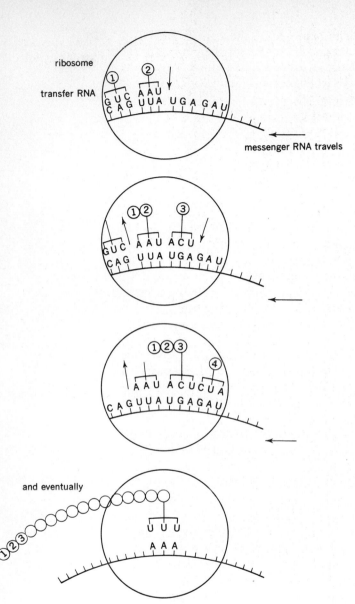

Figure 33–5. Messenger RNA directing proper sequence of transfer RNA's to form a protein.

The Triplet Code

The DNA code consists of four letters, A, T, C, and G, grouped in threes. It is believed that the code in the DNA molecule specifies individual amino acids. Since there are 20 primary amino acids that occur in nature, there must be at least 20 different coded groups in the DNA molecule.

The 20 primary amino acids and their abbreviations are

alanine—ala	glycine—gly	proline—pro
arginine—arg	histidine—his	serine—ser
asparagine—asn	isoleucine—ile	threonine—thr
aspartic acid—asp	leucine—leu	tryptophan—trp
cysteine—cys	lysine—lys	tyrosine—tyr
glutamic acid—glu	methionine—met	valine—val
glutamine—gln	phenylalanine—phe	

Decoding the Code

How can the DNA code be decoded? That is, how can we tell which amino acid is specified by a certain coded group?

In 1961 it was found that if a synthetic RNA composed only of uracil nucleotides was substituted for messenger RNA in a protein-synthesis system, a polypeptide was formed which contained only the amino acid phenylalanine. Since the synthetic messenger RNA contained only uracil, it must have the code group UUU. The corresponding group in the DNA molecule must be AAA. Thus we can say that the coded group AAA in the DNA molecule corresponds to the code UUU in messenger RNA, which in turn specifies the amino acid phenylalanine. Another synthetic messenger RNA, consisting solely of adenine nucleotides, produced a polypeptide containing only lysine. Thus the messenger RNA code for lysine is AAA and the corresponding DNA code must be TTT.

Table 33–1 indicates the amino acids and their coded groups.

Some experiments indicated that there may be more than one code group for a specific amino acid. That is, alanine is indicated by the messenger RNA codes GCU, GCC, GCA, and also GCG. Note that all of these code groups begin with the letters "GC."

Figure 33–6 indicates the synthesis of a polypeptide chain from messenger RNA, indicating the various amino acids specified by that messenger RNA.

Mutations

Suppose that one of the letters (nucleotides) in a DNA code group was substituted by another letter. Then the message would be miscopied and a mutation would occur. A mutation will also occur if one of the code letters is omitted or if one is added or if the order of the code letters is rearranged.

Although some mutations may be beneficial, most are harmful. Mutations occur because of exposure to radiation (in industry, medically, because of naturally occurring radiations and cosmic rays, or because of fallout) and possibly because of certain chemicals.

TABLE 33–1

Messenger RNA

First Letter	Second Letter	Third Letter			
		A	C	G	U
A	A	lys	asn	lys	asn
	C	thr	thr	thr	thr
	G	arg	ser	arg	ser
	U	ile	ile	met	ile
C	A	gln	his	gln	his
	C	pro	pro	pro	pro
	G	arg	arg	arg	arg
	U	leu	leu	leu	leu
G	A	glu	asp	glu	asp
	C	ala	ala	ala	ala
	G	gly	gly	gly	gly
	U	val	val	val	val
U	A	*	tyr	*	tyr
	C	ser	ser	ser	ser
	G	*	cys	trp	cys
	U	leu	phe	leu	phe

* End of chain.

Figure 33–6. Synthesis of a polypeptide from messenger RNA.

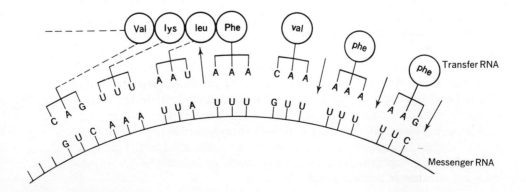

Consequences of Inherited Enzyme Deficiency

Most genetic diseases are caused by a defective gene, which results in a loss of activity of some enzyme. Even though the body has thousands of enzymes, the loss of only one may be disastrous.

Consider a simple enzymatic reaction in which compound X is changed to compound Y under the influence of enzyme x_1, and then compound Y is changed to compound Z under the influence of enzyme y_1.

$$X \xrightarrow{x_1} Y \xrightarrow{y_1} Z$$

If the enzyme y_1 is deficient, one of the following consequences might occur:

$X \xrightarrow{x_1} Y \not\xrightarrow{y_1} Z$
A. The body would be lacking in compound Z, with the resulting effect of that lack.

$X \xrightarrow{x_1} Y \not\xrightarrow{y_1} Z$
R
B. Compound Y might be changed to a harmful byproduct R.

Y
$X \xrightarrow{x_1} Y \not\xrightarrow{y_1} Z$
Y
C. Compound Y might accumulate and may be harmful if present in large amounts.

In the above cases, the genetic disease is caused by a lack of enzyme y_1, which in turn is due to a defective gene—one that might have been copied incorrectly, or changed, or even be lacking altogether. We shall see in the following paragraphs how a single simple change in the sequence of a code group in messenger RNA can lead to a genetic defect.

Common Genetic Diseases

The most common genetic disease in the United States is cystic fibrosis, which affects 1 individual in 2000. Another genetic disease, phenylketonuria, affects 1 in 10,000. Most states now require routine screening of newborn infants for this latter disease. Tay-Sachs disease affects 1 in 1000 among persons of Eastern European Jewish origin. One out of every 10 black persons in the United States carries a sickle cell gene. Today there are at least 1200 distinct inherited genetic diseases that have been identified.

Sickle Cell Anemia

Notice the arrangement for the 146 amino acids in the beta chain of normal hemoglobin, shown overleaf.

Val-His-Leu-Thr-Pro-[Glu]-Glu-Lys-Ser-Ala-Val-Thr-Ala-Leu-Try-Gly-Lys⌐

Val

⌐Val-Val-Leu-Leu-Arg-Gly-Leu-Ala-Glu-Gly-Gly-Val-Glu-Asp-Val-Asn⌐

Tyr

⌐Pro-Try-Thr-Gln-Arg-Phe-Phe-Glu-Ser-Phe-Gly-Asp-Leu-Ser-Thr-Pro⌐

Asp

⌐Leu-Val-Lys-Lys-Gly-His-Ala-Lys-Val-Lys-Pro-Asn-Gly-Met-Val-Ala⌐

Gly

⌐Ala-Phe-Ser-Asp-Gly-Leu-Ala-His-Leu-Asp-Asn-Leu-Lys-Gly-Thr-Phe⌐

Ala

⌐Asn-Glu-Pro-Asp-Val-His-Leu-Lys-Asp-Cys-His-Leu-Glu-Ser-Leu-Thr⌐

Phe

⌐Arg-Leu-Leu-Gly-Asn-Val-Leu-Val-Cys-Val-Leu-Ala-His-His-Phe-Gly⌐

Lys

⌐Val-Gly-Ala-Val-Val-Lys-Gln-Tyr-Ala-Ala-Gln-Val-Pro-Pro-Thr-Phe-Glu⌐

Ala

⌐Asn-Ala-Leu-Ala-His-Lys-Tyr-His

Each of these 146 amino acids was designated by a certain arrangement of three nucleotides in messenger RNA. Altogether there must have been 438 (3 × 146) nucleotides, arranged in the proper sequence in order to form this molecule of the beta chain of hemoglobin.

Look at amino acid number 6, the one with a box around it. This amino acid, glu, is glutamic acid. The messenger RNA code group for glutamic acid is either GAA or GAG. If the middle codon of this group is changed from A to U, the sequence becomes either GUA or GUG, both of which designate the amino acid valine, val. That is, if there is a change in only one of the nucleotides, from A to U, on the sixth codon of the 146 amino acid chain of the beta chain of hemoglobin, a different type of molecule is produced. This type of hemoglobin is called hemoglobin-S and causes the genetic disease sickle cell anemia.

The red blood cells normally have a concave shape when deoxygenated (see Figure 33–7, *A*). Red blood cells containing hemoglobin-S look like a sickle (see Figure 33–7, *B*). These sickle cells are more fragile than normal red blood cells, leading to anemia. They can also occlude capillaries, leading to thrombosis. The points and abnormal shapes of the sickle cells cause slowing and sludging of the red blood cells in the capillaries with resulting hypoxia of the tissues. This produces such symptoms as fever, swelling, and

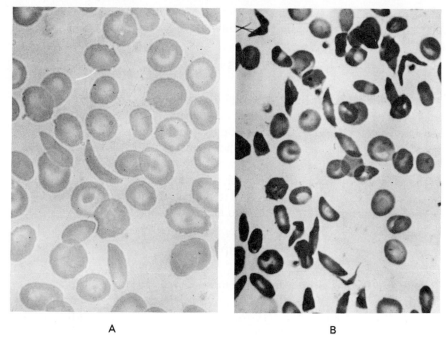

A   B

Figure 33–7. *A.* Normal red blood cells. *B.* Red blood cells of an individual with the sickle cell trait. (Courtesy of Dr. M. Eisenberger, Chicago, Ill.)

pain in various parts of the body. Eventually the spleen is affected. Many victims of severe sickle cell anemia die in childhood.

Sickle cell anemia is a hereditary condition found primarily among Latins and Blacks. Many of these people have the sickle cell trait but are relatively unaffected by it until there is a sharp drop in blood oxygen level, such as might be caused by strenuous exercise at high altitudes, underwater swimming, and drunkenness.

Phenylketonuria

Phenylketonuria, PKU, results when the enzyme phenylalanine hydroxylase is absent. A person having PKU cannot convert phenylalanine to tyrosine, and so the phenylalanine accumulates in the body, resulting in injury to the nervous system. In infants and in children up to age 6, an accumulation of phenylalanine leads to retarded mental development.

This disease may be readily diagnosed from a sample of blood or urine. Treatment consists of giving the patient a diet low in phenylalanine.

Galactosemia

Galactosemia results from the lack of the enzyme that catalyzes the formation of glucose from galactose. This disease may result in damage to the central nervous system, mental retardation, and even death. The disease may be controlled by the administration of a diet free of galactose.

Wilson's Disease

Wilson's disease is caused by the body's failure to eliminate excess Cu^{2+} ions because of lack of ceruloplasmin (see page 437) or because of a failure in the bonding of copper ions to the copper-bonding globulin, or because of both factors. In this disease copper accumulates in the liver, kidneys, and brain. There is also an excess of copper in the urine. If deposition of copper in the liver becomes excessive, cirrhosis may develop. In addition, accumulation of copper in the kidneys may lead to damage of the renal tubules, leading to increased urinary output of amino acids and peptides.

Albinism

Albinism is caused by a missing enzyme, tyrosinase, which is necessary for the formation of melanin, the pigment of the hair, skin, and eyes. Consequently, albinos have very white skin and hair. Although this disease is not serious, persons affected by it are very sensitive to sunburn.

Hemophilia

Hemophilia is caused by a missing protein, an antihemophilic globulin, which is important in the normal clotting process of the blood. Consequently, any cut may be life threatening to hemophiliacs, but the primary damage is the crippling effect of repeated episodes of internal bleeding into body joints.

Other Genetic Diseases

Niemann–Pick disease is caused by a lack of the enzyme sphingomyelinase, which causes an accumulation of sphingomyelin in the liver, spleen, bone marrow, and lymph nodes. This disease affects the brain and causes mental retardation and early death.

Gaucher's disease is caused by a lack of the enzyme glucocerebrosidase, which is necessary for the cleavage of glucocerebrosides into glucose and ceramide. This disease is characterized by the accumulation of glycolipids in the spleen and liver. In children, Gaucher's disease causes severe mental retardation and early death (see page 296). In adults, the spleen and liver enlarge progressively but the disease is compatible with long life.

Tay-Sachs disease is due to a lack of the enzyme hexosaminidase-A, leading to the accumulation of glycolipids in the brain and the eyes. Red spots show up in the retina and there is also muscular weakness. This disease is fatal to infants before the age of 4.

Summary

Originally it was believed that the cell's chromosomes transmitted information from one generation to the next. Then it was believed that the chromosomes contained genes, with each gene giving a particular characteristic to that individual. However, recent studies have shown that genes are composed of deoxyribonucleic acid (DNA), which stores and transfers hereditary characteristics.

The Watson-Crick model of DNA proposes a double-coiled chain consisting of two strands intertwined around one another. The chains were composed of alternating sugar (deoxyribose) and phosphates with the sugar being bonded to one of four different heterocyclic compounds, adenine, guanine, cytosine, and thymine (A, G, C, T). The model suggested that the adenine of one chain was always bonded to a thymine of the opposite chain and that cytosine was always bonded to guanine.

All DNA molecules have the same sequence of deoxyribose and phosphates in the ladder part of the chain. The difference lies in the order of adenine, thymine, guanine, and cytosine.

When a cell divides, the DNA molecule uncoils and each half acts as a template for the formation of a new chain.

During replication, the DNA molecule uncoils and both halves produce molecules identical to the original. However, when the DNA molecule synthesizes RNA, although the coil again unwinds, only one half acts as the template to produce RNA. There are three kinds of RNA: ribosomal RNA (rRNA), messenger RNA (mRNA), transfer RNA (tRNA), and possibly heterogeneous RNA (hRNA).

When messenger RNA is formed from DNA, guanine is still bonded to cytosine but adenine is bonded to uracil (U) rather than to thymine.

Messenger RNA moves through the cytoplasm to the ribosomes, where it acts as a template for the formation of protein. The ribosomes contain transfer RNA. Transfer RNA carries a specific amino acid to the messenger RNA. The identity of this amino acid is determined by the code on the end of the transfer RNA, and this code is specific for another code on the messenger RNA.

As the messenger RNA moves through the ribosome, the first of its coded groups picks up and holds a corresponding coded group on transfer RNA. The second, third, and succeeding groups on messenger RNA do likewise. The specific amino acids carried by the transfer RNA becomes bonded to one another to form the designated protein, and the transfer RNAs are released to find more of the specific amino acids and begin the procedure again.

The DNA code consists of four letters (A, T, C, and G) arranged in groups of three.

The three-letter coded groups specify the 20 primary amino acids occurring in nature.

If one of the code letters is changed or is missing, the information copied would be incorrect and a mutation would occur. Mutations occur because of exposure to radiation, both naturally occurring and that in industry and medicine, and possibly because of certain chemicals.

DNA directs the synthesis of enzymes (which are also proteins). If the message is transferred incorrectly, the proper enzyme will not be synthesized and so will be lacking in the body. This lack of a specific enzyme may lead to a genetic disease such as phenyl-ketonuria, galactosemia, albinism, pentosuria, hemophilia, and others.

Questions and Problems

1. What is the relationship between chromosomes, genes, and DNA?
2. What are nucleic acids?
3. Diagram briefly the Watson-Crick model of DNA.
4. What types of bonds are present in the DNA molecule? Are they all equal in strength?
5. How does DNA replicate?
6. What are the three types of RNA? What is the function of each?
7. What is the difference in nucleotides in messenger RNA and DNA?
8. How do mRNA and tRNA function in the formation of protein?
9. What is the most common genetic disease in the United States?
10. How many primary amino acids occur in nature?
11. Name the primary amino acids and indicate their abbreviations.
12. How many units are present in one unit of the DNA code?
13. Describe briefly how the DNA code was decoded.
14. What amino acid does each of the following coded groups of messenger RNA represent? ACG, GCA, UUU, GUU, CCC, CGA.
15. What is a mutation in terms of the DNA molecule?
16. What is sickle cell disease? What causes it?
17. How do normal blood cells differ from sickle cells in (a) shape and (b) arrangement of amino acids?
18. What causes the following genetic diseases?
    Phenylketonuria
    Hemophilia
    Galactosemia
    Albinism
    Wilson's disease
    Tay-Sachs disease.
    What are the symptoms of each disease?
19. Explain how a genetic disease might occur if an enzyme is deficient.

References

Baum, S. J., and Scaife, C. W.: *Chemistry: A Life Science Approach.* Macmillan Publishing Co., Inc., New York, 1975, Chap. 27.

Brady, R. O.: Hereditary fat-metabolism diseases. *Scientific American*, **229**:88–97 (Aug.), 1973.

Bronk, J. R.: *Chemical Biology.* Macmillan Publishing Co., Inc., New York, 1973, Chap. 5.

Butten, R. J., and Kohne, D. E.: Repeated segments of DNA. *Scientific American*, **222**:84–93 (Apr.), 1970.

Cerami, A., and Peterson, C. M.: Cyanate and sickle-cell disease. *Scientific American*, **232**:44–50 (Apr.), 1975.

Crick, F. H.: The genetic code. *Scientific American*, **207**:66–74 (Oct.), 1962.

———: The genetic code: III. *Ibid.*, **215**:55–60 (Oct.), 1966.

Friedmann, T.: Prenatal diagnosis of genetic disease. *Scientific American*, **225**:34–42 (Nov.), 1971.

Goodenough, V. W., and Levine, R. P.: The genetic activity of mitochondria and chloroplasts. *Scientific American*, **223**:22–29 (Nov.), 1970.

Hurwitz, J., and Furth, J. J.: Messenger RNA. *Scientific American*, **207**:41–49 (Feb.), 1962.

Kornberg, A.: Synthesis of DNA. *Scientific American*, **219**:64–78 (Oct.), 1968.

McEwen, B. S.: The brain as a target organ of endocrine hormones. *Hospital Practice*, **10**:95–104 (May), 1975).

Maniatis, T., and Ptashne, M.: A DNA operator-repressor system. *Scientific American*, **234**:64–76 (Jan.), 1976.

Mirsky, A. E.: The discovery of DNA. *Scientific American*, **218**:78–88 (June), 1968.

Montgomery, R.; Dryer, R. L.; Conway, T. W.; and Spector, A. A.; *Biochemistry*, C. V. Mosby Co., St. Louis, 1974, Chap. 12.

Ptashne, M., and Gilbert, W.: Genetic repressors. *Scientific American*, **222**:36–44 (June), 1970.

Raff, M. C.: Cell-surface immunology. *Scientific American*, **234**:30–39 (May), 1976.

Sobell, H. M.: How actinomycin binds to DNA. *Scientific American*, **231**:82–91 (Aug.), 1974.

Spector, D. H., and Baltimore, D.: The molecular biology of poliovirus. *Scientific American*, **232**:25–31 (May), 1975.

Exponential Numbers

Changing Exponential Numbers to Common Numbers

Mathematicians have developed a shorthand method for expressing very large or very small numbers. This system involves the use of a base number, 10, raised to some power. A power, or exponent, indicates how many times the base number, 10, is repeated as a factor. Thus:

1×10^2 (ten repeated as a factor 2 times) $= 1 \times 10 \times 10 = 100$
1×10^3 (ten repeated as a factor 3 times) $- 1 \times 10 \times 10 \times 10 = 1000$
1×10^5 (ten repeated as a factor 5 times) $= 1 \times 10 \times 10 \times 10 \times 10 \times 10 = 100,000$

Negative exponents are used to indicate numbers less than 1. A negative exponent indicates the reciprocal of the same number with a positive exponent:

$$1 \times 10^{-1} = 1 \times \frac{1}{10^1} = 0.1$$

$$1 \times 10^{-2} = 1 \times \frac{1}{10^2} = 0.01$$

$$1 \times 10^{-4} = 1 \times \frac{1}{10^4} = 0.0001$$

Note: The positive exponent indicates how many places the decimal point must be moved to the right (from the number 1). Also, note that the negative exponent indicates how many places the decimal point must be moved to the left (from the number 1).

Any number that is not an exact power of 10 may be expressed as a product of two numbers, one of which is a power of 10. The other number is always written with just one figure to the left of the decimal point.

Example No. 1. Express 6.2×10^3 as a common number.
A positive three $(+3)$ exponent indicates that the decimal point should be moved three places to the right. Thus

$$6.2 \times 10^3 = (6.200) = 6200$$

Example No. 2. Change 8.45×10^5 to common numbers.

A $+5$ exponent indicates that the decimal point should be moved five places to the right.

$$8.45 \times 10^5 = (8.45000) = 845,000$$

Example No. 3. Change 1.27×10^{-2} to common numbers.

A -2 exponent indicates that the decimal point should be moved two places to the left.

$$1.27 \times 10^{-2} = (001.27) = 0.0127$$

Example No. 4. Change 3.5×10^{-9} to common numbers.

$$3.5 \times 10^{-9} = (0000000003.5) = 0.0000000035$$

Changing Common Numbers to Exponential Numbers

When changing a common number to an exponential number, the decimal point is moved so that there is just one digit in front of it. The exponent corresponds to the number of places the decimal point must be moved. If the decimal point is moved to the left, the exponent is positive; if to the right, it is negative.

Example No. 5. Change 4000 to an exponential number.

The decimal point must be moved to the left three places in order for just one digit to remain before that decimal point. Three places to the left indicates an exponent of 3 $(+3)$.

$$4000 = (4000) = 4 \times 10^3$$

Example No. 6. Change 604 to exponential numbers.

$$604 = (604) = 6.04 \times 10^2$$

Example No. 7. Change 0.00037 to exponential numbers.

$$0.00037 = (0.00037) = 3.7 \times 10^{-4}$$

where the negative exponent indicates that the decimal point has been moved to the right.

Functional Groups

Class of Compound	Functional Group	Typical Aliphatic Example Formula	Name
alkane	$-\overset{\mid}{\underset{\mid}{C}}-\overset{\mid}{\underset{\mid}{C}}-$	$CH_3CH_2CH_3$	propane
alkene	$-\overset{\mid}{C}=\overset{\mid}{C}-$	$CH_3CH=CH_2$	propene
alkyne	$-C\equiv C-$	$CH_3C\equiv CH$	propyne
alcohol	$-OH$	CH_3OH	methanol (methyl alcohol)
ether	$-O-$	$CH_3CH_2OCH_2CH_3$	ethyl ether
aldehyde	$-C\overset{\displaystyle O}{\underset{\displaystyle H}{}}$	CH_3CHO	ethanal (acetaldehyde)
ketone	$\overset{\diagdown}{\underset{\diagup}{}}C=O$	$CH_3\overset{O}{\overset{\|}{C}}CH_3$	propanone (acetone)
acid	$-\overset{O}{\overset{\|}{C}}-OH$	$CH_3\overset{O}{\overset{\|}{C}}OH$	ethanoic acid (acetic acid)
amine	$-NH_2$	CH_3NH_2	methylamine
amide	$-\overset{O}{\overset{\|}{C}}-NH_2$	$CH_3\overset{O}{\overset{\|}{C}}-NH_2$	acetamide
ester	$-\overset{O}{\overset{\|}{C}}-O-$	$CH_3\overset{O}{\overset{\|}{C}}OCH_2CH_3$	ethyl acetate

Index

Ptyalin, 335, 337, 348
Pulmonary respiration, 90
Purine, 246, 319, 320, 322, 333, 397
Putrescine, 352, 396
Pyridine, 246
Pyridoxal, 454
Pyridoxal phosphate, 395
Pyridoxamine, 454
Pyridoxine, 334, 454
Pyrimidine, 246, 319, 321, 397
Pyrogallol, 257
Pyrrole, 246, 418
Pyruvic acid, 364, 366, 367, 378, 393, 397, 398, 452

Quark, 26
Quaternary ammonium salts, 224
Quinine, 252

Rachitic rosary, 448
Rad, 43
Radiation, 315, 418, 503
    biologic effects, 57
    measurement, 43
    sickness, 58
    sources, 59
    standards, 59
    types, 37
    units, 43
Radicals, 66, 68, 194
Radioactive material, 128
Radioactive tracers, 418
Radioactivity, 36, 41
Radioimmunoassay, 472
Radioisotopes, 41, 50
    in medicine, 47
Rancidity, 290
Rare gases, 18
Rat antidermatitis factor, 454
Ratio solution, 135
Rayon, 275
Reactants, 74
Reaction rates, 79
Red blood cells, 346, 396, 414
Reducing agents, 111
Reduction, 110, 267, 269, 270
Relative weight, 25
Relaxin, 487
Release-inhibiting factors, 485
Rem, 43
Renal albuminuria, 408

Renal diabetes, 409
Renal disease, 408, 416
Renal failure, 435
Renal stones, 476
Renal threshold, 356, 371, 372, 373, 480
Rennin, 335, 349
Replication of DNA, 416
Reserpine, 252
Resonance, 229
Resorcinol, 235
Respiration, 86, 424
Respiratory disease, 416
Reticuloendothelial cells, 346
Retinal, 443
Retinene isomerase, 338
Retinol, 443
Retinol equivalent, 444
Retrolental fibroplasia, 103
Reverse chloride shift, 426
Rhamnose, 257
Rheumatic fever, 239
Rhodopsin, 261, 445
Ribitol, 453
Riboflavin, 334, 452, 454
Ribonuclease, 310, 338
Ribonucleic acid, 495
Ribose, 264, 320, 333, 397, 456, 495
Ribsomal RNA, 499
Ribosomes, 498
Ribulose, 264
Rickets, 416, 448
RNA, 246, 264, 319, 320, 397, 467, 477, 481, 495
RNA polymerase, 324
Rochelle salts, 221
Roentgen, 43

Saccharic acid, 269
Salicylic acid, 238
Saline cathartic, 142
Salivary amylase, 337, 344, 348
Salivary digestion, 343
Salting out, 314
Salts, 165
    heavy metals, 314
Saponification, 289
Saturated and unsaturated hydrocarbons, 191, 200, 201, 228
Saturated fatty acids, 283
Saturated solutions, 134
Scanner, 42